Linear Systems

Linear Systems

TIME DOMAIN AND TRANSFORM ANALYSIS

Michael O'Flynn
Eugene Moriarty

San Jose State University

JOHN WILEY & SONS
New York   Chichester   Brisbane   Toronto   Singapore

Library of Congress Cataloging-in-Publication Data

O'Flynn, Michael, 1935–
  Linear systems.

  Bibliography: p.
  Includes index.
  1. Linear systems.   2. Transformations
(Mathematics)   I. Moriarty, Eugene.   II. Title.
QA402.039   1987   003   86-11975
ISBN 0-471-60373-2

10 9 8 7 6 5

Contents

*We suggest omitting starred chapters in an introductory course.

Preface

The text *Linear Systems: Time Domain and Transform Analysis* contains material that is suitable for a first course in linear systems as well as for a more advanced course. The second course could be taught at either the first year graduate level or the advanced undergraduate level. Methods of representation and analysis are developed for both discrete time and continuous time systems. The parallel consideration of discrete and continuous ideas provides an efficient way to profit from the similarities between these two fundamental representational modalities.

The prerequisite skills for understanding this material are those that most engineering students have mastered by their junior year. In particular, a basic knowledge of differential equations and dc/ac circuit analysis is necessary. There are also other requirements that are more important but more difficult to gauge. These include the readers' sincere intention to learn and their dedication to the subject matter. Intention, rather than mere attention, implies an active involvement with the text. The level of involvement will vary for each student as he or she progresses through the book and will differ depending on the perspective brought to the material.

The order of presentation of the subject matter does not proceed at a uniform level of difficulty. This unevenness, we believe, actually mirrors the learning process itself. The first three chapters (those dealing with time-domain analysis) will probably be the hardest for the student to assimilate. Then, the degree of sophistication drops in Chapter 4 where we begin transform analysis. The level of difficulty is incremented with the Fourier transform (Chapter 8). The final chapter on state variables uses ideas from the entire text. In the actual knowledge acquisition process, time-domain analysis techniques are mastered first, providing an introduction to the rigor of linear system concepts. For most students this introduction occurs during their first circuits course. Solving differential or difference equations yields descriptions and predictions of important variables within the system under consideration. Although these solutions are often not easy to obtain, with transform techniques the process is reduced to mere algebra. Transforms not only provide interesting insights into problems but also make solutions easier to determine.

There are two major parts of the text plus a final culminating chapter on state variables. The first portion consists of time-domain analysis and comprises the first three chapters. Chapter 1 presents basic background material for studying linear systems: important operations on continuous and discrete waveforms and the theory of singularity functions. Chapter 2 deals with the response of linear time-invariant systems to known or deterministic inputs. As a prelude to finding such responses we will develop the ideas that:

1. A system is often governed by a differential or difference equation.
2. A continuous system can be characterized by the impulse response, $h(t)$, which is the response to a delta function $\delta(t)$, and a discrete system can be characterized by the pulse response, $h(n)$, which is the response to the unit pulse, $\delta(n)$.
3. The zero-state system output is given by the convolution of the impulse response with the input in the continuous case and by the convolution summation of the input with the pulse response in the discrete case.

Chapter 3 extends the material of Chapter 2 to the case of systems with random inputs or a signal plus noise input. This is a starred chapter denoted, Chapter 3*, which indicates it should be omitted in an introductory course. First, a detailed treatment of correlation integrals and summations for finite energy and finite power (periodic) functions is given. Then, using only the concept of an average value, correlation integrals and summations are defined and interpreted for simple random waveforms. The material is intuitively challenging and presents a novel introduction to the world of ergodic random processes. The chapter culminates with the derivation and application of the input–output relations for autocorrelation and cross-correlation functions for linear systems with random inputs.

The second part of the text consists of transform analysis and comprises Chapters 4 through 9. Prior to the middle 1970s, many students graduated with a knowledge of the one-sided Laplace transform and its application to solving for complete responses in *RLC* circuits. They also had a nodding acquaintance with the Fourier transform as an extension of the Fourier series. Now a graduating senior must have facility in using the Z transform as well as the Laplace and Fourier transforms. Concise and precise presentations are therefore required. Chapter 4 covers the one-sided Laplace transform with an emphasis on solving for linear systems with deterministic causal inputs. Chapter 5, a starred chapter, develops the two-sided Laplace transform and concentrates on applications involving systems with random or signal plus noise inputs. Chapter 6 treats the one-sided Z transform and stresses discrete systems with causal inputs. Its starred counterpart, Chapter 7, deals with the two-sided Z transform and in particular its use in analyzing discrete systems with random or signal plus noise inputs. The frequency interpretation is given for continuous and discrete signals by using the Fourier transform and discrete Fourier transform in Chapters 8 and 9. Chapter 8 considers Fourier series and develops the Fourier transform from the exponential Fourier series. At the end of the chapter a number of Fourier analysis applications are examined. In Chapter 9 the discrete Fourier transform is studied and the decimation in time and the decimation in frequency approaches to the fast Fourier transform (FFT) are discussed.

The third and final part of the text is a last long chapter, Chapter 10. This chapter deals with state variables and focuses specifically on applications in control theory. The material in Chapter 10 does not properly belong in either the time-domain analysis section or the transform analysis section of the text. However, it employs many of the ideas explored in the first nine chapters and can function as a culminating experience in the study of linear systems. The general state equation formulation is developed in various realizations for both the continuous and discrete cases. The solution of the state equations is considered. Controllability and observability are studied and state variable feedback is discussed along with the fundamentals of observer theory.

A goal of *Linear Systems: Time Domain and Transform Analysis* is to develop intuitive and practical understanding of the essentials in linear systems analysis. The stress is on fundamentals that are illustrated with many examples and problems. General theories are best learned through many particulars. The philosophy of "learning by doing" provides the framework for our presentations. Although we believe that there exists a rough proportionality between the amount of knowledge acquired and the number of problems worked, there are many different ways to use the text. We encourage instructors to experiment. The sequence that we use for the undergraduate and graduate courses is indicated schematically:

First Course	Second Course
• *Chapter 1* Signal Operations and Singularity Functions	• Review of Laplace, Z, and Fourier Transforms • *Chapter 3** Linear Systems with Random Inputs
• *Chapter 2* Time-Domain Analysis of Linear Systems	• *Chapter 5** The Two-sided Laplace Transform
• *Chapter 4* The One-sided Laplace Transform	• *Chapter 7** The Two-sided Z Transform
• *Chapter 6* The One-sided Z Transform	• *Chapter 9* The Fast Fourier Transform
• *Chapter 8* The Fourier Transform	• *Chapter 10* State Variables
• *Chapter 9* or *Chapter 10* (as time permits) The Fast Fourier Transform or State Variables	

The authors wish to thank all those who have contributed to the development of this textbook: Stephen Gold, University of Southwest Louisiana; John Golzy, Ohio University; Abraham Haddad, Georgia Institute of Technology; Shlomo Karni, University of New Mexico; K. S. P. Kumar, University of Minnesota; S. C. Prabhakar, California State at Northridge; Martin Roden, California State University at Los Angeles; Lee Rosenthal, Fairleigh Dickinson University; and John Thomas, Princeton University. We appreciate the patience and critiques of the many students, both graduate and undergraduate, who have endured the rough drafts out of which this textbook evolved. We are grateful to our colleagues, especially our department chair, Professor Jim Freeman, who have offered advice and encouragement along the way. We would not have survived the manuscript preparation phase of our work without the superb assistance of our secretaries. Special thanks go to Lynn Hong, Julie Pierce, Lee Clark, and Susan Owen.

Michael O'Flynn

Eugene Moriarty

Chapter 1

Signal Operations and Singularity Functions

INTRODUCTION

Chapters 1 through 3 are concerned with the evaluation of the output of linear, time-invariant causal systems (LTIC) for different inputs. Both continuous and discrete systems are considered; they are subject to deterministic and random inputs.

A continuous system is one whose input $x(t)$ and output $y(t)$ are continuous time functions related by a rule as in Figure 1-1(a). A discrete system is one whose input $x(n)$ and output $y(n)$ are discrete time functions related by a rule as in Figure 1-1(b).

The case of systems with deterministic or known inputs is treated in Chapter 2, whereas random and signal plus random inputs are treated in Chapter 3. As a prerequisite to system analysis in both the time and transform domains (Chapters 2 through 9), it is essential to be able to:

1. represent both continuous and discrete signals
2. understand the important signal operations of time-scaling, reflecting, and time-shifting
3. physically interpret and intuitively and rapidly operate with singularity functions.

Tasks 1, 2, and 3 will be accomplished in Chapter 1.

1-1 CONTINUOUS AND DISCRETE WAVEFORMS

A continuous waveform $x(t)$ assigns a unique numerical value to $x(t)$ for all t, $-\infty < t < \infty$. A discrete waveform $x(n)$ assigns a unique numerical value to $x(n)$

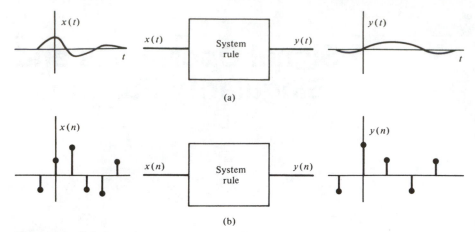

Figure 1-1   (a) A continuous system; (b) a discrete system.

for all integer n, $-\infty < n < \infty$. A number of waveforms that are prevalent throughout system theory will now be defined. A **waveform** is a function whose domain is from $-\infty$ to $+\infty$.

The Unit Step Function *u(t)*

$u(t)$ is defined as:

$$u(t) = 1, \qquad t > 0$$
$$= 0, \qquad t < 0$$

and its plot is shown in Figure 1-2(a).

The Rectangle Function ⊓(*t*)

$\sqcap(t)$ is defined as:

$$\sqcap(t) = 1, \qquad -0.5 < t < 0.5$$
$$= 0, \qquad \text{otherwise, except at } t = -0.5 \text{ and } 0.5$$

The rectangle function is normalized with unit area and is even, as plotted in Figure 1.2(b).

The Triangle Function Λ(*t*)

$\Lambda(t)$ is defined as:

$$\Lambda(t) = 1 - |t|, \qquad -1 < t < 1$$
$$= 0, \qquad \text{otherwise}$$

The triangle function is normalized with unit area and is even, as shown in Figure 1-2(c).

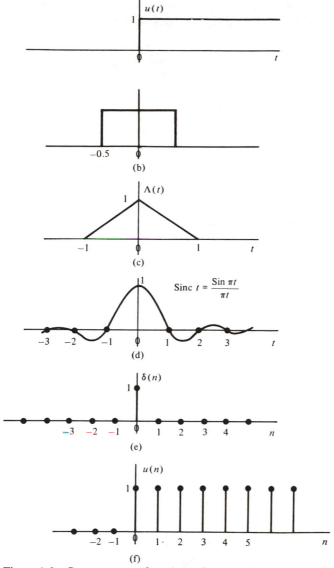

Figure 1-2 Some common functions of system theory.

The Sinc Function
Sinc (t) is defined as:

$$\text{Sinc } (t) = \frac{\text{Sin } \pi t}{\pi t}, \qquad -\infty < t < \infty$$

With investigation it can be shown (do so) that Sinc (t) has unit area, has a value of 1 at $t = 0$, and is even. The plot of Sinc (t) is illustrated in Figure 1.2(d).

The Unit Pulse Function δ(n)

The unit pulse function $\delta(n)$ is a discrete function defined by:

$$\delta(n) = 1, \qquad n = 0$$
$$= 0, \qquad \text{otherwise}$$

$\delta(n)$ is plotted in Figure 1-2(e). Any discrete function may be written as:

$$f(n) = \sum_{\substack{k=-\infty \\ \text{integer}}}^{\infty} f(k)\delta(n - k) \tag{1-1}$$

although more commonly we just give a formula for $f(n)$ for different values of n.

The Discrete Unit Step Function u(n)

Corresponding to its continuous counterpart, the discrete step function $u(n)$ is defined as:

$$u(n) = 1, \qquad n \geq 0$$
$$= 0, \qquad n < 0$$

The plot for $u(n)$ is shown in Figure 1-2(f).

An example will now be solved to develop familiarity with discrete functions.

EXAMPLE 1-1

Plot the following discrete functions, $f(n)$ versus n, and note the different notations:

(a) $f(n) = (0.5)^n u(n)$ or $\sum_{k=0}^{\infty} (0.5)^k \delta(n - k)$

(b) $f(n) = n(n - 1)2^n, \qquad n \geq 0$
$= 0, \qquad \text{otherwise}$

or, alternatively:

$$f(n) = n(n - 1)2^n u(n - 1) \quad \text{or} \quad n(n - 1)2^n u(n - 2) \tag{1-2}$$

(c) $f(n) = 2^n u(-n) + nu(n - 1)$

Solution.   The plot for each of the required functions is shown in Figure 1-3(a) through (c). In part (a) we note that since $u(n) = 0$ for $n < 0$, then $f(n)$ must be zero for $n < 0$. In part (b) since $n(n - 1)2^n$ is zero for $n = 0$ and $n = 1$, then $n(n - 1)2^n u(n)$ and $n(n - 1)2^n u(n - 2)$ are equivalent. In part (c) since $u(-n) = 0$ for $n > 0$ and is 1 for $n \leq 0$, then $2^n u(-n)$ assigns values to $f(n)$ for $n \leq 0$, and since $u(n - 1) = 0$ for $n < 1$ and is 1 for $n \geq 1$, then $nu(n - 1)$ assigns values to $f(n)$ for $n \geq 1$.

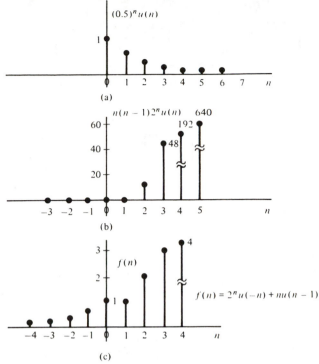

Figure 1-3   The discrete functions of Example 1-1.

1-2 SHIFTING, REFLECTING, AND TIME-SCALING OPERATIONS

There are a number of important operations on functions that occur repeatedly throughout system theory which will now be discussed. It is important to handle these operations fluently and quickly and to visualize them physically.

In Example 1-2 we use a test function that contains no special symmetric properties and thoroughly digest these operations for continuous functions.

EXAMPLE 1-2
For the test function $f(t)$ shown in Figure 1-4(a), find analytic expressions for and plot the following functions versus time:

(a) $f(2t)$, $f(\tfrac{1}{3}t)$
(b) $f(t-3)$, $f(t+4)$
(c) $f(-t)$, $f(-\tfrac{1}{2}t)$
(d) $f(3-t)$, $f(2t-3)$, $f(-3t-4)$

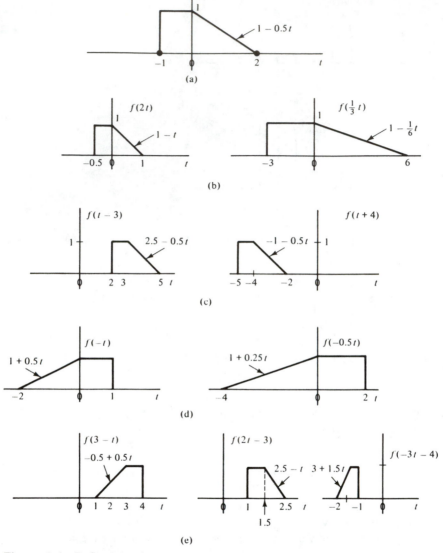

Figure 1-4 Reflecting, time-scaling, and shifting the continuous test function of Example 1-2.

Solution

(a) The Time-scaling Operation. Using the fundamental definition of a function, we say:

$$y(t) = f(2t) = 1 - 0.5(2t), \qquad 0 < 2t < 2$$

$$= 1, \qquad -1 < 2t < 0$$

$$= 0, \qquad \text{otherwise}$$

This leads to:

$$f(2t) = 1 - t, \qquad 0 < t < 1$$
$$= 1, \qquad -0.5 < t < 0$$
$$= 0, \qquad \text{otherwise}$$

$f(2t)$ is plotted in Figure 1-4(b) and may be described as $f(t)$ contracted by a factor of 2. This is the operation of *time-scaling*. It is left as an exercise for the reader to show:

$$y(t) = f(\tfrac{1}{3}t) = 1 - \tfrac{1}{6}t, \qquad 0 < t < 6$$
$$= 1, \qquad -3 < t < 0$$
$$= 0, \qquad \text{otherwise}$$

$y(t)$ is shown in Figure 1-4(b). We say $f(\tfrac{1}{3}t)$ is $f(t)$ widened or expanded by a factor of 3.

(b) The Shifting Operation. Using the fundamental definition of a function, we obtain:

$$f(t - 3) = 1 - 0.5(t - 3), \qquad 0 < t - 3 < 2$$
$$= 1, \qquad -1 < t - 3 < 0$$
$$= 0, \qquad \text{otherwise}$$

This leads to:

$$f(t - 3) = 2.5 - 0.5t, \qquad 3 < t < 5$$
$$= 1, \qquad 2 < t < 3$$
$$= 0, \qquad \text{otherwise}$$

$f(t - 3)$, as plotted in Figure 1-4(c), may be described as $f(t)$ *shifted 3 units to the right*. Similarly, we find:

$$f(t + 4) = -1 - 0.5t, \qquad -4 < t < -2$$
$$= 1, \qquad -5 < t < -4$$
$$= 0, \qquad \text{otherwise}$$

and $f(t + 4)$ is illustrated in Figure 1-4(c) and may be described as $f(t)$ *shifted 4 units to the left* on the time axis.

(c) The Reflecting Operation

$$f(-t) = 1 - 0.5(-t), \qquad 0 < -t < 2$$
$$= 1, \qquad -1 < -t < 0$$
$$= 0, \qquad \text{otherwise}$$

This becomes:

$$f(-t) = 1 + 0.5t, \qquad -2 < t < 0$$

$$= 1, \qquad 0 < t < 1$$

$$= 0, \qquad \text{otherwise}$$

$f(-t)$ is illustrated in Figure 1-4(d) and is said to be $f(t)$ reflected about the vertical axis. Similarly, it can be shown that:

$$f(-0.5t) = 1 + 0.25t, \qquad -4 < t < 0$$

$$= 1, \qquad 0 < t < 2$$

$$= 0, \qquad \text{otherwise}$$

$f(-0.5t)$ is plotted in Figure 1-4(d) and is said to be $f(t)$ *reflected* and then *widened* by 2 units or vice versa. The operations of reflecting and time-scaling are obviously commutative. (Check this.)

(d) This part contains combinations of the three basic operations— *reflecting, time-scaling,* and *shifting.*

$$f(3 - t) = 1 - 0.5(3 - t), \qquad 0 < 3 - t < 2$$

$$= 1, \qquad -1 < 3 - t < 0$$

$$= 0, \qquad \text{otherwise}$$

This leads to:

$$f(3 - t) = -0.5 + 0.5t, \qquad 1 < t < 3$$

$$= 1, \qquad 3 < t < 4$$

$$= 0, \qquad \text{otherwise}$$

$f(3 - t)$ versus t is shown in Figure 1-4(e) and may be visualized as $f(t)$ *reflected* and then *shifted* 3 units to the right. Also, we note the operations of shifting and reflecting are not commutative: if we shift first, $f(t)$ extends from $t = 2$ to $t = 5$ and on reflection extends from $t = -5$ to $t = -2$. This is different from the plot of $f(3 - t)$ in Figure 1-4(e).

The result obtained is most easily summarized as:

$$f(3 - t) = f[-(t - 3)]$$

$$\underset{①}{\uparrow} \qquad \underset{②}{\uparrow}$$

where ① indicates reflection and ② indicates subsequent shifting to the right. Now considering $f(2t - 3)$, we obtain with work:

$$f(2t - 3) = 1, \qquad 1 < t < 1.5$$

$$= 2.5 - t, \qquad 1.5 < t < 2.5$$

$$= 0, \qquad \text{otherwise}$$

$f(2t - 3)$ versus t is plotted in Figure 1-4(e) and may be visualized as $f(t)$ contracted by a factor 2 (or time-scaled by a factor 0.5) and then shifted 1.5 units to the right. Again, it could be demonstrated that if $f(t)$ is shifted and then contracted, a different result is obtained, unless we modify the time-scaling operation when shifting is performed first (check this and state the modification). The result we obtained is most easily summarized by:

$$f(2t - 3) = f[2(t - 1.5)]$$

① Contract by 2   ② Shift right 1.5 units to right

Finally, we consider $f(-3t - 4)$. With work this becomes:

$$f(-3t - 4) = 1 - 0.5(-3t - 4), \qquad 0 < -3t - 4 < 2$$

$$= 3 + 1.5t, \qquad -2 < t < -1.33$$

$$= 1, \qquad -1.33 < t < -1$$

$$= 0, \qquad \text{otherwise}$$

$f(-3t - 4)$ is most easily summarized by writing it as:

$$f(-3t - 4) = f[-3(t + 1.33)]$$

① Reflect   ② Contract   ③ Shift left 1.33 units
                      by 3

If the shifting operation is performed last, then the other two operations are commutative. Otherwise we must modify the definitions for time-scaling and reflecting to ensure commutativeness after shifting has occurred.

As a consequence of the solution of Example 1-2, we can enumerate for any general function $f(t)$ the operations involved when the argument is changed to $\pm at \pm b$ for any real positive numbers a and b.

$$f(at \pm b) = f\left[a\left(t \pm \frac{b}{a}\right)\right] \tag{1-3}$$

①   Contract
     by a
          ② Shift right $\dfrac{b}{a}$ units with $-$ sign

          Shift left $\dfrac{b}{a}$ units with $+$ sign

$$f(-at \pm b) = f\left[-a\left(t \mp \frac{b}{a}\right)\right] \tag{1-4}$$

①   Reflect and
     contract by a
          ② Shift left $\dfrac{b}{a}$ units with $+$ sign

          Shift right $\dfrac{b}{a}$ units with $-$ sign

The preceding notation may be used to indicate a signal existing for a finite duration of time.

EXAMPLE 1-3

Find a single analytic expression for the following using the unit step and rectangle functions.

(a) a pulse of amplitude 2 extending from $t = 6$ to $t = 6.2$ s
(b) the signal Sin t existing for negative time only
(c) the sinusoidal burst Cos $10^4\,t$ lasting from $t = 2$ s until 2.01 s

Solution

(a) If we consider that $\prod(t)$ is of width 1 and is centered at $t = 0$, then our required pulse is:

$$f(t) = 2\prod[5(t - 6.1)] \quad \text{or} \quad 2[u(t - 6) - u(t - 6.2)]$$

(b) $f(t) = $ Sin $t\,u(-t)$ where $u(t)$ the step function is reflected.
(c) Again, the rectangle or window function is useful:

$$f(t) = (\text{Cos } 10^4 t)\prod[100(t - 2.005)]$$

or
$$= \text{Cos } 10^4 t[u(t - 2) - u(t - 2.01)]$$

1-2-1 Shifting and Reflecting for Discrete Functions

A discrete waveform will either inherently exist or be formed by sampling a continuous time signal every τ seconds. Regardless of the sampling rate, it is common to assume the spacing between samples is unity and to take account of the actual spacing as a separate problem when all the necessary discrete operations are performed. For this reason we will rarely use the time-scaling operation when dealing with discrete waveforms. The use of the reflecting and shifting operations are analogous as for continuous functions. $f(n \pm k)$ versus n is the discrete function $f(n)$ shifted k units to the left for the positive sign and k units to the right for the negative sign where k is a positive integer.

$f(-n \pm k)$ versus n is visualized as $f[-(n \mp k)]$ and is the discrete function $f(n)$ reflected and shifted k units to the right for the positive sign of k and k units to the left for the negative sign on k in $f(-n \pm k)$.

EXAMPLE 1-4

Given

$$f(n) = (-0.6)^n u(n)$$

Find analytic expressions for and plot:

(a) $f(-n)$
(b) $f(n + 3)$
(c) $f(-n - 4)$

Solution

(a) $$g(n) = f(-n) = (-0.6)^{-n}u(-n)$$

$$= 0, \qquad n > 0$$

Therefore $$f(-n) = (-0.6)^{-n}, \qquad n \le 0$$

$f(n)$ is plotted in Figure 1-5(a) and $f(-n)$ in Figure 1-5(b). $f(-n)$ is $f(n)$ *reflected.*

(b) $f(n + 3) = (-0.6)^{n+3}u(n + 3)$

$$= (-0.6)^{n+3}, \qquad n \ge -3$$

$$= 0, \qquad \text{otherwise}$$

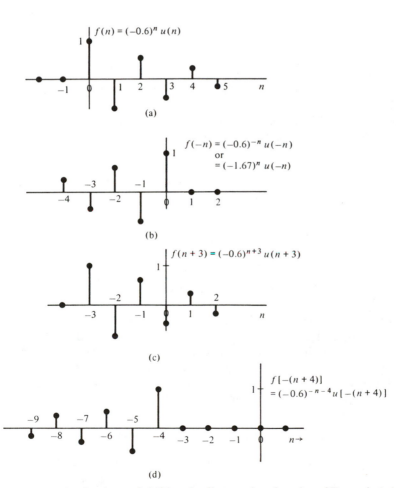

Figure 1-5   Reflecting and shifting the discrete time function of Example 1-4.

$f(n + 3)$ is shown in Figure 1-5(c) and is $f(n)$ shifted 3 units to the left.

(c) $f(-n - 4) = (-0.6)^{-n-4}u[-(n + 4)]$

$$= (-0.6)^{-(n+4)}, \qquad n \leq -4$$

$$= 0, \qquad \text{otherwise}$$

$f(-n - 4)$ is plotted in Figure 1-5(d) and is $f(n)$ reflected and shifted 4 units to the left.

As preparation for the evaluation of convolution and correlation integrals in Chapters 2 and 3, we solve another example.

EXAMPLE 1-5
Given

$$f(t) = 1 + t, \qquad -1 < t < 0$$

$$= 0, \qquad \text{otherwise}$$

(a) Find and plot $f(p - t)$ versus p for $t = -4, 0,$ and 2.
(b) Develop a rule for describing $f(p - t)$ versus p for all t, $-\infty < t < \infty$.
(c) Find and plot $f(t - p)$ versus p for $t = -4, 0,$ and 2.
(d) Develop a rule for describing $f(t - p)$ versus p for all t, $-\infty < t < \infty$.
(e) For what values of t is $\int_{-\infty}^{\infty} f(p) f(t - p)\, dp = 0$?

Solution

(a) $f(p - t)$ is $f(p)$ shifted t units to the right if $t > 0$ and "$-t$" units to the left if $t < 0$. The analytic formula is:

$$f(p - t) = 1 + (p - t), \qquad -1 < p - t < 0$$

$$= (1 - t) + p, \qquad -1 + t < p < t$$

$$= 0, \qquad \text{otherwise}$$

$f(p)$ is plotted in Figure 1-6(a), whereas $f[p - (-4)], f(p - 0),$ and $f(p - 2)$ are shown in Figure 1-6(b).
(b) $f(p - t)$ extends from $-1 + t$ to t on the p axis and since $f(p)$ is a pulse of finite duration $f(p - t)$ moves along the p axis from $p = -\infty$ to $p = +\infty$ as t varies from $-\infty$ to $+\infty$. Figure 1-6(c) indicates this.
(c) $f(t - p)$ versus p is interpreted by writing $f(t - p) = f[-(p - t)]$. This says $f(t - p)$ is $f(p)$ *reflected* and then *shifted* t units to the right if $t > 0$ or $-t$ units to the left if $t < 0$. The analytic formula for our pulse is:

$$f(t - p) = 1 + (t - p), \qquad -1 < t - p < 0$$

$$= 1 + t - p, \qquad t < p < t + 1$$

$$= 0, \qquad \text{otherwise}$$

$f(-4 - p), f(0 - p),$ and $f(2 - p)$ are plotted in Figure 1-6(d).

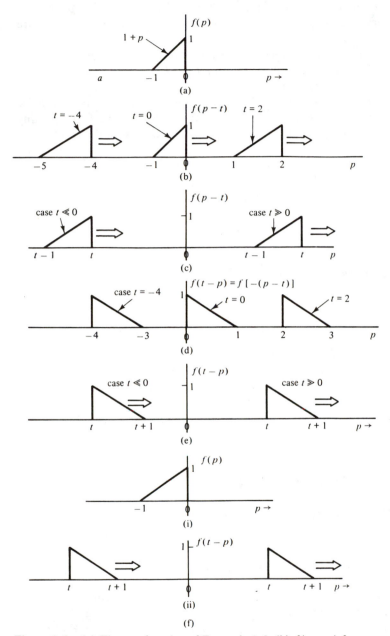

Figure 1-6   (a) The test function of Example 1-5; (b) $f(p-t)$ for $t = -4, 0,$ and 2; (c) $f(p-t)$ versus p; (d) $f(t-p)$ for $t = -4, 0,$ and 2; (e) $f(t-p)$ versus p; (f) determining when $f(p)f(t-p) = 0$.

(d) $f(t - p)$ extends from t to $t + 1$ on the p axis for any value of t. Since $f(-p)$ is a pulse of finite duration, $f(t - p)$ represents $f(-p)$ moving from $p = -\infty$ to $p = +\infty$ as t varies $-\infty < t < \infty$ as shown in Figure 1-6(e).

(e) We are now asked to interpret when $\int_{-\infty}^{\infty} f(p) f(t - p)\, dp$ is zero. This is so if the integrand product function $f(p) f(t - p)$ is zero. Observing Figure 1-6(f) we note that if $1 + t < -1$ (i.e., $t < -2$) or if $t > 0$, then the product is zero. For any $t < -2$ the front of the moving pulse $f(t - p)$ does not overlap the rear of the stationary pulse $f(p)$. For any $t > 0$ the rear of the moving pulse $f(t - p)$ has passed the front of the stationary pulse $f(p)$.

$$f(p) f(t - p) = 0, \qquad t < -2 \cup t > 0$$

where $\cup$ is the union operation. Therefore

$$\int_{-\infty}^{\infty} f(p) f(t - p)\, dp = 0, \qquad t < -2 \cup t > 0$$

Drill Set: Operating on Discrete Functions

1. Consider

$$f(n) = (0.7)^{n-1} [u(n - 1) - u(n - 4)]$$

(a) Find and plot $f(n - 4)$, $f(4 - n)$, and $f(-n - 3)$.
(b) Plot $f(k - n)$ versus k for $n = -3, 1, 4$.
(c) Plot $f(n - k)$ versus k for $n = -3, 1, 4$.
(d) For what values of n is $g(n) = \Sigma_k\ f(k) f(k - n)$ equal to zero?
(e) For what values of n is $g(n) = \Sigma_k\ f(k) f(n - k)$ equal to zero?

1-3 SINGULARITY FUNCTIONS

1-3-1 Definition of $\delta(t)$

The Dirac delta function is defined by:

$$\int_{-\infty}^{\infty} f(t)\delta(t)\, dt \triangleq f(0) \tag{1-5}$$

where $f(t)$ is any function that is continuous at $t = 0$. The requirements on $\delta(t)$ to obey Equation 1-5 for a function $f(t)$ which is changing rapidly (e.g., Cos $10^{10}t$) at $t = 0$ are very severe. Two engineering models for $\delta(t)$ are shown in Figure 1-7. Both $p_1(t)$ and $p_2(t)$ have the following properties for ϵ "small":

- The value at $t = 0$ is very large.
- The duration is very short.
- The area is one.

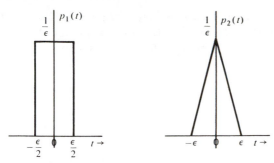

Figure 1-7 Two engineering models for $\delta(t)$.

The term "very short" in particular is a relative one, as in some applications "seconds" may be a brief duration, whereas in other applications "nanoseconds" may be of seemingly infinite duration.

EXAMPLE 1-6

(a) Show that $p_1(t) = \lim\limits_{\epsilon \to 0} \frac{1}{\epsilon} \sqcap (\frac{1}{\epsilon} t)$ is a mathematical model for $\delta(t)$.

(b) Discuss practical versions of $p_1(t)$ that are appropriate as far as sampling the functions e^{-4t} and $\text{Cos } 10^4 t$ are concerned.

Solution

(a) Consider any function $f(t)$ that is continuous at $t = 0$, and assume that in the range $-w/2 < t < w/2$, $f(t)$ varies only slightly from $f(0)$. Now

$$\int_{-\infty}^{\infty} f(t)p_1(t)\, dt = \int_{-\epsilon/2}^{\epsilon/2} f(t) \frac{1}{\epsilon}\, dt$$

$$= \frac{1}{\epsilon} f(0) \int_{-\epsilon/2}^{\epsilon/2} 1\, dt \qquad \text{for } \epsilon \ll w$$

$$= f(0)$$

Theoretically, what we have done here is to replace the order of taking a limit and integrating.

(b) The function e^{-4t} has a time constant $\tau = 0.25$, and if ϵ in $p_1(t)$ is chosen much less than 0.25 s (say, 0.05 s), then $p_1(t) = 20 \sqcap (20t)$ would be an acceptable model for $\delta(t)$.

The function $\text{Cos } 10^4 t$ may be expanded in a Taylor series about $t = 0$ to yield $\text{Cos } 10^4 t = 1 - 10^8 t^2/2! + 10^{16} t^4/4! + \cdots$ and $\text{Cos } 10^4 t \approx 1$ if $t \ll 2 \cdot 10^{-4}$. If ϵ is chosen such that $\epsilon \ll 10^{-4}$ (say, 10^{-6}), then $p_1(t) = 10^6 \sqcap (10^6 t)$ is an acceptable practical model of $\delta(t)$ as far as $\text{Cos } 10^4 t$ is concerned.

As a result of the definition and Example 1-6, we can say that any pulse with area 1 and a duration of ϵ seconds from $-\epsilon/2 < t < \epsilon/2$ serves as a practical model for $\delta(t)$ as far as a function $f(t)$ is concerned. We require $\epsilon \ll w$ where $f(t)$ is continuous at $t = 0$ and varies little over the range $-w/2 < t < w/2$ (subjective).

1-3-2 Notation and Mathematical Properties of Models

The delta function is indicated by a spike as shown in Figure 1-8(a). As a consequence of Equation 1-5 the theoretical properties of any model of $\delta(t)$ or $\delta(t)$ itself are:

- $\delta(0) \longrightarrow \infty$ (1-6)
- $\delta(t) \longrightarrow 0,$ $t \neq 0$ (1-7)
- Area is 1   (1-8)
- $\delta(t)$ is even

The fact that $\delta(t)$ is even is important when we extend our treatment to higher-order singularity functions and consider taking derivatives of models such as $p(t)$ in Figure 1-9.

EXAMPLE 1-7

Intuitively, show that $p(t) = \lim_{a \to 0^+} [2a/(a^2 + 4\pi^2 t^2)]$ satisfies all the properties of a delta function model.

Solution

- $p(0) = \lim_{a \to 0^+} \dfrac{2a}{a^2} = \infty$

- $p(t) = \lim_{a \to 0^+} \dfrac{2a}{0 + 4\pi^2 t^2} = 0,$ for $t \neq 0$

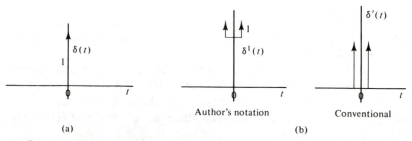

Author's notation   Conventional

(a)   (b)

Figure 1-8   The notation for $\delta(t)$ and $\delta'(t)$.

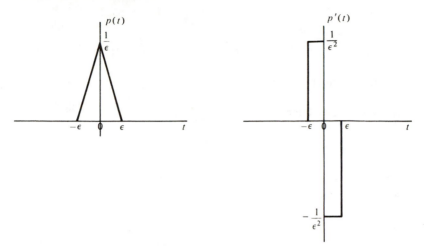

Figure 1-9   A delta function model and its derivative.

- From any conventional set of tables or using the appropriate trigonometric or hyperbolic substitution of a variable (which one?), it is easy to show:

$$\int_{-\infty}^{\infty} \frac{2a}{a^2 + 4\pi^2 t^2} \ dt = 1$$

- $p(t) = p(-t)$ and therefore $p(t)$ is even.

    In practice, delta functions will often appear in disguised model form and are only recognizable by checking the four basic properties. The interpretation of a number of models is required in Problem 1-7 at the end of the chapter.

1-3-3 Working Properties of Delta Functions

When operating with delta functions, we have four properties that are repeatedly used. These properties are:

Property 1

$$\int_{-\infty}^{\infty} f(t)\delta(t - a) \ dt \equiv f(a) \qquad (1\text{-}9)$$

This is called the "sampling property."

Property 2

$$f(t)\delta(t - a) \equiv f(a)\delta(t - a) \qquad (1\text{-}10)$$

This is called the "sifting property."

Property 3

$$\delta(at + b) \equiv \frac{1}{|a|} \delta\left(t + \frac{b}{a}\right) \tag{1-11}$$

This is called the "scaling property."

Property 4 (to be discussed later)

$$f(t) * \delta(t - a) \triangleq \int_{-\infty}^{\infty} f(u)\delta(t - u - a) \, du \tag{1-12}$$

$$= f(t - a)$$

This is called the "convolution property."

EXAMPLE 1-8

Prove Properties 1 through 3 and mechanically show Property 4 is true.

Solution

Property 1.   Using the change of variable $p = t - a$, we obtain:

$$\int_{-\infty}^{\infty} f(t)\delta(t - a) \, dt = \int_{-\infty}^{\infty} f(p + a)\delta(p) \, dp$$

$$= f(a)$$

by Equation 1-5. This property is often used in lieu of Equation 1-5 as the definition of a delta function located at a.

Property 2

$$f(t)\delta(t - a) \equiv f(a)\delta(t - a) \text{ if for any continuous } g(t)$$

$$\int_{-\infty}^{\infty} g(t) f(t)\delta(t - a) \, dt = \int_{-\infty}^{\infty} g(t) f(a)\delta(t - a) \, dt$$

Both sides obviously equal $g(a) f(a)$. Observing Property 2 we say that $\delta(t - a)$ sifts out the value of $f(t)$ at $t = a$.

Property 3.   In order to prove Property 3 we need to consider the two cases $a > 0$ and $a < 0$.

Case $a > 0$
Show

$$\int_{-\infty}^{\infty} f(t)\delta(at + b) \, dt = \int_{-\infty}^{\infty} \frac{1}{a} f(t)\delta\left(t + \frac{b}{a}\right) dt$$

Obviously, by definition the right-hand side (RHS) yields $(1/a) f(-b/a)$. For the left-hand side let:

$$p = at + b, \quad \text{then} \quad dt = \frac{1}{a} dp$$

and as $-\infty < t < \infty$, $-\infty < p < \infty$. The left-hand side now becomes:

$$\int_{-\infty}^{\infty} f(t)\delta(at + b)\, dt = \int_{-\infty}^{\infty} f\left(\frac{p-b}{a}\right)\frac{1}{a}\,\delta(p)\, dp$$

$$= \frac{1}{a}f\left(\frac{0-b}{a}\right)$$

$$= \frac{1}{a}f\left(-\frac{b}{a}\right) = \text{RHS}$$

Case $a < 0$
Now if we let $at + b = p$, $dt = -1/|a|\, dp$, and as $-\infty < t < \infty$, $\infty > p > -\infty$
and we obtain:

$$\int_{-\infty}^{\infty} f(t)\delta(at + b)\, dt = \int_{\infty}^{-\infty} f\left(\frac{p-b}{a}\right)\delta(p)\,\frac{-1}{|a|}\, dp$$

$$= \frac{1}{|a|}\int_{-\infty}^{\infty} f\left(\frac{p-b}{a}\right)\delta(p)\, dp$$

$$= \frac{1}{|a|}f\left(-\frac{b}{a}\right), \qquad \text{as before}$$

Property 4. We will mechanically prove the convolution property and discuss it in depth in the next chapter when we consider convolution. Consider $\int_{-\infty}^{\infty} f(u)\delta(t - u - a)\, du$. As far as integrating with respect to u for a fixed t, Property 3 says

$$\delta(t - u - a) \equiv \delta[\underset{\substack{\uparrow \\ a = -1}}{-u} + \underset{\substack{\searrow \\ b = t - a}}{(t - a)}] \equiv \delta(u - t + a)$$

Therefore

$$\int_{-\infty}^{\infty} f(u)\delta(t - u - a)\, du = \int_{-\infty}^{\infty} f(u)\delta[u - (t - a)]\, du$$

$$= f(u)\big|_{u=t-a}$$

$$= f(t - a)$$

Another example will be solved to demonstrate the simplicity of integration when delta functions are involved.

EXAMPLE 1-8
Evaluate:

(a) $\int_{-2}^{4} (t^2 + 5t)\delta(-3t + 5)\, dt$
(b) $\int_{-3}^{-1} (t^2 + 5t)\delta(-3t + 5)\, dt$
(c) $F(t) = \int_{-\infty}^{t} \delta(u - a)\, du$

Solution

(a) Using Property 3, we obtain:

$$\int_{-2}^{4} (t^2 + 5t)\delta(-3t + 5)\, dt = \int_{-2}^{4} (t^2 + 5t)\tfrac{1}{3}\,\delta(t - \tfrac{5}{3})\, dt$$

Since $t = \tfrac{5}{3}$ lies within the range $-2 < t < 4$,

$$\text{RHS} = \tfrac{1}{3}(t^2 + 5t)|_{t-5/3}$$

$$= \tfrac{25}{27} + \tfrac{25}{9}$$

$$= 3.7$$

(b) Using Property 3 the integral becomes $\int_{-3}^{-1} (t^2 + 5t)\tfrac{1}{3}\,\delta(t - \tfrac{5}{3})\, dt = 0$ because the delta function occurs outside the range of integration.

In general, we can summarize parts (a) and (b) by:

$$\int_{-\infty}^{t} (p^2 + 5p)\tfrac{1}{3}\,\delta\!\left(p - \frac{5}{3}\right) dp = 3.7u\!\left(t - \frac{5}{3}\right)$$

which indicates the solution is 3.7 when the upper limit of integration is greater than $\tfrac{5}{3}$ and is zero otherwise.

(c) $F(t) = \int_{-\infty}^{t} \delta(p - a)\, dp = u(t - a)$. The functions $\delta(t - a)$ and $u(t - a)$ form an integral–derivative pair. Alternatively:

$$\frac{d}{dt}u(t - a) = \frac{d}{dt}\!\left[\int_{-\infty}^{t} \delta(p - a)\, dp\right]$$

$$= \delta(t - a)$$

1-3-4 Derivatives of Delta Functions

$\delta'(t)$, called the derivative of a delta function or the unit doublet, is defined by:

$$\int_{-\infty}^{\infty} f(t)\delta'(t)\, dt = -f'(0) \tag{1-13}$$

for any $f(t)$ that possesses a derivative $f'(0)$ at $t = 0$. This result may be demonstrated using integration by parts:

$$\int_{-\infty}^{\infty} f(t)\delta'(t)\, dt = \int_{-\infty}^{\infty} f(t)\, d[\delta(t)]$$

$$= f(t)\delta(t)\,\Big|_{-\infty}^{\infty} - \int_{-\infty}^{\infty} \delta(t)\, f'(t)\, dt$$

$$= 0 - 0 - f'(0)$$

The notation for $\delta'(t)$ is shown in Figure 1-8(b); its advantages over the conventional notation, which is also shown, will be discussed shortly.

1-3-5 Models for $\delta'(t)$

Any function is said to be a mathematical model of the derivative of a delta function if when placed inside the integral of Equation 1-13 it yields $-f'(0)$ for any differentiable $f(t)$. It can be demonstrated that the derivative of any mathematical model of $\delta(t)$ is a mathematical model of $\delta'(t)$.

EXAMPLE 1-9

Show that the derivative of $p(t) = \lim_{\epsilon \to 0} [\frac{1}{\epsilon} \Lambda (\frac{1}{\epsilon} t)]$ is a model for $\delta'(t)$.

Solution

$p(t)$ and $p'(t)$ are shown in Figure 1-9. Using as our model $\lim_{\epsilon \to 0} p'(t)$, we can say for any differentiable $f(t)$ that:

$$\int_{-\infty}^{\infty} f(t)p_2'(t)\, dt = \lim_{\epsilon \to 0}\left[\int_{-\epsilon}^{0} \frac{1}{\epsilon^2}f(t)\, dt \right.$$

$$\left. + \int_{0}^{\epsilon} \frac{-1}{\epsilon^2}f(t)\, dt\right]$$

$$= \lim_{\epsilon \to 0}\left[\left(\frac{1}{\epsilon^2}f\left(\frac{-\epsilon}{2}\right)\epsilon\right) - \left(\frac{1}{\epsilon^2}f\left(\frac{\epsilon}{2}\right)\epsilon\right)\right]$$

$$= \lim_{\epsilon \to 0}\frac{f(-\epsilon/2) - f(\epsilon/2)}{\epsilon}$$

$$= -f'(0) \tag{1}$$

Alternatively, we could have obtained two other results instead of Equation (1). These are $\lim_{\epsilon \to 0} [f(0) - f(\epsilon)]/\epsilon$ or $\lim_{\epsilon \to 0} [f(-\epsilon) - f(0)]/\epsilon$ depending on whether over a small range of integration we use a forward, midpoint, or backpoint value times the range as an approximation for the integral. If $f(t)$ is differentiable all these three results define $-f'(0)$.

As was done for $\delta(t)$, we could now derive properties for $\delta'(t)$. Some of these are:

- $f(t)\delta'(t) \equiv -f'(0)\delta(t)$

- $\delta'(at + b) \equiv \frac{1}{a^2} \delta'\left(t + \frac{b}{a}\right), a > 0; \equiv \frac{-1}{a^2} \delta'\left(t + \frac{b}{a}\right), a < 0$

- $\int_{-\infty}^{t} \delta'(p - a)\, dp = \delta(t - a), \qquad$ if $t > a$ and 0 otherwise

$$= \delta(t - a)u(t - a)$$

1-3-6 Higher Derivatives of $\delta(t)$

By extending the definition of $\delta'(t)$, we define an nth-order derivative of $\delta(t)$ by:

$$\int_{-\infty}^{\infty} f(t)\delta^n(t)\, dt = (-1)^n f^n(0) \tag{1-14}$$

The notations for $\delta''(t)$ and $\delta'''(t)$ are shown in Figure 1-10(a). One great advantage of our notation for $\delta'(t)$, $\delta''(t)$, and so on is the facility of the notation in allowing us to denote clearly the presence of a delta function and derivatives of a delta function at the same point. For example, in Figure 1-10(b) there is the notation for $3\delta(t-4)$; in Figure 1-10(c), the notation for $-5\delta'(t-4)$; and in Figure 1-10(d), $f(t) = 3\delta(t-4) - 5\delta'(t-4) + \delta''(t-4)$ is denoted.

It is important to be relaxed and confident when handling singularity functions and to realize that their purpose is to greatly simplify analytic work. We now consider some applications.

1-3-7 Singularity Functions in Differentiation and Integration

EXAMPLE 1-10

Consider the function:

$$f(t) = \delta'(t+3) - 3\delta(t+3) + 4\delta(t+2)$$

Integrate $f(t)$ twice and graphically show the results. More formally, find

$$F(t) = \int_{-\infty}^{t} f(p)\, dp \quad \text{and} \quad G(t) = \int_{-\infty}^{t} F(p)\, dp$$

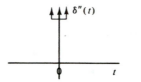
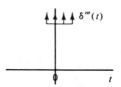

(a)

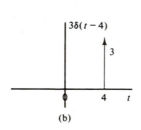

(b)

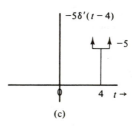

(c)

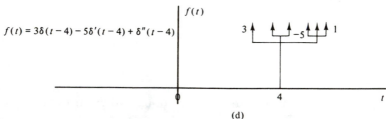

(d)

Figure 1-10 (a) Notation for $\delta''(t)$ and $\delta'''(t)$; (b) $3\delta(t-4)$; (c) $-5\delta'(t-4)$; (d) $3\delta(t-4) - 5\delta'(t-4) + \delta''(t-4)$.

Solution

$f(t)$ is plotted in Figure 1-11(a), $F(t)$ in Figure 1-11(b), and $G(t)$ in Figure 1-11(c). The more formal derivation will now be followed.

The Integral of $f(t)$

$$F(t) = \int_{-\infty}^{t} f(p)\, dp$$

$$= 0, \qquad t < -3$$

$$= \delta(t + 3), \qquad t = -3$$

$$= -3, \qquad -3 < t < -2$$

$$= -3 + 4, \qquad t > -2$$

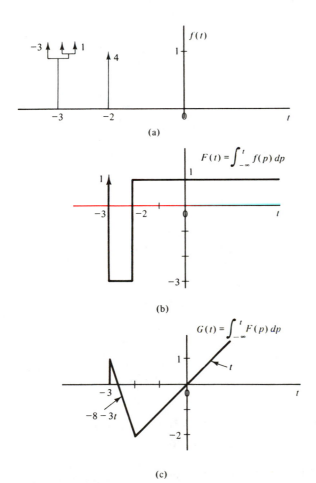

(a)

(b)

(c)

Figure 1-11 (a) The function $f(t)$ for Example 1-10; (b) $F(t)$ the integral of $f(t)$; (c) $G(t)$ the integral of $F(t)$.

Alternatively, using the results:

$$\int_{-\infty}^{t} \delta^{n}(p - a)\, dp = \delta^{n-1}(t - a)u(t - a) \quad \text{and} \quad \int_{-\infty}^{t} \delta(p - a)\, dp = u(t - a)$$

we may write:

$$F(t) = \delta(t + 3) - 3u(t + 3) + 4u(t + 2)$$

which on interpretation gives the same result.

The Integral of F(t)

$$G(t) = \int_{-\infty}^{t} F(p)\, dp$$

$$= 0, \qquad t < -3$$

$$= 1 - 3(t + 3), \qquad -3 < t < -2$$

$$= -8 - 3t, \qquad -3 < t < -2$$

$$= 1 - 3(t + 3) + 4(t + 2), \qquad t > -2 \text{ which becomes:}$$

$$= t, \qquad t > -2$$

Alternatively:

$$G(t) = \int_{-\infty}^{t} [\delta(p + 3) - 3u(p + 3) + 4u(p + 2)]\, dp$$

$$= u(t + 3) - 3(t + 3)u(t + 3) + 4(t + 2)u(t + 2)$$

using the result $\int_{-\infty}^{t} cu(p + a)\, dp = c(t + a)u(t + a)$. On interpretation, this yields the same result for each range.

We should carefully observe $f(t)$, $F(t)$, and $G(t)$ and note how we can quickly visualize the differentiation problem. If $G(t)$ is given we introduce the delta function at $t = -3$ when finding $F(t) = G'(t)$, and when we differentiate again we obtain the two delta functions plus the doublet.

EXAMPLE 1-11

Given

$$f(t) = 2e^{-3t}u(t)$$

Find and plot $f'(t)$ and $f''(t)$.

Solution.   Figure 1-12(a) to (c) shows $f(t)$, $f'(t)$, and $f''(t)$ plotted. Formally,

$$f'(t) = 2e^{-3t}\delta(t) - 6e^{-3t}u(t) = 2\delta(t) - 6e^{-3t}u(t)$$

$$f''(t) = 2\delta'(t) - 6e^{-3t}\delta(t) + 18e^{-3t}u(t)$$

$$= 2\delta'(t) - 6\delta(t) + 18e^{-3t}u(t)$$

We note these results agree with Figure 1-12(b) and (c).

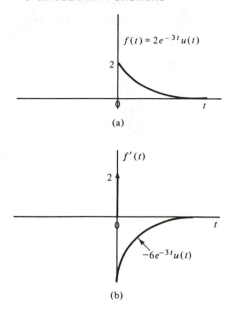

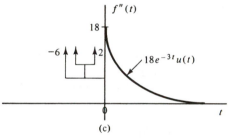

Figure 1-12 $f(t), f'(t)$, and $f''(t)$ for Example 1-11.

There is one common mistake that is often made, particularly when occurring in a more complex situation. The mistake in Example 1-11 is in effect saying

$$f'(t) = 2e^{-3t}\delta(t) - 6e^{-3t}u(t) \qquad \text{(which is fine)}$$

and then differentiating by parts:

$$f''(t) = 2e^{-3t}\delta'(t) - 6e^{-3t}\delta(t) - 6e^{-3t}\delta(t) + 18e^{-3t}u(t) \qquad \text{(which is incorrect)}$$

Using the properties:

$$f(t)\delta(t - a) = f(a)\delta(t - a) \quad \text{and} \quad f(t)\delta'(t) = -f'(0)\delta(t)$$

we correctly obtain:

$$f'(t) = 2\delta(t) - 6e^{-3t}u(t) \qquad \text{(correct)}$$

and $\qquad f''(t) = 2\delta'(t) - 6\delta(t) + 18e^{-3t}u(t) \qquad \text{(correct)}$

The circled term is in error because we cannot differentiate $f(t)\delta(t-a)$ by parts to obtain two terms. The delta function sifts out $f(a)$ and thus

$$\frac{d}{dt}[f(t)\delta(t-a)] = f(a)\delta'(t-a)$$

not $\frac{d}{dt}[f(t)\delta(t-a)] = f(a)\delta'(t-a) + \boxed{f'(a)\delta(t-a)}$ (error term)

1-3-8 The Language of "Density Functions" (A Challenging Comment)

Throughout engineering we continually talk about "density functions." In physics courses linear charge density $\rho_L(l)$ in coulombs per meter is a quantity that when integrated $\int_a^b \rho_L \, dl$ gives the total charge from a to b on the line. If $\rho_L(5) = 10$ and $\rho_L(3) = 2$, then the total charge in a small interval Δl about $l = 5$ is five times greater than in a small interval about $l = 3$. The presence of a delta function in a linear charge density indicates a point charge is present there. For example, if the "x axis" has a charge density:

$$\rho(x) = 2 + 0.5\delta(x-1) + 4\delta(x), \qquad -5 < x < 2$$
$$= 0, \qquad \text{otherwise}$$

This says that a point charge of 0.5 C is present at $x = 1$ and a point charge of 4 C is present at $x = 0$. The total charge between $x = -5$ and $x = 2$ is given by:

$$Q = \int_{-5}^{2} [2 + 0.5\delta(x-1) + 4\delta(x)] \, dx$$
$$= 2(7) + 0.5 + 4$$
$$= 18.5 \text{ C}$$

In our future studies we will consider three very famous density functions:

1. the energy spectral density function $E(f)$
2. the power spectral density function $S(f)$
3. the probability density function $f_X(\alpha)$

Each of these three functions, similar to linear charge density in physics, is such that when integrated from a to b it will give either:

1. the total signal energy content from a to b Hz,
2. the total signal power content from a to b Hz, or
3. the probability of assuming a value between a and b.

EXAMPLE 1-12

Given the power spectral density of a noise waveform is:

$$S(f) = 0.6\delta(f) + 0.2, \qquad -10 < f < 10$$

$$= 0, \qquad \text{otherwise}$$

Find:

(a) the total power contained between 6 and 8 Hz
(b) the total power contained below 2 Hz
(c) the total average power of the waveform

Solution. By definition the **power spectral density** is a function that measures the distribution of power with respect to frequency. The total average power contained between f_1 and f_2 is $\int_{-f_2}^{-f_1} S(f)\,df + \int_{f_1}^{f_2} S(f)\,df$ which must equal $2\int_{f_1}^{f_2} S(f)\,df$. If the density function is continuous, then $2S(f)\,\Delta f$ is the total output power of a bandpass filter of bandwidth Δf about f (where $\Delta f \to 0$). Solving the problem we obtain:

(a)
$$P_{av}(6 < f < 8) = 2\int_6^8 0.2\,df$$

$$= 0.8 \text{ W}$$

(b)
$$P_{av}(|f| < 2) = \int_{-2}^2 [0.6\delta(f) + 0.2]\,df$$

$$= 0.6 + 0.8$$

$$= 1.4 \text{ W}$$

(c)
$$P_{av} = \int_{-\infty}^{\infty} S(f)\,df$$

$$= \int_{-10}^{10} (0.6\delta(f) + 0.2)\,df$$

$$= 0.6 + 4$$

$$= 4.6 \text{ W}$$

Physically $S(f_1)$ as compared to $S(f_2)$ gives the relative power contained in a small bandwidth about f_1 as compared to a small bandwidth about f_2. The fact that $S_x(f)$ contains a delta function at $f = 0$ indicates there is a specific power content at $f = 0$ and that our waveform has a dc component. The ratio of the specific power content in a tiny bandwidth about $f = 0$ as compared to power content in a tiny bandwidth about any other frequency is infinite.

EXAMPLE 1-13

Given the density function associated with the values obtained when sampling a waveform is $f_X(\alpha) = 0.2\delta(\alpha + 1) + 0.3\delta(\alpha - 2) + \alpha \sqcap (\alpha - 0.5)$ where X describes the value obtained, find:

(a) the probability X is less than 0.6
(b) the probability $-1 < X < 4$

Solution.   By definition the probability density function for any random quantity yields on integration from α_1 to α_2 the probability $\alpha_1 < X \le \alpha_2$. Denoting the word probability by P and observing the density function in Figure 1-13, we obtain:

(a)
$$P[X < 0.6] = \int_{-\infty}^{0.6} [0.2\delta(\alpha + 1) + 0.3\delta(\alpha - 2)$$
$$+ \alpha \sqcap (\alpha - 0.5)] \, d\alpha$$
$$= 0.2 + 0 + \int_{0}^{0.6} \alpha \, d\alpha$$
$$= 0.2 + 0.18$$
$$= 0.38$$

(b)
$$P[-1 < X < 4] = \int_{-1^+}^{4} f_X(\alpha) \, d\alpha$$
$$= \int_{-1^+}^{4} 0.2\delta(\alpha + 1) + \int_{-1^+}^{4} 0.3\delta(\alpha - 2) \, d\alpha$$
$$+ \int_{0}^{1} \alpha \, d\alpha$$
$$= 0 + 0.3 + 0.5$$
$$= 0.8$$

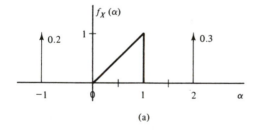

(a)

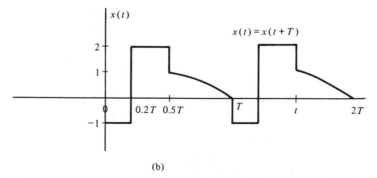

(b)

Figure 1-13   (a) The density function for Example 1-13; (b) a periodic waveform whose sampled values yield $f_X(\alpha)$.

Examples 1-12 and 1-13 were included to encourage intuitiveness. If they are too shocking, they can safely be ignored but the effort in considering them can be beneficial.

Drill Set: Singularity Functions

1. Given

$$f(t) = 8t^2 \prod (t)$$

Find and plot $f'(t) f''(t)$, and $f'''(t)$.

2. Given

$$f(t) = 2\delta(t + 2) - \prod (0.5t - 1.5) - 3\delta'(t - 2)$$

Find and plot $F(t)$ the integral of $f(t)$, and $G(t)$ the integral of $F(t)$.

1-3-9 The Discrete Function $\delta(n)$

The pulse function $\delta(n)$, called the "Kronecker delta," is much simpler to handle than its continuous counterpart since it is just a sequence of one number. However, algebraically many of its properties are analogous to those of $\delta(t)$. These are:

Property 1

$$\sum_{n=-\infty}^{\infty} f(n)\delta(n - k) \equiv f(k) \qquad (1\text{-}15)$$

Property 2

$$f(n)\delta(n - k) \equiv f(k)\delta(n - k) \qquad (1\text{-}16)$$

Property 3

$$\delta(an + b) \equiv \delta\left(n + \frac{b}{a}\right) \qquad (1\text{-}17)$$

We will only use this if a and b/a are integers.

Property 4 (to be discussed later)

$$f(n) * \delta(n - k) \triangleq \sum_{p=-\infty}^{\infty} f(p)\delta(n - p - k)$$

$$= f(n - k) \qquad (1\text{-}18)$$

EXAMPLE 1-14
Prove the discrete delta function properties (1-15) through (1-18).

Solution

(a) By definition: In $\sum_{n=-\infty}^{\infty} f(n)\delta(n-k)$ every value of $f(n)$ except $f(k)$ is multiplied by zero. Therefore

$$\sum_{-\infty}^{\infty} f(n)\delta(n-k) = f(k) \cdot 1$$

$$= f(k)$$

(b) $f(n)\delta(n-k) \equiv f(k)\delta(n-k)$ if:

$$\sum_n g(n)[f(n)\delta(n-k)] \equiv \sum_n g(n)[f(k)\delta(n-k)]$$

Both sides obviously yield $g(k)f(k)$.

(c) From part (a),

$$\sum_{n=-\infty}^{\infty} f(n)\delta(n) = f(0)$$

Therefore Property 3 is proved if:

$$\sum_n f(n)\delta(an+b) \equiv \sum_n f(n)\delta\left(n + \frac{b}{a}\right)$$

By definition the RHS is $f(-b/a)$. For the LHS if $a > 0$, we let

$$an + b = p \quad \text{and} \quad n = \frac{p-b}{a}$$

Therefore

$$\text{LHS} = \sum_{\text{all }p} f\left(\frac{p-b}{a}\right)\delta(p)$$

$$= f\left(-\frac{b}{a}\right)$$

We notice this property does not correspond exactly to:

$$\delta(at+b) = \frac{1}{|a|}\delta\left(t + \frac{b}{a}\right)$$

in the continuous case. Also, we require b/a to be an integer.

(d) Property 4 is the definition of the discrete convolution of a function with the discrete pulse function and will be encountered in the next chapter. By definition:

$$\sum_p f(p)\delta(n-p-k) = \sum_p f(p)\delta[-(p-n+k)]$$

$$= \sum_p f(p)\delta[p-(n-k)]$$

$$= f(n-k)$$

Readers who are completely unfamiliar with discrete functions must study this example with care and not be too mechanical. Consider specific sequences, plot them, and interpret the general results with specific numbers. Do the following drill set slowly and carefully.

Drill Set: Discrete Pulse Function $\delta(n)$

1. Given

$$f(n) = 2^n$$

Evaluate:

(a) $\sum_{-\infty}^{\infty} f(n)\delta(n + 2)$
(b) $\sum_{\infty}^{\infty} f(n)\delta(0.5n + 1.5)$
(c) $\sum_{p=-\infty}^{\infty} f(p)\delta(n - p + 2) = F(n)$ for $n = -4, 0,$ and 3.

Plot $F(n)$ for all n.

SUMMARY

This chapter introduced signals and operations on them that are in prevalent use throughout the time and transform analysis of linear time-invariant systems. Although all operations were discussed in the continuous or discrete time domain, they are also used frequently in the transform domain.

Some of the more commonly occurring and useful notational functions were first considered. Chief among these were $\prod(t)$, $u(t)$, $\mathrm{Sinc}(t)$, $\delta(n)$, and $u(n)$. The basic operations of reflecting, time-scaling, and time-shifting were then treated. These are operationally summarized by:

$$f(\pm at \pm b) = f\left[\pm a\left(t \pm \frac{b}{a}\right)\right]$$
$$\underset{①}{\quad} \underset{②}{\quad} \underset{③}{\quad}$$

or

$$f(\pm an \pm b) = f\left[\pm a\left(n \pm \frac{b}{a}\right)\right]$$
$$\underset{①}{\quad} \underset{②}{\quad} \underset{③}{\quad}$$

where ① indicates whether or not the function is reflected, ② indicates time-scaling, and ③ indicates shifting left or right by b/a. Since discrete functions are defined only where n is an integer, for them a and b/a must be integers.

Singularity functions were considered next. A delta function located at $t = a$ was defined by $\int_{-\infty}^{\infty} f(t)\delta(t - a) \, dt = f(a)$ and the unit pulse function at $n = k$ could analogously be defined as:

$$\sum_{n=-\infty}^{\infty} f(n)\delta(n - k) = f(k)$$

The main properties of continuous and discrete delta functions are tabulated in Table 1-1.

For the continuous case higher-order singularity functions were defined by $\int_{-\infty}^{\infty} f(t)\delta^n(t) \, dt \triangleq (-1)^n f^n(0)$. The use of singularity functions in integration and differentiation was stressed. Shown also in Table 1-1 are the definitions of integration and differentiation along with their discrete counterparts, that is, summations and differences.

TABLE 1-1

Continuous	Discrete
$\int_{-\infty}^{\infty} f(t)\delta(t - a) \, dt = f(a)$	$\sum_{-\infty}^{\infty} f(n)\delta(n - k) = f(k)$
$f(t)\delta(t - a) \equiv f(a)\delta(t - a)$	$f(n)\delta(n - k) = f(k)\delta(n - k)$
$\delta(at + b) \equiv \dfrac{1}{\|a\|}\delta\left(t + \dfrac{b}{a}\right)$	$\delta(an + b) \equiv \delta\left(n + \dfrac{b}{a}\right)$
$\int_{-\infty}^{\infty} f(p)\delta(t - p - a) \, dp = f(t - a)$	$\sum_{p} f(p)\delta(n - p - k) \equiv f(n - k)$
Integration	**Discrete Integration or Summation**
$F(t) = \int_{-\infty}^{t} f(p) \, dp$	$F(n) = \displaystyle\sum_{p=-\infty}^{n} f(p)$
Differentiation	**Differences**
$f'(t) = \dfrac{f(t + h) - f(t)}{h}$	$f(n) - f(n - 1)$
or	or
$\dfrac{f\left(t + \dfrac{h}{2}\right) - f\left(t - \dfrac{h}{2}\right)}{h}$	$f(n + 1) - f(n)$
	Approximate Derivative
or	$\dfrac{f(n\tau) - f[(n - 1)\tau]}{\tau}$
$\dfrac{f(t) - f(t - h)}{h}, \quad h \to 0$	or
	$\dfrac{f[(n + 1)\tau] - f(n\tau)}{\tau}, \quad \tau \text{ "small"}$

PROBLEMS

1-1. Plot the following functions:

(a) $u(n) + u(-n)$ (b) $u(n) + u(-n + 1)$

(c) $u(n) + u(-n - 1)$ (d) $(-1)^n u(n)$

(e) $2^n u(-n - 1) + n(n - 1)u(n)$ (f) $(0.5)^n u(-n - 1) + 0.5^n u(n)$

(g) $2^{-n} u(-n - 1) + n0.5^{n-1} u(n)$ (h) $u(\cos \pi n)$

(i) $e^{2t} u(-t) + e^{-2t} u(t)$ (j) $e^{-2t} u(-t) + e^{-2t} u(t)$

(k) $e^{2t} u(-t) + e^{2t} u(t)$ (l) $e^{-2t} u(-t) + e^{2t} u(t)$

(m) $u(t^2 - 16)$ (n) $\prod(t^2 - 9)$

(o) $u(\cos t)$ (p) $\prod(\cos t)$

1-2. Given

$$x(t) = e^t, \qquad t < 0$$

$$= t, \qquad 0 < t < 2$$

$$= 0, \qquad \text{otherwise}$$

Plot and find an analytic expression for the following:

(a) $x(t - 4)$ (b) $x(3 - t)$ (c) $x(4t + 2)$ (d) $x(-4t - 2)$

1-3. Given

$$f(n) = n(n - 1)0.5^{n-2} u(n)$$

Plot and find an analytic expression for the following:

(a) $f(2 - n)$ (b) $f(n + 5)u(n)$

1-4. Given

$$f(t) = t[u(t) - u(t - 2)] \quad \text{and} \quad g(t) = 2\prod(t)$$

Find for what values of t the product functions:

(a) $h(p) = f(p)g(t - p) = 0$ (b) $k(p) = f(p)g(p - t) = 0$

Specifically find and sketch $h(p)$ and $k(p)$ versus p for $t = 0.5$.

1-5. Consider the following discrete elements with their input–output definitions:

- the delay element

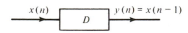

- the multiplier element

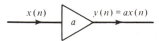

- the summer element

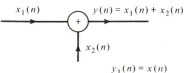

- the branching element

Given

$$x(n) = 0.5^n u(n)$$

Find an expression for and plot $y(n)$ for the following systems:

(a)   **(b)**

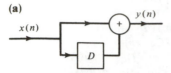

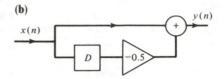

1-6. Given

$$f(n) = n[u(n) - u(n - 4)]$$

(a) Plot and evaluate $f(n - k)$ versus n for $k = -4, 1,$ and 3.
(b) Discuss the behavior of $f(n - k)$ versus n as k varies $-\infty < k < \infty$.
(c) Plot and evaluate $f(k - n)$ versus n for $k = -4, 1,$ and 3.
(d) Discuss the behavior of $f(k - n)$ versus n as k varies $-\infty < k < \infty$.
(e) Consider $h(n) = f(n) f(n - k)$. For what values of k is $h(n) = 0$?
(f) Consider $l(n) = f(n) f(k - n)$. For what values of k is $l(n) = 0$?

1-7. Check which of the following may be used as a weighted mathematical model of a delta function or a derivative of a delta function:

(a) $\lim\limits_{a \to 0} \left[\dfrac{1}{a^2} \Lambda^2 \left(\dfrac{t}{a} \right) \right]$

(b) $\lim\limits_{a \to 0} \dfrac{5a}{a^2 + 4\pi^2 t^2}$

(c) $\lim\limits_{a \to \infty} \dfrac{8\pi^2 at}{(a^2 + 4\pi^2 t^2)^2}$

(d) $\lim\limits_{a \to \infty} \dfrac{\operatorname{Sin} \pi at}{\pi at}$

(e) $\lim\limits_{a \to \infty} \dfrac{a}{\sqrt{1 + a^2 t^2}}$

(f) $\lim\limits_{a \to \infty} \dfrac{a}{\pi} \dfrac{1}{\operatorname{Cosh} at}$

(g) $\lim\limits_{a \to 0} \dfrac{2at}{a^2 + 4\pi^2 t^2}$

(h) $\int_{-\infty}^{\infty} e^{-j2\pi ft} \, df$

1-8. Evaluate:

(a) $\int_{-\infty}^{\infty} t \operatorname{Sin} t\delta \left(t - \dfrac{\pi}{2} \right) dt$

(b) $\int_{-5}^{4} 4t^3 \delta(-4t + 17) \, dt$

(c) $\int_{-4}^{5} 4t^3 \delta(-4t + 17) \, dt$

(d) $\int_{-4}^{5} 4t^3 \delta'(-4t + 17) \, dt$

1-9.

(a) Prove $f(t)\delta'(t) = -f'(0)\delta(t)$
(b) Is $\delta'(at + b) \equiv (1/|a|)\,\delta'(t + b/a)$? If the answer is no, what should the right-hand side be?

1-10.

(a) Plot:

$$f(t) = \delta'(t + 4) - 2\delta(t + 4) + t\delta(t + 1) + 2e^{-t}u(t + 1)$$

(b) Plot and evaluate (or vice versa):

$$F(t) = \int_{-\infty}^{t} f(u) \, du$$

(c) Plot and evaluate:

$$G(t) = \int_{-\infty}^{t} F(p)\, dp = \int_{-\infty}^{t}\left[\int_{-\infty}^{p} f(u)\right] dp$$

1-11. Consider the waveform $g(t)$:
(a) Plot and evaluate $g'(t)$. Use step functions in your analytic expression.
(b) Plot and evaluate $g''(t)$.
(c) Check your answer to (b) by integration:

$$g'(t) = \int_{-\infty}^{t} g''(p)\, dp$$

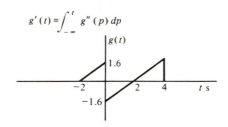

1-12. The probability $X \le \alpha$ is found by integrating the probability density function $f_x(\alpha)$ to obtain:

$$P[X \le \alpha] = \int_{-\infty}^{\alpha^+} f_X(p)\, dp$$

Given

$$f_X(\alpha) = 0.3\delta(\alpha + 1) + 0.4\delta(\alpha - 2) + 0.1\,[u(\alpha - 1) - u(\alpha - 4)]$$

Find:
(a) $P[X \le 2]$ (b) $P[X > 1]$ (c) $P[-1 < X < 2.6]$
(d) Find and plot the cumulative distribution function:

$$F_X(\alpha) = P[X \le \alpha] = \int_{-\infty}^{\alpha^+} f_X(p)\, dp$$

1-13.
(a) Plot $f(t) = 2e^{-(t-1)} u(t - 1) + t^2\delta(t - 2)$
(b) Evaluate $g(t) = (d/dt)[f(t)]$
(c) $K(t) = (d/dt)[g(t)]$
(d) Check $\int_{-\infty}^{t} g(l)\, dl = f(t)$

1-14. Evaluate and plot as a function of n:

(a) $F(n) = \sum_{k=-\infty}^{n} 2^k \delta(k - 3)$

(b) $G(n) = \sum_{k=-\infty}^{n} 2^k \delta(3 - k)$

(c) $H(n) = \sum_{p=0}^{\infty} 2^p (\tfrac{1}{4})^{n-p}$

Hint:

$$\sum_{0}^{\infty} \alpha^n = \frac{1}{1 - \alpha}, \quad \alpha \text{ real}, |\alpha| < 1$$

$$\sum_{p=0}^{n} \alpha^p = \frac{\alpha^{n+1} - 1}{\alpha - 1}, \quad \text{any real } \alpha$$

Linear Time-Invariant Systems with Deterministic Inputs

INTRODUCTION

Chapter 2 considers the time-domain analysis of linear, time-invariant, causal (LTIC) systems with continuous and discrete deterministic inputs. Although this chapter is quite extensive, it forms the foundation for linear system theory and the different transforms encountered throughout the text.

The case of a continuous system is treated in Sections 2-1 through 2-3 and that of a discrete system in Sections 2-4 through 2-6. Since the input and output of these systems are governed by a differential or difference equation, a review is given for solving linear differential equations with constant coefficients and the solution of linear difference equations with constant coefficients is covered. The concept of the system function $H(s)$ or $H(z)$ and its relation to the governing differential or difference equation is developed. As an extension of solving differential or difference equations, the response to an impulse function $\delta(t)$, called the impulse response $h(t)$ for a continuous system, and the response due to a unit pulse $\delta(n)$, called the pulse response $h(n)$ for a discrete system, are found. Then the general zero-state output of a continuous system is derived as the convolution integral of the input with the impulse response and as the convolution summation of the input with the pulse response for a discrete system. A thorough general treatment of convolution integrals and summations is given.

2-1 THE SYSTEM DIFFERENTIAL EQUATION AND SYSTEM FUNCTION $H(s)$

2-1-1 Review of Solving Differential Equations

As a prelude to the time-domain analysis of continuous, linear, time-invariant causal systems* (LTIC), a tutorial summary of solving linear differential

---

*The terms *linear, time-invariant,* and *causal* are assumed to be familiar but in Problem 2-5 they are defined and discussed.

equations with constant coefficients is considered. The output $y(t)$ and input $x(t)$ of a LTIC system are related by a linear differential equation of the form:

$$a_n y^n(t) + a_{n-1} y^{n-1}(t) + \cdots + a_0 y(t)$$
$$= b_0 x(t) + b_1 x'(t) + \cdots + b_m x^m(t) \qquad (2\text{-}1)$$

The right-hand terms are often lumped together as:

$$f(t) = b_0 x(t) + b_1 x^1(t) + \cdots + b_m x^m(t)$$

and called the forcing function. If the system differential equation is obtained from a passive network where the input $x(t)$ is a voltage or current source and $y(t)$ is a branch voltage or current, then all the b and a coefficients are constants. The following theorem summarizes the theory of linear differential equations:

Theorem. Given

$$a_n y^n(t) + a_{n-1} y^{n-1}(t) + \cdots + a_0 y(t) = f(t) \qquad (2\text{-}2)$$

and the initial conditions $y(0), \ldots, y^{n-1}(0)$, the complete response is of the form

$$y(t) = y_{ho}(t) + y_{fo}(t) \qquad (2\text{-}3)$$

where $y_{ho}(t)$ the homogeneous response is the solution to the differential equation with $f(t) = 0$ and contains n arbitrary constants and $y_{fo}(t)$ the forced response is that one particular solution to the differential equation that contains no part of $y_{ho}(t)$. The n arbitrary constants in Equation 2-3 may be found by applying the values for $y(0), y'(0), \ldots, y^{n-1}(0)$ to it.

There is much alternative terminology used. The terms *homogeneous, natural, free, complimentary function,* and *transient* [if $y_{fo}(t)$ does not approach zero for $t \to \infty$] *response* are used interchangeably. Also, the terms *forced, particular integral,* and, if appropriate, *final,* and *steady-state* (when are these appropriate?) *response* are used interchangeably. Our use and definition of forced response should be noted. In most classical mathematics books the particular integral or **forced response** is defined as "any solution to the differential equation." For example, if we consider the differential equation:

$$\frac{dy}{dt} + 4y(t) = 8$$

our definition specifically yields $y_{fo}(t) = 2$, but the classical definition allows:

$$y_{fo}(t) = 2 + 3e^{-4t} \quad \text{or} \quad y_{fo}(t) = 2 - 5e^{-4t}, \qquad \text{and so on}$$

as particular integrals because when substituted into the differential equation it is satisfied. A number of problems on solving differential equations are given at the end of the chapter but it is assumed the reader is already competent analytically and philosophically on the topic. A review problem will now be solved.

EXAMPLE 2-1
Solve:

$$y''(t) + 5y'(t) + 6y(t) = 3e^{2t}$$

given

$$y(0) = 2, \qquad y'(0) = 3$$

Solution

Step 1: $y_{ho}(t)$
The homogeneous equation is:

$$y''(t) + 5y'(t) + 6y(t) = 0$$

On guessing $y_{ho}(t) = Ae^{mt}$, we obtain:

$$Ae^{mt}(m^2 + 5m + 6) = 0$$

and no value of A (except zero) or e^{mt} (except $m = -\infty$) makes this equation zero via the Ae^{mt} term.

Any value of m satisfying the characteristic equation:

$$m^2 + 5m + 6 = 0$$

yields a solution for $y_{ho}(t)$.

Therefore $y_{ho}(t) = A_1 e^{-2t} + A_2 e^{-3t}$

Step 2: $y_{fo}(t)$
Consider:

$$y''(t) + 5y'(t) + 6y(t) = 3e^{2t}$$

Based on our knowledge of calculus we guess that

$$y_{fo}(t) = Ae^{2t}$$

and substitute this into the equation.

Therefore $(4A + 10A + 6A)e^{2t} \equiv 3e^{2t}$ and $y_{fo}(t) = 0.15e^{2t}$

Step 3: $y(t)$
Combining steps 1 and 2, we obtain as the form of the complete response:

$$y(t) = A_1 e^{-2t} + A_2 e^{-3t} + 0.15e^{2t}$$

Step 4: Applying the initial conditions
Applying $y(0) = 2$ and $y'(0) = 3$ yields the simultaneous equations:

$$2 - 0.15 = A_1 + A_2 \quad \text{or} \quad A_1 + A_2 = 1.85$$

$$3 - 0.3 = -2A_1 - 3A_2 \quad \text{or} \quad -2A_1 - 3A_2 = 2.7$$

with solution $A_1 = 8.25$ and $A_2 = -6.4$. Finally:

$$y(t) = 8.25e^{-2t} - 6.4e^{-3t} + 0.15e^{2t}$$

This is the one unique function that when you add its second derivative to five times its first derivative to six times the function the resultant is $3e^{2t}$ and $y(t)$ has an initial value of 2 and a slope of 3. The solution holds for all time, $-\infty < t < \infty$.

2-1-2 The System Function $H(s)$

Given a system with a governing linear differential equation with constant coefficients as in Equation 2-1, we abstractly define the system function $H(s)$ as:

$$H(s) \triangleq \frac{y_{f0}(t)}{x(t)}, \qquad \text{when } x(t) = e^{st} \tag{2-4}$$

In words, the system function is the forced response divided by the input when the input is e^{st}. We now find $H(s)$ corresponding to Equation 2-1. The differential equation with $x(t) = e^{st}$ is:

$$a_n y''(t) + a_{n-1} y^{n-1}(t) + \cdots + a_0 y(t) = (b_0 + sb_1 + s^2 b_2 + \cdots + s^m b_m)e^{st}$$

where if $s = \alpha$ the forcing term is similar to, say, $7.3e^{\alpha t}$. For the forced response we assume $y(t) = Ae^{st}$ where A is to be found. On substituting into Equation 2-1, we obtain:

$$(a_n s^n + a_{n-1} s^{n-1} + \cdots a_0) Ae^{st} = (b_0 + b_1 s + \cdots + b_m s^m)e^{st}$$

and A must be:

$$A = \frac{b_m s^m + b_{m-1} s^{m-1} + \cdots + b_0}{a_n s^n + a_{n-1} s^{n-1} + \cdots + a_0}$$

Now

$$H(s) = \frac{y_{f0}(t)}{e^{st}} = \frac{Ae^{st}}{e^{st}}$$

$$= \frac{b_m s^m + b_{m-1} s^{m-1} + \cdots + b_0}{a_n s^n + a_{n-1} s^{n-1} + \cdots + a_0} \tag{2-5}$$

We notice that $H(s)$ may be found by inspection of the system differential equation. For example, if the system is governed by:

$$y'(t) + 6y(t) = 2x(t)$$

$$H(s) = \frac{2}{s + 6}$$

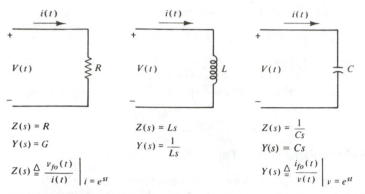

$$Z(s) = R$$
$$Y(s) = G$$
$$Z(s) \triangleq \left. \frac{v_{fo}(t)}{i(t)} \right|_{i\,=\,e^{st}}$$

$$Z(s) = Ls$$
$$Y(s) = \frac{1}{Ls}$$

$$Z(s) = \frac{1}{Cs}$$
$$Y(s) = Cs$$
$$Y(s) \triangleq \left. \frac{i_{fo}(t)}{v(t)} \right|_{v\,=\,e^{st}}$$

Figure 2-1   Exponential impedance and admittance for the network elements.

Obviously, finding $H(s)$ knowing the system equation provides no new insight, but in practice the procedure of finding $H(s)$ first, which is simple, provides a powerful means for finding the governing system equation.

The two most famous system functions for a system composed of resistors, inductors, and capacitors (plus possibly ideal transformers, dependent sources, and gyrators) are the impedance function $Z(s)$ and the admittance function $Y(s)$. These are defined as:

$$Z(s) = \left. \frac{v_{fo}(t)}{i(t)} \right|_{i(t)\,=\,e^{st}} \tag{2-6}$$

and

$$Y(s) = \left. \frac{i_{fo}(t)}{v(t)} \right|_{v(t)\,=\,e^{st}} \tag{2-7}$$

and their values are shown for a resistor, inductor, and capacitor in Figure 2-1. It is very easy to verify the well-known rules for combining impedances and admittances in series and parallel and for the voltage and current division of a forced exponential voltage or current. We now demonstrate briefly how to find a system's differential equation using the system function.

EXAMPLE 2-2

For the system shown in Figure 2-2(a) find:

(a) the governing system differential equation
(b) the forced response if
    (i) $x(t) = 3e^{-t}$
    (ii) $x(t) = 6$
    (iii) $x(t) = 2 \cos (4t - 40°)$
(c) the complete response when $x(t) = 6$ and $y(0) = 2$, $y'(0) = -1$

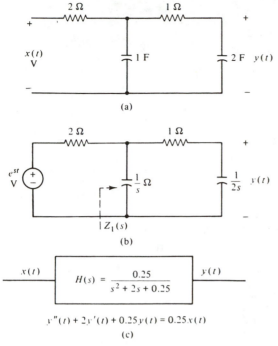

Figure 2-2 (a) The circuit of Example 2-2; (b) the exponential forced response model; (c) the system function and SDE.

Solution

(a) The forced response model for the system when $x(t) = e^{st}$ is shown in Figure 2-2b. Observing the circuit:

$$Z_1(s) = \frac{(1/s)(1 + 1/2s)}{1/s + 1 + 1/2s}$$

$$= \frac{2s + 1}{2s^2 + 3s}$$

and using voltage division twice:

$$H(s) = \frac{e^{st}(2s + 1)/(2s^2 + 3s)}{2 + [(2s + 1)/(2s^2 + 3s)]} \times \frac{1/2s}{1 + 1/2s} \div e^{st}$$

$$= \frac{0.25}{s^2 + 2s + 0.25} \quad \text{(with work)}$$

we obtain the governing system differential equation:

$$y''(t) + 2y'(t) + 0.25y(t) = 0.25x(t)$$

and the system representation of the circuit is shown in Figure 2-2(c).

(b) (i) if $x(t) = 3e^{-t}$

$$y_{f0}(t) = H(-1)3e^{-t}$$

$$= \frac{0.25}{1 - 2 + 0.25} 3e^{-t}$$

$$= -e^{-t}$$

(ii) $y_{f0}(t) = \dfrac{0.25}{0 + 0 + 0.25} 6$

$$= 6$$

(iii) $y_{f0}(t) = \text{Re}\left[\dfrac{0.25}{(j4)^2 + 2(j4) + 0.25} 2 \lfloor -40^0 e^{j4t}\right]$

where "Re" indicates the "real part of." With previous knowledge of manipulations from a first circuits course this yields:

$$y_{f0}(t) = 0.028 \text{ Cos } (4t + 167^0)$$

(c) The characteristic equation is:

$$m^2 + 2m + 0.25 = 0$$

Therefore $(m + 1)^2 = 0.75$

and $m = -0.13$ and -1.87. The homogeneous response is:

$$y_{h0}(t) = A_1 e^{-0.13t} + A_2 e^{-1.87t}$$

The forced response is $y(t) = 6$ and the complete response is of the form:

$$y(t) = A_1 e^{-0.13t} + A_2 e^{-1.87t} + 6$$

Applying the initial conditions $y(0) = 2$ and $y'(0) = -1$, we get:

$$A_1 + A_2 = -4$$

$$-0.13A_1 - 1.87A_2 = -1$$

which have solutions $A_1 = -4.87$ and $A_2 = 0.87$. Finally:

$$y(t) = -4.87e^{-0.13t} + 0.87e^{-1.87t} + 6$$

2-2 THE IMPULSE RESPONSE $h(t)$

If a LTIC system is governed by the differential equation:

$$a_n y^n(t) + a_{n-1} y^{n-1}(t) + \cdots + a_0 y(t) = b_0 x(t) + \cdots + b_m x^m(t) \qquad (2\text{-}8)$$

or characterized by the system function:

$$H(s) = \frac{b_m s^m + b_{m-1} s^{m-1} + \cdots + b_0}{a_n s^n + a_{n-1} s^{n-1} + \cdots + a_0}$$

we define the impulse response $h(t)$ as the response $y(t)$ when:

$$x(t) = \delta(t) \quad \text{and} \quad y(t) = 0, \quad -\infty < t < 0$$

$$h(t) = y(t) \mid_{x(t) = \delta(t)} \tag{2-9}$$

Later the impulse response will be found with ridiculous ease using the Laplace transform, but we now consider the classical time-domain derivation.

EXAMPLE 2-3

Find the impulse response $h(t)$ for the system governed by:

(a) $2y'(t) + 3y(t) = 5x(t)$
(b) $2y'(t) + 3y(t) = 5x(t) + 4x'(t)$

Solution

(a) We are required to solve:

$$2y'(t) + 3y(t) = 5\delta(t)$$

We want to obtain the one unique function $h(t)$ such that:

$$2h'(t) + 3h(t) = 5\delta(t)$$

We assume:

$$h(t) = Ae^{-1.5t}u(t) + 0\delta(t) \tag{1}$$

The motivation for this is that for $t > 0$, $\delta(t) = 0$ and we have the homogeneous equation.

Also, $h(t)$ cannot contain a delta function, because then $2h'(t)$ would contribute a weighted derivative of a delta function to the right-hand side. Substituting (1) into the equation yields:

$$2\frac{d}{dt}[Ae^{-1.5t}u(t)] + 3Ae^{-1.5t}u(t) \equiv 5\delta(t)$$

Equating the coefficients of $\delta(t)$, we obtain:

$$2Ae^{-1.5t}\Big|_{t=0}\delta(t) \equiv 5\delta(t)$$

or $2A\delta(t) = 5\delta(t)$ [using the sifting property of $\delta(t)$]

and $A = 2.5$

Therefore $h(t) = 2.5e^{-1.5t}u(t)$

Figure 2-3(a) shows the plots of $h(t)$, $3h(t)$, $2h'(t)$, and $3h(t) + 2h'(t)$ which is equal to $5\delta(t)$. $h(t)$ is that unique transient response that satisfies initial conditions at $t = 0^+$ created by $x(t) = \delta(t)$. Of course, observation of Figure 2-3 makes it clear that with a little experience this result can be determined by inspection.

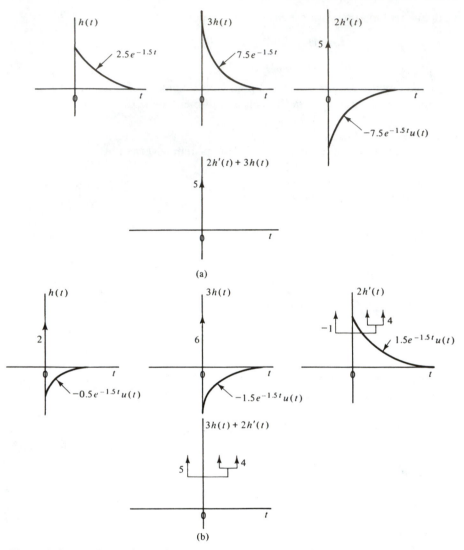

Figure 2-3 (a) $h(t)$, $3h(t)$, $2h'(t)$, and verification that $2h'(t) + 3h(t) = 5\delta(t)$ for the system of Example 2-3(a); (b) $h(t)$, $3h(t)$, $2h'(t)$, and verification that $2h'(t) + 3h(t) = 5\delta(t) + 4\delta'(t)$ for the system of Example 2-3(b).

(b) We are required to solve:

$$2y'(t) + 3y(t) = 5\delta(t) + 4\delta'(t)$$

We assume:

$$h(t) = Ae^{-1.5t}u(t) + B\delta(t) \qquad (1)$$

The weighted delta function must be present so that $2y'(t)$ contributes $4\delta'(t)$ to the right-hand side.

Obviously

$$2B = 4 \quad \text{and} \quad B = 2 \quad \text{(be absolutely sure)}$$

The singularity functions contributed when substituting (1) in the differential equation and equating coefficients are:

$$2[B\delta'(t) + A\delta(t)] + 3B\delta(t) \equiv 5\delta(t) + 4\delta'(t)$$

This gives

$$2A + 3B = 5, \qquad \text{coefficient of } \delta(t)$$

$$B = 2, \qquad \text{coefficient of } \delta'(t)$$

Therefore

$$A = -0.5$$

and

$$h(t) = -0.5e^{-1.5t}u(t) + 2\delta(t)$$

Figure 2-3(b) shows $h(t)$, $3h(t)$, $2h'(t)$, and $2h'(t) + 3h(t)$, and we note that $h(t)$ is the one unique transient solving this equation.

EXAMPLE 2-4

Find the impulse response for the second-order system:

$$y''(t) + 3y'(t) + 2y(t) = 3x(t) + 2x'(t)$$

Solution. We must find the unique solution for:

$$y''(t) + 3y'(t) + 2y(t) = 3\delta(t) + 2\delta'(t)$$

Since the characteristic equation is:

$$m^2 + 3m + 2 = 0$$

$$(m + 2)(m + 1) = 0$$

We assume:

$$y(t) = A_1 e^{-t}u(t) + A_2 e^{-2t}u(t) + 0\delta(t)$$

since if $y(t)$ contains a delta function, $y''(t)$ would contribute $\delta''(t)$ to the right-hand side. Equating coefficients rapidly, we obtain:

$$A_1 + A_2 = 2, \qquad \text{coefficient of } \delta'(t)$$

$$3(A_1 + A_2) - A_1 - 2A_2 = 3, \qquad \text{coefficient of } \delta(t), \text{ be sure}$$

or

$$A_1 + A_2 = 2$$

$$2A_1 + A_2 = 3$$

which yields $A_1 = 1$ and $A_2 = 1$

and

$$h(t) = (e^{-t} + e^{-2t})u(t)$$

2-2-1 Comment—Comparing, "Finding *h(t)*," and "Solving Differential Equations"

Obtaining $h(t)$ may be compared to solving a differential equation classically but we must be careful. For example, in Example 2-3 we sought the unique solution to:

$$2y'(t) + 3y(t) = 5\delta(t) + 4\delta'(t) \tag{1}$$

and obtained:

$$h(t) = -0.5e^{-1.5t}u(t) + 2\delta(t)$$

Classically, we can say this is equivalent to solving:

$$2y'(t) + 3y(t) = 0, \qquad t > 0 \tag{2}$$

subject to an initial condition at $t = 0^+$ caused by $x(t) = \delta(t)$. Substituting into:

$$2y'(t) + 3y(t) = 5\delta(t) + 4\delta'(t)$$

with $$y(t) = Ae^{-1.5t}u(t) + 2\delta(t)$$

and $$y'(t) = Ae^{-1.5(0)}\delta(t) - 1.5Ae^{-1.5t}u(t) + 2\delta'(t)$$

$$= A\delta(t) - 1.5Ae^{-1.5t}u(t) + 2\delta'(t)$$

we have that:

$$2A\delta(t) + 6\delta(t) = 5\delta(t)$$

which gives

$$A = -0.5 \quad \text{or} \quad y(0^+) = -0.5$$

We may replace (1) by:

$$2y'(t) + 3y(t) = 0, \qquad t > 0$$

with $$y(0^+) = -0.5$$

which yields:

$$h(t) = -0.5e^{-1.5t}, \qquad t > 0$$

Indeed, the solution:

$$y(t) = -0.5e^{-1.5t}$$

satisfies $2y'(t) + 3y(t) = 0$ with $y(0) = -0.5$ for all $-\infty < t < \infty$ but as far as our system is concerned, the equation is only valid for $t > 0$ and it omits information about possible singularity functions at $t = 0$. Sometimes in circuit analysis the initial conditions are obtained at $t = 0^+$ using the conservation of energy from 0^- to 0^+ and then our solutions are only valid for $t > 0$ and may not address the possible singularity functions at $t = 0$.

EXAMPLE 2-5

Solve for $y(t)$ the capacitor current in the circuit of Figure 2-4 when $x(t) = \delta(t)$.

Solution. Consider the circuit exactly at $t = 0$. We assume that the $\delta(t)$ amperes flow through the capacitor because otherwise there would be an infinite voltage across the capacitor. Since $i_c(t) = \delta(t)$ at $t = 0$, the voltage across the capacitor at $t = 0^+$ is:

$$v_c(0^+) = 2 \int_{0^-}^{0^+} \delta(\alpha)\, d\alpha$$

$$= 2 \text{ V}$$

The form of $y(t)$ for $t > 0$ is:

$$y(t) = Ae^{-(1/R_{eq}C)t}$$

where

$$R_{eq}C = \left(\frac{6}{5}\right)\left(\frac{1}{2}\right) = 0.6$$

Therefore

$$y(t) = Ae^{-1.67t} \qquad \text{for } t > 0$$

Considering the circuit as shown in Figure 2-4(b) at $t = 0^+$ and using KCL with an initial capacitor voltage of 2 V, we obtain:

$$y(0^+) = -1 - 0.67 = -1.67$$

and

$$A = -1.67$$

Finally:

$$y(t) = -1.67e^{-1.67t}u(t) \qquad \text{for } t > 0$$

This answer is correct for $t > 0$, but at $t = 0$ we should verify the existence of a $\delta(t)$ term and argue that:

$$y(t) = \delta(t) - 1.67e^{-1.67t}u(t)$$

The reader should check this solution by finding:

$$H(s) = e^{st} \frac{0.5s}{0.5 + 0.5s + 0.33} \div e^{st} = \frac{s}{s + 1.67}$$

and solving for $h(t)$ as in Examples 2-3 and 2-4.

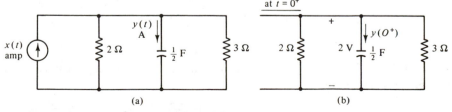

(a) (b)

Figure 2-4 (a) The circuit for Example 2-5; (b) the circuit at $t = 0^+$.

Drill Set: The Impulse Response

1. Given the solution to a differential equation is:

$$y(t) = 3e^{-2t} - e^{-t} + 5 \cos t$$

   where the forced response is $5 \cos t$ and the homogeneous response is $3e^{-2t} - e^{-t}$, find the differential equation with appropriate initial conditions.

2. Given the impulse response of a system is:

$$h(t) = 3e^{-t}u(t) - \delta(t)$$

   find the system differential equation and system function.

2-2-2 Representing a Function by Weighted Delta Functions

From the preceding section it is apparent that if the input to a LTIC system was the sum of weighted delta functions, the response could easily be found.

EXAMPLE 2-6

Given the input to a LTIC system with impulse response $h(t) = 4e^{-t}u(t)$ is:

$$x(t) = 3\delta(t) - 4\delta(t - 2)$$

find $y(t)$.

Solution.   Since the system is linear and time-invariant:

$$y(t) = 3h(t) - 4h(t - 2)$$
$$= 12e^{-t}u(t) - 16e^{-(t-2)}u(t - 2)$$

Now let us consider the representation of any function $f(t)$ as a string of weighted delta functions. Figure 2-5 parts (a) to (c) help us to visualize this in three stages. Part (a) shows $f(t)$ and for a "small" range of width ϵ, say, from $p\epsilon$ to $(p + 1)\epsilon$, $f(t)$ is approximately $f(p\epsilon)$. Therefore from $p\epsilon$ to $(p + 1)\epsilon$:

$$f(t) \approx f(p\epsilon) \prod \left[\frac{1}{\epsilon} [t - (p + 0.5)\epsilon] \right]$$

Part (b) shows $f(t)$ approximated by a sum of weighted rectangular pulses and we denote this by $f_A(t)$:

Therefore $$f(t) \approx f_A(t) = \sum_{p=-N}^{M} f(p\epsilon) \prod \left[\frac{1}{\epsilon} [t - (p + 0.5)\epsilon] \right] \qquad (2\text{-}10)$$

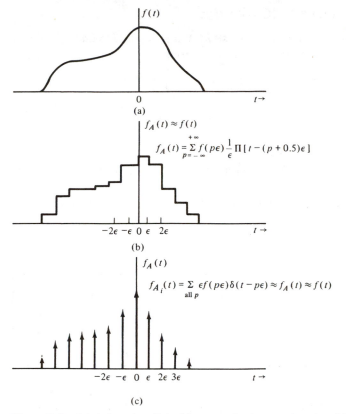

Figure 2-5   (a) A function $f(t)$; (b) a quantized approximation $f_A(t)$; (c) a weighted delta function approximation $f_{Ai}(t)$.

From the discussion of the models for a delta function in the last chapter:

$$f(p\epsilon) \prod \left[\frac{1}{\epsilon}\left[t - (p + 0.5)\epsilon\right]\right] \approx \epsilon f(p\epsilon)\delta(t - p\epsilon) \qquad \text{(important)} \qquad (2\text{-}11)$$

since the area of a pulse of width ϵ and height 1 is ϵ. Now from Equation 2-11 we can write:

$$f(t) \approx \sum_{\text{all } p} \epsilon f(p\epsilon)\delta(t - p\epsilon) \qquad (2\text{-}12)$$

where the symbol "$\approx$" indicates "approximately." This sum of weighted delta functions approximation of $f(t)$ is shown in Figure 2-5(c).

A very important subjective judgment is made when deciding on an acceptable value of ϵ. If the criterion is that each truncated rectangular segment is to appear as a weighted delta function as far as some system with impulse response $h(t)$ is concerned, then $h(t)$ must change slowly over any range ϵ. For a first-order system we would require $\epsilon \ll \tau$ where τ is the time constant.

2-2-3 The Output of a LTIC System

Given a LTIC system with impulse response $h(t)$, it is possible to find the output $y(t)$ in terms of the input $x(t)$ and $h(t)$. Representing:

$$x(t) = \sum_{\text{all } p} \epsilon x(p\epsilon)\delta(t - p\epsilon)$$

the following are true:

- The response to $\delta(t)$ is $h(t)$.
- The response to $\epsilon x(p\epsilon)\delta(t - p\epsilon)$ is $\epsilon x(p\epsilon)h(t - p\epsilon)$ using linearity and time invariance.
- The response to $\sum_p \epsilon x(p\epsilon)\delta(t - p\epsilon)$ is $\sum_{\text{all } p} \epsilon x(p\epsilon)h(t - p\epsilon)$ using linearity.

We have derived:

$$y(t) = \sum_{\text{all } p} \epsilon x(p\epsilon)h(t - p\epsilon)$$

$$= \int_{-\infty}^{\infty} x(u)h(t - u)\, du, \qquad \text{as } \epsilon \rightarrow 0 \qquad (2\text{-}13)$$

This is the famous convolution integral and it has some special notations. For any two functions $x(t)$ and $h(t)$ the convolution integral $r(t)$ is denoted by $x(t)*h(t)$ and defined as:

$$r(t) = x(t)*h(t)$$

$$\triangleq \int_{-\infty}^{\infty} x(u)h(t - u)\, du \qquad (2\text{-}14)$$

The response in Equation 2-13 for a LTIC system is called the **zero-state response** since when deriving $h(t)$ all the initial conditions at $t = 0^+$ are caused by applying an impulse at $t = 0$ and assuming the energy at $t = 0^-$ or prior to the impulse being applied is zero.

An example will now be solved incorporating all the ideas considered so far in the chapter.

EXAMPLE 2-7

Consider the simple first-order system shown in Figure 2-6(a) and find:

(a) the system function $H(s)$
(b) the system differential equation
(c) the impulse response $h(t)$
(d) the response $y(t)$ (zero-state) by convolution if $x(t) = u(t)$
(e) the complete response given $v_c(0^+) = 3$ with $x(t) = u(t)$

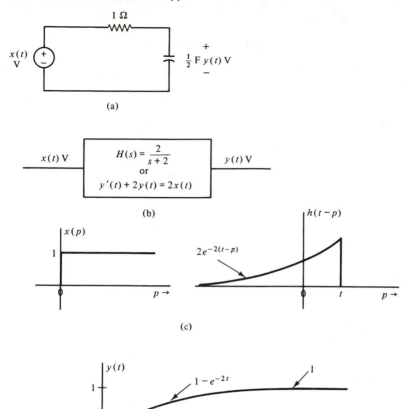

Figure 2-6 (a) The circuit of Example 2-7; (b) system representation; (c) $x(p)$ and $h(t - p)$ for convolution; (d) the output $y(t)$.

Solution

(a) The system function is:

$$H(s) = \frac{2/s}{1 + 2/s}$$

$$= \frac{2}{s + 2}$$

(b) The system differential equation is:

$$y'(t) + 2y(t) = 2x(t)$$

as indicated in Figure 2-6(b).

(c) We need to solve:

$$y'(t) + 2y(t) = 2\delta(t)$$

Obviously:

$$h(t) = 2e^{-2t}u(t)$$

(d) The zero-state output is:

$$y(t) = x(t)*h(t)$$

$$= \int_{-\infty}^{\infty} x(p)h(t - p)\, dp \qquad \text{for all } t$$

Figure 2-6(c) shows both $x(p)$ and $h(t - p)$ plotted as a function of p for a general value of t. We now systematically evaluate $y(t)$ for all $-\infty < t < \infty$.

Case $t < 0$
If $t < 0$, $h(t - p)$ is zero for $p > 0$ and hence $x(p)h(t - p) = 0$ for all p.

Therefore $y(t) = 0$

Case $t > 0$
For any $t > 0$ the product function $x(p)h(t - p)$ is nonzero for $0 < p < t$.

Therefore
$$y(t) = \int_0^t (1)2e^{-2(t-p)}\, dp$$

$$= 2e^{-2t} \int_0^t e^{2p}\, dp$$

$$= 2e^{-2t}[0.5(e^{2t} - 1)]$$

$$= 1 - e^{-2t}$$

We have found:

$$y(t) = (1 - e^{-2t})u(t)$$

and notice that as a complete response $y(t)$ has transient part $-e^{-2t}$ and steady-state part $u(t)$. This is the simple first circuit course problem of charging an uncharged capacitor. $y(t)$ is plotted in Figure 2-6(d).

(e) Since the initial energy is not zero the output may not be found by convolution because the transient part will be different.

However, $y_{fo}(t) = 1$

and $y(t) = Ae^{-2t} + 1, \qquad t > 0$

Since $y(0) = 3$, then $3 = A + 1$ and $A = 2$:

$$y(t) = (1 + 2e^{-2t})u(t)$$

In general, a complete response may be written as:

$$y_{com}(t) = y_{z.s}(t) + y_1(t)$$

$$= [x(t)*h(t)] + y_1(t) \qquad (2\text{-}15)$$

where the subscript z.s indicates the zero-state response and $y_1(t)$ is that transient component that must be added to $y_{z.s}(t)$ to give the correct complete response.

2-2-4 Comment—The Form of Terms in Convolution Integrals

Before we embark on detailed evaluation of convolution integrals there are a number of expected properties or results that we can visualize from our knowledge of differential equations and system theory.

The solution of a differential equation is the sum of the homogeneous and forced responses where the homogeneous response is found with the forcing term put equal to zero and the forced response is of the form of the forcing function. A LTIC system is governed by a linear differential equation with constant coefficients. Its impulse response $h(t)$ is that particular transient response with possible singularity functions evoked by a delta function input. The zero-state response is the convolution of the impulse response with the input. The zero-state response must consist of a transient part of the form of $h(t)$ and a forced part of the form of the input.*

When evaluating convolution integrals, we can anticipate the form of the result. For example:

$$6e^{-2t}u(t)*8e^{3t}u(t) = a_1e^{-2t}u(t) + b_1e^{3t}u(t)$$

because if a system has an impulse response $h(t) = 6e^{-2t}u(t)$ and an input $x(t) = 8e^{3t}u(t)$, then the output must contain a transient a_1e^{-2t} term and a forced b_1e^{3t} term. The constants will be found on completion of the integral. Similarly:

$$2e^{-t}u(t)*(4 + 3t)u(t) = c_1e^{-t}u(t) + (c_2 + c_3t)u(t)$$

subject to the same reasoning. The reader should interpret what terms are involved when $2e^{-t}u(t)*(3 + t)e^{-t}u(t)$ is evaluated.

2-3 THE EVALUATION AND PROPERTIES OF CONVOLUTION INTEGRALS

2-3-1 Some Convolution Integrals

In this section convolution is treated in its own right since it is prevalent throughout many fields of engineering. First, a number of problems will be solved and then convolution properties will be developed.

---

*This is the case if the input $x(t)$ is not of the form of the impulse response $h(t)$.

EXAMPLE 2-8

Given

$$x(t) = \sqcap(t - 0.5) \quad \text{and} \quad y(t) = \sqcap(t - 2)$$

Evaluate:

$$x(t)*y(t)$$

Solution.   Figure 2-7(a) shows $x(p)$ and $y(t - p)$ plotted as a function of p. To visualize $y(t - p)$ we write it as:

$$y(t - p) = \sqcap[t - p - 2]$$
$$= \sqcap[-(p - (t - 2))]$$

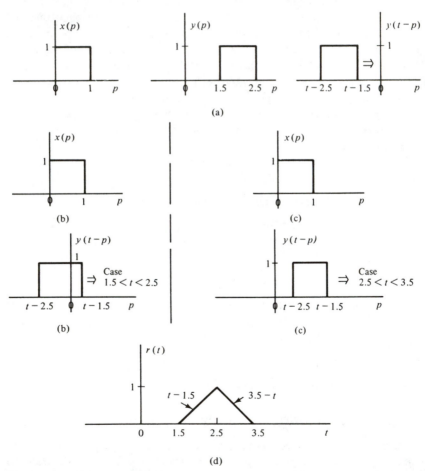

(a)

(b)

(c)

(b)

(c)

(d)

Figure 2-7   (a) $x(p)$, $y(p)$, and $y(t - p)$ for Example 2-8; (b) $x(p)$ and $y(t - p)$ when $1.5 < t < 2.5$; (c) $x(p)$ and $y(t - p)$ when $2.5 < t < 3.5$; (d) the convolution integral $r(t)$.

which is $\Pi(p)$ reflected and shifted $t - 2$ units to the right or $y(p)$ reflected and shifted t units to the right. We now systematically carry out the convolution starting at $t = -\infty$ and proceed with each different range until $t = \infty$.

Range 1: $t < 1.5$

If $t - 1.5 < 0$, then the front edge of the moving pulse $y(t - p)$ has not reached the rear end of the stationary pulse $x(p)$ which is at $p = 0$.

Therefore
$$r(t) = \int_{-\infty}^{\infty} x(p)y(t - p)\, dp$$
$$= 0$$

Range 2: $1.5 < t < 2.5$

If $1.5 < t < 2.5$, then the product function $x(p)y(t - p)$, as may be seen from Figure 2-7(b), is nonzero when $0 < p < t - 1.5$

and
$$r(t) = \int_{0}^{t-1.5} (1)(1)\, dp$$
$$= t - 1.5$$

Range 3: $2.5 < t < 3.5$

If $2.5 < t < 3.5$, then the product function $x(p)y(t - p)$, as may be seen in Figure 2-7(c), is nonzero when $t - 2.5 < p < 1$

and
$$r(t) = \int_{t-2.5}^{1} (1)(1)\, dp$$
$$= 3.5 - t$$

Range 4: $t > 3.5$
$$r(t) = 0$$

The convolution integral is shown in Figure 2-7(d).

EXAMPLE 2-9
Given
$$x(t) = \Pi(t - 0.5) \quad \text{and} \quad y(t) = \Pi[0.5(t - 4)]$$

Evaluate:
$$x(t)*y(t)$$

Solution.   Figure 2-8(a) shows $x(p)$ and $y(t - p)$ plotted as a function of p. $y(p)$ extends from $3 < p < 5$ and $y(t - p)$ from $-5 + t < p < -3 + t$.

Range $t < 3$
Since $t - 3 < 0$, the product function $x(p)y(t - p) = 0$.

Therefore
$$r(t) = 0$$

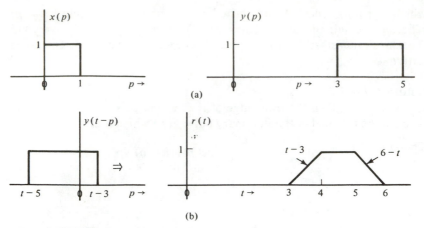

Figure 2-8   (a) $x(p)$ and $y(p)$ and $y(t - p)$ for Example 2-9; (b) the convolution integral $r(t)$.

Range $3 < t < 4$

$$r(t) = \int_0^{t-3} (1)(1)\, dp$$

$$= t - 3$$

Range $4 < t < 5$
Now the $y(t - p)$ function straddles $x(p)$

and
$$r(t) = \int_0^1 (1)(1)\, dp$$

$$= 1$$

Range $5 < t < 6$

$$r(t) = \int_{t-5}^1 (1)(1)\, dp$$

$$= 6 - t$$

Range $t > 6$
$$r(t) = 0$$

The convolution integral is illustrated in Figure 2-8(b), and it can be noted that the width of $r(t)$ equals the sum of the widths of the two individual functions.

In the next example we consider the case where the functions are more complicated and greater attention to detail is necessary.

EXAMPLE 2-10
Carefully set up the appropriate formulas for each range of integration when evaluating $x(t)*y(t)$ with $x(t) = 2, 0 < t < 1; x(t) = 3 - t, 1 < t < 3;$

$x(t) = 0$, otherwise, and $y(t) = 1.25(t - 1)$, $1 < t < 2.6$; $y(t) = 0$, otherwise.

Solution. Figure 2-9 shows $x(p)$, $y(p)$, and $y(t - p)$. The convolution integral may now be systematically set up as follows:

Range $t < 1$
$$r(t) = 0$$

Range $1 < t < 2$
The upper limit of integration is determined by noting that the width of $y(t)$ is 1.6, which is greater than 1, the range for which $x(p)$ maintains the value 2.

Therefore
$$r(t) = \int_0^{t-1} (2)1.25(t - p - 1) \ dp$$

Range $2 < t < 2.6$
$$r(t) = 2 \int_0^1 1.25(t - p - 1) \ dp + \int_1^{t-1} (3 - p)1.25(t - p - 1) \ dp$$

If $t > 2.6$, the lower limit will no longer be zero in the first part of integration.

Range $2.6 < t < 3.6$
$$r(t) = 2 \int_{t-2.6}^1 1.25(t - p - 1) \ dp + \int_1^{t-1} (3 - p)1.25(t - p - 1) \ dp$$

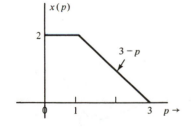

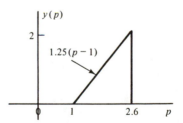

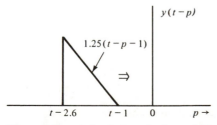

Figure 2-9 $x(p)$, $y(p)$, and $y(t - p)$ for Example 2-10.

Range $3.6 < t < 4$

$$r(t) = \int_{t-2.6}^{t-1} (3 - p)1.25(t - p - 1) \, dp$$

Range $4 < t < 5.6$

$$r(t) = \int_{t-2.6}^{3} (3 - p)1.25(t - p - 1) \, dp$$

Range $t > 5.6$

$$r(t) = 0$$

The reader should carefully note the necessity for five different ranges of integration. Also, the analytical "nastiness" of convolution is apparent and the time required to complete this problem with a careful check on the work is about 5 hours.

2-3-2 A Summary of Convolution

Consider two general functions $x(t)$ and $y(t)$ where $x(t)$ is nonzero from x_A to x_B and is of width $w_x = x_B - x_A$ and where $y(t)$ is nonzero from y_A to y_B and is of width $w_y = y_B - y_A$. Figure 2-10 shows a sketch of such general functions. The convolution integral is:

$$r(t) = x(t)*y(t) = \int_{-\infty}^{\infty} x(p)y(t - p) \, dp$$

We visualize $y(t - p)$ as a moving pulse that is $y(p)$ reflected and moved t units to the right. $y(t - p)$ extends from $p = -y_B + t$ to $-y_A + t$. $y(t - p)$ (if $w_y < \infty$) moves from $p = -\infty$ to $p = +\infty$ as t varies $-\infty < t < \infty$.

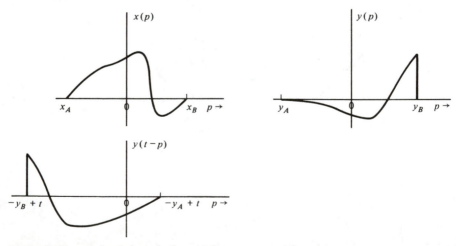

Figure 2-10 $x(p), y(p),$ and $y(t - p)$ for two general functions.

Observing Figure 2-10 we clearly note the following:

- $r(t) = 0$ if $-y_A + t < x_A$ or $t < x_A + y_A$.
- $r(t) = 0$ if $-y_B + t > x_B$ or $t > x_B + y_B$.
- The convolution integral extends from $x_A + y_A < t < x_B + y_B$ and is of resultant width $w = w_A + w_B$.
- For any specific problem, when determining the different analytic ranges of $x(p)$ and $y(t - p)$, we must carefully pick out each range from $t = x_A + y_A$ to $t = x_B + y_B$ for which the integral has a different result. Example 2-10 illustrated this procedure.

2-3-3 Properties of Convolution

In order to distinguish between the order of functions being convolved, we may use a double subscript notation. We denote:

$$x(t)*y(t) = r_{xy}(t) \triangleq \int_{-\infty}^{\infty} x(p)y(t - p) \, dp$$

and

$$y(t)*x(t) = r_{yx}(t) \triangleq \int_{-\infty}^{\infty} y(p)x(t - p) \, dp$$

Some of the more important properties of convolution are:

Commutative Law

$$x(t)*y(t) = y(t)*x(t)$$

or

$$r_{xy}(t) = r_{yx}(t) \tag{2-16}$$

Distributive Law

$$x(t)*[y(t) \pm z(t)] = [x(t)*y(t)] \pm [x(t)*z(t)] \tag{2-17}$$

or

$$r_{x(y\pm z)}(t) = r_{xy}(t) \pm r_{xz}(t)$$

Associative Law

$$x(t)*[y(t)*z(t)] = [x(t)*y(t)]*z(t) \tag{2-18}$$

The proofs of the commutative and distributive properties are straightforward but handling expressions similar to those in the associative property requires care and adherence to the fundamental definition.

EXAMPLE 2-11

Prove the commutative property and set up the proof of the associative property for convolution.

Solution

(a) We want to show:

$$\int_{-\infty}^{\infty} x(p)y(t - p)\, dp = \int_{-\infty}^{\infty} x(t - u)y(u)\, du$$

Considering the RHS we should feel motivated to let $t - u = s$; then $du = -ds$ and $+\infty > s > -\infty$ as $-\infty < u < \infty$.

$$\text{RHS} = \int_{\infty}^{-\infty} x(s)y(t - s)\,[-ds]$$

$$= \int_{-\infty}^{\infty} x(s)y(t - s)\, ds = \text{LHS}$$

(b) We require by transforming variables to show:

$$x(t)*[y(t)*z(t)] = [x(t)*y(t)]*z(t)$$

Initially it is difficult to write expressions for both sides of this equation and the t in the notation is confusing. Physically it is obvious what is implied. Convolving two functions $y(t)$ and $z(t)$ yields $r_{yz}(t)$, and convolving this with $x(t)$ yields the answer as a function of t. Mathematically, to obtain an expression for $x(t)*[y(t)*z(t)]$, we write:

$$r_{yz}(t) = \int_{-\infty}^{\infty} y(p)z(t - p)\, dp$$

$$x(t)*r_{yz}(t) = \int_{-\infty}^{\infty} x(u)r_{yz}(t - u)\, du$$

$$= \int_{-\infty}^{\infty} x(u)\left[\int_{-\infty}^{\infty} y(p)z(t - u - p)\, dp\right] du \qquad (1)$$

and this is an expression for the LHS.

Similarly, for the RHS $[x(t)*y(t)]*z(t)$ we write:

$$r_{xy}(t) = \int_{-\infty}^{\infty} x(r)y(t - r)\, dr$$

$$r_{xy}(t)*z(t) = \int_{-\infty}^{\infty} r_{xy}(m)z(t - m)\, dm$$

$$= \int_{-\infty}^{\infty}\left[\int_{-\infty}^{\infty} x(r)y(m - r)\, dr\right]z(t - m)\, dm \qquad (2)$$

and this is the RHS.

The proof of the associative law requires showing that (1) and (2) are identical. In doing so we assume the functions are such that the order of integration may be interchanged. The proof of this law is trivial using the Laplace or Fourier transform. These transforms, however, normally require that the functions x, y, and z are analytically transformable. In the computer era when many important design problems are not analytically transformable a deep knowledge of the time domain is a necessity.

2-3-4 Convolution with Delta Functions

When a function may be approximated by a weighted delta function great simplicity occurs in convolution. Two important results are:

$$x(t)*\delta(t - a) = x(t - a) \tag{2-19}$$

and

$$\delta(t - a)*\delta(t - b) = \delta(t - a - b) \tag{2-20}$$

EXAMPLE 2-12

Prove Equations 2-19 and 2-20.

Solution. By definition:

$$x(t)*\delta(t - a) = \int_{-\infty}^{\infty} x(p)\delta(t - p - a)\, dp$$

$$= \int_{-\infty}^{\infty} x(p)\frac{1}{|-1|}\delta(p - t + a)\, dp$$

$$= x(t - a)$$

$$\delta(t - a)*\delta(t - b) = \int_{-\infty}^{\infty} \delta(s - a)\delta(t - s - b)\, ds$$

$$= \delta(t - s - b)\,|_{s-a}$$

$$= \delta(t - a - b)$$

Physically this may be verified by convolving two rectangular models about a and b and obtaining a triangular model about $t = a + b$.

EXAMPLE 2-13

Obtain a good approximation for $f(t)*g(t)$ for the functions shown in Figure 2-11(a).

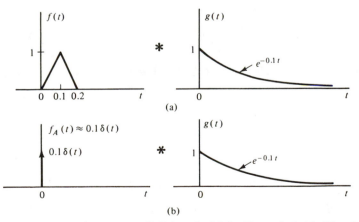

Figure 2-11 (a) A "narrow" $f(t)$ and $g(t)$ for Example 2-13; (b) a delta approximation and $g(t)$.

Solution.   Considering $f(t)$ and $g(t)$ we see that $f(t)$ which is of width 0.2 s is very "narrow" as compared to $g(t)$ which has a time constant of $\tau = 10$ s. Therefore as shown in Figure 2-11(b) as far as $g(t)$ is concerned $f(t)$ may be replaced by a delta function weighted by its area.

$$f(t) \approx 0.1\delta(t) \qquad [\text{really } 0.1\delta(t - 0.1)]$$

and using the commutative property:

$$e^{-0.1t}u(t)*0.1\delta(t) \approx 0.1e^{-0.1t}u(t)$$

This result is only substantially in error over the small range $0 < t < 0.2$.

2-3-5 Approximate Convolution

Figure 2-12(a) shows two functions $f_A(p)$ and $g_A(p)$ which have constant values over ranges of width 0.25. Let us find the convolution at $t = n(0.25)$ for all integer n. $g_A(t - p)$ is also shown.

$$r_{f_{A}g_{A}}(t) = f_A(t)*g_A(t) = \int_{-\infty}^{\infty} f_A(p)g_A(t - p)\, dp$$

$$r(t) = 0, \qquad t < 0.75 \cup t > 2.5$$

Now evaluating $r(t)$ at $t = 0.25n$ for $4 \le n \le 10$, we obtain:

$$r(1) = (2 \times 1)0.25$$

$$r(1.25) = [3 \times 1 + 2(-2)]0.25$$

$$r(1.5) = [1 \times 1 + 3(-2) + 2(2)]0.25$$

$$r(1.75) = [1 \times (-2) + 3 \times 2 + 2(-2)]0.25$$

$$r(2) = [1 \times 2 + 3(-2)]0.25$$

and $$r(2.25) = [1(-2)]0.25$$

$r_{f_{A}g_{A}} \div 0.25$ is plotted in Figure 2-12(b). The simplicity of convolving these quantized functions should be evident.

To conform with the convolution of quantized waveforms, we define the convolution product of sets. If the set:

$$(f) \triangleq (2, 3, 1)$$

and $$(g) \triangleq (1, -2, 2, -2)$$

we define:

$$(f)*(g) \triangleq [2 \times 1, 3 \times 1 + 2(-2), 1 \times 1 + 3(-2) + 2(2), 1(-2)$$

$$+ 3(2) + 2(-2), 1 \times 2 + 3(-2), 1(-2)]$$

$$= (2, -1, -1, 0, -4, -2)$$

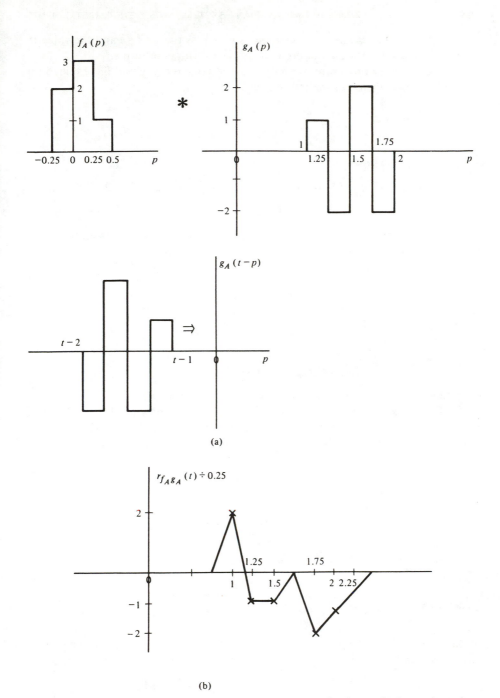

Figure 2-12 (a) Two quantized functions; (b) the normalized convolution evaluated at $t = p\Delta$ where $\Delta = 0.25$.

To obtain this the set (g) is reflected and slid across the set (f) and the value in the convolved set is the sum of products of the overlapping terms.

Again, being observant, we note that the same terms could be obtained for the coefficients when multiplying two series:

$$2 + 3z^{-1} + z^{-2}$$
$$\times\ \ 1 - 2z^{-1} + 2z^{-2} - 2z^{-3}$$

$$= (2 \times 1) + [3 \times 1 + 2(-2)]z^{-1} + [1 + 3(-2) + 2(2)]z^{-2}$$
$$+ [-2 + 3(2) + 2(-2)]z^{-3} + (-2 + 3(-2)z^{-4} - 2z^{-5}$$
$$= 2 - z^{-1} - z^{-2} + 0z^{-3} - 4z^{-4} - 2z^{-5}$$

Indeed, multiplication of numbers may be carried out by convolution. For example, 324×56 may be found as $(3, 2, 4)*(5, 6)$ to yield:

$$324$$
$$65\longrightarrow$$

$$\overline{15,\ 10 + 18,\ 20 + 12,\ 24} = 18{,}144$$

thousands hundreds tens units

and some old desk calculators used to work on this principle.

In general, given two functions $f(u)$ and $g(u)$ as shown in Figure 2-13(a), we may approximate them by $f_A(u)$ and $g_A(u)$ as shown in Figure 2-13(b) where the choice of Δ made subject to the criteria $f(u)$ and $g(u)$ may not vary by "much" over a range Δ. The discrete functions $f(n)$ and $g(n)$ in Figure 2-13(c) represent the quantized values of $f(t)$ and $g(t)$.

The convolution integral is found as:

$$r_{fg}(t) = \int_{-\infty}^{\infty} f(u)g(t - u)\, du$$
$$\approx \int_{-\infty}^{\infty} f_A(u)g_A(t - u)\, du$$

Therefore
$$r_{fg}(p\Delta) = \Delta \sum_{all\ i} f_i g_{p-i} \tag{2-21}$$

If as shown in Figure 2-13(c) $f(n)$ extends from M_1 to N_1 and $g(n)$ extends from M_2 to N_2, we find:

$$r_{fg}(p\Delta) = 0, \qquad p < M_1 + M_2$$
$$= 0, \qquad p > N_1 + N_2$$
$$= \Delta \sum_{all\ i} f_i g_{p-i}, \qquad \text{otherwise} \tag{2-22}$$

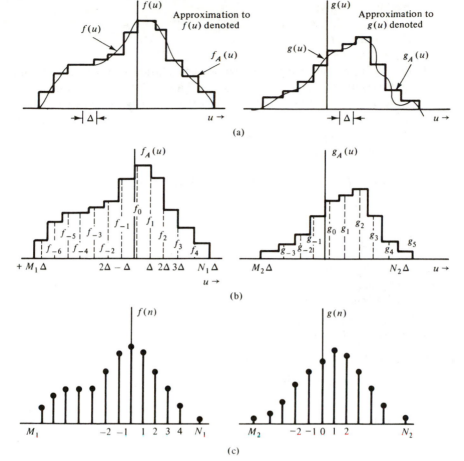

Figure 2-13   (a) Two general functions; (b) their quantized approximations; (c) their discrete approximations.

When to use the summation limits $M_1 < i < -M_2 + p$ or $-N_2 + p < i < N_1$ is left as an exercise.

Drill Set: Convolution Integrals

1. Given a LTIC system has impulse response

$$h(t) = 2e^{-t}u(t)$$

find the response $y(t)$ if:
(a) $x(t) = \prod (t - 0.5)$
(b) $x(t) = \prod [20(t - 0.025)]$ (a quick approximation will suffice)

2. Evaluate rapidly:
   (a) $\prod (t - 2.5) * 2\prod [0.2(t - 6)]$
   (b) $tu(t) * \delta(t - 2)$
   (c) $e^{-2t}u(t) * e^{t}u(-t)$
3. (a) Find two pulse functions whose convolution starts at $t = 1$ and increases linearly until it has the value 2 at $t = 1.5$. This constant value is maintained until $t = 3$ and then decreases linearly to the value 0 at $t = 3.5$.
   (b) Is the answer in part (a) unique?

2-4 THE SYSTEM DIFFERENCE EQUATION AND SYSTEM FUNCTION $H(\alpha)$ OR $H(z)$

2-4-1 Solving Difference Equations

As a prelude to the time-domain analysis of a discrete, linear, time-invariant causal system (LTIC) a summary of solving linear difference equations with constant coefficients will be given.

The output $y(n)$ and input $x(n)$ of a LTIC system are related by a linear difference equation of the form:

$$a_n y(n) + a_{n-1} y(n-1) + \cdots + a_{n-p} y(n-p)$$
$$= b_n x(n) + b_{n-1} x(n-1) + \cdots + b_{n-l} x(n-l) \qquad (2\text{-}23)$$

The right-hand terms of Equation 2-23 are often lumped together and called the forcing function $f(n)$. We confine ourselves to the case where the a's and b's are constant coefficients. Two methods of solving difference equations will be discussed:

1. a classical approach that is analogous to that used for solving differential equations in Section 2-1
2. a recursive or iterative approach such as that carried out by a digital computer

2-4-2 The Classical Solution of Difference Equations

Theorem. Given a pth-order difference equation as in Equation 2-23 plus the initial conditions $y(-1), y(-2), \ldots, y(-p)$, the complete response is of the form:

$$y(n) = y_{h0}(n) + y_{f0}(n) \qquad (2\text{-}24)$$

where $y_{h0}(n)$, the homogeneous response, is the solution to the difference equation with $f(n) = 0$ and contains p arbitrary constants and $y_{f0}(n)$, the forced response, is that one particular solution to the equation that contains no part of

$y_{ho}(n)$. The p arbitrary constants may be found by applying (as a rule) $y(-1)$, $y(-2), \ldots, y(-p)$ to Equation 2-24. Indeed, any other p values for $y(n)$ are equally adequate.

We now solve a number of problems widely utilizing the analogy to solving linear differential equations with constant coefficients.

EXAMPLE 2-14

Solve the following difference equations:

(a) $y(n) + 0.5y(n - 1) = 3, \qquad y(-1) = 4$
(b) $y(n) - 4y(n - 1) + 4y(n - 2) = 4(-3)^n + 3n$
   $y(-1) = 0$ and $y(-2) = 2$

Solution

(a) *Step 1: $y_{ho}(n)$*
The homogeneous equation is:

$$y(n) + 0.5y(n - 1) = 0$$

Let us try:

$$y(n) = C(\alpha)^n$$

since shifting it yields:

$$y(n - 1) = C(\alpha)^{n-1} = k(\alpha)^n$$

where $k = C\alpha^{-1}$. Substituting in the equation, we obtain:

$$C(\alpha)^n + 0.5\,C\alpha^{n-1} = 0$$

$$C\alpha^{n-1} (\alpha + 0.5) = 0$$

Only $C = 0$ or $\alpha^{n-1} = 0$ will satisfy this equation via the $C\alpha^{n-1}$ term. Therefore any value of α satisfying the characteristic equation:

$$\alpha + 0.5 = 0$$

will be a solution.

Thus $\qquad\qquad y_{ho}(n) = C(-0.5)^n \qquad$ for any C

is the most general solution.

Step 2: $y_{fo}(n)$
Considering

$$y(n) + 0.5y(n - 1) = 3$$

where it is implied $f(n) = 3$ for all n, corresponding to the continuous case, we try:

$$y_{fo}(n) = A \qquad \left[\text{This is the same as } \sum_{k=-\infty}^{\infty} A\delta(n - k)\right]$$

Substituting in the equation, we obtain:

$$A + 0.5A = 3$$

or $$A = 2$$

Therefore $$y_{fo}(n) = 2 \quad \text{for all } n$$

Step 3: $y_{com}(n)$
The form of the complete response is:

$$y(n) = 2 + C(-0.5)^n$$

Step 4: Applying the initial conditions

$$y(-1) = 4$$

Therefore $$4 = 2 + C(-0.5)^{-1}$$

$$-2C = 2$$

and $$C = -1$$

Finally:

$$y(n) = 2 - (-0.5)^n$$

is the one unique discrete function satisfying the difference equation subject to the initial condition. Figure 2-14 shows a plot of $y(n)$, $0.5y(n - 1)$, and $y(n) + 0.5y(n - 1)$. If discrete functions are somewhat new, study the figures and show in addition that $y(n) + 0.5y(n - 1)$ is zero for any homogeneous solution and $y(n) + 0.5y(n - 1) = 3$ for the forced solution. Also, the solution satisfies the difference equation for all integers $n, -\infty < n < \infty$.

(b) The difference equation to be solved is:

$$y(n) - 4y(n - 1) + 4y(n - 2) = 4(-3)^n + 3n$$

with initial conditions $y(-1) = 0, y(-2) = 2$

Step 1: $y_{ho}(n)$
Let us try:

$$y_{ho}(n) = C(\alpha)^n$$

Quickly we find the characteristic equation:

$$\alpha^2 - 4\alpha + 4 = 0$$

$$(\alpha - 2)^2 = 0$$

and $$y_{ho}(n) = C_1(2)^n + C_2 n(2)^n$$

using the analogy to repeated roots from a differential equation. The reader should verify this by substituting $C_2 n(2)^n$ into the homogeneous equation.

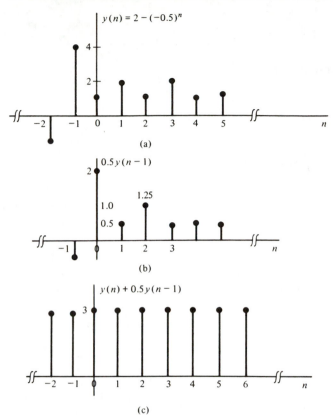

Figure 2-14 (a) The solution $y(n)$ for $y(n) + 0.5y(n - 1) = 3, y(-1) = 4$ in Example 2-14(a); (b) $0.5\,y(n - 1)$; (c) $y(n) + 0.5y(n - 1)$.

Step 2: $y_{f0}(n)$
Observing the difference equation, from our knowledge of differential equations, we guess $y_{f0}(n)$ of the form:

$$y_{f0}(n) = A(-3)^n + B + Cn$$

If a mistake is made here it is to omit the B term.
     Substituting into the equation, we obtain:

$$A[(-3)^n - 4(-3)^{n-1} + 4(-3)^{n-2}] + B(1 - 4 + 4)$$
$$+ C[n - 4(n - 1) + 4(n - 2)] \equiv 4(-3)^n + 3n$$

Now we equate coefficients:

Coefficient of $(-3)^n$

$$A(1 + \tfrac{4}{3} + \tfrac{4}{9}) = 4$$

Therefore $A = 1.44$

Coefficient of n

$$C = 3$$

Coefficient of Constant Term

$$B - 4C = 0$$

Therefore

$$B = 12$$

Finally:

$$y_{f0}(n) = 1.44(-3)^n + 12 + 3n$$

Step 3: $y_{com}(n)$

$$y(n) = 1.44(-3)^n + 12 + 3n + C_1(2)^n + C_2 n(2)^n$$

Step 4: Applying the initial conditions

The given initial conditions are $y(-1) = 0$ and $y(-2) = 2$. For our equation, since the initial condition $y(0)$ would be very easy to substitute into the solution, we will find it and use $y(0)$ and $y(-1)$. From our equation at $n = 0$:

$$y(0) - 4y(-1) + 4y(-2) = 4 + 0$$

This becomes:

$$y(0) - 4(0) + 4(2) = 4$$

Therefore

$$y(0) = -4$$

Substituting $y(0) = -4$ and $y(-1) = 0$ into our general solution from step 3 yields:

$$-4 = 1.44 + 12 - 0 + C_1$$

Thus

$$C_1 = -17.44$$

and for $y(-1)$:

$$0 = -0.48 + 12 - 3 - 8.72 - 0.5C_2$$

Thus

$$-0.5C_2 = 0.2$$

and

$$C_2 = -0.4$$

Finally:

$$y(n) = 1.44(-3)^n + 12 + 3n - 17.44(2)^n - 0.4n(2)^n$$

There are many other situations we could consider for difference equations. In particular, the case of conjugate imaginary and complex conjugate roots for the characteristic equation are somewhat important but will not be pursued here. The goal is to have a clear insight as to the basic conceptual simplicity of difference equations and to realize with practice they are as simple to solve as differential equations.

2-4-3 The Recursive or Iterative Solution of Difference Equations

The previous two examples will now be solved in a recursive manner. The general procedure should then be obvious.

EXAMPLE 2-15

Solve the following difference equations recursively for $n \geq 0$:

(a) $y(n) + 0.5y(n-1) = 3,$ $\qquad y(-1) = 4$
(b) $y(n) - 4y(n-1) + 4y(n-2) = 4(-3)^n + 3n,$ $\qquad y(-1) = 0,$
$y(-2) = 2$

Solution

(a) In general:

$$y(n) = -0.5y(n-1) + 3$$

and knowing $y(-1)$, all $y(n)$, $n \geq 0$ may be found in an iterative manner. We obtain:

$$y(0) = -0.5(4) + 3 = 1$$
$$y(1) = -0.5(1) + 3 = 2.5$$
$$y(2) = -0.5(2.5) + 3 = 1.75$$
$$y(3) = -0.5(1.75) + 3 = 2.125$$
$$y(4) = -0.5(2.125) + 3 = 1.94$$
$$y(5) = -0.5(1.94) + 3 = 2.03$$

and $\qquad y(n) = 2, \quad n \geq 5$

This result is seen to be identical for the values obtained by substituting into the complete response $y(n) = 2 - (-0.5)^n$ from Example 2-14(a).

(b) Considering:

$$y(n) - 4y(n-1) + 4y(n-2) = 4(-3)^n + 3n$$

We write:

$$y(n) = 4y(n-1) - 4y(n-2) + 4(-3)^n + 3n$$

and knowing $y(-1) = 0$ and $y(-2) = 2$, we obtain:

$$y(0) = 4(0) - 4(2) + 4 + 0 = -4$$
$$y(1) = 4(-4) - 4(0) + 4(-3) + 3 = -25$$
$$y(2) = 4(-25) - 4(-4) + 4(9) + 6 = -42$$

$$y(3) = 4(-42) - 4(-25) + 4(-27) + 9 = -167$$
$$y(4) = 4(-167) - 4(-42) + 4(81) + 12 = -164$$

and so on.

Difference equations are extremely simple to solve using a programmable calculator or computer and the reader should redo these problems by either means. Theoretically, the classical solution is interesting as the homogeneous and forced part of the response are distinguishable and the analogy to differential equations is apparent.

2-4-4 Finding the System Difference Equation

Figure 2-15 shows the basic block diagram representations for operations on discrete signals. These are the delay, multiplier, summing, and branching elements. Given the representation of a system in block diagram form, with practice, it is almost trivial to find the system difference equation. In Figure 2-16 three block diagram representations of systems are shown with the corresponding difference equation written underneath.

2-4-5 The Discrete System Function

For discrete systems we will not develop in depth the problem of finding a system difference equation via a system function. However, the theoretical time-domain definition of the discrete system function is important.

Given a system governed by:

$$a_n y(n) + a_{n-1} y(n - 1) + \cdots + a_{n-p} y(n - p)$$
$$= b_n x(n) + \cdots + b_{n-k} x(n - k) \qquad (2-25)$$

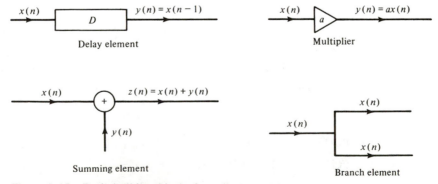

Figure 2-15 Basic building blocks for a discrete system.

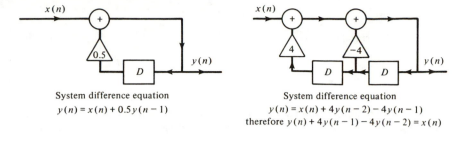

System difference equation
$$y(n) = x(n) + 0.5y(n-1)$$

System difference equation
$$y(n) = x(n) + 4y(n-2) - 4y(n-1)$$
therefore $y(n) + 4y(n-1) - 4y(n-2) = x(n)$

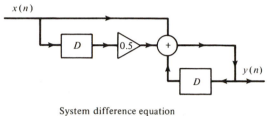

System difference equation
$$y(n) - y(n-1) = x(n) + 0.5x(n-1)$$

Figure 2-16 Block diagram representation of discrete systems.

We initially define the system function:

$$H(\alpha) \triangleq \left. \frac{y_{fo}(n)}{x(n)} \right|_{x(n) = \alpha^n} \tag{2-26}$$

In words, the system function is the forced response to α^n divided by α^n.

EXAMPLE 2-16

Find the system function for the systems defined by the following difference equations:

(a) $y(n) + 0.6y(n-1) = x(n)$
(b) $y(n) + 2y(n-1) + y(n-2) = x(n) - 0.5x(n-1)$

Solution

(a) $y(n) + 0.6y(n-1) = \alpha^n$

Try:

$$y_{fo}(\alpha) = A\alpha^n$$

Therefore $A\alpha^n + 0.6A\alpha^{n-1} \equiv \alpha^n$

or $A = \dfrac{1}{1 + 0.6\alpha^{-1}}$

$$H(\alpha) = \frac{\dfrac{\alpha^n}{1 + 0.6\alpha^{-1}}}{\alpha^n}$$

$$= \frac{1}{1 + 0.6\alpha^{-1}}$$

We may leave $H(\alpha)$ as the ratio of polynomials in terms of α^{-1} and see how it may be written by inspection of the difference equation or, alternatively:

$$H(\alpha) = \frac{\alpha}{\alpha + 0.6}$$

where the higher-order terms are associated with $x(n)$ and $y(n)$.

(b) $y(n) + 2y(n - 1) + y(n - 2) = \alpha^n - 0.5\alpha^{n-1}$

If we guess $y_{f0}(n) = A\alpha^n$ with a little work or inductively by inspection of part (a)

$$H(\alpha) = \frac{1 - 0.5\alpha^{-1}}{1 + 2\alpha^{-1} + \alpha^{-2}}$$

$$= \frac{\alpha^2 - 0.5\alpha}{\alpha^2 + 2\alpha + 1}$$

For the general system governed by equation:

$$a_n y(n) + a_{n-1} y(n - 1) + \cdots + a_{n-p} y(n - p)$$
$$= b_n x(n) + \cdots + b_{n-k} x(n - k)$$

$$H(\alpha) = \left. \frac{y_{f0}(n)}{x(n)} \right|_{x=\alpha^n}$$

$$= \frac{b_n + b_{n-1}\alpha^{-1} + \cdots + b_{n-k}\alpha^{-k}}{a_n + a_{n-1}\alpha^{-1} + \cdots + a_{n-p}\alpha^{-p}} \qquad (2\text{-}27)$$

and if $p \geq k$:

$$H(\alpha) = \frac{b_n\alpha^p + b_{n-1}\alpha^{p-1} + \cdots + b_{n-k}\alpha^{p-k}}{a_n\alpha^p + a_{n-1}\alpha^{p-1} + \cdots + a_{n-p}} \qquad (2\text{-}28)$$

To conform to the Z transform to be discussed in Chapter 6, we present a different version of the system function:

$$H(z) \triangleq \left. \frac{y_{f0}(n)}{x(n)} \right|_{x(n)=z^n}$$

The $H(z)$ associated with the general system governed by Equation 2-23 is:

$$H(z) = \frac{b_n z^p + b_{n-1} z^{p-1} + \cdots + b_{n-l} z^{p-l}}{a_n z^p + a_{n-1} z^{p-1} + \cdots + a_{n-p}} \qquad (2\text{-}29)$$

Equation 2-29 is the normal version of the system function given for a discrete system.

Drill Set: Difference Equations

1. For a system with system function:

$$H(z) = \frac{2z^2 + z}{z^2 + 0.5z - 0.5}$$

(a) Find the governing difference equation.
(b) Find the forced response when $x(n) = 2^n$. Try to develop a system approach with $z = 2$.
(c) Find the complete response when $x(n) = 3$ and $y(-1) = y(-2) = 1$ using the classical approach.
(d) Redo part (c) recursively and check the values for $y(0)$, $y(1)$, and $y(2)$ with the classical solution.

2-5 THE PULSE RESPONSE h(n)

If a LTIC system is governed by:

$$a_n y(n) + a_{n-1} y(n - 1) + \cdots + a_{n-p} y(n - p)$$
$$= b_n x(n) + \cdots + b_{n-l} x(n - l) \qquad (2\text{-}30)$$

or characterized by the system function:

$$H(z) = \frac{b_n z^p + \cdots + b_{n-l} z^{p-l}}{a_n z^p + \cdots + a_{n-p}}$$

we define the pulse response h(n) *as the response when* x(n) = δ(n) *and* y(-1), y(-2), . . . , y(-p) *all equal zero.* Later the pulse response will be found easily using the Z transform but it is instructive and enlightening to derive it directly in the time domain.

Before embarking on solving for pulse responses directly, let us philosophically summarize our results when solving differential or difference equations and obtaining impulse responses for continuous systems. (2-1 and 2-2)

- Solving the differential equation:

$$a_n y''(t) + a_{n-1} y^{n-1}(t) + \cdots + a_0 y(t) = f(t)$$

given any n initial conditions whether they are $y(0)$, $y'(0)$, . . . , $y^{n-1}(0)$, or indeed $y(-20)$, $y(-6)$, and so on leads to a solution that satisfies the differential equation (DE) for all time, $-\infty < t < \infty$.

- Solving for an impulse response $h(t)$ for a system differential equation such as:

$$a_n y''(t) + a_{n-1} y^{n-1}(t) + \cdots + a_0 y(t) = b_0 x(t) + \cdots + b_m x^m(t)$$

led to consideration of:

$$a_n y''(t) + a_{n-1} y^{n-1}(t) + \cdots + a_0 y(t) = b_0 \delta(t) + \cdots + b_m \delta^m(t)$$

where $y(t) = 0$, $t < 0$. This equation is contentious because of the discontinuities at $t = 0$. Essentially, the solution is the sum of a homogeneous solution plus a "forcing part" composed of singularity functions. If $m < n$ the forcing part is zero.

- So far in Chapter 2 we have considered solving the difference equation:

$$a_n y(n) + a_{n-1} y(n - 1) + \cdots + a_{n-p} y(n - p) = f(n)$$

given any p initial conditions whether they are $y(-1), \ldots, y(-p)$ or $y(-16)$, $y(17)$, and so on. The solution $y(n)$ satisfies the difference equation for all n, $-\infty < n < \infty$.

Now let us solve for pulse response in an analogous manner to impulse responses.

EXAMPLE 2-17

Find the pulse response $h(n)$ for the system defined by:

(a) $y(n) + 0.6y(n - 1) = 3x(n)$
(b) $y(n) + 0.6y(n - 1) = 3\dot{x}(n) + x(n - 1)$

Solution

(a) We are required to solve:

$$y(n) + 0.6y(n - 1) = 3\delta(n)$$

The form of $y(n)$ is $y(n) = y_{ho}(n)u(n) + A\delta(n)$. The term A must be zero because otherwise $0.6y(n - 1)$ would contribute a term $0.6A\delta(n - 1)$. Therefore $y(n) = C(-0.6)^n u(n)$ where C is to be found.
At $n = 0$ substituting $y(-1) = 0$, we have:

$$y(0) + 0.6(0) = 3 \quad \text{or} \quad y(0) = 3$$

This yields:

$$h(n) = 3(-0.6)^n u(n)$$

This one special homogeneous solution is the pulse response.

(b) We are required to solve:

$$y(n) + 0.6y(n - 1) = 3\delta(n) + \delta(n - 1)$$

The form of $y(n)$ is:

$$y(n) = A(-0.6)^n u(n) + B\delta(n)$$

since $y(n)$ must contain $B\delta(n)$ or it is impossible for $y(n) + 0.6y(n - 1)$ to contribute a $\delta(n - 1)$ term. In a manner analogous to the continuous case when finding impulse responses we must have:

$$0.6B = 1 \qquad \text{[Be very sure!]}$$

or $\qquad\qquad B = 1.67$

Therefore $\qquad y(n) = A(-0.6)^n u(n) + 1.67\delta(n)$

Considering our equation at $n = 0$, we require:

$$y(0) + 0.6(0) = 3$$

Therefore $\qquad\qquad A + B = 3 \quad \text{and} \quad A = 1.33$

Finally:

$$h(n) = 1.67\delta(n) + 1.33(-0.6)^n u(n)$$

or an equivalent alternate form is:

$$h(n) = 3\delta(n) + 1.33(-0.6)^n u(n - 1)$$

2-5-1 An Alternative Approach

$$y(n) + 0.6y(n - 1) = 3\delta(n) + \delta(n - 1)$$

may also be solved by superposition. The solution to $y(n) + 0.6y(n - 1) = \delta(n)$ is obviously $y(n) = (-0.6)^n u(n)$. The solution to $y(n) + 0.6y(n - 1) = \delta(n - 1)$ is:

$$y(n) = (-0.6)^{n-1} u(n - 1)$$

Therefore by superposition the solution to our equation is:

$$h(n) = 3(-0.6)^n u(n) + (-0.6)^{n-1} u(n - 1)$$
$$= 3\delta(n) + [3(-0.6)^n + (-0.6)^{n-1}]u(n - 1)$$
$$= 3\delta(n) + 1.33(-0.6)^n u(n - 1)$$

as before. The reader is exhorted to sketch $h(n)$ and $0.6h(n - 1)$ and to appreciate their sum yields $3\delta(n) + \delta(n - 1)$.

EXAMPLE 2-18

(a) Solve for $h(n)$ for the system whose governing equation is:

$$y(n) - 0.5y(n - 1) + 0.06y(n - 2) = 2x(n) + x(n - 1)$$

(b) Check your result recursively.

Solution

(a) Consider:

$$y(n) - 0.5y(n - 1) + 0.06y(n - 2) = 2\delta(n) + \delta(n - 1)$$

The solution will not contain an extra $\delta(n)$ term and will be of the form:

$$y(n) = [A_1(+0.2)^n + A_2(+0.3)^n]u(n)$$

where $$h(0) - 0 + 0 = 2$$

and $$h(1) - 0.5(2) + 0 = 1$$

which gives $h(1) = 2$. Applying $h(0) = 2$ and $h(1) = 2$ yields the simultaneous equations:

$$A_1 + A_2 = 2$$

and $$0.2A_1 + 0.3A_2 = 2$$

Solving gives $A_1 = -14$ and $A_2 = 16$

and $$h(n) = [-14(+0.2)^n + 16(+0.3)^n]u(n) \tag{1}$$

(b) $y(n) = 0.5y(n-1) - 0.06y(n-2)$, $n \geq 2$ with starting conditions: $y(0) = 2$, $y(1) = 2$. Proceeding iteratively:
$y(2) = 0.5(2) - 0.06(2) = 0.88$
$y(3) = 0.5(0.88) - 0.06(2) = 0.32$
$y(4) = 0.5(0.32) - 0.06(0.88) = 0.11$
and so on

These values agree with those obtained on substituting into the classical solution (1).

2-5-2 Some General Comments About $h(n)$

If a LTIC system is governed by:

$$a_n y(n) + \cdots + a_{n-p} y(n-p) = b_n x(n) + \cdots + b_{n-k} x(n-k)$$

then we obtain the following forms for $h(n)$:

Case $p > k$

$$h(n) = \left[\sum_{i=1}^{p} A_i(\alpha_i)^n \right] u(n)$$

when the roots of the characteristic equation are real and distinct; α_1, $\alpha_2, \ldots, \alpha_p$

Case $p = k$

$$h(n) = \left[\sum_{i=1}^{p} A_i(\alpha_i)^n \right] u(n) + B_0\delta(n)$$

Case $p < k$

$$h(n) = \left[\sum_{i=1}^{p} A_i(\alpha_i)^n \right] u(n) + B_0\delta(n) + \cdots + B_{\alpha-p}\delta[n - (k-p)]$$

The adjustment to the preceding cases for repeated real roots is simple, whereas conjugate imaginary and complex conjugate roots may be investigated when the need arises.

2-5-3 The Output of a LTIC System

Given a LTIC system with pulse response $h(n)$, it is possible to find the response to any input $x(n)$. If

$$x(n) = \sum_{k=-\infty}^{\infty} x(k)\delta(n - k)$$

then:

- The response to $\delta(n)$ is $h(n)$.

- The response to $x(k)\delta(n - k)$ is $x(k)h(n - k)$ using time invariance and linearity.

- The response to $x(n) = \sum_{k=-\infty}^{\infty} x(k)\delta(n - k)$ is $\sum_{k=-\infty}^{\infty} x(k)h(n - k)$ = $y(n)$.

We have derived in general that the response to $x(n)$ is:

$$y(n) = \sum_{k=-\infty}^{\infty} x(k)h(n - k)$$

$$= x(n)*h(n) \tag{2-31}$$

and this is the discrete convolution of $x(n)$ and $h(n)$. We have already discussed discrete convolution in Section 2-3 as the convolution set product of two sets of numbers. We call $y(n)$ the zero-state complete response as $h(n)$ is derived with $h(-1) = h(-2) = \cdots = h(-n) = 0$. As in the continuous case, we solve an example incorporating the ideas of discrete systems already examined.

EXAMPLE 2-19

Consider the simple first-order system shown in block diagram form in Figure 2-17(a). Find:

(a) the system difference equation
(b) the system function $H(z)$
(c) the pulse response $h(n)$
(d) the zero-state response when $x(n) = u(n)$
(e) the complete response when $y(-1) = 2$ and $x(n) = u(n)$

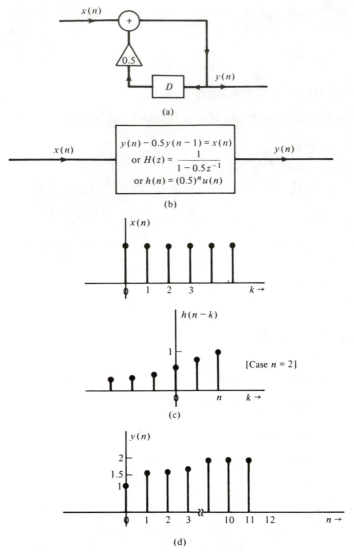

Figure 2-17 (a) The block diagram for the system of Example 2-19; (b) the system representation; (c) the system input $x(k)$ and $h(n - k)$ for convolution; (d) the system zero-state output.

Solution

(a) From Figure 2-17(a):

$$y(n) = 0.5y(n - 1) + x(n)$$

Therefore $$y(n) - 0.5y(n - 1) = x(n)$$

(b) $H(z) = \dfrac{y_{fo}(n)}{x(n)}\bigg|_{x(n)=z^n}$

$$= \frac{1}{1 - 0.5z^{-1}} = \frac{z}{z - 0.5}$$

This part is included to show the analogy to $H(s)$ which was much utilized for continuous systems. Philosophically, it will be useful to understand the meaning of $H(z)$ in the time domain when we consider the Z transform later.

(c) $y(n) - 0.5y(n - 1) = \delta(n)$

and $\qquad\qquad h(n) = A(0.5)^n u(n)$

where $\qquad\qquad h(0) = 1$

Therefore $\qquad h(n) = (0.5)^n u(n)$

(d) The zero-state response is:

$$y_{zs}(n) = (0.5)^n u(n) * u(n)$$

$$= \sum_{k=0}^{n} (0.5)^k (1) \text{ as may be seen in Figure 2-17c.}$$

$$= \sum_{k=0}^{\infty} (0.5)^k - \sum_{k=n+1}^{\infty} (0.5)^k$$

$$= 2 - 0.5^{n+1}(2)$$

$$= 2u(n) - 0.5^n u(n)$$

The first few terms of $y_{zs}(n)$ are:

$$(y) = (1, 1.5, 1.75, \ldots, 2, \ldots, 2)$$

and $y(n)$ is illustrated in Figure 2-17(d).

(e) Given:

$$y(-1) = 2 \quad \text{or} \quad y(0) = 0.5(2) + 1 = 2$$

then only the transient term changes

and $\qquad\qquad\qquad y(n) = A(0.5)^n + 2$

with $\qquad\qquad\qquad\qquad 2 = A + 2$

$$A = 0$$

$$y(n) = 2 \qquad \text{for all } n$$

It is interesting to note that $y(-1) = 2$ makes the transient term zero.

2-6 DISCRETE CONVOLUTION AND PROPERTIES OF CONVOLUTION SUMMATIONS

The discrete convolution of two functions is:

$$x(n)*y(n) = \sum_{\text{all } k} x(k)y(n-k)$$

and we may denote this by $r_{xy}(n)$. If $x(n)$ and $y(n)$ are of finite duration, we observe the product function $x(k)y(n-k)$ as a function of k and decide for what values of n it is zero. Convolution begins when the moving sequence $y[-(k-n)]$ which is $y(k)$ reflected and moved n units to the right arrives at the first value of $x(k)$. Convolution concludes when $y(n-k)$ clears the last value in the sequence $x(k)$. We now solve a few problems.

EXAMPLE 2-20

Evaluate:

(a) $\delta(n-3)*\delta(n+4)$
(b) $u(n)*u(n-2)$
(c) $2^n u(n)*u(n-4)$
(d) $2^n u(n)*(-\frac{1}{3})^n u(n)$

Solution

(a) The answer is obviously $\delta(n+1)$ as can be seen from Figure 2-18(a). However we now determine it formally:

$$\delta(n-3)*\delta(n+4) = \sum_{\text{all } k} \delta(k-3)\delta(n-k+4)$$

$$= 1, \quad \text{when } n+4 = 3 \text{ or } n = -1$$

$$= 0, \quad \text{otherwise}$$

Therefore $\delta(n-3)*\delta(n+4) = \delta(n+1)$

(b) $u(n)*u(n-2) = \sum_{\text{all } k} u(k)u(n-k-2)$

$$= \sum_{k} u(k)u[-(k-n+2)]$$

$$= 0, \quad \text{if } n-2 < 0$$

$$= \sum_{k=0}^{n-2} (1)(1), \quad n \geq 2$$

$$= (n-1)$$

Therefore $u(n)*u(n-2) = (n-1)u(n-2)$

We notice a slight operational difference between the continuous and

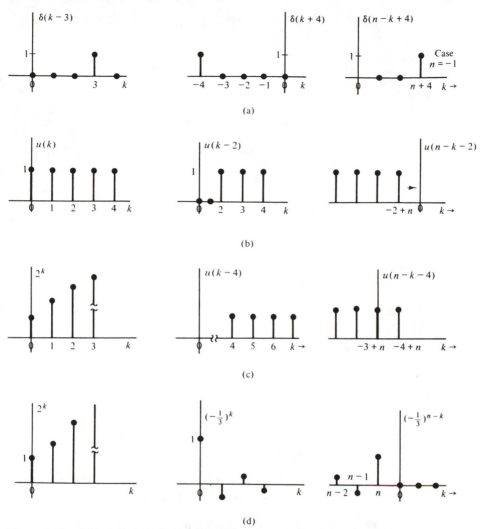

Figure 2-18 $f(k)$ and $g(n - k)$ for convolving the functions of Example 2-20(a) through (d).

discrete unit "step" functions:

$$u(t)*u(t) = tu(t)$$

$$u(n)*u(n) = (n + 1)u(n)$$

Figure 2-18(b) shows $u(k)$ and $u(n - k - 2)$ versus k.

(c) $2^n u(n)*u(n - 4) = \displaystyle\sum_{k} 2^k u(k)u(n - k - 4)$

$\qquad\qquad\qquad\quad = \displaystyle\sum_{k=0}^{n-4} 2^k(1)$

$$= 0, \qquad n < 4$$

$$= \sum_{k=0}^{n-4} 2^k, \qquad n \geq 4$$

The convolution sequence is:

$$r(n) = (\underset{\underset{n=0}{\uparrow}}{0}, 0, 0, 0, 1, 1 + 2, 1 + 2 + 2^2, \cdots)$$

Figure 2-18(c) shows $2^k u(k)$, $u(k - 4)$, and $u(n - k - 4)$ versus k.

(d) $2^n u(n) * (-\tfrac{1}{3})^n u(n) = \displaystyle\sum_{k=0}^{n} 2^k (-\tfrac{1}{3})^{n-k}$

$$= 0, \qquad n < 0$$

$$= (-\tfrac{1}{3})^n \sum_{k=0}^{n} (-6)^k, \qquad n \geq 0$$

The convolution sequence is:

$$r(n) = (1, (-\tfrac{1}{3})(1 - 6), \tfrac{1}{9}(1 - 6 + 36) + \cdots)$$

if n is large:

$$r(n) = (-\tfrac{1}{3})^n [(-6)^0 + (-6) + (-6)^2 + \cdots]$$

Figure 2-18(d) shows $2^k u(k)$, $(-\tfrac{1}{3})^k u(k)$, and $(-\tfrac{1}{3})^{n-k} u(n - k)$ versus k. Later we will find closed form expressions for parts (c) and (d).

2-6-1 Comment—An Intuitive Way to Evaluate Discrete Convolutions

It is possible to use our knowledge of difference equations and discrete systems to rapidly find formulas for convolution summations of "easy analytical functions."

We found that the solution of a linear difference equation was the sum of the homogeneous solution plus the forced response that has the same form as the forcing term or the input. The output of a LTIC system was obtained as the discrete convolution of the input $x(n)$ and the pulse response $h(n)$. Since this is the solution of the system difference equation with $x(n)$ as input, it must be the sum of a homogeneous term of the form of $h(n)$ and a forced response of the form of the input $x(n)$. Without doing any work we can make the following observations based on discrete system theory.

$$a^n u(n) * b^n u(n) = (c_1 a^n + c_2 b^n) u(n)$$

or $\qquad a^n u(n) * n u(n) = [c_1 a^n + c_2 + c_3 n] u(n)$

We now redo some of the previously solved problems and obtain closed form expressions for our answers.

EXAMPLE 2-21

Intuitively, obtain solutions for:

(a) $2^n u(n) * (-\frac{1}{3})^n u(n)$
(b) $2^n u(n) * u(n - 4)$
(c) $u(n) * u(n - 2)$

Solution

(a) We assume from our knowledge of discrete systems that:

$$r(n) = 2^n u(n) * (-\tfrac{1}{3})^n u(n) = c_1 2^n u(n) + c_2(-\tfrac{1}{3})^n u(n)$$

where from discrete convolution:

$$r(0) = 1 \quad \text{and} \quad r(1) = -\tfrac{1}{3} + 2 = +1.67$$

Substituting in our assumed form, we obtain:

$$1 = c_1 + c_2 \quad \text{and} \quad 1.67 = 2c_1 - 0.33c_2$$

Solving for c_1 and c_2, we have:

$$c_2 = \frac{0.33}{2.33} = 0.14 \quad \text{and} \quad c_1 = 0.86$$

Comparing

$$r(n) = [0.86(2)^n + 0.14(-0.33)^n]u(n) \tag{1}$$

to the result:

$$r(n) = [1, (-0.33)(-5), 0.11\,(1 - 6 + 36) + \cdots] \tag{2}$$

from Example 2-20(d), we note for both results (1) and (2):

$$r(0) = 1$$
$$r(1) = 1.69 \text{ from } (1) \text{ and } 1.65 \text{ from } (2)$$
$$r(2) = 3.44 + 0.001 = 3.45 \text{ from } (1)$$

and
$$r(2) = 0.11(31) = 3.41 \text{ from } (2)$$

which checks out well.

(b) We now assume:

$$r(n) = 2^n u(n) * u(n - 4) = c_1 2^{n-4} u(n - 4) + c_2 u(n - 4)$$

From discrete convolution we find:

$$r(4) = 1 \quad \text{and} \quad r(5) = 3$$

Substituting in our assumed form, we obtain:

$$1 = c_1 + c_2 \quad \text{and} \quad 3 = 2c_1 + c_2$$

Solving for c_1 and c_2, we have:

$$c_1 = 2 \quad \text{and} \quad c_2 = -1$$

Therefore $$r(n) = 2(2)^{n-4}u(n-4) - u(n-4) \qquad (1)$$

Comparing this to the solution by discrete convolution from Example 2-20(c), which was

$$r(n) = (0, 0, 0, 0, 1, 3, 7, 1 + 2 + 2^2 + 2^3, \text{etc.}) \qquad (2)$$

We note:

$$r(4) = 1 \qquad \text{for both (1) and (2)}$$
$$r(5) = 3 \qquad \text{for both (1) and (2)}$$
$$r(6) = 7 \qquad \text{for both (1) and (2)}$$

(c) Considering $u(n)*u(n-2)$, we note that the homogeneous and forcing terms are of the same form and therefore we assume:

$$r(n) = u(n)*u(n-2) = c_1 u(n-2) + c_2(n-2)u(n-2)$$

From discrete convolution we find:

$$r(2) = 1 \quad \text{and} \quad r(3) = 2$$

Substituting in our assumed form, we obtain:

$$1 = c_1 \quad \text{and} \quad 2 = c_1 + c_2$$

Solving for c_1 and c_2, we have:

$$c_1 = 1 \quad \text{and} \quad c_2 = 1$$

and $$r(n) = u(n-2) + (n-2)u(n-2) \qquad (1)$$

Comparing this to the solution by discrete convolution from Example 2-20(b), which was

$$r(n) = n - 1, \qquad n \geq 2 \qquad (2)$$

We note:

$$r(2) = 1, \qquad r(3) = 2, \qquad r(4) = 3, \text{ and so on for both}$$

Also, we may say:

$$r(n) = u(n-2) + (n-2)n(n-2)$$
$$= [1 + (n-2)]u(n-2)$$
$$= (n-1)u(n-2)$$

which is the same as (2).

2-6-2 Closed Form Expressions for Discrete Convolution Analytically

The solutions for parts (c) and (d) of Example 2-19 were not obtained in closed form. We now quote some results from evaluating sums of series and apply them to discrete convolution. These are:

$$\sum_{k=0}^{n} \alpha^k = \frac{\alpha^{n+1} - 1}{\alpha - 1} \tag{2-32}$$

and

$$\sum_{k=0}^{n} k\,\alpha^k = \alpha \frac{d}{d\alpha}\left[\frac{\alpha^{n+1} - 1}{\alpha - 1}\right] \tag{2-33a}$$

$$= \frac{\alpha\,[(\alpha - 1)(n + 1)\alpha^n - (\alpha^{n+1} - 1)]}{(\alpha - 1)^2} \tag{2-33b}$$

EXAMPLE 2-22

Prove Equation 2-32 for $|\alpha| < 1$ and show Equation 2-33a follows from it.

Solution

$$\sum_{0}^{n} \alpha^k = \sum_{0}^{\infty} \alpha^k - \sum_{n+1}^{\infty} \alpha^k$$

$$= \frac{1}{1 - \alpha} - \alpha^{n+1} \frac{1}{1 - \alpha}$$

$$= \frac{1 - \alpha^{n+1}}{1 - \alpha} \quad \text{or} \quad \frac{\alpha^{n+1} - 1}{\alpha - 1}$$

This follows from the fact that:

$$(1 - \alpha)^{-1} = 1 + \alpha + \alpha^2 + \cdots = \sum_{0}^{\infty} \alpha^p \qquad \text{for } |\alpha| < 1$$

Now since:

$$\frac{\alpha^{n+1} - 1}{\alpha - 1} = 1 + \alpha + \alpha^2 + \cdots + \alpha^n$$

on differentiation, we obtain:

$$\frac{d}{d\alpha}\left[\frac{\alpha^{n+1} - 1}{\alpha - 1}\right] = 1 + 2\alpha + 3\alpha^2 + \cdots + n\alpha^{n-1}$$

and

$$\alpha \frac{d}{d\alpha}\left[\frac{\alpha^{n+1} - 1}{\alpha - 1}\right] = \alpha + 2\alpha^2 + 3\alpha^3 + \cdots + n\alpha^n = \sum_{k=0}^{n} k\,\alpha^k$$

which proves Equation 2-33a. It can be shown that these results are also true for $|\alpha| > 1$.

It is fascinating to see how apparently divergent results for infinite sequences can be combined to give correct results for finite sequences. For example, let us consider evaluating $\Sigma_0^6(-3)^k$ and $\Sigma_{k=0}^3 k(-3)^k$.

$$\sum_0^6 (-3)^k = \sum_0^\infty (-3)^k - (-3)^7 \sum_0^\infty (-3)^k$$

$$= \frac{1}{1+3} - (-3)^7 \frac{1}{1+3} \qquad \begin{array}{l}\text{(Note how untrue}\\ \text{the separate}\\ \text{expressions are)}\end{array}$$

$$= \frac{1}{4}(1 - (-3)^7)$$

$$= 547$$

This result may be verified by adding the terms:

$$\sum_0^6 (-3)^k = [1 - 3 + 9 - 27 + 81 - 243 + 729] = 547$$

Using Equation 2-33b

$$\sum_0^3 k(-3)^k = \frac{(-3)[(-4)(4)(-3)^3 - ((-3)^4 - 1)]}{(-4)^2}$$

$$= \frac{+3[432 - 80]}{16}$$

$$= -66$$

This result may be verified by directly showing:

$$\sum_0^3 k(-3)^k = 0 + (-3) + 18 - 81 = -66$$

We now consider some convolution problems.

EXAMPLE 2-23

Evaluate in closed form using Equations 2-32 and 2-33(b)

(a) $4(-2)^n u(n) * 5(3)^n u(n)$
(b) $(2 + 3n)(-3)^n u(n) * u(n)$

Solution

(a) $4(-2)^n u(n) * 5(3)^n u(n) = \displaystyle\sum_{k=0}^n 4(-2)^k 5(3)^{n-k}$

$$= 20(3)^n \sum_0^n \left(-\frac{2}{3}\right)^k$$

$$= 20(3)^n \frac{(-\frac{2}{3})^{n+1} - 1}{-1.67} \text{ using Equation 2-32}$$

$$= -12\left[\left(-\frac{2}{3}\right)(-2)^n - (3)^n\right]$$

$$= 8(-2)^n + 12(3)^n, \qquad n \ge 0$$

(b) $(2 + 3n)(-3)^n u(n) * u(n) = \sum_0^n 2(-3)^k$

$$+ \sum_0^n 3k(-3)^k$$

$$= 2\frac{(-3)^{n+1} - 1}{-4}$$

$$+ 3\frac{(-3)[(-4)(n + 1)(-3)^n - ((-3)^{n+1} - 1)]}{16}$$

$$= 1.5(-3)^n + 0.5$$

$$+ \left(-\frac{9}{16}\right)[-4n(-3)^n - 4(-3)^n + 3(-3)^n + 1]$$

$$= 1.5(-3)^n + 0.5$$

$$- 0.56[-4n(-3)^n - (-3)^n + 1]$$

$$= -0.06 + 2.06(-3)^n + 2.24n(-3)^n, \quad n \ge 0$$

This material may be extended by finding expressions for $\Sigma_{k-0}^n k^2\alpha^k$, $\Sigma_{k-0}^n k^3\alpha^k$, and so on. Indeed, when using Z transforms in Chapters 6 and 7, we accomplish this routinely with partial fractions and a table of transforms.

2-6-3 Properties of Discrete Convolution

Some of the more important properties of discrete convolution are:

Commutative Law

$$x(n)*y(n) = y(n)*x(n) \qquad (2\text{-}34)$$

or
$$r_{xy}(n) = r_{yx}(n)$$

Distributive Law

$$x(n)*[y(n) \pm z(n)] = x(n)*y(n) \pm x(n)*z(n) \qquad (2\text{-}35)$$

or
$$r_{x(y \pm z)}(n) = r_{xy}(n) \pm r_{xz}(n)$$

Associative Law

$$[x(n)*y(n)]*z(n) = x(n)*[y(n)*z(n)] \qquad (2\text{-}36)$$

The proofs of these laws are very similar to the continuous case and we now solve an analogous problem to Example 2-12.

EXAMPLE 2-24

Prove the commutative law and set up the proof of the associative law for discrete convolution.

Solution

(a) We want to show:

$$\sum_{k=-\infty}^{\infty} x(k)y(n-k) = \sum_{p=-\infty}^{\infty} y(p)x(n-p)$$

Considering the RHS, let $n - p = s$

then

$$\text{RHS} = \sum_{s=-\infty}^{\infty} y(n-s)x(s)$$

$$= \text{LHS}$$

(b) We are required to show:

$$[x(n)*y(n)]*z(n) = x(n)*[y(n)*z(n)]$$

Carefully we obtain expressions for both sides as a function of n:

$$r_{xy}(n) = x(n)*y(n)$$

$$= \sum_{k} x(k)y(n-k)$$

Therefore $$\text{LHS} = r_{xy}(n)*z(n) = \sum_{p} r_{xy}(p)z(n-p)$$

$$= \sum_{p}\left[\sum_{k} x(k)y(p-k)\right]z(n-p)$$

$$(2\text{-}37)$$

Similarly, for the right-hand side:

Therefore $$\text{RHS} = x(n) * r_{yz}(n)$$

$$= \sum_{s} x(s)r_{yz}(n-s)$$

$$= \sum_{s} x(s)\sum_{m} y(m)\,z(n-s-m) \qquad (2\text{-}38)$$

It is now possible but difficult to show that the LHS and RHS are identical.

Drill Set: Discrete Convolution

1. For a linear time-invariant system with pulse response

$$h(n) = (-0.7)^n u(n)$$

Find the response if:
(a) $x(n) = \delta(n) - 2\delta(n - 1) + 3\delta(n - 4)$
(b) $x(n) = u(n)$

2. Evaluate rapidly and plot $r(n)$:
(a) $\delta(n + 3) * \delta(n - 2)$
(b) $nu(n - 1) * \delta(n - 2)$
(c) $2^n u(n) * (-0.5)^{n-2} u(n - 2)$

SUMMARY

A linear time-invariant system is governed by a linear differential equation:

$$a_n y^n(t) + a_{n-1} y^{n-1}(t) + \cdots + a_0 y(t) = b_0 x(t) + \cdots + b_m x^m(t) \qquad (1)$$

relating a continuous output and input, or a linear difference equation:

$$a_n y(n) + \cdots + a_{n-p} y(n - p) = b_n x(n) + \cdots + b_{n-k} x(n - k) \qquad (2)$$

relating a discrete output and input. The continuous case was treated thoroughly. The system function $H(s)$ was defined as the ratio of the forced response to the input when the input $x(t) = e^{st}$. For a system governed by equation (1)

$$H(s) = \frac{b_m s^m + b_{m-1} s^{m-1} + \cdots + b_0}{a_n s^n + a_{n-1} s^{n-1} + \cdots + a_0}.$$

For any passive circuit $H(s)$ is readily found by simple circuit analysis and represents the tool for rapidly finding the system differential equation. The impulse response $h(t)$ is the output of a zero-initial energy system to a delta function input. Impulse responses were found by assuming logical forms and obtaining the required coefficients. A side effect of all this was to obtain deeper insight into and more confidence with singularity functions. Since it is possible to approximate any continuous function as a weighted string of delta functions, the zero-state output $y(t)$ is derived as the input convolved with the impulse response. If the initial energy of a system is not zero, the output has different coefficients associated with its homogeneous part from those in the zero-state output.

In the discrete case the system function $H(z)$ was defined as the ratio of the forced response to the input when $x(n) = z^n$. For a system governed by (2)

$$H(z) = \frac{b_n z^p + \cdots + b_{n-k} z^{p-k}}{a_n z^p + a_{n-1} z^{p-1} + \cdots + a_{n-p}} \qquad \text{if } p \geq k.$$

In this chapter we did not use $H(z)$ as a vehicle for finding the system difference equation because we only modeled discrete systems with multiplying, delaying, summing, and branching blocks. In this case finding the system equation directly is simple. The pulse response $h(n)$ is the output of a system with all initial values $y(-1)$, $y(-2)$, and so on zero when the input is $\delta(n)$. Pulse responses were found by assuming logical forms for $h(n)$ and obtaining the required coefficients. The zero-state output for any input $x(n)$ was derived as the discrete convolution of $x(n)$ and $h(n)$. If the initial values $y(-1) \cdots y[-(n-p)]$ are not zero, then the output has different values associated with the coefficients of its homogeneous part from those in the zero-state output. In addition to using a classical approach, we can readily solve for difference equations, pulse responses, and complete responses using a recursive or iterative procedure.

PROBLEMS

2-1. Solve the following differential equations:
   (a) $y''(t) + 3y'(t) + 2y(t) = 2e^{-t} - t$
   $$y(0) = 2, \qquad y'(0) = -1$$
   (b) $y''(t) + 2y'(t) + y(t) = 2t \cos t$
   $$y(0) = 2, \qquad y'(0) = -1$$

2-2. Given the solution of a differential equation is:

$$y(t) = 4e^{-2t} + 2e^{-3t} + 3t$$

where $y_{\text{homo}}(t) = 4e^{-2t} + 2e^{-3t}$ and $y_{fo}(t) = 3t$

find the differential equation.

2-3. Find the impulse response for the systems governed by the following equations:
   (a) $3y'(t) + 5y(t) = 7x(t)$
   (b) $3y'(t) + 5y(t) = 7x(t) + 3x'(t)$
   (c) $3y'(t) + 5y(t) = 7x(t) + 3x'(t) + x''(t)$
   (d) $y''(t) + 2y'(t) + 2y(t) = 2x(t) + 3x'(t)$

2-4. Find the system function and hence the system differential equations for the following systems:

(a)

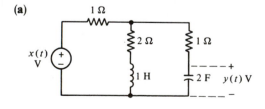

(b)

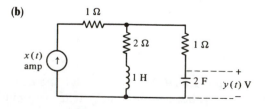

(c)

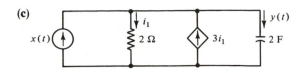

2-5. Linearity, time-invariance, and causality are defined as follows:
- A system is **linear** if when $x_1(t)$ yields $y_1(t)$ and $x_2(t)$ yields $y_2(t)$, then $ax_1(t) + bx_2(t) \Rightarrow ay_1(t) + by_2(t)$ for any $x_1(t)$ and $x_2(t)$ where a and b are constants.
- A system is **time-invariant** if when $x(t)$ yields $y(t)$, then $x(t - \tau)$ yields $y(t - \tau)$.
- A system is **causal** if the output cannot commence before the input or $h(t) = 0$, $t < 0$.

(i) Check whether systems defined by the following differential equations are linear, time-invariant or causal.
   (a) $y'(t) + 3y(t) = 2x(t) + x'(t)$
   (b) $y(t) = 2x(t) + x'(t)$
   (c) $y(t) = 2$
   (d) $y'(t) + 2y(t) = tx(t)$
   (e) $y'(t) + 2y^2(t) = x(t)$
   (f) $y'(t) * y(t) = x(t - 3)$
   (g) $y'(t) + y(t) = x(t + 3)$

(ii) Check whether systems defined by the following impulse responses are linear, time-invariant or causal.
   (a) $h(t) = 2e^{-5t}u(t)$ (b) $h(t) = 2e^{-5(t-2)}u(t - 2)$
   (c) $h(t) = 2e^{-5(t+1)}u(t + 1)$ (d) $h(t) = 2e^{-t}u(t) + e^t u(-t)$

2-6. Some other different type systems are defined as follows:
- A real system is defined by:

$$Im[h(t)] = 0$$

where Im denotes "the imaginary part of."
- A stable system is defined by:

$$\int_{-\infty}^{\infty} |h(t)|^2\, dt < \infty$$

Consider a second-order LTIC system governed by:

$$a_2 y''(t) + a_1 y'(t) + a_0 y(t) = b_0 x(t) + b_1 x'(t) + b_2 x''(t)$$

where the a's and b's are real. What are the conditions on the coefficients for the system to be real and stable? Assume $a_2 \neq 0$. Make thumbnail sketches of $h(t)$ versus t for all the different possible forms of $h(t)$.

2-7. For the three systems of Problem 2-4 find:
   (a) the zero-state response when (1) $x(t) = 2u(t)$ and (2) $x(t) = te^{-t}u(t)$
   (b) the complete response when (1) $x(t) = 2u(t)$ and (2) $x(t) = te^{-t}u(t)$ and all the initial inductor currents and capacitor voltages are 1 amp and 1 V, respectively (use current direction down and voltage polarity top to bottom).

2-8. Evaluate:
   (a) $x(t) * y(t)$
      where $x(t) = u(t - 1) - u(t - 3)$ and $y(t) = -2[u(t + 2) - u(t - 1)]$

   (b) $x(t) * y(t)$
     where $x(t) = [u(t - 1) - u(t - 3)]$ and $y(t) = [2u(t + 1) - 3u(t - 2) + u(t - 3)]$

   (c) $2e^{-2t}u(t) * (-3e^{-t})u(t)$ both directly and intuitively by assuming a solution

   (d) $2e^{-2t}u(t) * e^{t}u(-t)$

   (e) $2e^{-2t}u(t) * e^{-t}u(-t)$

   (f) $te^{-t}u(t) * e^{-t}u(-t)$

2-9. Set up the different ranges and integrals when evaluating the convolution integral $x(t) * y(t)$ for the functions shown.

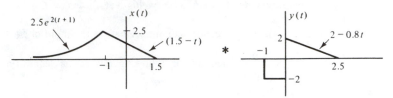

2-10. Solve the following difference equations classically:

   (a) $y(n) - 0.2y(n - 1) = 0, \qquad y(0) = 3$

   (b) $y(n) + \frac{1}{3}y(n - 1) - \frac{2}{9}y(n - 2) = \frac{2}{9}, \qquad y(-1) = 2, \quad y(-2) = 1$

   (c) $y(n) + \frac{1}{3}y(n - 1) - \frac{2}{9}y(n - 2) = n(\frac{1}{4})^{n}, \qquad y(0) = 1, \quad y(-1) = 2$

2-11. Find the pulse response $h(n)$ for the following systems:

   (a) $5y(n) - 2y(n - 1) = 3x(n)$

   (b) $5y(n) - 2y(n - 1) = 3x(n) - 2x(n - 1)$

   (c) $y(n) + \frac{1}{3}y(n - 1) - \frac{1}{9}y(n - 2) = 2x(n)$

2-12. Resolve Problem 2-11 recursively and verify your results for $n \le 4$.

2-13. In an analogous manner to the continuous case, define the terms: linear, time-invariant, causal, real, and stable for a discrete system.

2-14. Determine whether each of the following systems is linear, time-invariant, causal, or stable:

   (a) $y(n) + 2y(n) = x(n)$

   (b) $y(n) - 0.6y(n - 1) = x(n - 2)$

   (c) $h(n) = 0.5^{n-1}nu(n)$

   (d) $h(n) = 0.5^{n}u(n) + 0.5^{-n}u(-n - 1)$

2-15. For the three systems of Problem 2-11 find:

   (a) the zero-state response when (1) $x(n) = 3u(n)$ and (2) $x(n) = (2^{n} + n)u(n)$

   (b) the complete response when (1) $x(n) = 3u(n)$ and (2) $x(n) = (2^{n} + n)u(n)$ and the initial condition for the first two systems is $y(-1) = 2$ and for the third system is $y(-1) = y(-2) = 2$.

2-16. Evaluate (1) analytically (if the summation exists) and (2) by assuming the correct form.

   (a) $2^{n}u(n) * 3^{n}u(n)$ **(b)** $2^{n}u(n) * u(-n)$

   (b) $2^{n}u(n) * 3^{n}u(-n)$ **(c)** $a^{n}u(n) * b^{n}u(-n)$

2-17. Convolution integrals may be evaluated by assuming the form of the solution and obtaining the necessary initial conditions. Let us consider causal functions where the required initial conditions are $y(0^{+}), y'(0^{+}), y''(0^{+})$, and so on. If

$$y(t) = f(t)u(t) * g(t)u(t)$$

(a) Show $y'(t) = [f'(t)u(t) + f(0)\delta(t)] * g(t)u(t)$

(b) Show $y'(0^+) = f'(0^+)g(0^+) + f(0^+)g'(0^+)$

(c) Find an expression for $y''(0^+)$

2-18. By assuming a solution and obtaining the required initial conditions, evaluate:

(a) $3e^{-2t}u(t) * 2e^{-t}u(t)$

(b) $2e^{-t}u(t) * (2 + 3t)u(t)$

(c) Check your answers analytically.

2-19. Discrete Frequency Response

(a) Given a discrete system with system function H(z) show the forced response to:

i. $e^{j\omega n}$ is $H(e^{j\omega}) e^{j\omega n}$

ii. $\cos \omega n$ is $|H(e^{j\omega})| \cos[\omega n + \angle H(e^{j\omega n})]$

(b) Plot the waveforms:

i. $\cos 2\pi n$,   ii. $\cos \pi n$,   iii. $\cos \dfrac{\pi}{2}n$, iv. $\cos \dfrac{\pi}{10}n$

(c) Given the following systems plot $|H(e^{j\omega})|$ vs ω, $-\alpha < \omega < \alpha$

i. $H(z) = \dfrac{z}{z-T}$ ii. $H(z) = \dfrac{z^2(z + 0.5)}{(z - 1\angle 1 \cdot 1)(z - 1\angle 1 \cdot 1)}$

iii. $H(z) = \dfrac{z}{z+T}$

(d) Prove,

i. $|H(e^{j\omega})| = |H(e^{j(\omega + \pi)})|$

ii. $\angle H(e^{j\omega}) = -\angle H(e^{-j\omega})$

iii. $\angle H(e^{j\omega}) = \angle H(e^{j(\omega + 2\pi)})$

Linear Time-invariant Systems with Random Inputs

INTRODUCTION

Chapter 2 treated the time-domain analysis of LTIC continuous and discrete systems with deterministic inputs. A system is characterized by its system equation or system function. The **impulse response** $h(t)$ or **pulse response** $h(n)$ is the solution of the system differential or difference equation when the input is $\delta(t)$ or $\delta(n)$, respectively, with initial conditions equal to zero. The zero-state response to an input $x(t)$ or $x(n)$ is the continuous or discrete convolution of the input with the impulse or pulse response.

Chapter 3 considers the time-domain analysis of LTIC continuous and discrete systems with random or deterministic signal plus random inputs. Achieving the goals of analyzing systems with random inputs involves much preparation before concluding with the famous Wiener-Khinchine results of Section 3-5. This procedure includes the following stages:

1. the evaluation and interpretation of correlation integrals for continuous and discrete finite energy functions
2. the evaluation and interpretation of correlation integrals for infinite energy, finite power, periodic functions
3. a tutorial introduction to statistics of and correlation integrals for finite power noise waveforms. Although a thorough treatment of this topic would involve knowledge of probability theory and random processes, the ideas of how correlation functions measure randomness for noise waveforms are intuitively discussed relying only on the simple concept of what an average value is. It is stimulating, simple, and very proper to think statistically of previously mechanically used time averages.

Stages 1 through 3 are considered in Sections 3-1 through 3-4. With this procedure the input–output relations between the correlation functions of input and output noise waveforms for a system with a random input are examined in Section 3-5. In addition, the case of an input deterministic signal plus an independent random waveform is also treated. Sections 3-4 and 3-5 are quite challenging, because we are developing very powerful results without the often *required prerequisite* of three units of probability theory.

3-1 CORRELATION INTEGRALS FOR CONTINUOUS FINITE ENERGY WAVEFORMS

3-1-1 Evaluation of Correlation Integrals

The definition of the correlation of $f(t)$ with $g(t)$ also called the **cross-correlation** of $f(t)$ and $g(t)$ is denoted and defined as:

$$C_{fg}(t) = f(t) \oplus g(t)$$

$$\triangleq \int_{-\infty}^{\infty} f(u)g(u + t)\, du \qquad (3\text{-}1a)$$

For the purpose of evaluation it is somewhat easier to use the equivalent definition:

$$C_{fg}(t) = \int_{-\infty}^{\infty} f(u - t)g(u)\, du \qquad (3\text{-}1b)$$

The reader should show that Equations 3-1a ad 3-1b are identical by using an appropriate substitution of variable.

From 3-1a or 3-1b the correlation of $f(t)$ with itself which is called the **autocorrelation** function is:

$$C_{ff}(t) = f(t) \oplus f(t)$$

$$\triangleq \int_{-\infty}^{\infty} f(u) f(u \pm t)\, du \qquad (3\text{-}2)$$

where either the plus or minus sign may be used. Similarly, the cross-correlation of $g(t)$ and $f(t)$ is:

$$C_{gf}(t) = \int_{-\infty}^{\infty} g(u) f(u + t)\, du \qquad (3\text{-}3a)$$

or

$$C_{gf}(t) = \int_{-\infty}^{\infty} f(u)g(u - t)\, du \qquad (3\text{-}3b)$$

The reason for giving two equivalent definitions for correlation is that the 3-3a definition is standard in noise and communication theory, whereas for dealing with deterministic waveforms the 3-3b definition may be simpler for evaluation purposes.

A number of problems will now be solved in detail.

EXAMPLE 3-1
   Given

$$x(t) = 2e^{-3t}u(t)$$

Find its autocorrelation function.

Solution.   Figure 3-1(a) shows a plot for $x(u)$ and $x(u - t)$ versus u for t both a positive and negative number.

Range of $-\infty < t < 0$
For a negative value of t the product function $x(u)x(u - t)$ is nonzero $0 < u < \infty$ because $x(u)$ is zero for $-\infty < u < 0$.

Therefore
$$C_{xx}(t) = \int_0^\infty (2e^{-3u})(2e^{-3(u-t)}) \, du$$

$$= 4e^{3t} \int_0^\infty e^{-6u} \, du$$

$$= e^{3t}(-\tfrac{2}{3} e^{-6u} \big|_0^\infty)$$

$$= 0.67e^{3t}$$

Range of $0 < t < \infty$
For a positive value of t the product function $x(u)x(u - t)$ is nonzero for $t < u < \infty$.

Therefore
$$C_{xx}(t) = \int_t^\infty (2e^{-3u})(2e^{-3(u-t)}) \, du$$

$$= 4e^{3t} \int_t^\infty e^{-6u} \, du$$

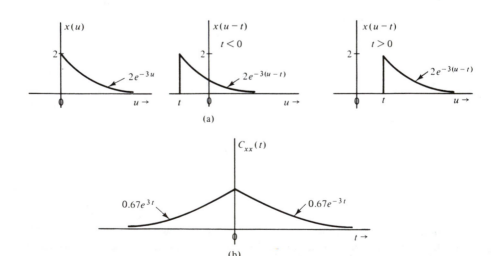

(a)

(b)

Figure 3-1   (a) $x(u)$ and $x(u - t)$ for Example 3-1; (b) the autocorrelation function.

$$= 4e^{3t}(0 + \tfrac{1}{6}e^{-6t})$$

$$= 0.67e^{-3t}$$

This autocorrelation function, which is plotted in Figure 3-1(b), shows that it is an even function of t represented as:

$$C_{xx}(t) = 0.67e^{-3|t|}$$

EXAMPLE 3-2

For the simple constant functions $x(t) = \sqcap(t - 0.5)$ and $y(t) = \sqcap(t - 2)$, quickly and as intuitively as possible evaluate and plot (or vice versa):

(a) the convolution integral $r(t)$
(b) the auto- or self-correlation functions $C_{xx}(t)$ and $C_{yy}(t)$
(c) the cross-correlation functions $C_{xy}(t)$ and $C_{yx}(t)$

Solution.   The functions $x(u)$, $y(u)$, $x(u - t)$, $x(t - u)$, $y(u - t)$, and $y(t - u)$ are shown as a function of u for some general t in Figure 3-2(a).

(a) The convolution integral as treated in Chapter 2 is:

$$r(t) = \int_{-\infty}^{\infty} x(u)y(t - u)\, du$$

Observing Figure 3-2, we find that $r(t)$ is zero for $t - 1.5 < 0$ or $t < 1.5$ and increases linearly from $t = 1.5$ to $t = 2.5$ where $r(2.5) = 1$. $r(t)$ then decreases linearly from $t = 2.5$ to $t = 3.5$. $r(t)$ is plotted in Figure 3-2(b).

(b) The autocorrelation function for $x(t)$ is:

$$C_{xx}(t) = \int_{-\infty}^{\infty} x(u)x(u - t)\, du$$

We see from $x(u - t)$ in Figure 3-2(a) that $C_{xx}(t)$ is zero for $t + 1 < 0$ or $t < -1$ and increases linearly from $t = -1$ to $t = 0$ where $C_{xx}(0) = 1$. $C_{xx}(t)$, shown in Figure 3-2(b), then decreases linearly from $t = 0$ to $t = 1$. In a completely analogous manner we obtain the same result for $C_{yy}(t)$ as for $C_{xx}(t)$.

(c) The cross-correlation of $x(t)$ with $y(t)$ may be expressed as:

$$C_{xy}(t) = \int_{-\infty}^{\infty} y(u)x(u - t)\, du$$

which means that $x(u - t)$ moves from $u = -\infty$ to $u = +\infty$ as t varies from $-\infty$ to $+\infty$. Had we used the definition:

$$C_{xy}(t) = \int_{-\infty}^{\infty} x(u)y(u + t)\, du$$

we would have to consider $y(u + t)$ moving from $u = +\infty$ to $u = -\infty$ as t varies, $-\infty < t < \infty$. Considering Figure 3-2(a) we say $C_{xy}(t) = 0$ for

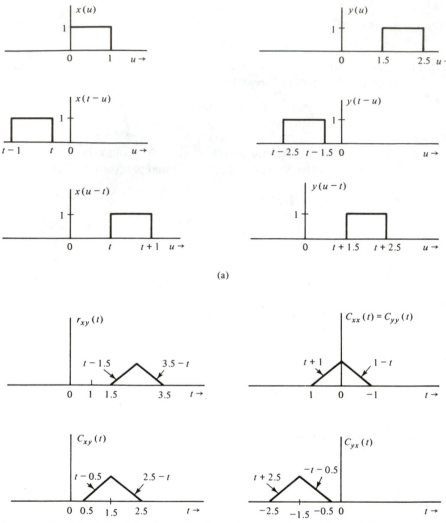

Figure 3-2 (a) $x(u)$, $y(u)$, $x(u - t)$, $x(t - u)$, $y(u - t)$, and $y(t - u)$ for Example 3-2; (b) the convolution and correlation integrals.

$t + 1 < 1.5$ or $t < 0.5$. From $t = 0.5$ to $t = 1.5$ $C_{xy}(t)$ increases linearly and $C_{xy}(1.5) = 1$. From $t = 1.5$ to $t = 2.5$ $C_{xy}(t)$, plotted in Figure 3-2(b), decreases linearly.

The cross-correlation of $y(t)$ with $x(t)$ may be defined as:

$$C_{yx}(t) = \int_{-\infty}^{\infty} x(u)y(u - t) \, du$$

Considering Figure 3-2(a) we see that $C_{yx}(t)$ is zero for $t + 2.5 < 0$ or $t < -2.5$ and $C_{yx}(t)$, as shown in Figure 3-2(b), increases linearly from

$t = -2.5$ to $t = -1.5$ where it has a value of 1 and then decreases linearly to have a value of zero at $t = -0.5$. We note that $C_{yx}(t)$ is $C_{xy}(t)$ reflected and indeed, we will see this is a general property for any two functions, $C_{xy}(t) = C_{yx}(-t)$. Attention should be paid as to how the analytic formulas were rapidly written down. If difficulty arises in visualizing these results, we should practice them until we feel more confident. It is instructive to redo the problem analytically. For example, for $r_{xy}(t)$ we obtain for the range $1.5 < t < 2.5$:

$$r(t) = \int_0^{t-1.5} 1 \, du = t - 1.5$$

This agrees with our result and we should proceed with other ranges until we are very confident.

EXAMPLE 3-3

Carefully set up the appropriate formulas for the different ranges of integration when evaluating $x(t) \oplus y(t)$ for the functions shown in Figure 3-3(a).

Solution

$$C_{xy}(t) = \int_{-\infty}^{\infty} y(u)x(u - t) \, du$$

Figure 3-3(b) shows a plot of $x(u - t)$ for a general t. The different ranges of t and the appropriate integrals for $C_{xy}(t)$ are:

Range of $t < -3$
If $t + 2 < -1$ or $t < -3$, then the integrand is zero and $C_{xy}(t) = 0$.

Range of $-3 < t < -2$

$$C_{xy}(t) = \int_{-1}^{t+2} [1 - 0.5(u - t)] \, du$$

Range of $-2 < t < -1$

$$C_{xy}(t) = \int_{-1}^{0} [1 - 0.5(u - t)] \, du + \int_0^{t+2} e^{-u} [1 - 0.5(u - t)] \, du$$

Range of $-1 < t < 0$

$$C_{xy}(t) = \int_t^0 [1 - 0.5(u - t)] \, du + \int_0^{t+2} e^{-u} [1 - 0.5(u - t)] \, du$$

Range of $t > 0$

$$C_{xy}(t) = \int_t^{t+2} e^{-u} [1 - 0.5(u - t)] \, du$$

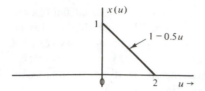

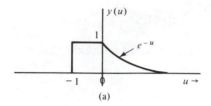

(a)

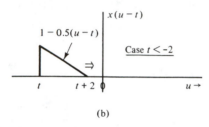

(b)

Figure 3-3 (a) The functions for Example 3-3; (b) $x(u - t)$ versus u for a general t.

3-1-2 Properties of Correlation Integrals

Figure 3-4 lists many properties of correlation integrals along with a few illustrative proofs. The two listed properties involving bounds will now be proved.

EXAMPLE 3-4

Prove:

(a) $C_{xx}(0) \geq C_{xx}(\tau)$ for any τ (3-4)

(b) $|C_{xy}(\tau)| \leq \sqrt{C_{xx}(0)C_{yy}(0)}$ (3-5)

Solution. These proofs involve inequalities of positiveness.

(a) By definition:

$$\int_{-\infty}^{\infty} [x(t) - x(t - \tau)]^2 \, dt \geq 0$$

Therefore

$$\int_{-\infty}^{\infty} x^2(t) \, dt + \int_{-\infty}^{\infty} x^2(t - \tau) \, dt$$

$$- 2 \int_{-\infty}^{\infty} x(t)x(t - \tau) \, dt \geq 0$$

or

$$2C_{xx}(0) \geq 2C_{xx}(\tau)$$

Operation	Properties	Illustrated proofs
Correlation		Proof 1
$C_{xx}(t) = \int_{-\infty}^{\infty} x(p)x(p-t)\,dp$	1. $C_{xx}(t) = C_{xx}(-t)$	$C_{xx}(t) = \int_{-\infty}^{\infty} x(p)x(p-t)\,dp$
denoted	2. $C_{xx}(0) \geq C_{xx}(t)$ for any t	let $u = p - t$
$C_{xx}(t) = x(t) \oplus x(t)$		$= \int_{-\infty}^{\infty} x(u+t)x(u)\,du$
		$C_{xx}(-t) = \int_{-\infty}^{\infty} x(p)x[p-(-t)]\,dp$
		$= C_{xx}(t)$
Cross-correlation	1. $x(t) \oplus y(t) \neq y(t) \oplus x(t)$	Proof of 4
$C_{xy}(t) = \int_{-\infty}^{\infty} x(p)y(p+t)\,dp$	2. $x(t) \oplus [y(t) + z(t)] = x(t) \oplus y(t)$	$C_{xy}(t) = \int_{-\infty}^{\infty} y(p)x(p-t)\,dp$
$= \int_{-\infty}^{\infty} y(p)x(p-t)\,dp$	$\qquad + x(t) \oplus z(t)$	let $p - t = s$
	3. $x(t) \oplus [y(t) \oplus z(t)] = [x(t) \oplus y(t)] \oplus z(t)$	$= \int_{-\infty}^{\infty} x(s)y(s+t)\,ds$
$C_{xy}(t) = x(t) \oplus y(t)$	4. $C_{xy}(t) = C_{yx}(-t)$	$C_{yx}(-t) = \int_{-\infty}^{\infty} x(p)y[p-(-t)]\,dp$
$C_{yx}(t) = \int_{-\infty}^{\infty} x(p)y(p-t)\,dp$	5. $|C_{xy}(t)| \leq \sqrt{C_{xx}(0)C_{yy}(t)}$	$= \int_{-\infty}^{\infty} x(p)y(p+t)\,dp$
		$= C_{xy}(t)$

Figure 3-4 Properties of correlation integrals for finite energy continuous functions.

(b) This property, known as the Cauchy-Schwartz inequality, has what will seem to be an abstract proof due to our previous limited exposure to bounds. By definition:

$$\int_{-\infty}^{\infty} [y(t) - kx(t - \tau)]^2 \, dt \geq 0 \qquad \text{for any real } k$$

On expanding:

$$\int_{-\infty}^{\infty} y^2(t) \, dt + k^2 \int_{-\infty}^{\infty} x^2(t - \tau) \, dt - 2k \int_{-\infty}^{\infty} y(t)x(t - \tau) \, dt \geq 0$$

we obtain the quadratic in k:

$$k^2 C_{xx}(0) - 2k C_{xy}(\tau) + C_{yy}(0) \geq 0 \tag{1}$$

Differentiating with respect to k, we find the quadratic to be a minimum when

$$k = \frac{C_{xy}(\tau)}{C_{xx}(0)}$$

Substituting this value of k into (1), we get:

$$\frac{C_{xy}^2(\tau)}{C_{xx}(0)} - 2\frac{C_{xy}^2(\tau)}{C_{xx}(0)} + C_{yy}(0) \geq 0$$

or

$$-C_{xy}^2(\tau) + C_{yy}(0)C_{xx}(0) > 0$$

and

$$\sqrt{C_{xx}(0)C_{yy}(0)} \geq |C_{xy}(\tau)|$$

Part (a) of this problem is a special case of part (b) when $y(t) = x(t)$.

When we obtain a somewhat abstract result such as this, we should always solve a specific simple problem for which the property is applicable.

EXAMPLE 3-5

For the pulse waveforms $x(t) = \sqcap(t)$ and $y(t) = \sqcap(0.5t)$ evaluate $C_{xx}(\tau)$, $C_{yy}(\tau)$, $C_{xy}(\tau)$, and $C_{yx}(\tau)$ and note that the inequality:

$$|C_{xy}(\tau)| \leq \sqrt{C_{xx}(0)C_{yy}(0)}$$

is satisfied for all τ.

Solution. Figure 3-5(a) shows the plots of $x(t)$, $y(t)$, $x(t - \tau)$, and $y(t - \tau)$. Figure 3-5(b) shows $C_{xx}(\tau)$, $C_{yy}(\tau)$, $C_{xy}(\tau)$, and $C_{yx}(\tau)$. For these simple constant functions the results should be clear. Observing the results we note that $C_{xx}(0) = 1$, $C_{yy}(0) = 2$ and that the maximum value of $C_{xy}(\tau)$ which occurs for all τ, $-0.5 < \tau < 0.5$, is 1. C_{xy} max $= 1 < \sqrt{2}$ which satisfies our inequality. Otherwise the left-hand side is less than 1.

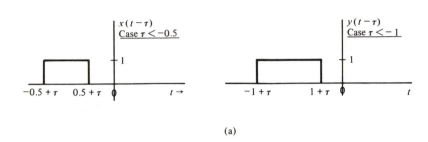

(a)

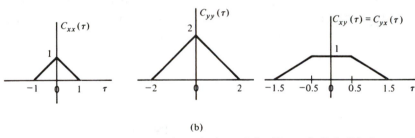

(b)

Figure 3-5 (a) $x(t)$, $y(t)$, $x(t-\tau)$, and $y(t-\tau)$ for Example 3-5; (b) the correlation integrals.

3-2 CORRELATION SUMMATIONS FOR DISCRETE FINITE ENERGY FUNCTIONS

The correlation of $f(n)$ and $g(n)$ is defined as:

$$C_{fg}(n) = f(n) \oplus g(n)$$

$$= \sum_{p} f(p)g(p+n) \qquad (3\text{-}6a)$$

It is easy to show (do so) that this is equivalent to:

$$C_{fg}(n) = \sum_{p} g(p)f(p-n) \qquad (3\text{-}6b)$$

The auto- or self-correlation of $f(n)$ with itself is:

$$C_{ff}(n) = \sum_{p} f(p)f(p \pm n) \qquad (3\text{-}7)$$

and the cross-correlation of $g(n)$ with $f(n)$ is:

$$C_{gf}(n) = \sum_p g(p) f(n + p) \tag{3-8a}$$

or

$$= \sum_p f(p)g(p - n) \tag{3-8b}$$

EXAMPLE 3-6
Given

$$f(n) = 2\delta(n + 1) + 4\delta(n) - 3\delta(n - 1)$$

and

$$g(n) = -3\delta(n + 1) + \delta(n)$$

evaluate $C_{ff}(n)$, $C_{gg}(n)$, $C_{fg}(n)$, and $C_{gf}(n)$.

Solution. The plots of $f(p)$, $g(p)$, $f(p - n)$, and $g(p - n)$ are shown as functions of p for some n in Figure 3-6(a).

$C_{ff}(n)$
$C_{ff}(n)$ is zero if $n + 1 < -1$ or $n < -2$ and if $n - 1 > 1$ or $n > 2$.
Considering $f(p - n)$ as a function that moves down the p axis from $p = -\infty$ to $+\infty$ as n varies from $-\infty$ to ∞, we obtain:

$$C_{ff}(-2) = 2 \times -3 = -6$$
$$C_{ff}(-1) = 4 \times -3 + 2 \times 4 = -4$$
$$C_{ff}(0) = -3 \times -3 + 4 \times 4 + 2 \times 2 = 29$$
$$C_{ff}(1) = -3 \times 4 + 4 \times 2 = -4$$
$$C_{ff}(2) = -3 \times 2 = -6$$

$C_{ff}(n)$ is shown in Figure 3-6(b). Corresponding to discrete correlation, the correlation of the two sets of numbers is defined as:

$$(2, 4, -3) \oplus (2, 4, -3) = (2 \times -3, 4 \times -3 + 2 \times 4,$$
$$-3 \times -3 + 4 \times 4 + 2 \times 2,$$
$$-3 \times 4 + 4 \times 2, -3 \times 2)$$
$$= (-6, -4, 29, -4, -6)$$

This is nothing more than interpreting the definition:

$$C_{ff}(n) = \sum_p f(p) f(p - n)$$

$C_{gg}(n)$
$C_{gg}(n)$ is zero for $n < -1$ and for $n - 1 > 0$ or $n > 1$.

$$C_{gg}(-1) = \sum_p g(p)g(p - 1) = -3 \times 1 = -3$$

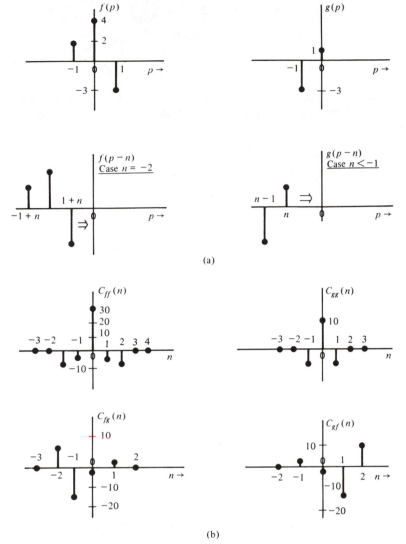

(a)

(b)

Figure 3-6 (a) $f(p)$, $g(p)$, $f(p-n)$, and $g(p-n)$ versus p for Example 3-6; (b) the correlation summations.

$$C_{gg}(0) = 1 \times 1 + -3 \times -3 = 10$$

and $$C_{gg}(1) = 1 \times -3 = -3$$

or using the correlation product set of numbers:

$$(-3, 1) \oplus (-3, 1) = (-3, 10, -3)$$

$C_{gg}(n)$ is shown in Figure 3-6(b).

$C_{fg}(n)$

$$C_{fg}(n) = f(n) \oplus g(n)$$

$$= \sum_p g(p) f(p - n)$$

Observing $g(p)$ and $f(p - n)$, we note that their product is zero if $n + 1 < -1$ or $n < -2$ and if $n - 1 > 0$ or $n > 1$. Evaluating term by term, we obtain:

$$C_{fg}(-2) = -3 \times -3 = +9$$

$$C_{fg}(-1) = 1 \times -3 + -3 \times 4 = -15$$

$$C_{fg}(0) = 1 \times 4 + -3 \times 2 = -2$$

$$C_{fg}(1) = 1 \times 2 = 2$$

Alternatively, defining the set product $(f) \oplus (g)$ to agree with this, we write the g set first and move the f set across it:

$$(2, 4, -3) \oplus (-3, 1) \triangleq (+9, -3 - 12, 4 - 6, 2)$$

$$= (+9, -15, -2, 2)$$

We must also note that 9 equals $C_{fg}(-2)$ or that the cross-correlation starts at $n = -2$. $C_{fg}(n)$ is plotted in Figure 3-6(b).

$C_{gf}(n)$

$$C_{gf}(n) = \sum_p f(p) g(p - n)$$

Observing $f(p)$ and $g(p - n)$, we note that their product is zero if $n < -1$ and if $n - 1 > 1$ or $n > 2$. Evaluating term by term, we obtain:

$$C_{gf}(-1) = 2 \times 1 = 2$$

$$C_{gf}(0) = 4 \times 1 + 2 \times -3 = -2$$

$$.C_{gf}(1) = -3 \times 1 + 4 \times -3 = -15$$

and
$$C_{gf}(2) = -3 \times -3 = +9$$

$C_{gf}(n)$ is shown in Figure 3-6(b), and we note that $C_{gf}(n) = C_{fg}(-n)$ (or vice versa).

Note: The problem could also be solved analytically. In Chapter 2 we found that:

$$\delta(n - l) * \delta(n - m) = \delta(n - l - m)$$

Similarly, $\delta(n - l) \oplus \delta(n - m)$ where $f(n) = \delta(n - l)$ and $g(n) = \delta(n - m)$ using $C_{fg}(n) = \sum g(p) f(p - n)$ yields the value 1 when $m + n = l$ or $n = l - m$.

Therefore
$$\delta(n - l) \oplus \delta(n - m) = \delta(n - m + l) \tag{3-9}$$

In understanding this problem we should be very careful since the notation is tricky. A graphical sketch will make it clearer.

Redoing the cross-correlating analytically, we have:

$$C_{fg}(n) = [2\delta(n+1) + 4\delta(n) - 3\delta(n-1)] \oplus [-3\delta(n+1) + \delta(n)]$$

$$= -6\delta(n) - 12\delta(n+1) + 9\delta(n+2) + 2\delta(n-1)$$

$$+ 4\delta(n) - 3\delta(n+1)$$

$$= 9\delta(n+2) - 15\delta(n+1) - 2\delta(n) + 2\delta(n-1)$$

which is what we obtained earlier.

EXAMPLE 3-7

Given

$$f(n) = (0.5)^n u(n)$$

and

$$g(n) = u(n)$$

Find $C_{gg}(n)$, $C_{ff}(n)$, and $C_{fg}(n)$.

Solution

$C_{gg}(n)$

We note that $g(n) = u(n)$ is not a finite energy function as $\sum_{n=0}^{\infty} 1 = \infty$. $C_{gg}(n)$ will not exist since it will be infinite for all n.

$C_{ff}(n)$

Figure 3-7(a) shows $f(p)$ and $f(p - n)$.

For $n \geq 0$

$$C_{ff}(n) = \sum_{p=n}^{\infty} (0.5)^p (0.5)^{p-n}$$

$$= (0.5)^{-n} \sum_{p=n}^{\infty} (0.5)^{2p}$$

$$= (0.5)^{-n} [(0.5)^{2n} + (0.5)^{2(n+1)} + (0.5)^{2(n+2)} + \cdots]$$

$$= (0.5)^{-n} \left[\frac{(0.5)^{2n}}{1 - (0.5)^2} \right]$$

$$= 1.33(0.5)^n, \qquad n \geq 0$$

For $n < 0$

$$C_{ff}(n) = 1.33(0.5)^{-n}$$

since $C_{ff}(n)$ is an even function. $C_{ff}(n)$ is shown in Figure 3-7(b). This result for $n < 0$ should be verified by setting up the summation.

$C_{fg}(n)$

Although $g(n)$ is not a finite energy function, $C_{fg}(n)$ will exist because the product $g(p) f(p - n)$ is a finite energy function. Observing Figure 3-7(a), we have:

For $n \geq 0$

$$C_{fg}(n) = \sum_{n}^{\infty} 1(0.5)^{p-n}$$

$$= (0.5)^{-n} \sum_{n}^{\infty} (0.5)^{p}$$

$$= (0.5)^{-n} \left[\frac{(0.5)^{n}}{1 - 0.5} \right]$$

$$= 2, \, n \geq 0$$

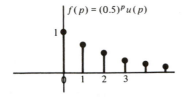

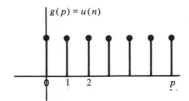

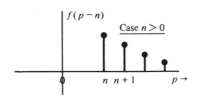

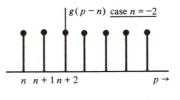

(a)

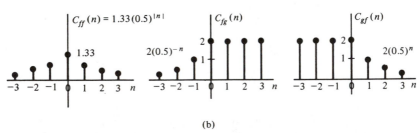

(b)

Figure 3-7   (a) $f(p)$, $g(p)$, $f(p - n)$, and $g(p - n)$ versus p for Example 3-7; (b) the correlation summations.

For n < 0

$$C_{fg}(n) = \sum_{0}^{\infty} (0.5)^{p-n}$$

$$= 0.5^{-n}[2]$$

$$= 2(0.5)^{-n}, \qquad n < 0$$

$C_{fg}(n)$ is plotted in Figure 3-7(b). Using the fact that $C_{gf}(-n) = C_{fg}(n)$, we plot $C_{gf}(n)$.

3-2-1 Closed Form Expressions for Correlation Summations

It is important to find expressions for the correlation integrals of sequences and to interpret them. We do this by means of presenting another example.

EXAMPLE 3-8

Given:

$$f(n) = A_1(\alpha)^n u(n)$$

and

$$g(n) = A_2(\beta)^n u(n)$$

Evaluate (a) $C_{ff}(n)$, (b) $C_{fg}(n)$, and (c) $C_{gf}(n)$

Solution

(a)

range n ≥ 0

$$C_{ff}(n) = \sum_{0}^{\infty} f(k) f(k+n),$$

$$= A_1^2 \sum_{k=0}^{\infty} \alpha^{2k} \alpha^n \qquad \text{(make a sketch)}$$

$$= A_1^2 \alpha^n \frac{1}{1 - \alpha^2}, \qquad \text{if } |\alpha| < 1$$

$$= \frac{A_1^2}{1 - \alpha^2} \alpha^n$$

range n < 0

$$C_{ff}(n) = \sum_{-n}^{\infty} f(n) f(k+n),$$

$$= \frac{A_1^2}{1 - \alpha^2} \alpha^{-n}$$

Therefore
$$C_{ff}(n) = \frac{A_1^2}{1 - \alpha^2} \alpha^{|n|}$$

(b)

range $n \geq 0$

$$C_{fg}(n) = \sum_0^\infty f(k)g(k + n),$$

$$= A_1 A_2 \sum_0^\infty \alpha^k \beta^{k+n} \qquad \text{(make a sketch)}$$

$$= A_1 A_2 \beta^n \sum_0^\infty (\alpha\beta)^k$$

$$= A_1 A_2 \beta^n \frac{1}{1 - \alpha\beta}, \qquad \text{if } |\alpha\beta| < 1$$

range $n < 0$

$$C_{fg}(n) = \sum_{-n}^\infty f(k)g(k + n),$$

$$= A_1 A_2 \beta^n \sum_{-n}^\infty (\alpha\beta)^k$$

$$= A_1 A_2 \beta^n [(\alpha\beta)^{-n} + (\alpha\beta)^{-n+1} + \cdots]$$

$$= A_1 A_2 \beta^n \left[(\alpha\beta)^{-n} \frac{1}{1 - \alpha\beta}\right], \qquad \text{if } |\alpha\beta| < 1$$

$$= \frac{A_1 A_2}{1 - \alpha\beta} \alpha^{-n},$$

We summarize:

$$\alpha^n u(n) \oplus \beta^n u(n) = \frac{A_1 A_2}{1 - \alpha\beta} \beta^n, \qquad n \geq 0$$

$$= \frac{A_1 A_2}{1 - \alpha\beta} \alpha^{-n}, \qquad n \leq 0$$

When we compare or correlate the waveforms $f(n)u(n)$ and $g(n)u(n)$ we find for $n > 0$ that the result is proportional to $g(n)$ as β^n predominates and for $n < 0$ the result is proportional to $f(n)$ as $\alpha^{|n|}$ predominates.

(c) By symmetry $C_{gf}(n)$ is found as

$$\beta^n u(n) \oplus \alpha^n u(n) = \frac{A_1 A_2}{1 - \alpha\beta} \alpha^n, \qquad n \geq 0$$

$$= \frac{A_1 A_2}{1 - \alpha\beta} \beta^{-n}, \qquad n \leq 0$$

Reflection on this problem might lead us to speculate on the results of more difficult problems. For example, $(A_1 + A_2 n)\alpha^n u(n) \oplus A_3 \beta^n u(n)$ should result in a solution as follows:

$$(A_1 + A_2 n)\alpha^n u(n) \oplus A_3 \beta^n u(n) = C_1 \beta^n u(n) + (C_2 + C_3 n)\alpha^{-n} u(-n)$$

where the constants may be found by evaluating the correlation summations or just obtaining appropriate initial conditions.

3-2-2 Properties of Discrete Correlation Functions

All the properties of Figure 3-4 for correlation integrals now carry over for correlation summations. Chief among these are:

$$C_{xx}(n) = C_{xx}(-n) \tag{3-10}$$

$$C_{xx}(0) \geq C_{xx}(n) \tag{3-11}$$

$$C_{xy}(n) = C_{yx}(-n) \tag{3-12}$$

and $$|C_{xy}(n)| \leq \sqrt{C_{xx}(0)C_{yy}(0)} \tag{3-13}$$

plus the distributive and associative rules. We prove a number of these properties.

EXAMPLE 3-9

Prove:

(a) $C_{xx}(n) = C_{xx}(-n)$
(b) $C_{xy}(n) = C_{yx}(-n)$

Solution

(a) $$C_{xx}(n) \triangleq \sum_p x(p)x(p - n)$$

Let $p - n = l$

Therefore $$C_{xx}(n) = \sum_l x(l + n)x(l)$$

$$= C_{xx}(-n)$$

(b) $$C_{xy}(n) = \sum_p x(p)y(p + n)$$

or $$= \sum_p y(p)x(p - n)$$

Now

$$C_{yx}(n) = \sum_s x(s)y(s - n)$$

Thus $$C_{yx}(-n) = \sum_s x(s)y(s + n) = C_{xy}(n)$$

EXAMPLE 3-10

Prove:

(a) $C_{xx}(0) \geq C_{xx}(n)$

(b) $|C_{xy}(n)| \leq \sqrt{C_{xx}(0)C_{yy}(0)}$

Solution

(a) These proofs involve inequalities of positiveness concerning sequences of numbers. Schwarz's inequality states for two sequences of numbers or discrete functions that:

$$\sum_n [f(n) \pm g(n)]^2 \geq 0$$

To prove part (a) from this inequality we say:

$$\sum_p [x(p) - x(p - n)]^2 \geq 0$$

Therefore $$\sum_p x^2(p) + \sum_p x^2(p - n) \geq 2 \sum_p x(p)x(p - n)$$

or $$2C_{xx}(0) \geq 2C_{xx}(n)$$

(b) By definition:

$$\sum_p [y(p) - kx(p - n)]^2 \geq 0$$

which in a similar manner to Example 3-4 yields a quadratic in k;

$$k^2 C_{xx}(0) - 2kC_{xy}(n) + C_{yy}(0) \geq 0 \tag{1}$$

Solving for k for the quadratic to be a minimum, we obtain:

$$k = \frac{C_{xy}(n)}{C_{xx}(0)}$$

and it is left to show that on substituting this value of k into (1), we obtain the famous Cauchy-Schwartz inequality which is our correlation property:

$$|C_{xy}(n)| \leq \sqrt{C_{xx}(0)C_{yy}(0)}$$

We typically have more familiarity with continuous functions than with discrete functions. To expand our confidence with discrete sequences, we must by trial and error use strings of numbers until we feel that these results are realistic when applied to specific problems.

Referring to Example 3-6 where $(f) = (2, 4, -3)$ and $(g) = (-3, 1)$, we found

$$(C_{ff}) = (-6, -4, 29, -4, -6), \quad (C_{gg}) = (-3, 10, -3)$$

$$\text{and } (C_{fg}) = (9, 15, -2, 2).$$

these are plotted in Figure 3-6. From Figure 3-6 we note that $C_{ff}(0) = 29$, $C_{gg}(0) = 10$, and $C_{fg}(n)$ is maximum at $n = -1$ where $C_{fg}(-1) = 15$. The inequality states that:

$$15 \leq \sqrt{290}$$

which is true for this $n = -1$ and obviously for all other n.

3-3 CORRELATION FUNCTIONS FOR PERIODIC WAVEFORMS (INFINITE ENERGY–FINITE POWER)

3-3-1 Periodic Waveforms

The definitions for correlation integrals will now be given for finite power continuous waveforms. The special case of periodic waveforms is first considered. A general periodic waveform is defined by:

$$x(t) = \sum_{n=-\infty}^{\infty} g(t - nT_0)$$

where $g(t)$ is zero for all t except for all or part of some range T_0 of the time axis. For example, given $g(t) = t, 0 < t < 1$, and is zero otherwise:

$$x(t) = \sum_{n=-\infty}^{\infty} g(t - 4n)$$

is a periodic function of period 4, which is shown in Figure 3-8(a). Obviously, a periodic function $x(t)$ with period T_0 is a nonfinite energy waveform as its energy

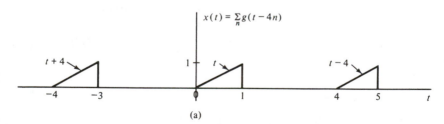

(a)

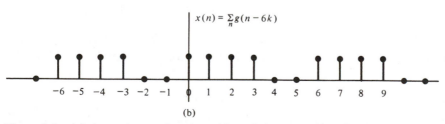

(b)

Figure 3-8   (a) A continuous function with period $T = 4$; (b) a discrete function with period $T = 6$.

$\int_{-\infty}^{\infty} x^2(t)\, dt = \infty$, but it is a finite power waveform as the total average power:

$$P_{av} = \lim_{T \to \infty} \frac{1}{2T} \int_{-T}^{T} x^2(t)\, dt = \frac{1}{T_0} \int_{\alpha}^{\alpha+T_0} x^2(t)\, dt \qquad (3\text{-}14)$$

exists.

The general definitions of correlation functions for finite power waveforms (periodic or not) are:

$$R_{xx}(\tau) \triangleq \lim_{T \to \infty} \frac{1}{2T} \int_{-T}^{T} x(t)x(t - \tau)\, dt \qquad (3\text{-}15)$$

which is also denoted $x(\tau) \oplus x(\tau)$ and called the autocorrelation function of $x(\tau)$:

$$R_{xy}(\tau) = \lim_{T \to \infty} \frac{1}{2T} \int_{-T}^{T} x(t)y(t + \tau)\, dt \qquad (3\text{-}16a)$$

$$= \lim_{T \to \infty} \frac{1}{2T} \int_{-T}^{T} y(t)x(t - \tau)\, dt \qquad (3\text{-}16b)$$

which is also denoted $x(\tau) \oplus y(\tau)$ and called the cross-correlation of $x(\tau)$ with $y(\tau)$,

and
$$R_{yx}(\tau) \triangleq \lim_{T \to \infty} \frac{1}{2T} \int_{-T}^{T} y(t)x(t + \tau)\, dt \qquad (3\text{-}17a)$$

or
$$= \lim_{T \to \infty} \frac{1}{2T} \int_{-T}^{T} x(t)y(t - \tau)\, dt \qquad (3\text{-}17b)$$

which is denoted by $y(\tau) \oplus x(\tau)$ and called the cross-correlation of $y(\tau)$ with $x(\tau)$.

If the functions are periodic with a period T_0 seconds, then the correlation functions are periodic with a period T_0 and the limits of integration may be taken over any period from $t = \alpha$ to $t = \alpha + T_0$ and then divided by T_0.

A discrete function $f(n) = \sum_{p=-\infty}^{\infty} f(p)\delta(n - p)$ is a nonfinite energy function if $\sum_p f^2(p) = \infty$, and it is a finite power waveform if:

$$\lim_{N \to \infty} \frac{1}{2N + 1} \sum_{p=-N}^{N} f^2(p) < \infty \qquad (3\text{-}18)$$

A general discrete periodic function with period N_0 is defined by:

$$x(n) = \sum_{k=-\infty}^{\infty} g(n - kN_0)$$

where N_0 is an integer and $g(n)$ is zero for all n except over all or part of a range $n_0 < n \le n_0 + N_0$. The discrete periodic function defined by:

$$x(n) = \sum_{k} g(n - 6k)$$

where
$$g(n) = 1, \qquad 0 \le n \le 3$$
$$= 0, \qquad \text{otherwise}$$

is illustrated in Figure 3-8(b). Obviously, this is a finite power waveform with total average power of 0.67 W. A few correlation integrals will now be evaluated for periodic functions.

EXAMPLE 3-11

Given the waveform:

$$g(t) = 1, \qquad 0 < t < 1$$

$$= -1, \qquad -1 < t < 0$$

$$= 0, \qquad \text{otherwise}$$

consider the two periodic waveforms:

$$x(t) = \sum_{n=-\infty}^{\infty} g(t - 2n)$$

and
$$y(t) = \sum_{n=-\infty}^{\infty} g(t - 1 - 2n)$$

Evaluate $R_{xx}(\tau)$ and $R_{xy}(\tau)$.

Solution. Figure 3-9(a) shows a plot of $x(t)$ and $x(t - \tau)$ for a general τ in the ranges $-1 < \tau < 0$ and $0 < \tau < 1$ and for $y(t)$.

$R_{xx}(\tau)$
For this periodic function with period 2:

$$R_{xx}(\tau) = \tfrac{1}{2} \int_0^2 x(t)x(t - \tau)\, dt$$

Range $0 < \tau < 1$

$$R_{xx}(\tau) = \tfrac{1}{2}\left[\int_0^\tau (1)(-1)\, dt + \int_\tau^1 (1)(1)\, dt \right.$$

$$\left. + \int_1^{1+\tau}(1)(-1)\, dt + \int_{1+\tau}^2 (-1)(-1)\, dt \right]$$

$$= 1 - 2\tau \qquad \text{(with work)}$$

Range $-1 < \tau < 0$
With work or using symmetry, we have:

$$R_{xx}(\tau) = 1 + 2\tau$$

and since $R_{xx}(\tau)$ is periodic with period 2 we can now find a formula for any value of t. $R_{xx}(\tau)$ is shown in Figure 3-9(b).

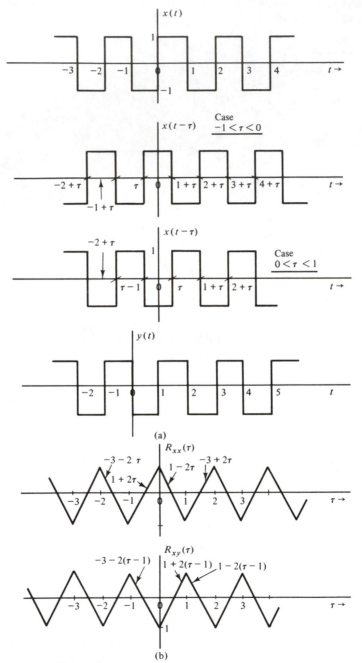

Figure 3-9 (a) The periodic functions $x(t)$, $x(t - \tau)$, and $y(t)$ for Example 3-11; (b) $R_{xx}(\tau)$ and $R_{xy}(\tau)$.

The result for this problem could also have been obtained by inspection,

$$R_{xx}(0) = R_{xx}(2n)$$

$$= \tfrac{1}{2} \int_0^2 1^2 \, dt = 1 \qquad \text{for any } \tau = n \text{ integer}$$

$$R_{xx}(1) = R_{xx}(1 + 2n)$$

$$= \tfrac{1}{2} \int_0^2 (1)(-1) \, dt = -1 \qquad \text{for any } \tau = n \text{ integer}$$

Since we are dealing with constant functions the value changes linearly. For $0 < \tau < 1$ we obtain $R_{xx}(\tau) = 1 - 2\tau$ as before.

$R_{xy}(\tau)$

We now intuitively derive the result:

$$R_{xy}(\tau) = \tfrac{1}{2} \int_0^2 y(t)x(t - \tau) \, dt$$

$y(t)$ is illustrated in Figure 3-9(a) and is $x(t)$ shifted 1 unit to the right.

Therefore $\qquad\qquad\qquad R_{xy}(1) = R_{xx}(0)$

and, in general:

$$R_{xy}(\tau + 1) = R_{xx}(\tau)$$

or $\qquad\qquad\qquad R_{xy}(\tau) = R_{xx}(\tau - 1)$

$R_{xy}(\tau)$ is shown in Figure 3-9(b). The reader may solve this analytically for the range $0 < \tau < 1$ to show that the same answer is obtained.

EXAMPLE 3-12

Given the discrete waveforms:

$$g_1(n) = 2\delta(n) - \delta(n - 1)$$

and $\qquad\qquad\qquad g_2(n) = \delta(n) + \delta(n - 1)$

consider the two periodic waveforms:

$$x(n) = \sum_{k=-\infty}^{\infty} g_1(n - 3k)$$

and $\qquad\qquad\qquad y(n) = \sum_{k=-\infty}^{\infty} g_2(n - 3k)$

Evaluate $R_{xx}(n)$ and $R_{xy}(n)$ using analogous formulas to Equations 3-15 through 3-17.

Solution. Figure 3-10(a) shows a plot of $x(p)$ and $x(p - n)$ for $n = 1$ and of $y(p)$ versus p.

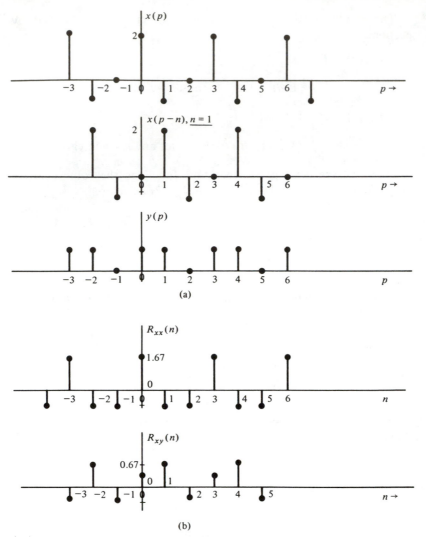

Figure 3-10   (a) $x(p)$, $x(p - n)$, and $y(p)$ for Example 3-12; (b) $R_{xx}(n)$ and $R_{xy}(n)$.

$R_{xx}(n)$

Since $x(p)$ is periodic with period 3, then $R_{xx}(n)$ will be periodic with period 3

and
$$R_{xx}(n) = \tfrac{1}{3} \sum_{p=0}^{2} x(p)x(p - n)$$

Observing Figure 3-10(a), we obtain:

$$R_{xx}(0) = \tfrac{1}{3} [2^2 + (-1)^2 + 0^2] = 1.67$$

$$R_{xx}(1) = \tfrac{1}{3} [2 \times 0 + -1 \times 2 + 0 \times -1] = -0.67$$

and
$$R_{xx}(2) = \tfrac{1}{3} [2 \times -1 + -1 \times 0 + 0 \times 2] = -0.67$$

Since $R_{xx}(n + 3) = R_{xx}(n)$, the autocorrelation function is now known for all n. $R_{xx}(n)$ is plotted in Figure 3-10(b).

$R_{xy}(n)$

$$R_{xy}(n) = \frac{1}{3}\left[\sum_{p=0}^{2} y(p)x(p - n)\right]$$

Observing Figure 3-10(a), we have:

$$R_{xy}(0) = \frac{1}{3}[1 \times 2 + 1 \times -1 + 0 \times 0] = 0.33$$

$$R_{xy}(1) = \frac{1}{3}[1 \times 0 + 1 \times 2 + 0 \times -1] = 0.67$$

and $\quad\quad R_{xy}(2) = \frac{1}{3}[1 \times -1 + 1 \times 0 + 0 \times 2] = -0.33$

$R_{xy}(n)$ is periodic with period 3, as illustrated in Figure 3-10(b).

3-3-2 Properties of Correlation Functions for Periodic Waveforms

Some of the main properties of correlation functions for continuous or discrete periodic waveforms are:

- $R_{xx}(\tau)$ and $R_{xx}(n)$ are even and periodic.   (3-19)
- $R_{xx}(0) \geq R_{xx}(\tau)$ and $R_{xx}(0) \geq R_{xx}(n)$.   (3-20)
- $R_{xx}(0)$ represents the total average power of the waveform.   (3-21)
- $R_{yx}(\tau) = R_{xy}(-\tau)$ and $R_{yx}(n) = R_{xy}(-n)$.   (3-22)
- $R_{xy}(\ ) \leq \sqrt{R_{xx}(0)\, R_{yy}(0)}$.   (3-23)

The blank arguments may be either τ or n.

3-4 STATISTICS AND CORRELATION FUNCTIONS FOR FINITE POWER NOISE WAVEFORMS

This section contains a tutorial consideration of noise waveforms and how correlation functions impart information about randomness. A rigorous treatment of random processes or noise is first studied in most curricula following a course on linear systems. Here we discuss the subject informally and intuitively.

3-4-1 Terminology and Definitions

A **continuous random process** is an ensemble or large collection of waveforms $x_1(t), x_2(t), \ldots, x_n(t)$, and so on from which it is possible to obtain statistics of a member or group of members. Two typical members of a continuous process $x(t)$

are shown in Figure 3-11(a). Before dealing with the concept of "ensemble" averages of a random process, we will consider our assumed previously held intuitive ideas of what is an "average value."

Average Values.   We now informally arrive at the mathematical definition of "averages" by using an example.

EXAMPLE 3-13

Consider the 10 numbers; 6.0, 4.0, −2.4, 3.1, −4.0, 6.0, 3.1, 1.5, 6.0, −2.4, and find:

(a) the average value of the numbers denoted $\bar{n}$
(b) the average value of the square of the numbers denoted $\overline{n^2}$
(c) the average of "4 times a number minus 3" denoted $\overline{4n - 3}$.

Solution

(a) $\bar{n} = \dfrac{3(6.0) + 2(-2.4) + 2(3.1) + 4.0 + (-4.0) + 1.5}{10} = 2.09$

(b) $\overline{n^2} = \dfrac{3(6.0)^2 + 2(-2.4)^2 + 2(3.1)^2 + (4.0)^2 + (-4.0)^2 + (1.5)^2}{10} = 17.3$

(c) $\overline{(4n - 3)} = \dfrac{\begin{array}{c}3[4(6.0) - 3] + 2[4(-2.4) - 3] + 2[4(3.1) - 3] \\ + (16 - 3) + (-16 - 3) + (6 - 3)\end{array}}{10} = 5.36$

It is assumed that the preceding calculations for average are intuitively logical and acceptable to us and that the reader could evaluate without effort $(2n^2 - 5n + 4)$ and with some thought $(6n^2 - 4\bar{n})$.

Mathematical Definition of Averages.   When we talk about such averages as, "the average height of a person," "the average value obtained on a die roll," "the average value of Sin^2 t," we are discussing assumed mathematical or statistical averages. It is assumed we can perform an identical experiment a large number of times N where different outcomes occur with statistical regularity as N increases. Then for any numerical quantity we define its average as:

$$\bar{n} = \lim_{N \to \infty} \frac{1}{N} \sum_{i=1}^{N} n_i \qquad (3\text{-}24)$$

where n_i is the value obtained on the ith trial.

EXAMPLE 3-14

Based on assumptions, if possible:

(a) Find the average value obtained on a die roll.
(b) Find the average value of Sin^2 t.

(c) Find the average height of a person.

(d) Give a statistical interpretation for the answers of Example 3-13.

Solution

(a) Observing Equation 3-24, we write:

$$\bar{n} = \frac{n_1(1) + n_2(2) + n_3(3) + n_4(4) + n_5(5) + n_6(6)}{n_1 + n_2 + n_3 + n_4 + n_5 + n_6}$$

where $N = \Sigma_{i-1}^6 \, n_i$ approaches infinity. Assuming the die is fair $n_i/N = \frac{1}{6}$ for all i and $\bar{n} = 3.5$.

(b) In a first circuits course without thinking statistically we mechanically accepted that:

$$\overline{\text{Sin}^2 \, t} = \lim_{T \to \infty} \frac{1}{2T} \int_{-T}^{T} \text{Sin}^2 \, t \, dt$$

Since $\text{Sin}^2 \, t$ is periodic with period π (or $n\pi$), we say:

$$\overline{\text{Sin}^2 \, t} = \frac{1}{\pi} \int_0^{\pi} \text{Sin}^2 \, t \, dt$$

$$= \frac{1}{\pi} \left[\int_0^{\pi} 0.5(1 - \text{Cos } 2t) \, dt \right]$$

$$= 0.5$$

Now we must interpret what we did in accordance with the definition for $\bar{n}$ in Equation 3-24. This follows from the definition of an integral as a limiting form of a summation and is equivalent to saying that if we carry out the experiment of choosing a point uniformly on the time axis, say, t_i, then

$$n_i = \text{Sin}^2 (t_i) \quad \text{and} \quad \overline{\text{Sin}^2 \, t} = \lim_{N \to \infty} \frac{1}{N} \sum_{i-1}^{N} \text{Sin}^2 \, t_i$$

The word "uniformly" is very important and means that the likelihood of choosing a value of t in any range Δt, say, $0 \le t \le \Delta t$ is identical to any other range Δt, say, $\pi - \Delta t \le t \le \pi$.

(c) Our definition Equation 3-24 implies that based on statistical data we might estimate the average height of a person but great care would have to be exercised.

(d) In order to change Example 3-13 to a problem involving statistical averages, we could restate it as follows: "Consider the experiment of choosing 1 of the 10 numbers, 6.0, 4.0, −2.4, 3.1, −4.0, 6.0, 3.1, 1.5, 6.0, and −2.4 where each has equal likelihood of being selected and find the average value, the average value of the square, and the average of "4 times n minus 3" for the number chosen.

Averages for a Random Process.   Let us reconsider the typical members of a continuous and a discrete random process as shown in Figure 3-11(a) and (b). We now define **first-order ensemble statistics** for a random process as follows: The **average of the process** at time t_1 for a continuous process and at time k for a discrete process is:

$$\overline{x(t_1)} = \lim_{N \to \infty} \frac{1}{N} \sum_{i=1}^{N} x_i(t_1) \tag{3-25a}$$

or

$$\overline{x(k)} = \lim_{N \to \infty} \frac{1}{N} \sum_{i=1}^{N} x_i(k) \tag{3-25b}$$

This means to estimate the average at t_1 for a continuous process we add the values of a number of members at time t_1 and divide by the number. We expect to obtain a stable limit as the number of members chosen gets large.

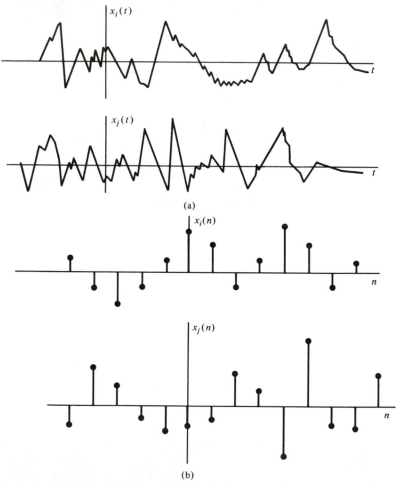

(a)

(b)

Figure 3-11   (a) A typical member waveform of a continuous ergodic random process; (b) a typical member waveform of a discrete ergodic random process.

Any first-order statistic is found in a similar manner. For example, the average of the square values at t_1 or k called the **mean square value** is defined on an ensemble basis as:

$$\overline{x^2(t_i)} = \lim_{N \to \infty} \frac{1}{N} \sum_{i=1}^{N} x_i^2(t_i) \tag{3-26a}$$

or

$$\overline{x^2(k)} = \lim_{N \to \infty} \frac{1}{N} \sum_{i=1}^{N} x_i(k) \tag{3-26b}$$

or any first-order statistic.

A **second-order statistic** of a random process is one involving values at two times t and s. The most famous second-order statistic is the autocorrelation function for the process $x(t)$ or $x(n)$ defined as:

$$R_{xx}(t, s) \triangleq \overline{x(t)x(s)} = \lim_{N \to \infty} \frac{1}{N} \sum_{p=1}^{N} x_p(t)x_p(s) \tag{3-27a}$$

or

$$R_{xx}(k, m) \triangleq \overline{x(k)x(m)} = \lim_{N \to \infty} \frac{1}{N} \sum_{p=1}^{N} x_p(k)x_p(m) \tag{3-27b}$$

Similarly, the reader should be able to write out definitions for second-order ensemble statistics such as $\overline{[6x(t) - 5x(s)]}$, $\overline{7x^2(k)x(m)}$, and so on (do so).

3-4-2 Types of Random Processes

We now define two very important types of **random process:** stationary and ergodic.

Stationary Random Processes.   A random process is **stationary** or stationary of order n if its first-order statistics are time-independent and higher-order statistics do not depend on the times involved but only on the differences between them. For example, if $x(t)$ or $x(n)$ are stationary, we must have:

$$\overline{x(t_1)} = \overline{x(t_2)} = \overline{x(t)} = \overline{x} \tag{3-28a}$$

$$\overline{x(k)} = \overline{x(n)} = \overline{x} \tag{3-28b}$$

$$R_{xx}(t, t + t) = \overline{x(t)x(t + t)} = R_{xx}(\tau) \qquad \text{for any } t \tag{3-29}$$

$$R_{xx}(n, n + k) = \overline{x(n)x(n + k)} = R_{xx}(k) \qquad \text{for any } n \tag{3-30}$$

Also, for example:

$$\overline{6x^2(2)x(4)x(7)} = \overline{6x^2(n)x(n + 2)x(n + 5)} \qquad \text{for any } n$$

Ergodic Random Processes.   In this text we confine ourselves to the case of ergodic random processes and the situation where each member is a finite power waveform. A random process is **ergodic** if any ensemble average equals the corresponding time average for any member. This is a very important process since the analysis of any member may be used to obtain complete statistical information about the process.

Some significant first-order time averages, with their notation for an ergodic random process are:

$$\widetilde{x(t)} \triangleq \lim_{T \to \infty} \frac{1}{2T} \int_{-T}^{T} x_p(t) \, dt \tag{3-31a}$$

or
$$\widetilde{x(n)} \triangleq \lim_{N \to \infty} \frac{1}{2N + 1} \sum_{n=-N}^{N} x_p(n) \tag{3-31b}$$

for any p. $\widetilde{x(t)}$ or $\widetilde{x(n)}$ is the average or mean of a general member of a continuous or discrete process. The wavy bar is used to distinguish a time average from an ensemble average.

$$\widetilde{x^2(t)} \triangleq \lim_{T \to \infty} \frac{1}{2T} \int_{-T}^{T} x_p^2(t) \, dt \tag{3-32a}$$

or
$$\widetilde{x^2(n)} \triangleq \lim_{N \to \infty} \frac{1}{2N + 1} \sum_{n=-N}^{N} x_p^2(n) \tag{3-32b}$$

$\widetilde{x^2(t)}$ or $\widetilde{x^2(n)}$ is the average of the squared value or more commonly called the mean square value. Similarly, the reader should be able to write formulas for any first-order statistic such as $\widetilde{[4x(t) - 3]}$, $\widetilde{[7x^3(n) + x(n)]}$, or $\widetilde{(x(t) - \bar{x})^2}$ called the **variance**. For any ergodic process

$$\overline{x(t)} = \widetilde{x(t)} = \bar{x} \tag{3-33a}$$

$$\overline{x^2(n)} = \widetilde{x^2(n)} = \overline{x^2}, \tag{3-34}$$

Also, for example, we must have:

$$\widetilde{[4x^3(n) + x(n)]} = \widetilde{4x^3(n)} + \widetilde{x(n)} = \widetilde{4x^3} + \bar{x}$$

The most important second-order time average for an ergodic random process is the autocorrelation function $R_{xx}(t)$ or $R_{xx}(k)$:

$$\widetilde{x(t)x(t \pm t)} \triangleq \lim_{T \to \infty} \frac{1}{2T} \int_{-T}^{T} x_p(t)x_p(t \pm t) \, dt \tag{3-35a}$$

or
$$\widetilde{x(n)x(n \pm k)} \triangleq \lim_{N \to \infty} \frac{1}{2N + 1} \sum_{n=-N}^{N} x_p(n)x_p(n \pm k). \tag{3-35b}$$

For an ergodic random process we have:

$$\overline{x(t)x(t \pm t)} = \widetilde{x(t)x(t \pm t)} = R_{xx}(\tau) \tag{3-36a}$$

and
$$\overline{x(n)x(n \pm k)} = \widetilde{x(n)x(n \pm k)} = R_{xx}(k) \tag{3-36b}$$

The only waveforms for which we can actually carry out the evaluation of time averages are periodic waveforms because then the limits of integration may be taken over any one period. For a noise waveform which is a member of an ergodic random process we must approximate time averages by chosing T or N large (subjective). The choice of T or N is of the utmost practical importance. We will, however, consider random processes where first- and second-order time averages may be predicted using symmetry and visualizing time averages statistically as was done in Example 3-14.

We now familiarize ourselves with the concept of what a random process is and with the prediction of first- and second-order statistics.

3-4-3 The Evaluation of Statistics for Ergodic Processes

EXAMPLE 3-15

Construct two typical member waveforms for each of the following random processes.

(a) Every second a noise voltage generator puts out a voltage that assumes all values from 0 to 10 V with the same chance of being in any fixed voltage interval. Assume each noise waveform has a uniform phase shift between 0 and 1 s.

(b) A discrete waveform assumes the value 0 or 1 every second. Assume there is a 70% chance of retaining the previous value.

Solution

(a) Two typical member waveforms of the random process are shown in Figure 3-12(a). We may visualize the generation of a member as follows. Start the waveform at, say, $t = -30 - \phi_i$ where ϕ_i is uniformly chosen between 0 and 1. Then choose a number uniformly between 0 and 10 and let the member take on this value from $-30 - \phi_i$ to $-29 - \phi_i$. Continually choose values uniformly between 0 and 10 to generate $x_i(t)$ for each succeeding interval. On reflection it should be clear why each member is analytically different while still retaining common statistics with other members. (e.g., the average value is 5).

(b) Two typical members of our discrete process are shown in Figure 3-12(b). A model for generating a specific member could be composed of a coin and an urn with seven red and three black balls. Start the process at $n = -20$ by tossing the coin with heads for the value 1 and tails for the value 0. Then continually draw a ball from the urn replacing it after each draw. When a red ball is obtained retain the previous waveform value, otherwise assume the other value. Our typical members reflect the tendency to stay with the previous value. Again, each member of the process is analytically different but shares common statistics (e.g., the average value of each waveform is 0.5).

EXAMPLE 3-16

For the random process of Example 3-15(b) evaluate the average, mean square, and variance of a typical member. Use symmetry and intuition. The variance is denoted by σ_x^2 and defined as $\sigma_x^2 \triangleq \overline{(x(n) - \overline{x(n)})^2}$.

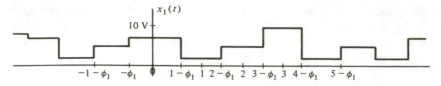

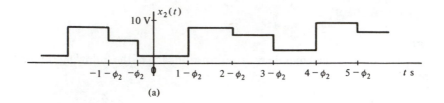

(a)

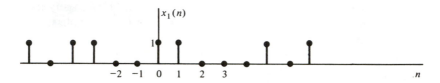

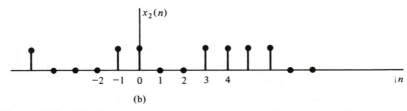

(b)

Figure 3-12 (a) Two typical members of the random process of Example 3-15(a); (b) two typical members of the random process of Example 3-15(b).

Solution. Observing the member $x_1(n)$, we note from symmetry that the value $x(n) = 1$ occurs half the time and the value $x(n) = 0$ the other half.

Therefore $$\overline{x(n)} = \tfrac{1}{2}(1 + 0) = 0.5$$

Similarly, the value $x^2(n) = 1$ occurs half the time and the value $x^2(n) = 0$ the other half.

Therefore $$\overline{x^2(n)} = \tfrac{1}{2}(1 + 0) = 0.5$$

Also $[x(n) - 0.5]^2 = (1 - 0.5)^2$ half the time and $(0 - 0.5)^2$ the other half,

Therefore $$\sigma_x^2 = \tfrac{1}{2}(0.5^2 + 0.5^2) = 0.25$$

EXAMPLE 3-17

Intuitively find the autocorrelation functions for the following two random processes:

(a) A continuous random process where a member is formed by generating with equal chances a pulse of value 2 or 1 every b seconds.
(b) The discrete random waveform of Example 3-15(b) where the value 0 or 1 is assumed every second and there is a 70% chance of retaining a value.

Solution

(a) A typical waveform of the process is shown in Figure 3-13(a). The formula:

$$R_{xx}(\tau) = \lim_{T \to \infty} \frac{1}{2T} \int_{-T}^{T} x(t)x(t - \tau) \, dt$$

cannot be evaluated since $x(t)$ is not deterministic. However, we will find $R_{xx}(\tau)$ from the statistical interpretation as the average of sampling two values spaced τ seconds apart. We now do this for different values of τ.

$\tau = 0$

$$R_{xx}(0) = \tfrac{1}{2}[2^2 + 1^2] = 2.5$$

$\tau \geq b$.   We now obtain product values of $(2)(2)$, $2(1)$, $1(2)$, or $(1)(1)$, and each of these products from symmetry should occur one-quarter of the time. (Stop: Be very sure!)

Therefore $R_{xx}(\tau) = \tfrac{1}{4}(4 + 2 + 2 + 1) = 2.25$

$0 < \tau < b$.   Since the waveform is always constant, we assume that $R_{xx}(\tau)$ decreases linearly from $R_{xx}(0) = 2.5$ to $R_{xx}(b) = 2.25$.

Thus $R_{xx}(\tau) = A_1 + A_2\tau$

where $2.5 = A_1$ and $2.25 = 2.5 + A_2 b$.

Therefore $R_{xx}(\tau) = 2.5 + \dfrac{2.25 - 2.5}{b}\tau$

$$= \dfrac{2.5(b - 0.1\tau)}{b}$$

By symmetry, $R_{xx}(\tau)$, shown in Figure 3-13(b), will be an even function of τ. We may now interpret the information conveyed by this autocorrelation function. $R_{xx}(\tau)$ varies from its initial value equal to the mean square value to a final value equal to the average value squared. We say the autocorrela-

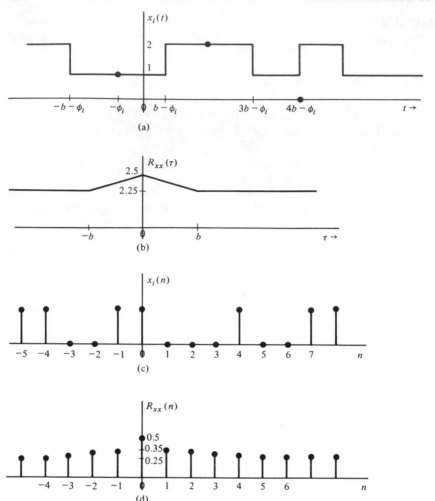

Figure 3-13 (a) A typical member of the random process of Example 3-17(a); (b) its autocorrelation function; (c) a typical member of the random process of Example 3-17(b); (d) its autocorrelation function.

tion function is a measure of the similarity of the waveform $x(t - \tau)$ to $x(t)$. If $\tau = 0$ the two waveforms are identical and the integral of the product is a maximum. When τ is close to zero there is still a close similarity. In this case for $\tau > b$ there is no similarity in the sense that by knowing $x(t)$ we can make no predictions as to whether the chances of $x(t - \tau)$ being 2 is better than its chances of being 1.

(b) A typical waveform of the process is shown in Figure 3-13(c). The formula:

$$R_{xx}(n) = \lim_{N \to \infty} \frac{1}{2N + 1} \sum_{k=-N}^{N} x(k)x(k - n)$$

cannot be evaluated. We now try to find $R_{xx}(n)$ for a few specific values of n from the statistical concept of the average of $x(k)x(k-n)$.

$n = 0$

$$R_{xx}(0) = \lim_{N \to \infty} \frac{1}{2N+1} \sum_k x^2(k)$$

Since 1^2 occurs about half the time and 0^2 the other half for many samples, we obtain:

$$R_{xx}(0) = \tfrac{1}{2}(1^2 + 0^2) = 0.5$$

and this is the mean square value.

$n = \pm 1$.   If $n = +1$

$$R_{xx}(1) = \lim \frac{1}{2N+1} \sum_k x(k)x(k-1)$$

Since there is a 70% chance of retaining a value we argue that on many samples (for different k) the product $(1)(1)$ or $(0)(0)$ occurs 70% of the time and the product $(1)(0)$ or $(0)(1)$ occurs 30% of the time. By symmetry, 1^2 and 0^2 occur about 35% of the time and $(1)(0)$ and $(0)(1)$ about 15%. From our concept of average we have:

$$R_{xx}(1) = [0.35 \times 1^2 + 0.35 \times 0^2 + 0.15(0) + 0.15(0)]$$

Therefore $R_{xx}(1) = 0.35$

and similarly, $R_{xx}(-1) = 0.35$

$|n| \gg 1$.   If $n \gg 1$ we can say that the chances of $x(k)x(k-n)$ assuming the four different possible products $(1)(1)$, $(0)(0)$, $(1)(0)$, and $(0)(1)$ are all 25%.

Therefore $R_{xx}(n) \approx 0.25$

which is the average value squared.

$R_{xx}(n)$, as shown in Figure 3-13(d), varies from its initial value equal to the mean square value to a final value equal to the mean or average value squared. We again say the autocorrelation function is a measure of the similarity of $x(k-n)$ to $x(k)$. In this case since there is a 70% chance of retaining a value, $R_{xx}(n)$ is always greater than $R_{xx}(\infty) = 0.25$, which implies that knowing $x(k-n)$ is 1, there is a greater than 50% chance that $x(k)$ is 1. The fact that $R_{xx}(n)$ is different from $\overline{x(n)}^2$ indicates similarity (or lack of it), which means $x(k)$ is more predictable knowing $x(k-n)$ than for the case $x(k-n)$ is not known. For example, in this problem knowing $x(k-1) = 1$ means that there is a 70% chance that $x(k) = 1$ as opposed to a 50% chance if we did not know it.

In later chapters a frequency interpretation will be given to randomness by considering the power spectral density $S_{xx}(\omega)$ which is the Fourier transform of the autocorrelation function $R_{xx}(\tau)$ of a continuous process and the power spectral density $S_{xx}(\omega)$ which is the discrete Fourier transform of a discrete autocorrelation function $R_{xx}(n)$.

Cross-correlation Functions.   For any two continuous noise waveforms $x(t)$ and $y(t)$ the cross-correlation integrals are:

$$R_{xy}(\tau) = \lim_{T\to\infty} \frac{1}{2T} \int_{-T}^{T} y(t)x(t - \tau)\, dt \tag{3-37}$$

and

$$R_{yx}(\tau) = \lim_{T\to\infty} \frac{1}{2T} \int_{-T}^{T} x(t)y(t - \tau)\, dt \tag{3-38}$$

called the correlation of $x(\tau)$ with $y(\tau)$ and $y(\tau)$ with $x(\tau)$, respectively. If $x(t)$ and $y(t)$ are not similar, we would expect that:

$$R_{xy}(\tau) = R_{yx}(\tau) = x(t)y(t)$$

However, if similarities exist in the sense that if $x(t - \tau)$ is known, then $y(t)$ is more predictable or vice versa, then $R_{xy}(\tau)$ will be different from $\overline{x(t)}\,\overline{y(t)}$. We would expect similarities in cases where $y(t)$ is the output of a system and $x(t)$ is the input. For two discrete noise waveforms:

$$R_{xy}(n) \triangleq \lim_{N\to\infty} \frac{1}{2N + 1} \sum_{k} y(k)x(k - n) \tag{3-39}$$

and

$$R_{yx}(n) \triangleq \lim_{N\to\infty} \frac{1}{2N + 1} \sum_{k} x(k)y(k - n) \tag{3-40}$$

If $x(n)$ and $y(n)$ are not "similar," then we should expect $R_{xy}(n) = R_{yx}(n) = \overline{x(n)}\,\overline{y(n)}$.

EXAMPLE 3-18

(a) Given the noise waveform $x(t)$ of Example 3-17(a) is the input to a system that amplifies the input by a factor of 10 and causes a time delay of $0.25b$, as shown in Figure 3-14(a), find the cross-correlation of the input and output and the output autocorrelation function. $R_{xx}(\tau)$ was found as:

$$R_{xx}(\tau) = \frac{2.5[b - 0.1|\tau|]}{b}, \qquad 0 < |\tau| < b$$

and $R_{xx}(\tau) = 2.25,$ otherwise

(b) Find the output autocorrelation and the cross-correlation of the input and output waveforms when the inputs to a summing device [i.e., $y(n) = x_1(n) + x_2(n)$] are two discrete waveforms, each defined as taking on

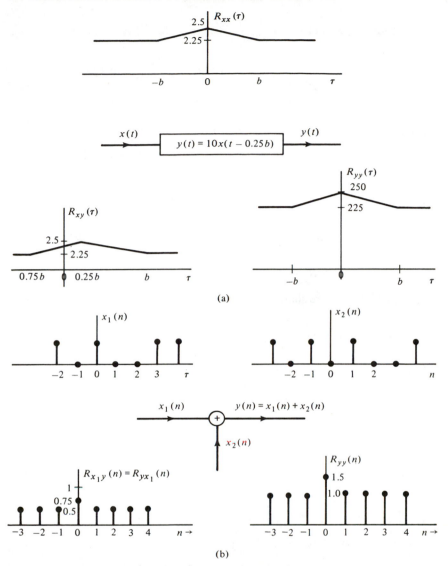

Figure 3-14 (a) The input autocorrelation function, system definition, and input–output cross-correlation and output autocorrelation function for Example 3-18(a); (b) the input waveforms, a summing device, and the output autocorrelation and input–output cross-correlation for Example 3-18(b).

the value 0 or 1 for each integer where previous values have no effect on the following ones and each value has equal chance.

Solution

(a) $$y(t) = 10x(t - 0.25b)$$

Therefore $$R_{xy}(\tau) = \lim_{T\to\infty} \frac{1}{2T} \int_{-T}^{T} 10x(t - 0.25b)x(t - \tau)\, dt$$

$$= \lim_{T\to\infty} \frac{1}{2T} \int_{-T}^{T} 10x(p)x(p - \tau + 0.25b)\, dt$$

$$= 10R_{xx}(\tau - 0.25b)$$

Using the formula for $R_{xx}(\tau)$, for Example 3-17a, we obtain:

$$R_{xy}(\tau) = \frac{25[b - 0.1(\tau - 0.25b)]}{b}$$

Since $$y(t) = 10x(t - 0.25b)$$

$$R_{yy}(\tau) = \lim_{T\to\infty} \frac{1}{2T} \int_{-T}^{T} 10x(t - 0.25b)\, 10x(t - \tau - 0.25b)\, dt$$

$$= 100R_{xx}(\tau)$$

$R_{xy}(\tau)$ and $R_{yy}(\tau)$ are shown in Figure 3-14(a).

(b) $y(n) = x_1(n) + x_2(n)$

$R_{yy}(n)$

$$R_{yy}(n) = \lim_{N\to\infty} \frac{1}{2N + 1} \sum_{k} y(k)y(k - n)$$

$$= \lim_{N\to\infty} \frac{1}{2N + 1} \sum_{k} ([x_1(k) + x_2(k)][x_1(k - n) + x_2(k - n)]$$

$$= R_{x_1 x_1}(n) + R_{x_2 x_2}(n) + R_{x_1 x_2}(n) + R_{x_2 x_1}(n)$$

Obviously, $R_{x_1 x_1}(n) = R_{x_2 x_2}(n)$ and on consideration:

$$R_{x_1 x_1}(0) = 0.5(1^2 + 0^2) = 0.5$$

and $$R_{x_1 x_1}(n) = 0.25(1 + 0 + 0 + 0) = 0.25, \qquad n \neq 0$$

In addition:

$$R_{x_1 x_2}(n) = R_{x_2 x_1}(n)$$

$$= \overline{x_1(n)x_2(n)}$$

$$= 0.25$$

Finally:

$$R_{yy}(n) = 2R_{xx}(n) + 2(0.25)$$

when $n = 0$:

$$R_{yy}(n) = 1 + 0.5$$

$$= 1.5, \qquad n = 0$$

when $n \neq 0$:

$$R_{yy}(n) = 1$$

This output autocorrelation function is shown in Figure 3-14(b).

$R_{x_1y}(n)$ and $R_{x_2y}(n)$

$$R_{x_1y}(n) = \lim_{N \to \infty} \frac{1}{2N + 1} \sum_{k=-N}^{N} [x_1(k) + x_2(k)]x_1(k - n)$$

$$= R_{x_1x_1}(n) + R_{x_2x_1}(n)$$

$$R_{x_1y}(0) = 0.5 + 0.25$$

$$= 0.75$$

and $\qquad R_{x_1y}(n) = 0.5, \qquad n \neq 0$

Obviously:

$$R_{x_2y}(n) = R_{x_1y}(n)$$

$R_{x_1y}(n)$ is plotted in Figure 3-14(b).

3-5 LINEAR SYSTEMS WITH RANDOM AND SIGNAL PLUS NOISE INPUTS

In this section we develop formulas for relating the input and output autocorrelation functions and the cross-correlation of the input and output when an ergodic noise waveform is the input to a linear time-invariant system. In addition, we examine the case of a deterministic signal plus noise input.

In Chapter 2 we derived the output of a LTIC system to a deterministic input $x(t)$ as:

$$y(t) = h(t)*x(t)$$

$$= \int_{-\infty}^{\infty} h(p)x(t - p)\, dp \quad \text{or} \quad \int_{-\infty}^{\infty} x(u)h(t - u)\, du$$

or for the discrete case:

$$y(n) = h(n)*x(n)$$

$$= \sum_{p} h(p)x(n - p) \quad \text{or} \quad \sum_{l} x(l)h(n - l)$$

We now utilize these formulas for the case of a random input.

3-5-1 Correlation Functions for a System with Random Inputs

The Cross-correlation $R_{xy}(\tau)$.

Table 3-1(a) shows the cases of a LTIC continuous or discrete system with a finite power noise waveform input that we assume is a member of an ergodic random process. The term **ergodic** means that all time averages are identical for each member waveform. Let us first consider the continuous case. The cross-correlation of the input and output is defined as:

$$R_{xy}(\tau) = \lim_{T \to \infty} \frac{1}{2T} \int_{-T}^{T} y(t)x(t - \tau)\, dt$$

$$= \lim_{T \to \infty} \frac{1}{2T} \int_{-T}^{T} \left[\int_{-\infty}^{\infty} h(l)x(t - l)\, dl \right] x(t - \tau)\, dt$$

Now interchanging the order of integration, we obtain:

$$R_{xy}(\tau) = \int_{-\infty}^{\infty} h(l) \left[\lim_{T \to \infty} \frac{1}{2T} \int_{-T}^{T} x(t - l)x(t - \tau)\, dt \right] dl$$

$$= \int_{-\infty}^{\infty} h(l) R_{xx}(l - \tau)\, dl$$

$$= R_{xx}(\tau) \oplus h(\tau) \tag{3-41a}$$

Since $R_{xx}(l - \tau) = R_{xx}(\tau - l)$ due to evenness:

$$R_{xy}(\tau) = \int_{-\infty}^{\infty} h(l) R_{xx}(\tau - l)\, dl$$

$$= h(\tau) * R_{xx}(\tau) \tag{3-41b}$$

Of course, $R_{yx}(\tau) = R_{xy}(-\tau)$ but we will also derive it.

$$R_{yx}(\tau) = \lim_{T \to \infty} \frac{1}{2T} \int_{-T}^{T} x(t)y(t - \tau)\, dt$$

$$= \lim_{T \to \infty} \frac{1}{2T} \int_{-T}^{T} x(t) \left[\int_{-\infty}^{\infty} h(s)x(t - \tau - s)\, ds \right] dt$$

Proceeding as before, we have:

$$R_{yx}(\tau) = \int_{-\infty}^{\infty} h(s) \left[\lim_{T \to \infty} \frac{1}{2T} \int_{-T}^{T} x(t)x(t - \tau - s)\, dt \right] ds$$

$$R_{yx}(\tau) = \int_{-\infty}^{\infty} h(s) R_{xx}(\tau + s)\, ds$$

$$= h(\tau) \oplus R_{xx}(\tau) \tag{3-42a}$$

TABLE 3-1 LINEAR SYSTEM WITH RANDOM AND SIGNAL PLUS NOISE INPUTS

Continuous	Discrete

(a) Random inputs

Continuous:

$x(t)$ with $R_{xx}(\tau)$ → $\boxed{h(t) \text{ and } C_{hh}(\tau)}$ → $y(t)$

$$R_{xy}(\tau) = R_{xx}(\tau) \oplus h(\tau) = R_{xx}(\tau)*h(\tau)$$
$$R_{yx}(\tau) = h(\tau) \oplus R_{xx}(\tau) = R_{xx}(\tau)*h(-\tau)$$
$$R_{yy}(\tau) = R_{xx}(\tau) \oplus C_{hh}(\tau) = R_{xx}(\tau)*C_{hh}(\tau)$$

Discrete:

$x(n)$ with $R_{xx}(n)$ → $\boxed{h(n) \text{ and } C_{hh}(n)}$ → $y(n)$

$$R_{xy}(n) = R_{xx}(n) \oplus h(n) = R_{xx}(n)*h(n)$$
$$R_{yx}(n) = h(n) \oplus R_{xx}(n) = R_{xx}(n)*h(-n)$$
$$R_{yy}(n) = R_{xx}(n) \oplus C_{hh}(n) = R_{xx}(n)*C_{hh}(n)$$

(b) Deterministic signal $f(t)$ plus zero mean noise $n(t)$

Continuous:

$x(t) = f(t) + n(t)$ → $\boxed{h(t) \text{ and } C_{hh}(\tau)}$ → $y(t) = g(t) + m(t)$

$R_{nn}(\tau)$ known
$R_{fn}(\tau) = 0$

$$g(t) = f(t)*h(t)$$
$$R_{mm}(\tau) = R_{nn}(\tau)*C_{hh}(\tau)$$

Deterministic signal $f(k)$ plus zero mean noise $n(k)$

Discrete:

$x(k) = f(k) + n(k)$ → $\boxed{h(k) \text{ or } C_{hh}(k)}$ → $y(k) = g(k) + m(k)$

$R_{nn}(k)$ known
$R_{fn}(k) = 0$

$$g(k) = f(k)*h(k)$$
$$R_{mm}(k) = R_{nn}(k)*C_{hh}(k)$$

Alternatively, we could obtain:

$$R_{yx}(\tau) = \int_{-\infty}^{\infty} h(s)R_{xx}(-\tau - s)\, ds$$

$$= R_{xy}(-\tau) \tag{3-42b}$$

The Output-Autocorrelation $R_{yy}(\tau)$

$$R_{yy}(\tau) = \lim_{T \to \infty} \frac{1}{2T} \int_{-T}^{T} y(t)y(t - \tau)\, dt$$

$$= \lim_{T \to \infty} \frac{1}{2T} \int_{-T}^{T} \left[\int_{-\infty}^{\infty} h(u)x(t - u)\, du \right.$$

$$\left. \cdot \int_{-\infty}^{\infty} h(s)x(t - \tau - s)\, ds \right] dt$$

Interchanging the order of integration and taking a limit, we obtain:

$$R_{yy}(\tau) = \int_{-\infty}^{\infty} h(u) \left[\lim_{T \to \infty} \frac{1}{2T} \int_{-T}^{T} x(t - u)x(t - \tau - s)\, dt \right] du \int_{-\infty}^{\infty} h(s)\, ds$$

$$= \int_{-\infty}^{\infty} h(u)R_{xx}(u - \tau - s)\, du \int_{-\infty}^{\infty} h(s)\, ds$$

Let $u - s = p$ and:

$$R_{yy}(\tau) = \int_{-\infty}^{\infty} h(p + s)R_{xx}(p - \tau)\, dp \int_{-\infty}^{\infty} h(s)\, ds$$

Now interchanging the order of integration, we have:

$$R_{yy}(\tau) = \int_{-\infty}^{\infty} \left[\int_{-\infty}^{\infty} h(s)h(p + s)\, ds \right] R_{xx}(p - \tau)\, dp$$

$$= \int_{-\infty}^{\infty} C_{hh}(p)R_{xx}(p - \tau)\, dp$$

$$[\text{where } C_{hh}(p) = h(p) \oplus h(p)]$$

$$= R_{xx}(\tau) \oplus C_{hh}(\tau) \tag{3-43a}$$

Since $C_{hh}(\tau)$ and $R_{xx}(\tau)$ are even, we can also write:

$$R_{yy}(\tau) = C_{hh}(\tau) * R_{xx}(\tau) \tag{3-43b}$$

$$= R_{xx}(\tau) \oplus C_{hh}(\tau) \tag{3-43c}$$

This relationship between the output and input autocorrelation functions for a LTIC system with a random input is one of the most important results in communication theory and is called the **Wiener-Khinchine theorem.** There are many different ways for obtaining these results and we should demonstrate versatility in deriving them. The reader should redo these derivations in a different manner and obtain the same result.

In a completely analogous way, we can show that for a discrete system with pulse response $h(n)$ and input $x(n)$ with autocorrelation $R_{xx}(n)$ that:

$$R_{xy}(n) = R_{xx}(n) \oplus h(n) \qquad (3\text{-}44a)$$

or

$$= h(n)*R_{xx}(n) \qquad (3\text{-}44b)$$

and that:

$$R_{yy}(n) = C_{hh}(n)*R_{xx}(n) \qquad (3\text{-}45a)$$

or

$$= C_{hh}(n) \oplus R_{xx}(n) \qquad (3\text{-}45b)$$

or

$$= R_{xx}(n) \oplus C_{hh}(n) \qquad (3\text{-}45c)$$

These results are tabulated in Table 3-1(a). We should note the similarity of the use of the transfer impulse or pulse correlation functions $C_{hh}(\tau)$ and $C_{hh}(n)$ for noise inputs to that of the impulse and pulse responses for deterministic inputs.

Deterministic input:

$$y(t) = h(t)*x(t) \quad \text{or} \quad y(n) = h(n)*x(n)$$

Noise input:

$$R_{yy}(\tau) = C_{hh}(\tau)*R_{xx}(\tau) \quad \text{or} \quad R_{yy}(n) = C_{hh}(n)*R_{xx}(n)$$

3-5-2 A Linear System with a Deterministic Signal plus Noise Input

Table 3-1(b) shows a linear system with input:

$$x(t) = f(t) + n(t) \quad \text{or} \quad x(k) = f(k) + n(k)$$

where $f(t)$ and $f(k)$ are deterministic signals and $n(t)$ and $n(k)$ are zero mean noise waveforms with $R_{fn}(\tau)$ and $R_{fn}(k) = 0$. The deterministic outputs are:

$$g(t) = f(t)*h(t) \quad \text{and} \quad g(k) = f(k)*h(k) \qquad (3\text{-}46)$$

respectively, and the output noise autocorrelation functions are:

$$R_{mm}(\tau) = R_{nn}(\tau)*C_{hh}(\tau) \quad \text{and} \quad R_{mm}(k) = R_{nn}(k)*C_{hh}(k) \qquad (3\text{-}47)$$

These results are tabulated in Table 3-1(b), and a few examples demonstrating them will now be solved.

EXAMPLE 3-19

Consider a linear system with pulse response $h(n) = (0.5)^n u(n)$ that has as its input a noise waveform with autocorrelation function $R_{xx}(n) = 2\delta(n)$. Find the cross-correlation function of the input and output and the output autocorrelation function, as well as the mean, mean square value, and variance of the output noise.

Solution

$$C_{hh}(n) = h(n) \oplus h(n)$$

$$= \sum_{k-n}^{\infty} 0.5^k 0.5^{k-n}$$

$$= 0.5^{-n}\left[\sum_{k-n}^{\infty} 0.5^{2k}\right]$$

$$= 1.33(0.5^n), \qquad n > 0$$

Therefore

$$C_{hh}(n) = 1.33(0.5^{|n|})$$

Thus

$$R_{yy}(n) = 1.33(0.5^{|n|})*2\delta(n)$$

$$= 2.67(0.5^{|n|})$$

The cross-correlation between the input and output is:

$$R_{xy}(n) = h(n)*R_{xx}(n)$$

$$= 0.5^n u(n)*2\delta(n)$$

$$= 2(0.5^n)u(n)$$

Considering the output autocorrelation function, we note that:

$$\overline{y^2(n)} = R_{yy}(0)$$

$$= 2.67$$

$$\overline{y(n)} = 0 \quad \text{and} \quad \sigma_y^2 = 2.67$$

EXAMPLE 3-20

Consider a linear system with impulse response $h(t) = 2e^{-3t}u(t)$ and with a deterministic input $f(t) = 3 \cos 2t$ plus uncorrelated white noise with a mean square value of 100 whose autocorrelation function for this system may be approximated by $R_{nn}(\tau) = 4\delta(\tau)$ (white noise). Find the output signal and noise autocorrelation function and the input and output signal to noise ratios.

Solution

The Output Signal
From a first circuits course:

$$g(t) = Re\left[\frac{2}{j2 + 3} 3\underline{|0°} e^{j2t}\right]$$

$$= \frac{6}{\sqrt{13}} \cos(2t - 33°)$$

$$= 1.66 \cos(2t - 33°)$$

The Output Noise Autocorrelation

$$R_{mm}(\tau) = C_{hh}(\tau) * 4\delta(\tau)$$

$C_{hh}(\tau)$

$$h(\tau) = 2e^{-3t}u(t)$$

and

$$C_{hh}(\tau) = \int_{\tau}^{\infty} (2e^{-3p})(2e^{-3(p-\tau)}) \, dp$$

$$= 4e^{+3\tau} \left[-\tfrac{1}{6}e^{-6p} \big|_{\tau}^{\infty} \right]$$

$$= 0.67e^{-3\tau}, \qquad \tau > 0$$

Therefore

$$C_{hh}(\tau) = 0.67e^{-3|\tau|}$$

and

$$R_{mm}(\tau) = 0.67e^{-3|\tau|} * 4\delta(\tau)$$

$$= 2.67e^{-3|\tau|}$$

The **signal to noise ratio** is defined as the ratio of the mean square value of the signal divided by the mean square value of the noise and denoted by S/N.

Therefore

$$\frac{S}{N}\bigg|_{\text{input}} = \frac{(3/\sqrt{2})^2}{100}$$

$$= 0.045$$

$$\frac{S}{N}\bigg|_{\text{output}} = \frac{(1.67/\sqrt{2})^2}{R_{mm}(0)}$$

$$= \frac{1.66}{2.67}$$

$$= 0.52$$

which is substantially better than at the input.

SUMMARY

The goal of the chapter was to extend the material of Chapter 2 involving systems with deterministic inputs to consideration of systems with ergodic noise inputs or deterministic signal plus noise inputs. This was achieved without the prerequisite of probability theory. In a subsequent course, probability theory and random processes should be treated in detail.

As a prelude to handling noise inputs, Sections 3-2 through 3-4 discussed auto- and cross-correlation for deterministic finite energy waveforms and periodic waveforms whether continuous or discrete. Periodic waveforms represent finite power waveforms that may be handled analytically. Section 3-5 considered the auto- and cross-correlation of ergodic noise waveforms:

$$R_{xx}(\tau) = \lim_{T \to \infty} \frac{1}{2T} \int_{-T}^{T} x(t)x(t - \tau)dt, \quad R_{xx}(n) = \lim_{N \to \infty} \frac{1}{2N + 1} \sum_{k} x(k)x(k - n)$$

and

$$R_{xy}(\tau) = \lim_{T \to \infty} \frac{1}{2T} \int_{-T}^{T} y(t)x(t - \tau)dt, \quad R_{xx}(n) = \lim_{N \to \infty} \frac{1}{2N + 1} \sum_{k} y(k)x(k - n)$$

Since noise waveforms do not possess analytic formulas these functions must in practice be estimated from a finite section or finite number of values using fast Fourier transform algorithms. We considered some very simple noise waveforms where by using the concept of a statistical average for different sampled values of the waveforms we were able to predict the correlation integrals. This led to an intuitive interpretation of how correlation functions measure similarity or randomness of waveforms. Later, these interpretations are examined with respect to power content in terms of frequency via power spectral densities.

Section 3-5 considered the evaluation of output autocorrelation functions and the cross-correlation of the input and output waveforms for LTIC continuous and discrete systems with ergodic noise inputs. The somewhat tricky manipulations involving time averages are algebraically equivalent to the ensemble average manipulations that are considered in the theory of random processes. Initially they appear quite awesome and it is better not to deal with them for the first time when also tackling ensemble concepts. The similarity of the role of the transfer correlation functions $C_{hh}(\tau)$ and $C_{hh}(n)$ to that of the impulse or pulse response of a system when finding deterministic outputs was stressed. For noise inputs we found that the output autocorrelation is:

$$R_{yy}(\tau) = C_{hh}(\tau) * R_{xx}(\tau) \quad \text{or} \quad R_{yy}(n) = C_{hh}(n) * R_{xx}(n)$$

as opposed to the output expressions:

$$y(t) = h(t) * x(t) \quad \text{and} \quad y(n) = h(n) * x(n)$$

for deterministic inputs.

PROBLEMS

3-1. Evaluate:
    (a) $e^{2t}u(t) \oplus e^{-2t}u(t)$
    (b) $e^{-2t}u(t) \oplus e^{t}u(t)$
    (c) $e^{-at}u(t) \oplus e^{-bt}u(t)$ How must b be related to a for this correlation integral to exist?
    (d) $e^{t}u(-t) \oplus u(t)$
    (e) $u(t) \oplus u(-t)$
    (f) $u(-t) \oplus u(t)$
    (g) $f(t) \oplus \delta(t - a)$
    (h) $\delta(t - a) \oplus f(t)$

3-2. Given:

$$f(t) = 0, \qquad t < t_{f_1} \cap t > t_{f_2}$$

and

$$g(t) = 0, \qquad t < t_{g_1} \cap t > t_{g_2}$$

and both functions are nonzero otherwise, find when:

(a) $f(t) * g(t) = 0$
(b) $f(t) \oplus g(t) = 0$
(c) $g(t) \oplus f(t) = 0$

3-3. In Chapter 2 based on our knowledge of linear systems we reasoned that:

$$c_1 e^{\alpha_1 t} u(t) * c_2 e^{\alpha_2 t} u(t) = [k_1 e^{\alpha_1 t} + k_2 e^{\alpha_2 t}] u(t)$$

The cross-correlation function $f(t) \oplus g(t)$ is described as a measure of the similarity of the two signals $f(t)$ and $g(t)$. Now evaluate $c_1 e^{\alpha_1 t} u(t) \oplus c_2 e^{\alpha_2 t} u(t)$ and interpret your results for $\alpha_2 = \alpha_1$ and $\alpha_2 \neq \alpha_1$ in terms of the similarity of $f(t)$ and $g(t)$.

3-4. Given:

$$f(t) = -t[u(t + 2) - u(t)] + 2e^{-t}[u(t) - u(t - 1)]$$

and

$$g(t) = (2 - t)[u(t + 1) - u(t - 2)]$$

set up each different range for evaluating $f(t) \oplus g(t)$. Do not leave unit step notation inside an integral.

3-5. Evaluate:

(a) $0.6^n u(n) \oplus (-0.5)^n u(n)$ (b) $(-0.5)^n u(n) \oplus 0.6^n u(n)$
(c) $a^n u(n) \oplus b^n u(n)$. How must b be related to a for this correlation summation to exist?
(d) $2^n u(-n) \oplus u(n)$ (e) $u(-n) \oplus u(n)$
(f) $f(n) \oplus \delta(n - a)$ (g) $\delta(n - a) \oplus \delta(n - b)$

3-6. Given:

$$f(n) = 0, \qquad n < n_{11} \cap n > n_{12}$$

and

$$g(n) = 0, \qquad n < n_{21} \cap n > n_{22}$$

and both functions are nonzero otherwise, find when:

(a) $f(n) * g(n) = 0$
(b) $f(n) \oplus g(n) = 0$
(c) $g(n) \oplus f(n) = 0$

3-7. In Chapter 2 based on our knowledge of discrete linear systems we reasoned that if the pulse responses $h(n) = c_1[a^n]u(n)$ and the input is $x(n) = c_2[b^n]u(n)$, then the output $y(n) = x(n) * h(n)$ must be of the form:

$$y(n) = c_1[a^n]u(n) + c_2[b^n]u(n), \qquad \text{(if } b \neq a)$$

Now evaluate:

$$k_1(a^n)u(n) \oplus k_2(b^n)u(n)$$

and interpret your results for $a = b$ and $a \neq b$ in terms of the similarity of the two signals.

3-8. Consider the periodic functions $x(t)$ and $y(t)$ defined as:

$$x(t) = \sum_{\substack{n=-\infty \\ \text{integer}}}^{\infty} g_1(t - 3n)$$

where

$$g_1(t) = 2, \quad 0 < t < 1$$
$$= 0, \quad \text{otherwise}$$

and

$$y(t) = \sum_{\substack{n=-\infty \\ \text{integer}}}^{\infty} g_2(t - 3n)$$

where

$$g_2(t) = -1, \quad -1 < t < 0$$
$$= 0, \quad \text{otherwise}$$

(a) Plot $x(t)$ and $y(t)$.
(b) Evaluate $R_{xx}(\tau) = x(\tau) \oplus x(\tau)$ and $R_{yy}(\tau) = y(\tau) \oplus y(\tau)$.
(c) Evaluate $R_{xy}(\tau)$ and $R_{yx}(\tau)$.

3-9. If in Problem 3-8 $y(t)$ is changed to:

$$y(t) = \sum_{n=-\infty}^{\infty} g_2(t - 2n)$$

evaluate $C_{xy}(\tau)$.

3-10. Consider the periodic functions $x(n)$ and $y(n)$ defined as:

$$x(n) = \sum_{\substack{k=-\infty \\ \text{integer}}}^{\infty} g_1(n - 4k)$$

where

$$g_1(n) = n[u(n) - u(n - 3)]$$

and

$$y(n) = \sum_{\substack{k=-\infty \\ \text{integer}}}^{\infty} g_2(n - 4k)$$

where

$$g_2(n) = (-1)^n[u(n) - u(n - 4)]$$

(a) Plot $x(n)$ and $y(n)$.
(b) Evaluate $R_{xx}(n) = x(n) \oplus x(n)$ and $R_{yy}(n) = y(n) \oplus y(n)$.
(c) Evaluate $R_{xy}(n)$ and $R_{yx}(n)$.

3-11. Prove for two discrete functions $x(n)$ and $y(n)$ that:
(a) $C_{xx}(n) = C_{xx}(-n)$
(b) $C_{xx}(0) \geq C_{xx}(n)$
(c) $C_{xy}(n) = C_{yx}(n)$
(d) $|C_{xy}(n)| \leq \sqrt{C_{xx}(0)C_{yy}(0)}$
(e) Verify all these relations with the result for $a^n u(n) \oplus b^n u(n)$.

3-12. Extend the proofs of the previous problems to the case where $x(n)$ and $y(n)$ are discrete periodic functions.

3-13. Consider an ergodic discrete random process with autocorrelation function $R_{xx}(n)$ and mean value $\overline{x(n)}$. If a member of a new process $y(n)$ is formed by adding two members of $x(n)$ and then subtracting a third member:

$$y_p(n) = x_i(n) + x_j(n) - x_k(n)$$

find a formula for $R_{yy}(n)$ in terms of $R_{xx}(n)$ and $\overline{x(n)}$.

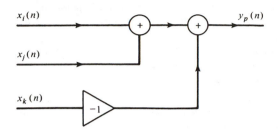

3-14. Consider the following random process. Every 3 s a pulse of amplitude ±2 V and duration 1 s is generated and the whole waveform is then given a phase shift uniformly chosen between 0 and 3 s. A sketch of a typical member of this process is shown. Find:

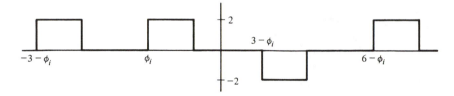

(a) the average, mean-square, and variance of any member.
(b) the autocorrelation function:

$$R_{xx}(\tau) = \overline{x_i(t)x_i(t + \tau)}$$

and plot your answer as a function of τ.
(c) If a new random process is formed by synchronously adding two members of $x(t)$ to find a member of $y(t)$, find and plot $R_{yy}(\tau)$, $R_{xy}(\tau)$, and $R_{yx}(\tau)$.
(d) If a new random process is formed by synchronously subtracting two members of $x(t)$ to find a member of $z(t)$, find and plot $R_{zz}(\tau)$, $R_{xz}(\tau)$, and $R_{zx}(\tau)$. Is your result surprising? At last, $a - b$ may equal $b - a$ and both not be identically zero.

3-15. Consider the following discrete random processes:
1. every second a member takes on the value ±2 V and the chances of retaining a previous value are 50%.
2. every second a member takes on the value ±2 V and the chances of retaining a previous value are 80%
3. every second a member takes on the value ±2 V and the chances of retaining a previous value are 20%
4. the trivial process where every second a member changes its value between +2 V and −2 V

(a) Sketch part of a typical member for each of the preceding processes.
(b) Find $\overline{x(t)}$, $\overline{x^2(t)}$, and σ_x^2 for each process.

(c) Find and plot $R_{xx}(\tau)$ for each process. For processes 2 and 3 just find $R_{xx}(0)$, $R_{xx}(\pm 1)$, and $R_{xx}(\pm\infty)$. If you are brave try to find $R_{xx}(\pm 2)$.

3-16. A random process is defined as a white noise process if its autocorrelation function can be approximated by a weighted delta function as far as a system is concerned. Given the input to a system with impulse response $h(t) = 2e^{-10t}u(t)$ is the deterministic signal $f(t) = 10 \cos 500t$ plus noise $n(t)$ with an autocorrelation function $R_{nn}(\tau) = 50e^{-a|\tau|}$. Find:

(a) the output signal and noise autocorrelation function if $a \gg 10$

(b) the input and output signal to noise ratios if $a \gg 10$.

(c) repeat parts (a) and (b) if $R_{nn}(\tau) = 50e^{-20|\tau|}$

3-17. Given the input to a system where the pulse response $h(n) = (-0.6)^n u(n)$ is $x(n)$. $x(n)$ is a member of a random process that assumes the value ± 2 V every second and the chances of retaining a previous value is always 50%. Find the output autocorrelation function and total average power of the input and the output.

The One-sided
Laplace Transform

INTRODUCTION

Chapters 1 and 2 treated the time-domain analysis of LTIC systems with deterministic continuous and discrete inputs. Chapter 3 intuitively discussed how the autocorrelation function measures randomness for a noise waveform and covered the analysis of systems with random inputs.

In Chapters 4 through 9 we are embarking on the transform theory. The most widely used transforms in electrical engineering will be considered. These are:

the one-sided Laplace transform (Chapter 4)

the two-sided Laplace transform* (Chapter 5)

the one-sided Z transform (Chapter 6)

the two-sided Z transform* (Chapter 7)

the Fourier transform (Chapter 8)

the discrete and fast Fourier transforms (Chapter 9)

All of these transforms will be handled in a similar manner: (1) The definition will be given and used by forming a small basic set of transforms; (2) The main properties and theorems of the transform will be enumerated; (3) These properties and theorems will be demonstrated by applying them to known results and utilized to extend the current vocabulary of transforms; (4) The problem of evaluating inverse transforms will be treated in detail; (5) The principal applications of the transform will be discussed.

4-1 DEFINITION AND EVALUATION
OF SOME TRANSFORMS

4-1-1 Definition

The one-sided Laplace transform of a real function $f(t)$ is defined as:

$$F(s) = \int_{0^-}^{\infty} f(t)e^{-st}\, dt \tag{4-1}$$

where $s = \sigma + j\omega$ is complex. $F(s)$ exists for all complex s for which Equation 4-1 converges. Therefore the condition for existence is that:

$$\int_{0^-}^{\infty} |f(t)e^{-st}|\, dt = \int_{0^-}^{\infty} |f(t)e^{-\sigma t}|\, dt < \infty \tag{4-2}$$

This leads to an allowable region of convergence $\sigma_1 < \mathrm{Re}(s) = \sigma < \sigma_2$ in the s plane.

We should immediately comment on the lower limit of integration $t = 0^-$ in the definition since some authors use $t = 0^+$ or $t = 0$. All three lower limits are equivalent if $f(t)$ does not contain a singularity function at $t = 0$. For any continuous or piecewise continuous function

$$\lim_{\epsilon \to 0} \int_{-\epsilon}^{\epsilon} f(t)\, dt = \lim_{\epsilon \to 0} \int_{-\epsilon}^{0} f(t)\, dt = 0$$

Indeed, when singularity functions do occur a certain amount of fudging often takes place. Our position is that:

$$\int_{-\epsilon}^{\epsilon} \delta(t)\, dt = 1$$

and

$$\int_{0}^{\epsilon} \delta(t)\, dt = 0.5 \qquad \text{(if it must be interpreted)}$$

and

$$\int_{0^+}^{\epsilon} \delta(t)\, dt = 0$$

if $\delta(t)$ is to retain its meaning from the fundamental definition. When a text states:

$$\int_{0}^{\infty} \delta(t)\, dt = \int_{0^+}^{\infty} \delta(t) = 1$$

it implies the delta function is shifted to the right or its model applies to a situation where it commences at $t = 0$ or 0^+.

A number of different short-hand notations are used to denote a Laplace transform. $F(s)$, $L[f(t)]$, and $\overline{f(t)}$ will all be used interchangeably. We now evaluate some Laplace transforms.

EXAMPLE 4-1

Find the Laplace transform of the following functions and discuss the region of convergence for which the transform exists:

(a) $f_1(t) = 6$
(b) $f_2(t) = 5e^{-2t}$
(c) $f_3(t) = 4e^{2t}$
(d) $f_4(t) = e^{j4t}$
(e) $f_5(t) = \text{Cos } 4t$
(f) $f_6(t) = \text{Sin } 4t$
(g) $f_7(t) = \delta(t)$
(h) $f_8(t) = d/dt[2e^{-3t}u(t)]$
(i) $f_9(t) = d/dt\,(2e^{-3t})$
(j) $f_{10}(t) = t$

Solution

(a) $F_1(s) = \displaystyle\int_0^\infty 6e^{-st}\, dt$

$$= \frac{6e^{-st}}{-s}\Big|_0^\infty$$

$$= 0\,|_{\text{if } \sigma > 0} + \frac{6}{s}$$

we note $e^{-s\infty}$ only approaches zero if $\text{Re}(s) = \sigma > 0$.

Therefore $\qquad\qquad L[6] = \dfrac{6}{s}, \qquad \sigma > 0$

(b) $F_2(s) = \displaystyle\int_0^\infty 5e^{-2t}e^{-st}\, dt$

$$= \frac{5e^{-(s+2)t}}{-(s+2)}\Big|_0^\infty$$

$$= 0 + \frac{5}{s+2}$$

if $\lim_{s\to\infty} 5e^{-(s+2)t} = 0$ and this is so if $\sigma > -2$.

Therefore $\qquad\qquad L[5e^{-2t}] = \dfrac{5}{s+2}, \qquad \text{Re}(s) > -2$

(c) $F_3(s) = \displaystyle\int_0^\infty 4e^{2t}e^{-st}\, dt$

$$= \frac{4}{s-2}, \qquad \text{Re}(s) > 2$$

using an analogous argument as in part (b).

(d) $f_4(t) = e^{j4t}$

This is a complex function of time and we must interpret what we mean by its Laplace transform. When for convenience we talk about the transform of a complex function $f(t) = f_1(t) + j f_2(t)$ we define:

$$L[f(t)] = F_1(s) + jF_2(s)$$

where $\sigma > $ maximum (σ_1, σ_2).

$$F_4(s) = \int_0^\infty e^{(j4-s)t}\, dt$$

$$= \frac{e^{(j4-s)t}}{(j4 - s)} \Bigg|_0^\infty$$

$$= 0 \Bigg|_{\text{if } \sigma>0} + \frac{1}{s - j4}$$

Therefore $\qquad L[e^{j4t}] = \dfrac{s}{s^2 + 16} + j\dfrac{4}{s^2 + 16}, \qquad \text{Re}(s) > 0$

(e) and **(f)**. From part (d) we conclude that:

$$L[\text{Cos } 4t] = \frac{s}{s^2 + 16}, \qquad \text{Re}(s) > 0$$

$$L[\text{Sin } 4t] = \frac{4}{s^2 + 16}, \qquad \text{Re}(s) > 0$$

and the reader should verify these results directly using integration by parts.

(g) $F_7(s) = \displaystyle\int_{0^-}^\infty \delta(t)e^{-st}\, dt \qquad$ (here we must use 0^-)

$$= 1, \qquad \text{for all } s$$

The lower limit of 0^- made this result very clear without essentially modifying the meaning of $\delta(t)$.

(h) $\qquad f_8(t) = \dfrac{d}{dt}[2e^{-3t}u(t)]$

$$= 2e^{-3t}\delta(t) - 6e^{-3t}u(t)$$

$$= 2\delta(t) - 6e^{-3t}u(t)$$

Therefore $F_8(s) = 2 - \dfrac{6}{s + 3}, \qquad \text{Re}(s) > \max(-\infty, -3)$

$$= \frac{2s}{s + 3}, \qquad \text{Re}(s) > -3$$

(i) $\qquad f_9(t) = \dfrac{d}{dt}(2e^{-3t})$

$\qquad\qquad\qquad = -6e^{-3t}$

Therefore $F_9(s) = \displaystyle\int_0^\infty -6e^{-3t}e^{-st}\,dt$

$\qquad\qquad\quad = \dfrac{-6}{s+3}, \qquad \sigma > -3$

Comparing parts (h) and (i) we note that we must be careful when a singularity function is present at $t = 0$. The derivative theorem of the next section will treat this situation thoroughly for the transform of a derivative.

(j) $F_{10}(s) = \displaystyle\int_0^\infty te^{-st}\,dt$

Integrating by parts, we find that:

$$F_{10}(s) = \int_0^\infty t\,d\left(-\frac{1}{s}e^{-st}\right)$$

$$= \left.\frac{te^{-st}}{-s}\right|_0^\infty + \int_0^\infty \frac{1}{s}e^{-st}(1)\,dt$$

$$= 0 - 0 + \frac{1}{s^2}, \qquad \text{Re}(s) > 0$$

The first zero term occurs because $\lim_{t\to\infty} te^{-st} = 0$ for any $\text{Re}(s) > 0$. The general result $\lim_{t\to\infty}[t^n e^{-\alpha t}] = 0$ for any $\alpha > 0$ is very important in determining the types of functions that have Laplace transforms. We can show this by expanding in a McLaurin's series:

$$t^n e^{-\alpha t} = \frac{t^n}{e^{\alpha t}} = \frac{t^n}{1 + \alpha t + (\alpha t)^2/2! + (\alpha t)^3/3! + \cdots}$$

$$= \frac{1}{t^{-n} + \alpha t^{-n+1} + \cdots + \alpha^n/n! + \alpha^{n+1}t/(n+1)! + \cdots}$$

We see that the denominator approaches infinity for large t and $\lim_{t\to\infty} t^n e^{-\alpha t} = 0$ for any $\alpha > 0$.

Based on Example 4-1 we may stop and form a short table of Laplace transforms. This is Table 4-1 and it also includes a plot for the region of convergence. Observing Table 4-1 some facts are evident. The region of convergence is always $\sigma > \sigma_1$ and if $F(s)$ is the ratio of two polynomials σ is to the right of all the poles.

In Table 4-1 every table entry except $\delta(t)$ can be written with $u(t)$ after it.

TABLE 4-1 A SHORT TABLE OF LAPLACE TRANSFORMS

$f(t)$	Plot	$F(s)$	Region of Convergence
$e^{-\alpha t}$	$\alpha < 0$ $\alpha > 0$	$\dfrac{1}{s + \alpha}$	$\sigma = \mathrm{Re}(s)$
1		$\dfrac{1}{s}$	
$e^{j\beta t}$	Complex	$\dfrac{1}{s - j\beta} = \dfrac{1}{s^2 + \beta^2} + j\,\dfrac{\beta}{s^2 + \beta^2}$	
Cos βt		$\dfrac{s}{s^2 + \beta^2}$	
Sin βt		$\dfrac{\beta}{s^2 + \beta^2}$	
$\delta(t)$		1	
t		$\dfrac{1}{s^2}$	

For example, we could write $e^{\alpha t}u(t)$, $u(t)$, for 1, $e^{j\beta t}u(t)$, and so on, and the results are identical.

4-2 IMPORTANT THEOREMS OF THE LAPLACE TRANSFORM

Table 4-2 lists some of the more important one-sided Laplace transform theorems. These will be used repeatedly throughout applications, and they have very similar counterparts for the Z and Fourier transforms. They should be thoroughly understood and considered as basic vocabulary. We will now prove a number of these, demonstrating their use.

TABLE 4-2   ONE-SIDED LAPLACE TRANSFORM THEOREMS

Given $f(t) \leftrightarrow F(s)$,	$\sigma > \sigma_1$,	$g(t) \leftrightarrow G(s)$,	$\sigma > \sigma_2$
Function		**Transform**	**Theorem's name**
$af(t) + bg(t)$		$aF(s) + bG(s)$ $\sigma > \max(\sigma_1, \sigma_2)$	Linearity
$f(at)$, $a > 0$		$\dfrac{1}{a} F\left(\dfrac{s}{a}\right)$, $\sigma > a\sigma_1$	Time-scaling
$f(t - a)u(t - a)$, $a > 0$		$e^{-as}F(s)$, $\sigma > \sigma_1$	Time-shifting
$e^{\alpha t}f(t)$		$F(s - a)$, $\sigma > \sigma_1 - \alpha$	"s-shifting"
$f(t)*g(t)$ for causal functions		$F(s)G(s)$, $\sigma > \max(\sigma_1, \sigma_2)$	Convolution
$f(t)g(t)$		$F(s)*G(s)$, $\sigma > (\sigma_1 + \sigma_2)$	Complex convolution (see Chapter 5)
$tf(t)$		$-F'(s)$, $\sigma > \sigma_1$	"s derivative"
$t^n f(t)$		$(-1)^n F^n(s)$, $\sigma > \sigma_1$	
$\displaystyle\int_{0^-}^{\infty} tf(t)\, dt$		$-F'(0)$, $\sigma > \sigma_1$	Moment
$\displaystyle\int_{0^-}^{\infty} t^n f(t)\, dt$		$(-1)^n F^n(0)$	
$f'(t)$		$sF(s) - f(0^-)$	
$f^n(t)$		$s^n F(s) - \displaystyle\sum_{p-0}^{n-1} s^{n-p-1}f^p(0^-)$	Derivative

Note: If $f'(t) \cdots f^n(t)$ contains no singularity functions, then $f^p(0^-)$ may be replaced by $f^p(0^+)$ for all p.

$\displaystyle\int_{0^-}^{t} f(\alpha)\, d\alpha$		$\dfrac{F(s)}{s}$, $\sigma > \sigma_1$	Integration
$f(0^+) = \lim\limits_{s\to\infty} sF(s)$			Initial value
$f(\infty) = \lim\limits_{s\to 0} sF(s)$			Final value

EXAMPLE 4-2

(a) Prove the time-scaling theorem.
(b) Given:

$$e^{-t}u(t) \leftrightarrow \frac{1}{s + 1}$$

Use the theorem to verify the transform of $e^{-3t}u(t)$.

Solution
(a) By definition:

$$L[f(at)] = \int_{0^-}^{\infty} f(at)e^{-st}\, dt$$

Let $at = p$;

Therefore $t = \frac{1}{a}p, \quad dt = \frac{1}{a}dp \quad$ and as $\quad 0^- < t < \infty, \ 0^- < p < \infty.$

$$L[f(at)] = \int_{0^-}^{\infty} f(p)e^{-(s/a)p}\left(\frac{1}{a}\right)dp$$

$$= \frac{1}{a}F\left(\frac{s}{a}\right)$$

(b) Given:

$$e^{-t}u(t) \leftrightarrow \frac{1}{s + 1}$$

we denote:

$$f(t) = e^{-t}u(t)$$

Therefore $e^{-3t}u(t) = f(3t) \qquad$ [since $u(3t) = u(t)$]

$$f(3t) \leftrightarrow \frac{1}{3}\frac{1}{s/3 + 1} = \frac{1}{s + 3}$$

This is a result we know to be true.

EXAMPLE 4-3

Prove the time-shifting and s-shifting theorems.

Solution.   By definition:

$$L[f(t - a)u(t - a)] = \int_{0^-}^{\infty} f(t - a)u(t - a)e^{-st}\, dt$$

$$= \int_{a}^{\infty} f(t - a)e^{-st}\, dt$$

Let $t - a = p$

$$dt = dp \qquad \text{as } a < t < \infty, 0 < p < \infty$$

The RHS becomes:

$$\text{RHS} = \int_0^\infty f(p)e^{-(p+a)s}\, dp$$

$$= e^{-as}F(s)$$

We note, of course, that in both the time-scaling and time-shifting theorems a must be positive or our function would be reflected in one case and shifted left in the other.

For the "s-shifting" theorem, by definition we have:

$$e^{\alpha t}f(t) \longleftrightarrow \int_{0^-}^\infty e^{\alpha t}f(t)e^{-st}\, dt$$

$$= \int_{0^-}^\infty f(t)e^{-(s-\alpha)t}\, dt$$

$$= F(s - \alpha)$$

EXAMPLE 4-4

Use the basic transforms of Table 4-1 plus the shifting theorems to find the Laplace transforms of:

(a) $(t - 2)u(t - 2)$
(b) $tu(t - 2)$
(c) $e^{-5t}u(t - 4)$
(d) te^{-3t}
(e) $e^{-2t} \text{Cos } 4t$

Solution
(a) Obviously:

$$(t - 2)u(t - 2) \longleftrightarrow e^{-2s}\frac{1}{s^2}$$

(b) $tu(t - 2) = (t - 2)u(t - 2) + 2u(t - 2)$

Therefore $\qquad\qquad tu(t - 2) \longleftrightarrow e^{-2s}\frac{1}{s^2} + 2e^{-2s}\frac{1}{s}$

$$= \left(\frac{2s + 1}{s^2}\right)e^{-2s}$$

(c) To use the shifting theorem, we write:

$$e^{-5t}u(t - 4) = e^{-20}e^{-5(t-4)}u(t - 4)$$

Therefore $\qquad L[e^{-5t}u(t - 4)] = \left[e^{-20}\frac{1}{s + 5}\right]e^{-4s}$

(d) $tu(t) \leftrightarrow 1/s^2$

and by the s-shifting theorem:

$$te^{-3t} \leftrightarrow \frac{1}{(s+3)^2}$$

(e) $\text{Cos } 4t \leftrightarrow s/(s^2 + 4^2)$

and

$$e^{-2t} \text{Cos } 4t \leftrightarrow \frac{s+2}{(s+2)^2 + 4^2}$$

$$= \frac{s+2}{s^2 + 4s + 20}$$

Two very important transforms are:

$$L[e^{-\alpha t} \text{Cos } \beta t] = \frac{s + \alpha}{(s + \alpha)^2 + \beta^2}$$

and

$$L[e^{-\alpha t} \text{Sin } \beta t] = \frac{\beta}{(s + \alpha)^2 + \beta^2}$$

These will allow us to find rapidly the inverse Laplace transform of any $F(s)$ with a denominator polynomial consisting of two complex conjugate roots.

EXAMPLE 4-5
  (a) Prove the convolution theorem.
  (b) Evaluate $4tu(t)*3e^{-2t}u(t)$.

Solution

(a) $L[f(t)*g(t)] = L\left[\int_{-\infty}^{\infty} f(p)g(t-p)\,dp\right]$

Since $f(t)$ and $g(t)$ are causal functions, the limits become 0 to t or 0 to ∞ as $g(t-p) = 0$ for $t < p < \infty$.

Therefore $L[f(t)*g(t)] = \int_0^{\infty}\left[\int_0^{\infty} f(p)g(t-p)\,dp\right]e^{-st}\,dt$

Assuming it is permissible to interchange the order of integration, then

$$\text{RHS} = \int_0^{\infty} f(p)\left[\int_0^{\infty} g(t-p)e^{-st}\,dt\right]dp$$

Now letting $u = t - p$

$$du = dt, \qquad \text{and as } 0 < t < \infty, \; -p < u < \infty.$$

and $\text{RHS} = \int_0^{\infty} f(p)\left[\int_{-p}^{\infty} g(u)e^{-s(u+p)}\,du\right]dp$

$$= \int_0^\infty f(p)e^{-sp}\left[\int_0^\infty g(u)e^{-su}\,du\right]dp$$

since $g(u)$ causal

$$= F(s)G(s)$$

This is very important derivation and should be thoroughly understood.

(b) $\overline{4tu(t)*3e^{-2t}u(t)} = \dfrac{4}{s^2}\dfrac{3}{s+2}$

$$= \frac{12}{s^2(s+2)}; \qquad \sigma > 0$$

The convolution theorem is the most important theorem in system analysis. If it is easier to find the inverse transform of $F(s)G(s)$ than to convolve $f(t)$ and $g(t)$, then we have found a simple way to convolve functions. The limitation on this will be that the functions must have Laplace transforms that are the ratio of two polynomials. If such is not the case we will normally use the fast Fourier transform and a computer to evaluate the convolution integral.

4-2-1 The Derivative Theorem

We now prove and discuss the derivative theorems and pay special attention to the occurrence of singularity functions in derivatives.

Derivation of the Derivative Theorem.   We wish to derive the derivative theorem:

if $\qquad\qquad\qquad\qquad f(t) \rightarrow F(s)$

then $\qquad\qquad\qquad L[f'(t)] = sF(s) - f(0^-)$ $\qquad\qquad$ (4-3)

Derivation

$$L[f'(t)] = \int_{0^-}^\infty f'(t)e^{-st}\,dt$$

$$= \int_{0^-}^\infty e^{-st}\,d[f(t)]$$

$$= e^{-st}f(t)\,|_{0^-}^\infty - \int_{0^-}^\infty f(t)(-se^{-st})\,dt$$

$$= 0 - f(0^-) + sF(s)$$

In order to appreciate this theorem we will solve a number of examples to indicate the necessity of using $f(0^-)$ when $f(t)$ is not continuous at $t = 0$.

EXAMPLE 4-6

Use the derivative theorem to find the following transforms and in each case check your answer by differentiating $f(t)$ and finding its transform directly.

(a) $d/dt\ [u(t)]$
(b) $d/dt\ [3e^{-2t}u(t)]$
(c) $d/dt\ [3e^{-2t}]$
(d) $d/dt\ [2\ \text{Cos}\ 4t\ u(t)]$

Solution

(a) Using the derivative theorem, we obtain:

$$L\left[\frac{d}{dt}\ u(t)\right] = s\frac{1}{s} - 0$$

$$= 1$$

Since $[(d/dt)\ u(t)] = \delta(t)$, we verify:

$$L[\delta(t)] = 1$$

(b) $L\left[\dfrac{d}{dt}\ (3e^{-2t}u(t))\right] = s\left[\dfrac{3}{s+2}\right] - 0$

$$= \frac{3s}{s+2} \tag{1}$$

using the derivative theorem. From Chapter 2:

$$\frac{d}{dt}[3e^{-2t}u(t)] = -6e^{-2t}u(t) + 3\delta(t) \qquad [\textit{Note:}\ 3e^{-2(0)} = 3.]$$

and finding the transform directly, we have:

$$L[-6e^{-2t}u(t) + 3\delta(t)] = \frac{-6}{s+2} + 3$$

$$= \frac{3s}{s+2} \tag{2}$$

which verifies our result in (1).

(c) $L\left[\dfrac{d}{dt}\ (3e^{-2t})\right] = s\left(\dfrac{3}{s+2}\right) - 3$

$$= \frac{3s}{s+2} - \frac{3s+6}{s+2}$$

$$= \frac{-6}{s+2}$$

Since $[d/dt(3e^{-2t})] = -6e^{-2t}$, we verify:

$$L[-6e^{-2t}] = \frac{-6}{s+2}$$

(d) $L\left[\dfrac{d}{dt}(2\cos 4t\, u(t))\right] = \dfrac{2s^2}{s^2+16} - 0$

$$= \frac{2s^2}{s^2+16} \tag{1}$$

using the derivative theorem. Differentiating, we obtain:

$$\frac{d}{dt}(2\cos 4t\, u(t)) = -8\sin 4t\, u(t) + 2\delta(t)$$

and its Laplace transform is:

$$L[-8\sin 4t + 2\delta(t)] = \frac{-32}{s^2+16} + 2$$

$$= \frac{2s^2}{s^2+16} \tag{2}$$

as found by using the derivative theorem in (1).

4-2-2 Equivalent Versions When Derivatives Have No Singularities

If $f(t)$ and its derivative are continuous at $t = 0$, then the derivative theorem may use $f(0^+), f'(0^+)$, and so on instead of $f(0^-), f'(0^-)$, and so on, and

$$L[f'(t)] = sF(s) - f(0^+) \qquad \text{if } f'(t) \text{ is continuous at } t = 0 \tag{4-4}$$

Reconsidering Example 4-6, we see that Equation 4-4 gives the correct result for only part (c). There seems to be too much controversy about this theorem, which really stems from its main applications which are:

1. solving differential equations
2. finding impulse responses
3. finding complete responses when switching occurs at $t = 0$

These applications are treated in Section 4-4.

4-2-3 Extensions of the Derivative Theorem

EXAMPLE 4-7

Use induction to extend the formula for $f'(t)$ to find the Laplace transform of $f''(t), f'''(t)$, and $f^n(t)$.

Solution

$$L[f''(t)] = L\left[\frac{d}{dt}f'(t)\right]$$

$$= s[sF(s) - f(0^-)] - f'(0^-)$$

$$= s^2 F(s) - sf(0^-) - f'(0^-) \tag{4-5}$$

Using the derivative theorem for $d/dt\,[f'(t)]$:

$$L[f'''(t)] = L\left[\frac{d}{dt}f''(t)\right]$$

$$= s[s^2 F(s) - sf(0^-) - f'(0^-)] - f''(0^-)$$

$$= s^3 F(s) - s^2 f(0^-) - sf'(0^-) - f''(0^-) \tag{4-6}$$

Inductively:

$$L[f^n(t)] = s^n F(s) - s^{n-1} f(0^-) - s^{n-2} f'(0^-) - \cdots - f^{n-1}(0^-)$$

$$= s^n F(s) - \sum_{p-0}^{n-1} s^{n-p-1} f^p(0^-) \tag{4-7}$$

One of the main applications of the one-sided Laplace transform is solving linear differential equations. We will demonstrate by an example the setting up of the solution but will have to wait until after the section on inverse transforms to complete the problem.

EXAMPLE 4-8

Set up the solution to:

$$y''(t) + 4y'(t) + 3y(t) = t$$

given:
$$y(0) = 2, \qquad y'(0) = -3$$

Solution. Taking the Laplace transform of both sides of the equation, we obtain:

$$s^2 Y(s) - 2s + 3 + 4sY(s) - 4(2) + 3Y(s) = \frac{1}{s^2}$$

$$(s^2 + 4s + 3)Y(s) = \frac{1}{s^2} + 2s + 5$$

Therefore
$$Y(s) = \frac{1 + s^2(2s + 5)}{(s^2 + 4s + 3)s^2}$$

and
$$y(t) = L^{-1}[Y(s)]$$

which we will evaluate when we encounter inverse transforms in Section 4-3.

Drill Set: Laplace Transforms and Theorems

1. Evaluate the Laplace transform and state the region of convergence for:
   - (a) Sinh $\alpha t\, u(t)$,   Cosh $\alpha t\, u(t)$
   - (b) $t^2 \sin t\, u(t)$, $(t^2 + 3)u(t - 4)$
2. (a) Prove the initial and final value theorems.
   - (b) Use the initial value theorem to verify the initial value of 4 Cos $5t$ $u(t)$. What happens if you apply it to $2\delta(t)$?
   - (c) Use the final value theorem to verify the final value of $te^{-2t}u(t)$. What happens if you apply it to Cos $4t\, u(t)$?

4-3 THE INVERSE LAPLACE TRANSFORM

The definition of the inverse Laplace transform (see Appendix A) is:

$$f(t) = \frac{1}{2\pi j} \int_{\substack{\sigma - j\infty \\ C_1}}^{\sigma + j\infty} F(s)e^{st}\, ds, \qquad \sigma > \sigma_1,$$

where C_1 is a straight line with path defined by:

$$z(\theta) = \sigma + j\theta, \quad -\infty < \theta < \infty$$

Using Jordan's lemma, we obtain:

$$f(t) = \frac{1}{2\pi j} \oint_C F(s)e^{st}\, ds \tag{4-8}$$

where C encloses all the poles of $F(s)$.

$$\dot{f}(t) = \Sigma\,[\text{residues of the poles of } F(s)e^{st}], \qquad t > 0 \tag{4-9}$$

$$= 0, \qquad t < 0 \tag{4-10}$$

Also, the uniqueness theorem for the Laplace transform is:

$$f(t) = \frac{1}{2\pi j} \int_{\sigma - j\infty}^{\sigma + j\infty} \left[\int_{0^-}^{\infty} f(p)e^{-sp}\, dp \right] e^{st}\, ds \tag{4-11}$$

which symbolically is:

$$L^{-1}[L[f(t)]] = f(t)$$

and this says that there is a unique transform corresponding to a causal time function and vice versa. Developing expression (4-8) using the complex variable theory will be deferred until Chapter 5. We will find inverse transforms by representing $F(s)$ in a form where $f(t)$ may be found by referring to Tables 4-1 and 4-2. To do this we apply two techniques.

Technique 1.   Intuitively adjust $F(s)$ so that the theorems and tables will allow us to state $f(t)$. This is very simple through the case of second-order polynomials in the denominator.

Technique 2.   Use "partial fraction theory" to expand $F(s)$ into terms whose inverse transforms are easily recognizable.

Technique 1 is really only used to increase our fluency and confidence.

EXAMPLE 4-9

Find the inverse transforms of the following, making adjustments to facilitate using Tables 4-1 and 4-2:

(a) $\dfrac{3s + 2}{4s + 5}$

(b) $\dfrac{3s + 2}{s^2}$

(c) $\dfrac{3s + 2}{s^2 + 16}$

(d) $\dfrac{3s + 2}{s^2 - 16}$

(e) $\dfrac{3s + 2}{s^2 + 2s + 5}$

(f) $\dfrac{3s + 2}{s^2 + 8s + 4}$

Solution

(a) $\dfrac{3s + 2}{4s + 5} = \dfrac{0.75(4s + 5)}{4s + 5} + \dfrac{2 - 3.75}{4s + 5}$

$\qquad\qquad = 0.75 + \dfrac{-0.44}{s + 1.25}$

Therefore $f(t) = 0.75\delta(t) - 0.44e^{-1.25t}u(t)$

(b) $\dfrac{3s + 2}{s^2} = \dfrac{3}{s} + \dfrac{2}{s^2}$

Thus $f(t) = (3 + 2t)u(t)$

(c) $\dfrac{3s + 2}{s^2 + 16} = \dfrac{3s}{s^2 + 16} + \dfrac{2}{4}\dfrac{4}{s^2 + 16}$

Therefore $f(t) = (3 \cos 4t + 0.5 \sin 4t)u(t)$

(d) $\dfrac{3s + 2}{s^2 - 16}$ = $(3 \text{ Cosh } 4t + 0.5 \text{ Sinh } 4t)u(t)$ (from Drill Set)

$$= [1.5(e^{4t} + e^{-4t}) + 0.25(e^{4t} - e^{-4t})]u(t)$$

$$= [1.75e^{4t} + 1.25e^{-4t}]u(t)$$

(e) $\dfrac{3s + 2}{s^2 + 2s + 5}$ = $\dfrac{3s + 2}{(s + 1)^2 + 4}$

$$= \dfrac{3(s + 1)}{(s + 1)^2 + 2^2} - \dfrac{1}{(s + 1)^2 + 2^2}$$

Thus $f(t) = [3e^{-t} \text{ Cos } 2t - 0.5e^{-t} \text{ Sin } 2t]u(t)$.
To show appreciation of the solution, the reader should immediately
write out the inverse transform of the general expression:

$$\dfrac{\bar{A}_1 s + A_2}{(s + \alpha)^2 + \beta^2}$$

where A_1, A_2, α, and β are constants.

(f) $\dfrac{3s + 2}{s^2 + 8s + 4}$ = $\dfrac{3s + 2}{(s + 4)^2 - (\sqrt{12})^2}$

$$= \dfrac{3(s + 4)}{(s + 4)^2 - (3.46)^2} - \dfrac{10}{(s + 4)^2 - (3.46)^2}$$

$$= 3e^{-4t} \text{ Cosh } 3.46t - \dfrac{10}{3.46} e^{-4t} \text{ Sinh } 3.46t$$

$$= 1.5e^{-4t} [e^{3.46t} + e^{-3.46t}] - 1.45e^{-4t}[e^{3.46t} - e^{-3.46t}]$$

$$= 1.5(e^{-0.54t} + e^{-7.46t}) - 1.45(e^{-0.54t} - e^{-7.46t})$$

$$= (0.05e^{-0.54t} + 2.95e^{-7.46t})u(t)$$

Rarely will a problem with two real roots in the denominator be
solved in this manner since a partial fraction solution will be easier to use.

We have included this example because, with the tremendous urgency to
teach transforms today, the result is a need for more drill and knowledge of
fundamentals. Students learning this material should be able to apply it, study it
in depth, and almost feel as if they were inventing the logical approach.

4-3-1 Inverse Transforms Using Partial Fractions

It is assumed that the reader is already familiar with partial fractions from algebra. We will systematically cover, via examples, and summarize, in general, the evaluation of inverse transforms for a function $F(s) = N(s)/D(s)$ which is the ratio of two polynomials where the order of $D(s)$ is *higher* than $N(s)$.

Case of Only First-Order Real Poles
Given:

$$F(s) = \frac{N(s)}{(s - \alpha_1)(s - \alpha_2) \cdots (s - \alpha_n)} \qquad (4\text{-}12)$$

where $N(s)$ does not have a root at α_i, $i = 1, \ldots, n$; then $F(s)$ may be expanded in partial fractions as:

$$F(s) = \frac{A_1}{(s - \alpha_1)} + \frac{A_2}{(s - \alpha_2)} + \cdots + \frac{A_n}{(s - \alpha_n)} \qquad (4\text{-}13)$$

To find A_i, multiply both sides by $(s - \alpha_i)$ and let $s = \alpha_i$. This yields:

$$A_i = (s - \alpha_i)F(s) \Big|_{s=\alpha_i} \qquad (4\text{-}14)$$

EXAMPLE 4-10
Find the inverse Laplace transform of:

$$F(s) = \frac{s^2 + 2s + 2}{(s + 1)(s + 2)(s + 3)}$$

Solution

$$F(s) = \frac{A_1}{s + 1} + \frac{A_2}{s + 2} + \frac{A_3}{s + 3}$$

$$= \frac{(s^2 + 2s + 2)/(s + 2)(s + 3)}{s + 1} \Bigg|_{s=-1}$$

$$+ \frac{(s^2 + 2s + 2)/(s + 1)(s + 3)}{s + 2} \Bigg|_{s=-2}$$

$$+ \frac{(s^2 + 2s + 2)/(s + 1)(s + 2)}{s + 3} \Bigg|_{s=-3}$$

$$= \frac{0.5}{s + 1} + \frac{-2}{s + 2} + \frac{2.5}{s + 3}$$

and by table recognition:

$$f(t) = (0.5e^{-t} - 2e^{-2t} + 2.5e^{-3t})u(t)$$

Case of Higher-Order Real Poles

EXAMPLE 4-11

Express

$$F(s) = \frac{3s^2 + 4s + 2}{(s + 1)^2(s + 3)}$$

in partial fractions and hence find its inverse transform.

Solution. We should check the factors of the numerator:

$$3(s^2 + 1.33s + 0.67) = 3[(s + 0.67)^2 + 0.67 - 0.67^2]$$
$$= 3(s + 0.67 + j0.47)(s + 0.67 - j0.47)$$

Therefore none of the zeros of $F(s)$ cancels the poles. We write:

$$\frac{3s^2 + 4s + 2}{(s + 1)^2(s + 3)} = \frac{A_1}{s + 3} + \frac{A_2}{s + 1} + \frac{A_3}{(s + 1)^2}$$

The reader should be philosophically convinced why the term $A_3/(s + 1)^2$ suffices and $(A_3 + A_4 s)/(s + 1)^2$ is not required. By inspection:

$$A_1 = \frac{3(9) + 4(-3) + 2}{(-3 + 1)^2} = 4.25$$

and multiplying both sides by $(s + 1)^2$, and letting $s = -1$, we obtain A_3 as:

$$A_3 = \frac{3s^2 + 4s + 2}{s + 3}\bigg|_{s=-1} = 0.5$$

Therefore
$$\frac{3s^2 + 4s + 2}{(s + 1)^2(s + 3)} = \frac{4.25}{s + 3} + \frac{A_2}{s + 1} + \frac{0.5}{(s + 1)^2}$$

which is true for all s. Letting $s = 0$, we have:

$$\frac{2}{3} = \frac{4.25}{3} + A_2 + 0.5$$

and
$$A_2 = 0.67 - 1.42 - 0.5 = -1.25$$

Therefore
$$F(s) = \frac{4.25}{s + 3} - \frac{1.25}{s + 1} + \frac{0.5}{(s + 1)^2}$$

and by table recognition:

$$f(t) = [4.25e^{-3t} - 1.25e^{-t} + 0.5te^{-t}]u(t)$$

This was the quickest way to do the problem, but we did avoid the theory of a second-order pole. Backtracking we found that:

$$\frac{3s^2 + 4s + 2}{(s + 1)^2(s + 3)} = \frac{4.25}{s + 3} + \frac{A_2}{s + 1} + \frac{0.5}{(s + 1)^2}$$

Multiplying by $(s + 1)^2$ yields:

$$\frac{3s^2 + 4s + 2}{s + 3} = \frac{(4.25)}{s + 3}(s + 1)^2 + A_2(s + 1) + 0.5$$

Now if we differentiate with respect to s and then let $s = -1$, we obtain:

$$A_2 = \left[\frac{d}{ds}\frac{3s^2 + 4s + 2}{s + 3}\right]\Bigg|_{s=-1}$$

or

$$A_2 = \frac{d}{ds}\left[(s + 1)^2 F(s)\right]\Bigg|_{s=-1}$$

The reader should check that we get the same result $A_2 = -1.25$ with this formula.

We now consider the general theory of a higher-order pole in $F(s)$:
Given:

$$F(s) = \frac{N(s)}{D(s)} = \frac{N(s)}{(s - \alpha)^p D_1(s)} \tag{4-15}$$

where α is not a root of $N(s)$, we can expand in partial fractions as:

$$F(s) = \frac{A_1}{s - \alpha} + \frac{A_2}{(s - \alpha)^2} + \cdots + \frac{A_p}{(s - \alpha)^p} + K(s) \tag{4-16}$$

or

$$(s - \alpha)^p F(s) = A_1(s - \alpha)^{p-1} + A_2(s - \alpha)^{p-2} + \cdots$$
$$+ A_{p-1}(s - \alpha) + A_p + K(s)(s - \alpha)^p$$

Substituting $s = \alpha$, we have:

$$A_p = (s - \alpha)^p F(s)\Big|_{s=\alpha} \tag{4-17}$$

Now continually differentiating and then substituting $s = \alpha$, we obtain:

first

$$A_{p-1} = \frac{d}{ds}[(s - \alpha)^p F(s)]\Big|_{s=\alpha} \tag{4-18}$$

then

$$A_{p-2} = \frac{1}{2}\frac{d^2}{ds^2}[(s - \alpha)^p F(s)]\Big|_{s=\alpha} \tag{4-19}$$

and, in general:

$$A_{p-m} = \frac{1}{m!}\frac{d^m}{ds^m}[(s - \alpha)^p F(s)]\Big|_{s=\alpha} \tag{4-20}$$

for $m = 0$ to $p - 1$. Finally, for a general integer n where $n \leq p$:

$$A_n = A_{p-(p-n)} = \frac{1}{(p - n)!}\frac{d^{p-n}}{ds^{p-n}}[(s - \alpha)^p F(s)]\Big|_{s=\alpha} \tag{4-21}$$

The symbol $m!$ denotes factorial m and is defined as $m! = m \times (m - 1) \times \cdots \times 2 \times 1$ for m a positive integer (e.g., $4! = 24$, $6! = 720$, etc.)

Case of Conjugate Imaginary Poles in F(s). Given:

$$F(s) = \frac{N(s)}{D(s)} = \frac{N(s)}{(s^2 + \beta^2)D_1(s)}$$

We know that $f(t)$ will contain $(A_1 \cos \beta t + A_2 \sin \beta t)u(t)$ if $s^2 + \beta^2$ is not a factor of $N(s)$. Using our intuitive skills, we now solve a problem quickly and then more slowly develop the theory for conjugate imaginary roots.

EXAMPLE 4-12

Find the inverse transform of:

$$F(s) = \frac{2s + 3}{(s + 1)(s^2 + 4)}$$

Solution

$$F(s) = \frac{0.2}{s + 1} + \frac{As + B}{s^2 + 4}$$

Therefore $\quad \dfrac{2s + 3}{(s + 1)(s^2 + 4)} = \dfrac{0.2(s^2 + 4) + (As + B)(s + 1)}{(s + 1)(s^2 + 4)}$

and equating coefficients, we get:

$$0.2 + A = 0$$

$$A + B = 2$$

$$0.8 + B = 3$$

which gives $A = -0.2$, $B = 2.2$, and a check on our work:

$$\frac{2s + 3}{(s + 1)(s^2 + 4)} = \frac{0.2}{s + 1} + \frac{-0.2s + 2.2}{s^2 + 4}$$

and with confidence:

$$f(t) = [0.2e^{-t} - 0.2 \cos 2t + 1.1 \sin 2t]u(t)$$

We now want to develop the theory of conjugate imaginary roots in the denominator. Using partial fractions, we obtain:

$$\frac{2s + 3}{(s + 1)(s^2 + 4)} = \frac{A_1}{s + 1} + \frac{A_2}{s + j2} + \frac{A_3}{s - j2}$$

where $\quad A_1 = 0.2$

$$A_2 = \frac{2(-j2) + 3}{(1 - j2)(-j4)} = \frac{(3 - j4)(-8 + j4)}{80}$$

$$= -0.1 + j0.55$$

$$A_3 = \frac{2(j2) + 3}{(1 + j2)(j4)} = A_2^* = -0.1 - j0.55$$

Therefore
$$F(s) = \frac{0.2}{s + 1} + \frac{-0.1 + j0.55}{s + j2} + \frac{-0.1 - j0.55}{s - j2}$$

$$f(t) = 0.2e^{-t} + (-0.1 + j0.55)e^{-j2t} + (-0.1 - j0.55)e^{j2t}$$

$$= 0.2e^{-t} + [-0.1 \text{ Cos } 2t + 0.55 \text{ Sin } 2t + j \text{ terms}]$$

$$+ [-0.1 \text{ Cos } 2t + 0.55 \text{ Sin } 2t - j \text{ terms}]$$

$$= [0.2e^{-t} - 0.2 \text{ Cos } 2t + 1.1 \text{ Sin } 2t]u(t)$$

as the imaginary terms cancel to yield zero. Observing our calculation we note that from $A_2 = -0.1 + j0.55 = a + jb$ we can immediately write the contribution of both $A_2/(s + j2)$ and $A_2^*/(s - j2)$ as $2a \text{ Cos } 2t + 2b \text{ Sin } 2t$.

General Summary of Conjugate Imaginary Poles.   Given:

$$F(s) = \frac{N(s)}{(s^2 + \beta^2)D_1(s)}$$

it is possible to expand this equation to write:

$$F(s) = \frac{a + jb}{s + j\beta} + \frac{a - jb}{s - j\beta} + K(s) \tag{4-22}$$

where
$$a + jb = (s + j\beta)F(s)|_{s = -j\beta} \tag{4-23}$$

and then:

$$f(t) = [2a \text{ Cos } \beta t + 2b \text{ Sin } \beta t]u(t) + L^{-1}[K(s)]$$

Case of Complex Conjugate Poles.   Given:

$$F(s) = \frac{N(s)}{D(s)} = \frac{N(s)}{[(s + \alpha)^2 + \beta^2]D_1(s)}$$

where $(s + \alpha)^2 + \beta^2$ is not a factor of $N(s)$, we know $f(t)$ will contain $[A_1 e^{-\alpha t} \text{ Cos } \beta t + A_2 e^{-\alpha t} \text{ Sin } \beta t]u(t)$. Proceeding as in the case of conjugate imaginary poles, we solve a problem and then develop the general theory.

EXAMPLE 4-13

Given:

$$F(s) = \frac{2s + 3}{s^3 + 5s^2 + 9s + 5}$$

Find $f(t)$ rapidly and then use an alternative approach to develop the theory of complex conjugate roots.

Solution. $D(s)$ has a root at $s = -1$.

Therefore
$$\frac{2s + 3}{s^3 + 5s^2 + 9s + 5} = \frac{2s + 3}{(s + 1)(s^2 + 4s + 5)}$$

$$= \frac{0.5}{s + 1} + \frac{As + B}{s^2 + 4s + 5}$$

$$\frac{2s + 3}{(s + 1)(s^2 + 4s + 5)} = \frac{0.5(s^2 + 4s + 5) + (As + B)(s + 1)}{(s + 1)(s^2 + 4s + 5)}$$

Equating coefficients for s^2, s, and the constant term, we have:

$$0.5 + A = 0$$

$$2 + A + B = 2$$

and
$$2.5 + B = 3$$

Solving, we obtain $A = -0.5$, $B = 0.5$, and a check on our work.

Therefore
$$\frac{2s + 3}{s^3 + 5s^2 + 9s + 5} = \frac{0.5}{s + 1} + \frac{-0.5s + 0.5}{s^2 + 4s + 5}$$

$$= \frac{0.5}{s + 1} + \frac{-0.5(s + 2)}{(s + 2)^2 + 1}$$

$$+ 1.5 \frac{1}{(s + 2)^2 + 1}$$

and confidently we state:

$$f(t) = [0.5e^{-t} - 0.5e^{-2t} \cos t + 1.5e^{-2t} \sin t] u(t)$$

Alternative
We now want to develop the theory of complex conjugate roots:

$$\frac{2s + 3}{(s + 1)(s + 2 + j)(s + 2 - j)} = \frac{A_1}{s + 1} + \frac{A_2}{s + 2 + j} + \frac{A_3}{s + 2 - j}$$

$$A_1 = 0.5$$

$$A_2 = \frac{2(-2 - j) + 3}{(-1 - j)(-j2)} = \frac{(-1 - j2)(-2 - j2)}{8}$$

$$= -0.25 + j0.75$$

$$A_3 = \frac{2(-2 + j) + 3}{(-1 + j)(j2)} = A_2^* = -0.25 - j0.75$$

Therefore
$$F(s) = \frac{0.5}{s + 1} + \frac{-0.25 + j0.75}{s + 2 + j} + \frac{-0.25 - j0.75}{s + 2 - j}$$

and
$$f(t) = 0.5e^{-t} + (-0.25 + j0.75)e^{-2t}e^{-jt}$$
$$+ (-0.25 - j0.75)e^{-2t}e^{jt}$$
$$= [0.5e^{-t} + e^{-2t}(-0.5 \cos t + 1.5 \sin t)]u(t)$$

using the theory of conjugate imaginary roots.

General Summary for Complex Conjugate Roots.   Given:

$$F(s) = \frac{N(s)}{[(s + \alpha)^2 + \beta^2]D_1(s)}$$

it is possible to expand this equation to:

$$F(s) = \frac{a + jb}{(s + \alpha + j\beta)} + \frac{a - jb}{(s + \alpha - j\beta)} + K(s) \qquad (4\text{-}24)$$

where $a + jb = (s + \alpha + j\beta)F(s)|_{s=-\alpha-j\beta}$ $\qquad (4\text{-}25)$

and $f(t) = e^{-\alpha t}[2a \cos \beta t + 2b \sin \beta t]u(t) + L^{-1}[K(s)]$

The next logical inverse transforms to consider are the cases of terms such as $(s^2 + \beta^2)^p$ and $[(s + \alpha)^2 + \beta^2]^p$ in the denominator of $F(s)$. For example, $(s^2 + \beta^2)^p$ gives rise to a term of the form:

$$f(t) = (a_1 + a_2t + \cdots + a_pt^p) \cos \beta t + (b_1 + b_2t + \cdots + b_pt^p) \sin \beta t$$

Rarely, however, do we evaluate these coefficients by hand.

Drill Set: Inverse Transforms

1. Find the inverse transforms of the following both directly and by using partial fractions:

   (a) $\dfrac{6s + 3}{s^2 + 9}$

   (b) $\dfrac{6s + 3}{s^2 - 9}$

   (c) $\dfrac{6s + 3}{s^2 + 2s + 1}$

   (d) $\dfrac{6s + 3}{s^2 + 2s + 0.5}$

   (e) $\dfrac{6s + 3}{s^2 + 2s + 8}$

4-4 APPLICATIONS OF THE LAPLACE TRANSFORM

In this section we now consider three important applications of the one-sided Laplace transform. These are:

1. the solution of linear differential equations and finding impulse responses
2. finding complete responses in a circuit
3. system analysis

4-4-1 The Solution of Linear Differential Equations

The procedure which is very straightforward will be demonstrated by a few examples.

EXAMPLE 4-14
  Solve:

$$y''(t) + 4y'(t) + 3y(t) = 2$$

given:

$$y(0) = 2, \qquad y'(0) = 1$$

Solution.   Taking the Laplace transform of both sides yields:

$$s^2 Y(s) - s(2) - 1 + 4[sY(s) - 2] + 3Y(s) = \frac{2}{s}$$

$$(s^2 + 4s + 3)Y(s) = \frac{2}{s} + 2s + 9$$

$$Y(s) = \frac{2 + (2s + 9)s}{s(s + 1)(s + 3)}$$

$$= \frac{0.67}{s} + \frac{2.5}{s + 1} + \frac{-1.17}{s + 3}$$

and $\qquad\qquad y(t) = 0.67 + 2.5e^{-t} - 1.17e^{-3t}$

Note: We should be very clear that this solution satisfies the differential equation for all t, $-\infty < t < \infty$. Whether we say $y(0^+) = 2$, $y'(0^+) = 1$, or $y(0^-) = 2$, $y'(0^-) = 1$, or $y(0) = 2$, $y(3) = 0.67 + 2.5e^{-3} - 1.17e^{-9}$ we obtain the same result. All solutions pertaining to a differential equation with any given initial conditions generate a solution that is continuous for all time.

EXAMPLE 4-15

Find the impulse response $h(t)$ for the system governed by:

$$y'(t) + 3y(t) = 2x(t) + x'(t)$$

Solution.   We want to solve:

$$y'(t) + 3y(t) = 2\delta(t) + \delta'(t)$$

Taking the Laplace transform of both sides yields [(using $y(0^-) = 0$)]:

$$(s + 3)Y(s) = 2 + s$$

Therefore
$$Y(s) = \frac{s + 2}{s + 3}$$

$$= 1 - \frac{1}{s + 3}$$

and
$$h(t) = \delta(t) - e^{-3t}u(t)$$

Since the impulse response is the solution obtained when $x(t) = \delta(t)$ and $y^p(0^-) = 0$ for all p, we obtain that unique discontinuous solution that is zero for $t < 0$.

4-4-2 Finding Complete Responses in a Circuit

Figure 4-1(a) shows the time-domain definitions of the circuit elements. We now derive the relations between the Laplace transform of the voltage and the Laplace transform of the current for each of the elements.

For an element with resistance R obviously $V(s) = RI(s)$. For an element with inductance L henries.

$$v(t) = L\frac{di}{dt}$$

Therefore
$$V(s) = L[sI(s) - i_L(0^-)]$$

$$= LsI(s) - Li_L(0^+)$$

We note for an inductor $i_L(0^+) = i_L(0^-)$. The Laplace transform model relating $V(s)$ to $I(s)$ is shown in Figure 4-2(b) and consists of a zero initial energy inductor in series with a voltage source. For an element with capacitance C farads:

$$v(t) = \frac{1}{C}\int_0^t i(\alpha)\, d\alpha + v_c(0^+)$$

assuming by the conservation of energy that

$$v_c(0^+) = v_c(0^-)$$

and
$$V(s) = \frac{1}{Cs}I(s) + \frac{v_c(0^+)}{s}$$

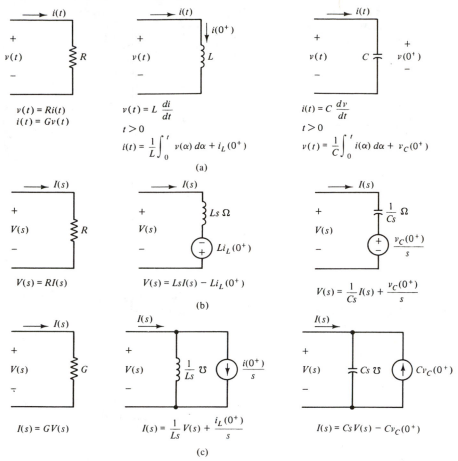

$$v(t) = Ri(t)$$
$$i(t) = Gv(t)$$

$$v(t) = L \frac{di}{dt}$$
$$t > 0$$
$$i(t) = \frac{1}{L} \int_0^t v(\alpha) \, d\alpha + i_L(0^+)$$

$$i(t) = C \frac{dv}{dt}$$
$$t > 0$$
$$v(t) = \frac{1}{C} \int_0^t i(\alpha) \, d\alpha + v_C(0^+)$$

(a)

$$V(s) = RI(s)$$

$$V(s) = LsI(s) - Li_L(0^+)$$

(b)

$$V(s) = \frac{1}{Cs}I(s) + \frac{v_C(0^+)}{s}$$

$$I(s) = GV(s)$$

$$I(s) = \frac{1}{Ls}V(s) + \frac{i_L(0^+)}{s}$$

$$I(s) = CsV(s) - Cv_C(0^+)$$

(c)

Figure 4-1 (a) Time-domain definitions of the network elements; (b) Laplace transform models for loop analysis; (c) Laplace transform models for nodal analysis.

The Laplace transform model relating $V(s)$ and $I(s)$, shown in Figure 4-1(b), consists of a zero initial energy capacitor in series with a voltage source.

In an analogous manner the relation between $I(s)$ and $V(s)$ may be found and the transform models are shown in Figure 4-1(c). For the inductor and capacitor the models consist of parallel combinations of zero initial energy elements and current sources describing the initial conditions. Based on our knowledge of circuit analysis it is obvious we will use the series models for loop analysis and the parallel models for nodal analysis.

EXAMPLE 4-16

Find $v(t)$ for $t > 0$ in the network of Figure 4-2(a).

Solution. At $t = 0^-$ just before the switch moves from position 1 we assume steady-state conditions. Using dc forced response analysis, we

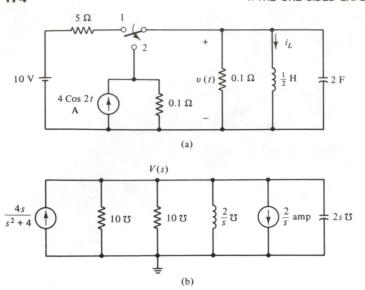

(a)

(b)

Figure 4-2 (a) The circuit for Example 4-17. At $t = 0$ the switch moves from position 1 to position 2; (b) the Laplace transform model for nodal analysis.

have:

$$i_L(0^-) = \frac{10}{5} = 2 \text{ amp}$$

and

$$v_c(0^-) = 0 \text{ V}$$

By the conservation of energy $i_L(0^+) = 2$ and $v_c(0^+) = 0$. Figure 4-2(b) shows a Laplace transform model suitable for nodal analysis incorporating the initial conditions.

KCL at Node 1

$$\left(2s + 20 + \frac{2}{s}\right) V(s) = \frac{4s}{s^2 + 4} - \frac{2}{s}$$

$$V(s) = \frac{4s^2 - 2(s^2 + 4)}{2(s^2 + 10s + 1)(s^2 + 4)}$$

$$V(s) = \frac{2s^2 - (s^2 + 4)}{(s + j2)(s - j2)(s + 0.1)(s + 9.9)}$$

$$= \frac{c_1}{s + j2} + \frac{c_1^*}{s - j2} + \frac{c_2}{s + 0.1} + \frac{c_3}{s + 9.9}$$

where

$$c_1 = \frac{2(-j2)^2 - 0}{(-j4)(-j2 + 0.1)(-j2 + 9.9)}$$

$$= 0.1 + j0.015 \qquad \text{(with work)}$$

$$c_2 = \frac{2(-0.1)^2 - [(-0.1)^2 + 4]}{[(-0.1)^2 + 4](9.8)} \approx -0.11$$

$$c_3 = \frac{2(-9.9)^2 - [(-9.9)^2 + 4]}{[(-9.9)^2 + 4](-9.8)} = -0.08$$

Finally:

$$v(t) = [0.2 \, \text{Cos} \, 2t + 0.03 \, \text{Sin} \, 2t - 0.11e^{-0.1t} - 0.08e^{-9.9t}]u(t)$$

The initial value $v(0^+) = 0$ checks reasonably well although more care is expected on calculations made by a student than by a harried professor.

EXAMPLE 4-17

Set up an expression for $V_2(s)$ the Laplace transform of $v_2(t)$ in the circuit of Figure 4-3(a).

Solution. The solution is very straightforward but messy to do algebraically. Figure 4-3(b) shows the Laplace transform model and the governing nodal KCL equations are:

$$\begin{pmatrix} 1.5 + \dfrac{2}{s} & -1 \\[2mm] -1 & \dfrac{s}{3} + 2 \end{pmatrix} \begin{pmatrix} V_1(s) \\[2mm] V_2(s) \end{pmatrix} = \begin{pmatrix} \dfrac{3}{s} - \dfrac{2}{s} - \dfrac{2s}{s^2 + 1} \\[2mm] \dfrac{2}{3} + \dfrac{2s}{s^2 + 1} \end{pmatrix}$$

$$V_2(s) = \cfrac{\begin{vmatrix} 1.5 + \dfrac{2}{s} & \dfrac{1}{s} + \dfrac{2s}{s^2 + 1} \\[2mm] -1 & \dfrac{2}{3} + \dfrac{2s}{s^2 + 1} \end{vmatrix}}{\begin{vmatrix} 1.5 + \dfrac{2}{s} & -1 \\[2mm] -1 & \dfrac{s}{3} + 2 \end{vmatrix}}$$

$$= \frac{(1.5s + 2)[2s/3 + 2s^2/(s^2 + 1)] + s[1 + 2s^2/(s^2 + 1)]}{(1.5s + 2)(s^2/3 + 2s) - s^2}$$

$$= \frac{(1.5s + 2)(2s^3 + 2s + 6s^2) + 3s(s^2 + 1 + 2s^2)}{(s^2 + 1)[(1.5s + 2)(s^2 + 6s) - 3s^2]}$$

$$= \frac{(1.5s + 2)(2s^2 + 6s + 2) + 3(3s^2 + 1)}{(s^2 + 1)[(1.5s + 2)(s + 6) - 3s]}$$

$$= \frac{N(s)}{(s + j)(s - j)(s + 2.65 + j0.9)(s + 2.67 - j0.9)}$$

and with substantial work $v_2(t)$ may be found.

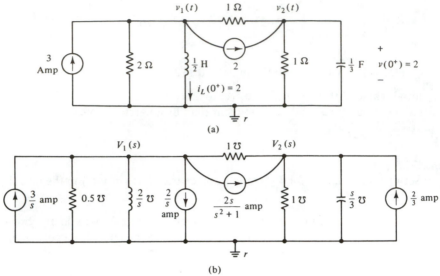

Figure 4-3 (a) The circuit for Example 4-18 with given initial conditions; (b) the Laplace transform model for nodal analysis.

Observing Example 4-17, we need to realize our analytical limitations. Doing an average second-order complete response problem like Example 4-17 and checking the answer is correct requires from 7 to 10 hours of work by an average junior or senior student. Obviously, such problems should be solved by computer means although it is instructive to do a few analytically.

4-4-3 System Analysis

The System Function H(s).   Given a linear time-invariant causal system (LTIC) is governed by:

$$a_n y^n(t) + \cdots + a_0 y(t) = b_0 x(t) + \cdots + b_m x^m(t) \qquad (4\text{-}26)$$

we define the system function $H(s)$ as:

$$H(s) = \frac{Y(s)}{X(s)} \qquad (4\text{-}27)$$

which is the Laplace transform of the output divided by the Laplace transform of the input with all the initial derivatives $y^p(0^-)$ equal to zero. Applying the definition to 4-26; we obtain:

$$(a_n s^n + a_{n-1} s^{n-1} + \cdots + a_0) Y(s) = (b_0 + b_1 s + \cdots + b_m s^m) X(s)$$

and
$$H(s) = \frac{b_0 + b_1 s + \cdots + b_m s^m}{a_0 + a_1 s + \cdots + a_n s^n} \qquad (4\text{-}28)$$

This is identical to the function obtained when $H(s)$ was defined as the forced

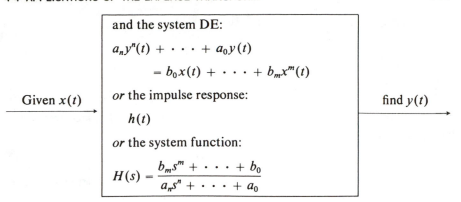

Given $x(t)$

and the system DE:

$$a_n y''(t) + \cdots + a_0 y(t)$$
$$= b_0 x(t) + \cdots + b_m x^m(t)$$

or the impulse response:

$$h(t)$$

or the system function:

$$H(s) = \frac{b_m s^m + \cdots + b_0}{a_n s^n + \cdots + a_0}$$

find $y(t)$

s Domain

$$H(s) \triangleq \frac{Y(s)}{X(s)} = \frac{b_0 + b_1 s + \cdots + b_m s^m}{a_0 + a_1 s + \cdots + a_n s^n} \tag{1}$$

$$Y(s) = H(s)X(s) \tag{2}$$

$$y_{zs}(t) = L^{-1}[H(s)X(s)] \tag{3}$$

$$h(t) = L^{-1}[H(s)]$$

Time Domain

$$H(s) \triangleq \frac{y_{f0}(t)}{x(t)}\bigg|_{x(t)-e^{st}} = \frac{b_0 + b_1 s + \cdots + b_m s^m}{a_0 + a_1 s + \cdots + a_n s^n} \tag{1'}$$

and $\quad a_n h^n(t) + \cdots + a_0 h(t) = b_0 \delta(t) + b_1 \delta'(t) + \cdots + b_m \delta^m(t) \tag{2'}$

from which $h(t)$ may be found.

Then $\qquad\qquad\qquad y_{zs}(t) = x(t) * h(t). \tag{3'}$

Figure 4-4 The transform and time-domain results for an LTIC system.

response when $x(t)$ is e^{st} divided by $x(t)$ in Chapter 2. The method for finding $H(s)$ and the system governing equation is identical in format to that used in Chapter 2. This is not surprising since the Laplace transform is the synthesis of a function into a string of exponentials.

The Zero-State Output.   Consider a system such as that shown in Figure 4-4. The Laplace transform of the zero-state output is:

$$Y(s) = H(s)X(s)$$

and $\qquad\qquad\qquad y(t) = L^{-1}[H(s)X(s)] \tag{4-29}$

which using the transform theory we find by taking partial fractions and referring to Tables 4-1 and 4-2. This corresponds to finding $x(t) * h(t)$ in the time domain.

H(s) and h(t).   If the input is $x(t) = \delta(t)$,

then
$$Y(s) = H(s)L[\delta(t)]$$
$$= H(s)$$

and
$$y(t) = L^{-1}[H(s)] \qquad (4\text{-}30)$$

Therefore the impulse response and system function are a Laplace transform pair:

$$h(t) \leftrightarrow H(s) \qquad (4\text{-}31)$$

Figure 4-4 summarizes the solution of a LTIC system using transform and time-domain analysis.

EXAMPLE 4-18
    Consider the circuit of Figure 4-5(a). Find:

(a) the system function
(b) the system differential equation
(c) the impulse response
(d) the zero-state output when $x(t) = tu(t)$
(e) the output when $x(t) = tu(t)$ and $v_{1/2F}(0^+) = 2, \quad v_{1/4F}(0^+) = 3$

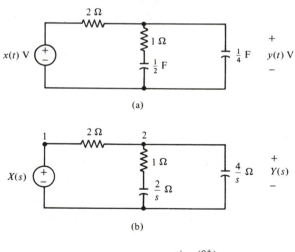

(a)

(b)

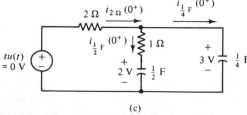

(c)

Figure 4-5   (a) The circuit for Example 4-18; (b) the Laplace transform model; (c) the initial conditions at $t = 0^+$.

Solution

(a) The transform model for the circuit is shown in Figure 4-5(b).

KCL at Node 2

$$\left(0.5 + \frac{1}{1 + 2/s} + \frac{s}{4}\right) Y(s) = 0.5 X(s)$$

Therefore $$H(s) = \frac{2(s + 2)}{s^2 + 8s + 4}$$

$$= \frac{2(s + 2)}{[(s + 4)^2 - 12]}$$

$$= \frac{2(s + 2)}{(s + 0.54)(s + 7.46)}$$

(b) The system differential equation is:

$$y''(t) + 8y'(t) + 4y(t) = 4x(t) + 2x'(t)$$

(c) The impulse response is:

$$h(t) = L^{-1}\left[\frac{2s + 4}{(s + 0.54)(s + 7.46)}\right]$$

$$= (0.43e^{-0.54t} + 1.56e^{-7.46t})u(t)$$

and we can check our answer with the initial value theorem.

(d) $$X(s) = \frac{1}{s^2}$$

hence $$Y(s) = \frac{2s + 4}{(s + 0.54)(s + 7.46)s^2}$$

$$= \frac{A_1}{s + 0.54} + \frac{A_2}{s + 7.46} + \frac{A_3}{s} + \frac{A_4}{s^2}$$

$$= \frac{1.43}{s + 0.54} + \frac{0.03}{s + 7.46} + \frac{A_3}{s} + \frac{1.0}{s^2}$$

To find A_3 we let $s = -2$

and $$0 = -0.98 + 0.005 - 0.5 A_3 - 0.25$$

and $A_3 = -1.45$.

Therefore $$y(t) = [1.43e^{-0.54t} + 0.03e^{-7.46t} - 1.45 + t]u(t)$$

(e) If the initial energy is not zero, then $y(t)$ is of the form:

$$y(t) = (A_1 e^{-0.54t} + A_2 e^{-7.46t} - 1.45 + t)u(t)$$

and we can find $y(0^+)$ and $dy/dt(0^+)$. Figure 4-5(c) shows the circuit at $t = 0^+$ and $i_{1/2F}(0^+) = 1A$, $i_{2\Omega}(0^+) = -1.5$ which by KCL gives $i_{1/4F}(0^+) = -2.5A$ and $dy/dt(0^+) = -10V/s$. Now incorporating these into $y(t)$, we have:

$$3 = A_1 + A_2 - 1.45$$

and $$-10 = -0.5A_1 - 7.5A_2 + 1.0$$

Solving for A_1 and A_2, we obtain:

$$A_1 = 3.20 \quad \text{and} \quad A_2 = 1.25.$$

Finally,

$$y(t) = [3.20e^{-0.54t} + 1.25e^{-7.46t} - 1.45 + t]u(t)$$

Alternatively, we could have used our Laplace transform model for the capacitors incorporating the initial conditions. The reader should redo part (e) in this manner.

Proving Properties of Time Operations.   A number of properties involving convolution and singularity functions that were challenging in the time domain are readily handled in the transform domain. Some of these will now be proved.

EXAMPLE 4-19

Prove the following using the Laplace transform and assuming all functions are causal:

(a) $f(t)*\delta(t - a) = f(t - a)$
(b) $\delta(t - a)*\delta(t - b) = \delta(t - a - b)$
(c) $f(t)*[g(t)*h(t)] = [f(t)*g(t)]*h(t)$

Solution

(a) $$\overline{f(t)*\delta(t - a)} = F(s)e^{-as}$$

Therefore $$f(t)*\delta(t - a) = L^{-1}[F(s)e^{-as}]$$

$$= f(t - a)$$

(b) $\overline{\delta(t - a)*\delta(t - b)} = (e^{-as})(e^{-bs})$

$$= e^{-(a+b)s}$$

Taking the inverse, we obtain:

$$\delta(t - a)*\delta(t - b) = \delta(t - a - b)$$

(c) $$\overline{f(t)*[g(t)*h(t)]} = \overline{f(t)} \ \overline{[g(t)*h(t)]}$$

$$= F(s)[G(s)H(s)]$$

$$= [F(s)G(s)]H(s) = L[\text{LHS}]$$

Therefore $$f(t)*[g(t)*h(t)] = \overline{F(s)G(s)} \ \overline{H(s)}$$

$$= [f(t)*g(t)]*h(t) = \text{RHS}$$

SUMMARY

Chapter 4 covered the Laplace transform in a fundamental manner. This is basic "bread-and-butter" material as far as systems, communications, control theory, plus many assorted fields are concerned and great fluency is expected.

The one-sided Laplace transform was defined and appreciated by forming a small vocabulary of time functions and transforms. The great theorems of the Laplace transform—the time-scaling, time- and s-shifting, convolution, derivative, integral, and initial and final value theorems were introduced. In particular, we focused on the derivative theorem and the necessity for a lower limit of $t = 0^-$ to cope with singularity functions at the origin.

The uniqueness of a time function and its transform was noted and inverse transforms were evaluated by referring to Tables 4-1 and 4-2. Knowing the transforms for 1, t, $e^{\alpha t}$, Cos t, Sin t, Cos h t, and Sin h t, we can handle quickly the case of finding the inverse transform of any function with at most a second-order polynomial in the denominator with the aid of the s-shifting theorem in particular. When the denominator is of the order of more than 2, partial fractions should be used. The properties of the partial fraction coefficients for conjugate imaginary and complex conjugate roots are important and we must proceed rapidly from finding a term such as $(a + jb)/(s + \alpha + j\beta)$ to quoting the time function

$$e^{-\alpha t}[2a \text{ Cos } \beta t + 2b \text{ Sin } \beta t]u(t)$$

Finally, some of the main applications of Laplace transforms were discussed. These were: solving differential equations incorporating the initial conditions, finding complete responses in circuits using a Laplace transform model incorporating the initial conditions, and the transform analysis of LTIC systems with deterministic inputs.

With effort the material is straightforward and philosophical understanding plus quick confident algebraic manipulations are assumed in later work.

PROBLEMS

4-1. Write out the Laplace transforms for $e^{\alpha t}$, 1, $e^{j\beta t}$, Cos βt, Sin βt, Cos hβt, Sin hβt, t, and t^n. Using these transforms plus the theorems, evaluate as quickly as possible the Laplace transforms for:

(a) $3e^{-4t}$

(b) e^{-j2t}

(c) 5 Cos $2t$

(d) 4 Sin $3t$

(e) 6 Cos $(2t - 40°)$

(f) 5 Cos h$4t$

(g) Sin h$3t$

(h) t^2

(i) $5t^2 + 3t$

(j) 4 $tu(t - 2)$

(k) $[3(t - 2)^2 + 4(t - 2) + 3]u(t - 2)$

(l) $(3t^2 - 4t + 3)u(t - 2)$

(m) $t^5 e^{-2t}$

(n) $3e^{-4t}$ Cos $2t$

(o) $5e^{-t}$ Sin $2t$

(p) $2te^{-4t}$ Cos $2t$

(q) $\int_{0^-}^{t} 2e^{-p}\, dp$

(r) $\int_{-4}^{t} 2e^{-p}\, dp$

(s) $4e^{-2t}u(t)*tu(t)$

(t) $\int_0^t 2pe^{-(t-p)}\,dp$

(u) $\int_0^\infty \dfrac{2}{p}e^{-(t-p)}u(t-p)\,dp$

4-2. Evaluate as quickly as possible the inverse Laplace transforms for:

(a) $\dfrac{4s^2-1}{3s+2}$

(b) $\dfrac{2s+1}{s^2+6s+1}$

(c) $\dfrac{2s+1}{s^2+6s+11}$

(d) $\dfrac{2s+1}{s^2+16}$

(e) $\dfrac{s^2-4}{s^3+3s^2+3s+2}$

(f) $\dfrac{2s+1}{s^3+3s^2+3s+2}$

(g) $\dfrac{2s+1}{(s+4)^4}$

(h) $\dfrac{3s^2+2}{(s^2+4)^2}$

(i) $e^{-5s}\dfrac{1}{(2s+1)^6}$

4-3. Given the Laplace transforms of the causal functions $f(t)$ and $g(t)$ have regions of convergence $\sigma > \sigma_1$ and $\sigma > \sigma_2$, respectively, where $\sigma_1 > \sigma_2$, find the regions of convergence for:

(a) $f(t) \pm g(t)$
(b) $f(t-a)u(t-a)$
(c) $e^{at}f(t)$
(d) $t^n f(t)$
(e) $f(t)g(t)$
(f) $f(t)*g(t)$

(g) $\int_{0^-}^t f(p)\,dp$

What is the condition for stability in each case.

4-4. For the systems shown find:
(a) the system function $H(s)$ and the system differential equation
(b) the impulse response as $h(t) = L^{-1}[H(s)]$
(c) the zero-state response $y(t) = L^{-1}[H(s)X(s)]$ where $x(t) = tu(t)$
(d) adjust your answer to part (c) to find the complete response when $y(0^+) = 2$

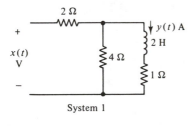

System 1

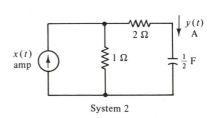

System 2

4-5. Solve. (a) $y''(t) + 4y(t) = 2$
$\qquad y(0) = 3, \qquad y'(0) = 2$
(b) $y''(t) + 4y(t) = 2u(t)$
$\qquad y(0^+) = 3, \qquad y'(0^+) = 2$

Carefully discuss the difference between the two problems. When you use $y(0^+)$ and $y'(0^+)$ for the Laplace transform, what assumption do you make about your solution?

4-6. Given the response of a system is:

$$y(t) = (2 + t) e^{-t} + 6 \cos (2t + 40°)$$

where $y_{\text{homo}}(t) = (2 + t) e^{-t}$ and $y_{f0}(t) = 6 \cos (2t + 40°)$

find the system differential equation. Is the solution unique?

4-7. Find the complete response indicated for the circuits shown using a Laplace transform model incorporating the initial energy for inductors or capacitors

(a)

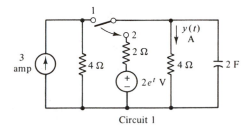

Circuit 1

At $t = 0$ the switch moves from position 1 to position 2.

(b)

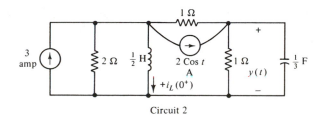

Circuit 2

$$i_L(0^+) = 2, \qquad y(0^+) = 1$$

4-8. Repeat Problem 4-7 in the following manner:
  (a) Find $H(s)$ and the form of $y_{\text{homo}}(t)$.
  (b) Find $y_{f0}(t)$ from the appropriate part of $L^{-1}[H(s)X(s)]$.
  (c) Use Kirchhoff's laws and the energy at $t = 0^+$ to find $y(0^+)$ and, if needed, $y'(0^+)$.
  (d) Apply the initial conditions to obtain the complete response.

4-9. (a) Find the Laplace transform of:

$$x(t) = \sum_{n=0}^{\infty} \delta(t - nT)$$

  (b) Show that the function:

$$f(t) = \sum_{0}^{\infty} f_T(t - nT)$$

where $f_T(t) = 0, \qquad t < 0 \cap t > T$

may be written as:

$$f(t) = f_T(t) * \sum_{n=0}^{\infty} \delta(t - nT)$$

(c) Using the convolution theorem, verify the Laplace transform of the semi-periodic function $f(t)$ is:

$$F(s) = \frac{F_T(s)}{1 + e^{-Ts}}$$

(d) Find the inverse transform of:

$$G(s) = \frac{1 - e^{-s}}{s(1 - e^{-4s})}, \qquad \text{Re}(s) > 0$$

and plot $g(t)$.

(e) Plot the pole-zero diagram of $G(s)$ and verify your result finding $g(t)$ by using the residue theory.

4-10. Given:

$$g(t) = \sum_{n=0}^{\infty} f(t - 2n)$$

where

$$f(t) = 1, \qquad 0 < t < 1$$
$$= 0, \qquad \text{otherwise}$$

is the input to a system with system function $H(s) = 2/(s + 3)$. Find and plot the zero-state output $y(t)$.

4-11. (a) Find $\theta(s) = Y(s)/X(s)$ for the system shown.

(b) For what values of K is the system stable?

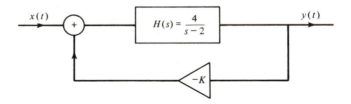

The Two-sided Laplace Transform

INTRODUCTION

Chapter 5 treats the two-sided or bilateral Laplace transform whose main application is the solving of LTIC continuous systems with random or signal plus random inputs. The chapter considers the transform analysis of the continuous material of Chapter 3.

The material is traversed by means of what is now our standard treatment of any transform. The different stages are:

1. The transform is defined and a number of transforms are evaluated. We utilize to the fullest our knowledge of one-sided transforms to evaluate two-sided ones.
2. The properties and theorems are given and attention is focused on the transform of convolution and correlation integrals.
3. The inverse transform is treated either by using previously mastered partial fraction techniques referring to tables, or using the residue theory from complex variables.
4. LTIC systems are solved for the cases of random and signal plus noise inputs.

5-1 DEFINITION AND EVALUATION OF SOME TRANSFORMS

The **two-sided** or **bilateral Laplace transform** of a real function $f(t)$ is defined as:

$$F_B(s) \triangleq \int_{-\infty}^{\infty} f(t)e^{-st}\, dt \tag{5-1}$$

for complex $s = \sigma + j\omega$. $F_B(s)$ if it exists will do so for a region of the complex s plane, $\sigma_1 < \text{Re}(s) < \sigma_2$, called the **region of convergence**. Normally, the subscript "B" is excluded and it will be clear from the context whether the one- or two-sided Laplace transform is being used. Again, alternate notations, $\mathcal{L}[f(t)]$, $\overline{f(t)}$ and $f(t) \leftrightarrow F(s)$ will be used.

A number of transforms will now be evaluated and the concept of the region of convergence will be expanded.

EXAMPLE 5-1

Find the two-sided Laplace transforms of the following functions and state the region of convergence:

(a) $f_1(t) = 3e^{2t}u(t)$
(b) $f_2(t) = 3e^{2t}u(-t)$
(c) $f_3(t) = 3e^{-2t}u(t) + 4e^{t}u(-t)$
(d) $f_4(t) = 3e^{-2t}u(t) - 4e^{t}u(t)$
(e) $f_5(t) = -3e^{-2t}u(-t) + 4e^{t}u(-t)$
(f) $f_6(t) = A_1e^{\alpha t}u(t) + A_2e^{\beta t}u(-t)$ in general for all appropriate real α and β.

Solution

(a) $f_1(t) = 3e^{2t}u(t)$

Therefore
$$F_1(s) = \int_{-\infty}^{0} 0e^{-st}\,dt + \int_{0}^{\infty} 3e^{2t}e^{-st}\,dt$$

$$= \frac{3}{s-2}, \qquad \text{Re}(s) > 2$$

$f_1(t)$, $F_1(s)$, and the region of convergence are shown in Figure 5-1(a).

(b) $f_2(t) = 3e^{2t}u(-t)$

$$F_2(s) = \int_{-\infty}^{0} 3e^{2t}e^{-st}\,dt$$

$$= \int_{-\infty}^{0} 3e^{(2-s)t}\,dt$$

$$= \frac{3}{2-s}e^{(2-\sigma)t}e^{-j\omega t}\,\Big|_{-\infty}^{0}, \qquad \text{writing } s = \sigma + j\omega$$

$$= \frac{3}{2-s} - 0, \qquad \text{if } \sigma < 2$$

since then $e^{(2-\sigma)(-\infty)} = 0 \qquad \text{for } 2 - \sigma > 0$

Therefore
$$F_2(s) = -\frac{3}{s-2}, \qquad \text{Re}(s) < 2$$

$f_2(t)$, $F_2(s)$, and the region of convergence are shown in Figure 5-1(b).

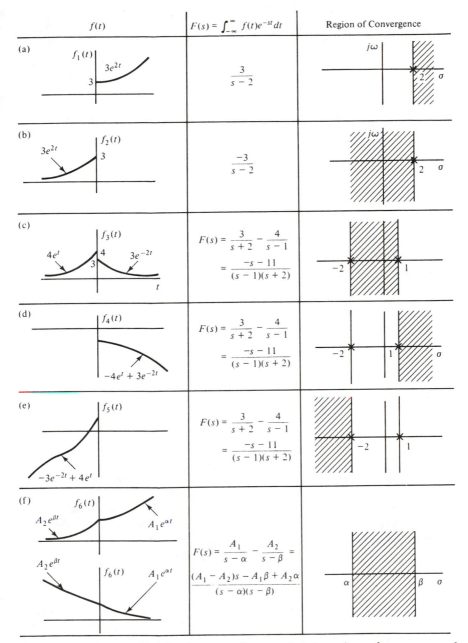

The table columns are: $f(t)$ | $F(s) = \int_{-\infty}^{\infty} f(t)e^{-st}\,dt$ | Region of Convergence

(a) $f_1(t)$, $3e^{2t}$, 3

$$\frac{3}{s-2}$$

(b) $3e^{2t}$, $f_2(t)$, 3

$$\frac{-3}{s-2}$$

(c) $f_3(t)$, $4e^t$, 4, $3e^{-2t}$, 3, t

$$F(s) = \frac{3}{s+2} - \frac{4}{s-1}$$
$$= \frac{-s-11}{(s-1)(s+2)}$$

(d) $f_4(t)$, $-4e^t + 3e^{-2t}$

$$F(s) = \frac{3}{s+2} - \frac{4}{s-1}$$
$$= \frac{-s-11}{(s-1)(s+2)}$$

(e) $f_5(t)$, $-3e^{-2t} + 4e^t$

$$F(s) = \frac{3}{s+2} - \frac{4}{s-1}$$
$$= \frac{-s-11}{(s-1)(s+2)}$$

(f) $f_6(t)$, $A_2 e^{\beta t}$, $A_1 e^{\alpha t}$, $A_2 e^{\beta t}$, $f_6(t)$, $A_1 e^{\alpha t}$

$$F(s) = \frac{A_1}{s-\alpha} - \frac{A_2}{s-\beta} =$$
$$\frac{(A_1 - A_2)s - A_1\beta + A_2\alpha}{(s-\alpha)(s-\beta)}$$

Figure 5-1 The two-sided Laplace transforms and their regions of convergence for Example 5-1.

(c) $f_3(t) = 3e^{-2t}u(t) + 4e^{+t}u(-t)$

$$F_3(s) = \int_{-\infty}^{0} 4e^t e^{-st}\, dt + \int_{0}^{\infty} 3e^{-st}e^{-2t}\, dt$$

$$= \frac{4}{1-s} e^{(-1-s)t} \Big|_{-\infty}^{0} + \frac{3}{-s-2} e^{(-2-s)t} \Big|_{0}^{\infty}$$

$$= \frac{-4}{s-1} + 0 \Big|_{\sigma<1} \text{if} + 0 \Big|_{\sigma>-2} \text{if} + \frac{3}{s+2}$$

$$= \frac{-s-11}{(s-1)(s+2)}; \qquad -2 < \text{Re}(s) < +1$$

For this function which exists for both positive and negative time we note that the behavior for negative time puts an upper bound on an allowable σ and the function behavior for positive time puts a lower bound on an allowable σ. Therefore we obtain the strip of convergence $\sigma_1 < \sigma < \sigma_2$. $f_3(t)$, $F_3(s)$, and the region of convergence are plotted in Figure 5-1(c).

(d) $f_4(t) = (3e^{-2t} - 4e^{+t})u(t)$

$$F_4(s) = \frac{3}{s+2} + 0 \Big|_{\text{if } \sigma>-2} - \frac{4}{s-1} + 0 \Big|_{\text{if } \sigma>+1}$$

$$= \frac{-s-11}{(s-1)(s+2)}, \qquad \text{Re}(s) > 1$$

We notice the bilateral transform expression is the same as in part (c) but the region of convergence is different. $f_4(t)$, $F_4(s)$, and the region of convergence are shown in Figure 5-1(d).

(e) $f_5(t) = -3e^{-2t}u(-t) + 4e^t u(-t)$

$$F_5(s) = \frac{3}{s+2} + 0 \Big|_{\text{if } \sigma<-2} - \frac{4}{s-1} + 0 \Big|_{\text{if } \sigma<1}$$

Therefore $\qquad F_5(s) = \dfrac{-s-11}{(s-1)(s+2)}, \qquad \text{Re}(s) < -2$

Since $e^{-2t}u(-t)$ is an increasing exponential of negative time its convergence factor $\sigma < -2$ determines the region of convergence. Again, the Laplace transform expression is identical to parts (c) and (d) but the region of convergence is different. $f_5(t)$, $F_5(s)$, and the region of convergence are plotted in Figure 5-1(e).

(f) $f_6(t) = A_1 e^{\alpha t}u(t) + A_2 e^{\beta t}u(-t)$

It can quite easily be shown or by now be clear that:

$$F_6(s) = \frac{A_1}{s-\alpha} + 0 \Big|_{\text{if } \sigma>\alpha} - \frac{A_2}{s-\beta} + 0 \Big|_{\text{if } \sigma<\beta}$$

$$= \frac{(A_1 - A_2)s - A_1\beta + A_2\alpha}{(s-\alpha)(s-\beta)} \qquad \text{if } \sigma > \alpha \cap \sigma < \beta$$

The bilateral Laplace transform will exist for all α and β such that $\alpha <$ β and then the region of convergence is $\alpha < \text{Re}(s) < \beta$. The different possible situations for $f_6(t)$ are shown in Figure 5-1(f).

5-1-1 Two-sided Transforms Using One-sided Transforms

The evaluation of two-sided Laplace transforms involves the same amount of work as doing two one-sided Laplace transforms and indeed a table of one-sided Laplace transforms may be used to find two-sided ones. We will now develop this technique. Given:

$$f(t) = f_1(t)u(t) + f_2(t)u(-t)$$

$$F(s) = F_1(s) + \int_{-\infty}^{0} f_2(t)e^{-st}\, dt$$

Letting $t = -p$, we obtain:

$$dt = -dp, \qquad -\infty < t < 0 \text{ gives } \infty > p > 0$$

$$F(s) = F_1(s) + \int_{\infty}^{0} f_2(-p)e^{ps}\,(-dp)$$

and

$$F(s) = F_1(s) + \mathcal{L}[f_2(-t)u(t)]\Big|_{s=-s} \qquad (5\text{-}2)$$

If $f_2(-t)u(t)$ has a region of convergence $\sigma > \sigma_2'$, then $f_2(t)u(-t)$ has a region of convergence $\sigma < -\sigma_2'$.

EXAMPLE 5-2

Using Equation 5-2, find the two-sided Laplace transform of:

(a) $f(t) = e^{2t}u(-t)$
(b) $f(t) = 3e^{-2t}u(t) + te^{-t}u(-t)$

Solution

(a) $F(s) = \mathcal{L}[e^{-2t}u(t)]\Big|_{s=-s}$

$$= \frac{1}{s+2}\Big|_{s=-s} = \frac{-1}{s-2}, \qquad \sigma < 2$$

(b) $f(t) = 3e^{-2t}u(t) + te^{-t}u(-t)$

Using Equation 5-2, we obtain:

$$F(s) = \frac{3}{s+2} + \mathcal{L}[-te^{t}u(t)]\Big|_{s=-s}$$

$$= \frac{3}{s+2} + \frac{-1}{(s-1)^2}\Big|_{s=-s}, \qquad \sigma > -2 \cap \sigma < -1$$

$$= \frac{3}{s+2} - \frac{1}{(1+s)^2}$$

$$= \frac{3s^2 + 5s + 1}{(s+2)(s-1)^2}, \qquad -2 < \sigma < -1$$

Example 5-1 should now be repeated using Equation 5-1.

There are a few very important functions for which a two-sided Laplace transform does not exist. The functions $f_1(t) = 1$, $f_2(t) = \cos(\omega_0 t + \phi)$, and $f(t) = \sum_{n=-\infty}^{\infty} g(t - nT)$ a periodic function, do not possess a Laplace transform as each function for $t < 0$ requires $\sigma < 0$, whereas the function behavior for $t > 0$ requires $\sigma > 0$. Therefore, no value of s exists for which the transform converges. These functions are not to be confused with the causal functions:

$$f_1(t) = u(t), \quad f_2(t) = \cos(\omega_0 t + \phi)u(t), \quad \text{and} \quad f_3(t) = \left[\sum_{n=-\infty}^{n=\infty} g(t - nT) \right] u(t)$$

Finally, we are rarely interested in the two-sided Laplace transforms of functions whose transforms are not the ratio of two polynomials in s. As we will see, the most common functions for which we find two-sided Laplace transforms are correlation functions for finite energy waveforms and autocorrelation functions for ergodic noise waveforms. Both these type functions are *even* and for positive or negative time may often be represented by the product of polynomials and exponentials.

5-2 IMPORTANT THEOREMS OF BILATERAL LAPLACE TRANSFORMS

Table 4-2 of Chapter 4 listed an extensive set of theorems and properties of the one-sided Laplace transform. Some important theorems for the bilateral Laplace transform are given in Table 5-1. Since the convolution and correlation theorems are of the utmost importance when applying this material to systems with random inputs we will prove and demonstrate these. Discussion of the complex convolution theorem is deferred until after the inverse transform is considered.

EXAMPLE 5-3
Prove the convolution theorem and comment on the region of convergence.

Solution. The convolution theorem states that if:

$$f(t) \leftrightarrow F(s), \qquad \sigma_{f_1} < \sigma < \sigma_{f_2}$$

and

$$g(t) \leftrightarrow G(s), \qquad \sigma_{g_1} < \sigma < \sigma_{g_2}$$

then

$$f(t)*g(t) \leftrightarrow F(s)G(s)$$

TABLE 5-1

Theorem	Time function	Two-sided Laplace transform
Linearity	$ax(t) + by(t)$	$aX(s) + bY(s)$
		$\sigma > \max(\sigma_{x1}, \sigma_{y1}), \ \sigma < \min(\sigma_{x2}, \sigma_{y2})$
Time-scaling	$x(at)$	$\dfrac{1}{\|a\|} X\left(\dfrac{s}{a}\right), \qquad \begin{array}{ll} a\sigma_1 < \sigma < a\sigma_2, & a > 0 \\ a\sigma_2 < \sigma < a\sigma_1, & a < 0 \end{array}$
Shifting	$x(t - a)$	$e^{-as} X(s), \qquad \sigma_1 < \sigma < \sigma_2$
Convolution	$x(t)*y(t)$	
	$= \displaystyle\int_{-\infty}^{\infty} x(p)y(t-p)\,dp$	$X(s)Y(s), \qquad (\sigma_{x_1} < \sigma < \sigma_{x_2}) \cap (\sigma_{y_1} < \sigma < \sigma_{y_2})^a$
Cross-correlation	$x(t) \oplus y(t)$	
		$Y(s)X(-s), \qquad (\sigma_{y_1} < \sigma < \sigma_{y_2}) \cap (-\sigma_{x_2} < \sigma < -\sigma_{x_1})^a$
	$= \displaystyle\int_{-\infty}^{\infty} y(p)x(p-t)\,dp$	
Autocorrelation	$x(t) \oplus x(t)$	
		$X(s)X(-s), \qquad \max(\sigma_1, -\sigma_2) < \sigma < \min(\sigma_2, -\sigma_1)^a$
	$= \displaystyle\int_{-\infty}^{\infty} x(p)x(p-t)\,dp$	

For causal functions all one-sided Laplace transform theorems carry over.

$^a \overline{x(t)} = X(s), \qquad \sigma_1 < \sigma < \sigma_2 \quad \text{or} \quad \sigma_{x_1} < \sigma < \sigma_{x_2}$
$\overline{y(t)} = Y(s), \qquad \sigma_{y_1} < \sigma < \sigma_{y_2}$

By definition:

$$\mathcal{L}[f(t)*g(t)] = \int_{-\infty}^{\infty} \left[\int_{-\infty}^{\infty} f(p)g(t-p)\,dp\right] e^{-st}\,dt$$

Interchanging the order of integration, we obtain:

$$\mathcal{L}[f(t)*g(t)] = \int_{-\infty}^{\infty} f(p)\left[\int_{-\infty}^{\infty} g(t-p)e^{-st}\,dt\right] dp$$

Letting $t - p = u$:

$$dt = du, \qquad -\infty < t < \infty \qquad \text{gives} \qquad -\infty < u < \infty$$

$$\mathcal{L}[f(t)*g(t)] = \int_{-\infty}^{\infty} f(p)\left[\int_{-\infty}^{\infty} g(u)e^{-s(u+p)}\,du\right] dp$$

$$= \int_{-\infty}^{\infty} f(p)e^{-sp}G(s)\,dp$$

$$= G(s)F(s) \tag{5-3}$$

If we write $R(s) = G(s)F(s)$, then $R(s)$ will converge for all s for which both $G(s)$ and $F(s)$ converge. For example, if $F(s)$ converges $-3 < \sigma < 1$ and $G(s)$ converges $-2 < \sigma < 4$, then $R(s)$ would converge $-2 < \sigma < 1$. In general, $R(s)$ converges over the intersection of the points

$$(\sigma_{f_1} < \sigma < \sigma_{f_2}) \cap (\sigma_{g_1} < \sigma < \sigma_{g_2}) \tag{5-4}$$

EXAMPLE 5-4

Find the transforms of the following functions and denote the regions of convergence if the transform exists:

(a) $f_1(t) = e^{-2t}u(t)*e^t u(-t)$
(b) $f_2(t) = e^{2t}u(t)*e^{-t}u(-t)$
(c) $e^{-2|t|}*u(t)$

Solution

(a) Let $x(t) = e^{-2t}u(t)$

and then $X(s) = \dfrac{1}{s+2}$, $\sigma > -2$

Let $y(t) = e^t u(-t)$

and then $Y(s) = -\dfrac{1}{s-1}$, $\sigma < 1$

By the convolution theorem:

$$\mathcal{L}[x(t)*y(t)] = \frac{-1}{(s+2)(s-1)}, \qquad \sigma > -2 \cap \sigma < +1$$

$$= \frac{-1}{(s+2)(s-1)}, \qquad -2 < \sigma < +1$$

(b) Let $x(t) = e^{2t}u(t)$

and then $X(s) = \dfrac{1}{s-2}$, $\sigma > 2$

Let $y(t) = e^{-t}u(-t)$

and then $Y(s) = \dfrac{-1}{s+1}$, $\sigma < -1$

By the convolution theorem:

$$\mathcal{L}[x(t)*y(t)] = \frac{-1}{(s-2)(s+1)}, \qquad \sigma > 2 \cap \sigma < -1$$

Since $\sigma > 2 \cap \sigma < -1 = \phi$ the Laplace transform does not exist.

(c) Let $x(t) = e^{-2|t|} = e^{-2t}u(t) + e^{2t}u(-t)$ and this has a Laplace transform:

$$X(s) = \frac{1}{s+2} - \frac{1}{s-2} = \frac{-4}{(s+2)(s-2)}, \qquad -2 < \sigma < 2$$

Let $y(t) = u(t)$

and then
$$Y(s) = \frac{1}{s}, \qquad \sigma > 0$$

By the convolution theorem:

$$\mathcal{L}[x(t)*y(t)] = \frac{-4}{s(s+2)(s-2)}, \qquad -2 < \sigma < 2 \cap \sigma > 0$$

$$= \frac{-4}{s(s+2)(s-2)}, \qquad 0 < \sigma < 2$$

Our anticipation of evaluating inverse two-sided transforms should be mounting since we now have an alternative to convolution.

EXAMPLE 5-5

Prove the correlation theorems:

(a) $\mathcal{L}[x(t) \oplus y(t)] = Y(s)X(-s)$
(b) $\mathcal{L}[x(t) \oplus x(t)] = X(s)X(-s)$

Solution

(a) By definition:

$$\mathcal{L}[x(t) \oplus y(t)] = \int_{-\infty}^{\infty} \left[\int_{-\infty}^{\infty} y(p)x(p-t)\, dp \right] e^{-st}\, dt$$

Interchanging the order of integration, we obtain:

$$\mathcal{L}[x(t) \oplus y(t)] = \int_{-\infty}^{\infty} y(p) \left[\int_{-\infty}^{\infty} x(p-t)\, e^{-st}\, dt \right] dp$$

Substitute $p - t = l$

$$\mathcal{L}[x(t) \oplus y(t)] = \int_{-\infty}^{\infty} y(p) \left[\int_{+\infty}^{-\infty} x(l)e^{-s(p-l)}\, (-dl) \right] dp$$

$$= \int_{-\infty}^{\infty} y(p)e^{-sp}\, dp \int_{-\infty}^{\infty} x(l)e^{sl}\, dl = Y(s)X(-s) \qquad (5\text{-}5)$$

where $C(s) = Y(s)X(-s)$ converges for all s for which both $Y(s)$ and $X(-s)$ converge. To interpret the region of convergence for $X(-s)$, we note that if $X(s)$ contains a pole at $s = -s_1$, then $X(-s)$ contains a pole at $s = s_1$. With a little more thought we conclude that if the region of convergence for $X(s)$ is $\sigma_{x_1} < \sigma < \sigma_{x_2}$, then the region of convergence for $X(-s)$ is $\sigma_{x_1} < -\sigma < \sigma_{x_2}$ or $-\sigma_{x_2} < \sigma < -\sigma_{x_1}$ and the region of convergence for $C(s)$ is the intersection of the points:

$$(\sigma_{y_1} < \sigma < \sigma_{y_2}) \cap (-\sigma_{x_2} < \sigma < -\sigma_{x_1}) \qquad (5\text{-}6)$$

(b) In the case $y(t) = x(t)$ we find by the cross-correlation theorem that:

$$\mathcal{L}[x(t) \oplus x(t)] = X(s)X(-s) \qquad (5\text{-}7)$$

If the region of convergence for $X(s)$ is $\sigma_1 < \sigma < \sigma_2$, then $X(s)X(-s)$ has a region of convergence $\max(\sigma_1, -\sigma_2) < \sigma < \min(\sigma_2, -\sigma_1)$ if it exists.

EXAMPLE 5-6

Find the Laplace transforms of the following correlation functions:

(a) $e^{-2t}u(t) \oplus e^{-2t}u(t) = C_{xx}(t)$
(b) $e^{-t}u(t) \oplus e^{2t}u(-t) = C_{xy}(t)$
(c) $e^{2t}u(-t) \oplus e^{-t}u(t) = C_{yx}(t)$

Solution

(a) $e^{-2t}u(t) \leftrightarrow \dfrac{1}{s+2}$

$$\mathcal{L}[e^{-2t}u(t) \oplus e^{-2t}u(t)] = \frac{1}{(s+2)} \frac{1}{(-s+2)}$$

$$= \frac{-1}{(s+2)(s-2)}, \qquad \sigma > -2 \cap \sigma < 2$$

$$= \frac{-1}{s^2 - 4}, \qquad -2 < \sigma < +2$$

(b) $x(t) = e^{-t}u(t) \leftrightarrow \dfrac{1}{s+1}, \qquad \sigma > -1,$

$y(t) = e^{2t}u(-t) \leftrightarrow \dfrac{-1}{s-2}, \qquad \sigma < -2$

Therefore $\qquad \mathcal{L}[e^{-t}u(t) \oplus e^{2t}u(-t)]$

$$= \frac{-1}{s-2} \frac{1}{-s+1}$$

$$= \frac{1}{(s-2)(s-1)}, \qquad \sigma < 2 \cap \sigma < 1$$

$$= \frac{1}{(s-2)(s-1)}, \qquad \sigma < 1$$

(c) $\mathcal{L}[e^{2t}u(-t) \oplus e^{-t}u(t)] = \dfrac{1}{s+1}\left(\dfrac{-1}{s-2}\right)\Bigg|_{s=-s}$

$$= \frac{1}{(s+1)(s+2)}, \qquad \sigma > -1 \cap \sigma > -2$$

$$= \frac{1}{(s+1)(s+2)}, \qquad \sigma > -1$$

The inverse Laplace transform will provide an alternative approach for correlation. For example, when finding $x(t) \oplus y(t)$ we have the choice of evaluating:

$$[x(t) \oplus y(t)] = \int_{-\infty}^{\infty} y(p)x(p - t) \, dp \qquad \text{or for example}$$

$$[e^{2t}u(-t) \oplus e^{-t}u(t)] = \mathcal{L}^{-1}\left[\frac{1}{(s + 1)(s + 2)}\right], \qquad \sigma > -1$$

Close examination of Example 5-6(a), (b), and (c) would lead to general relations concerning symmetry of the poles for the Laplace transform of correlation functions, which will be developed later in the chapter.

5-3 THE INVERSE TWO-SIDED LAPLACE TRANSFORM

The uniqueness theorem for the inverse two-sided Laplace transform states:

$$\mathcal{L}^{-1}[\mathcal{L}(f(t))] = f(t) \tag{5-8}$$

and in this section we discuss two techniques for finding inverses. These are:

1. the use of partial fraction expansions plus table reference.
2. the classical evaluation from the formal definition of the inverse using the residue theory from complex variables.

5-3-1 Inverse Transforms Using Partial Fraction

We confine our discussion to Laplace transforms which are the ratio of two polynomials in s:

$$F(s) = \frac{N(s)}{D(s)} = \frac{b_m s^m + b_{m-1} s^{m-1} + \cdots + b_0}{a_n s^n + a_{n-1} s^{n-1} + \cdots + a_0}$$

where the order of $D(s)$ is at least one order higher than $N(s)$. There is associated with $F(s)$ a region of convergence, $\sigma_1 < \sigma < \sigma_2$, where $D(s)$ contains two consecutive poles at $s = \sigma_1 + j\omega_1$ and $s = \sigma_2 + j\omega_2$ and we assume any cancellation of poles by zeros has been carried out. From Section 5-1 we know that a pole of $F(s)$ at $s = -s_1$ where $\text{Re}(-s_1) < \sigma_1$ contributes $Ae^{-s_1 t}u(t)$ if s_1 is real and a pole of $F(s)$ at $s = -s_2$ contributes $Be^{-s_2 t}u(-t)$ if s_2 is real and $-s_2 > \sigma_2$. We now use previously learned partial fraction theory to find inverses. A short general table of two-sided transforms is given in Table 5-2, which we will utilize when finding inverses.

TABLE 5-2   SOME FUNDAMENTAL TWO-SIDED LAPLACE TRANSFORMS

F(s)	Region of convergence	f(t)
$\dfrac{A}{s + \alpha}$	$\sigma > -\alpha$	$Ae^{-\alpha t}u(t)$
$\dfrac{A}{(s + \alpha)}$	$\sigma < -\alpha$	$-Ae^{-\alpha t}u(t)$
$\dfrac{A}{(s + \alpha)^2}$	$\sigma > -\alpha$	$Ate^{-\alpha t}u(t)$
$\dfrac{A}{(s + \alpha)^2}$	$\sigma < -\alpha$	$-Ate^{-\alpha t}u(-t)$
$\dfrac{A_1 s + A_2}{(s + \alpha)^2 + \beta^2}$	$\sigma > -\alpha$ α real, β real and positive	$e^{-\alpha t}\left[A_1 \cos \beta t + \dfrac{A_2 - \alpha A_1}{\beta} \sin \beta t\right]u(t)$
$\dfrac{A_1 s + A_2}{(s + \alpha)^2 + \beta^2}$	$\sigma < -\alpha$	$-e^{-\alpha t}\left[A_1 \cos \beta t + \dfrac{A_2 - \alpha A_1}{\beta} \sin \beta t\right]u(-t)$

EXAMPLE 5-7

Find the inverse Laplace transforms of the following functions using partial fractions:

(a) $F_1(s) = \dfrac{3s + 2}{(s + 1)^2(s - 2)}$, $-1 < \sigma < 2$

(b) $F_2(s) = \dfrac{3s + 2}{(s + 1)^2(s - 2)}$, $\sigma < -1$

(c) $F_3(s) = \dfrac{3s + 2}{(s + 1)^2(s - 2)}$, $\sigma > 2$

(d) $F_4(s) = \dfrac{2s + 3}{(s + 4)(s^2 + 2s + 3)}$, $-4 < \sigma < -1$

Solution

(a) $F_1(s) = \dfrac{3s + 2}{(s + 1)^2(s - 2)}$, $-1 < \sigma < 2$

$\quad = \dfrac{A}{s + 1} + \dfrac{B}{(s + 1)^2} + \dfrac{C}{s - 2}$

where $C = 8/(3)^2 = 0.89$ and $B = -1/-3 = 0.33$. Letting $s = 0$, we find:

$$\frac{2}{-2} = A + 0.33 + \frac{0.89}{-2}$$

Therefore $\quad A = -1 - 0.33 + 0.44$

$$= -0.89$$

$$F_1(s) = \frac{-0.89}{s+1} + \frac{0.33}{(s+1)^2} + \frac{0.89}{s-2}, \qquad -1 < \sigma < 2$$

and from Table 5-2:

$$f_1(t) = (-0.89e^{-t} + 0.33te^{-t})u(t) - 0.89e^{2t}u(-t)$$

$F_1(s)$ and $f_1(t)$ are shown in Figure 5-2(a).

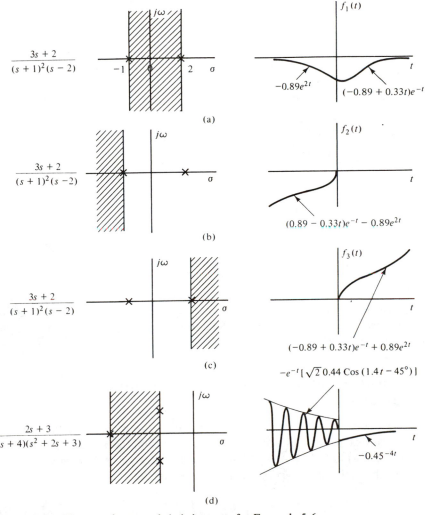

Figure 5-2 The transforms and their inverses for Example 5-6.

(b)
$$F_2(s) = \frac{3s + 2}{(s + 1)^2(s - 2)}, \qquad \sigma < -1$$

$$= \frac{-0.89}{s + 1} + \frac{0.33}{(s + 1)^2} + \frac{0.89}{s - 2}, \qquad \sigma < -1$$

Therefore $f_2(t) = [0.89e^{-t} - 0.33te^{-t} - 0.89e^{2t}]u(-t)$

$F_2(s)$ and $f_2(t)$ are shown in Figure 5-2(b).

(c)
$$F_3(s) = \frac{3s + 2}{(s + 1)^2(s - 2)}, \qquad \sigma > 2$$

Therefore $f_3(t) = [-0.89e^{-t} + 0.33te^{-t} + 0.89e^{2t}]u(t)$

$F_3(s)$ and $f_3(t)$ are shown in Figure 5-2(c).

(d) $F_4(s) = \dfrac{2s + 3}{(s + 4)(s^2 + 2s + 3)}, \qquad -4 < \sigma < -1$

$$= \frac{2s + 3}{(s + 4)[(s + 1)^2 + 2]}$$

$$= \frac{2s + 3}{(s + 4)(s + 1 + j1.4)(s + 1 - j1.4)}$$

$$= \frac{-0.45}{s + 4} + \frac{A_1}{s + 1 + j1.4} + \frac{A_1^*}{s + 1 - j1.4}, \qquad -4 < \sigma < -1$$

where $A_1 = \dfrac{2(-1 - j1.4) + 3}{(3 - j1.4)(-j2.83)} = 0.22 + j0.22$ and

$$f_4(t) = -0.45e^{-4t}u(t) - e^{-t}(0.44\,\mathrm{Cos}\,1.4t + 0.44\,\mathrm{Sin}\,1.4t)u(-t)$$

$F_4(s)$ and $f_4(t)$ are plotted in Figure 5-2(d).

From Example 5-7 it is seen that all the work required for evaluating inverse two-sided transforms by partial fractions was already mastered for the one-sided case. Now poles to the right of σ contribute to the function for negative time and pick up a minus sign.

5-3-2 Inverse Two-sided Laplace Transforms Using Residues

The Appendix on complex variables lists some of the more important results pertaining to system analysis and they will be utilized in this section. The inverse Laplace transform of $F(s) = N(s)/D(s)$ where the order of $D(s)$ is more than $N(s)$, with the region of convergence $\sigma_1 < \sigma < \sigma_2$ is defined as:

$$f(t) = \mathcal{L}^{-1}[F(s)] = \frac{1}{2\pi j} \int_{\substack{\sigma - j\infty \\ C}}^{\sigma + j\infty} F(s)e^{+st}\,ds \qquad (5-9)$$

where C is the straight line $s(\theta) = \sigma + j\theta$, $-\infty < \theta < \infty$. The contour C is shown in Figure 5-3(a). We now discuss the evaluation of Equation 5-9 for the cases of t positive and negative.

For $t > 0$
Consider closing the contour C to the left with contour C_1 as shown in Figure 5-3(b). For $t > 0$, $F(s)e^{+st} \to 0$ at all points on C_1 and by Jordan's lemma $\int_{C_1} F(s)e^{+st}\, ds = 0$.

Therefore
$$f(t) = \frac{1}{2\pi j} \int_{\sigma - j\infty}^{\sigma + j\infty} F(s)e^{st}\, ds + \frac{1}{2\pi j} \int_{C_1} F(s)e^{st}\, ds$$

$$= \frac{1}{2\pi j} \oint_{C + C_1} F(s)e^{st}\, ds$$

$$= \Sigma \, [\text{residues of the poles of } F(s)e^{st} \text{ to the left of } \sigma] \qquad (5\text{-}10)$$

For $t < 0$
Consider closing the contour C to the right with contour C_2 as shown in Figure 5-3(c). For $t < 0$, $F(s)e^{st} \to 0$ at all points on C_2 and by Jordan's lemma $\int_{C_2} F(s)e^{st}\, ds = 0$.

Therefore
$$f(t) = \frac{1}{2\pi j} \int_{\sigma - j\infty}^{\sigma + j\infty} F(s)e^{st}\, ds + \frac{1}{2\pi j} \int_{C_2} F(s)e^{st}\, ds$$

$$= \frac{1}{2\pi j} \oint_{C + C_2} F(s)e^{st}\, ds$$

$$= -\Sigma \, [\text{residues of the poles of } F(s)e^{st} \text{ to the right of } \sigma] \qquad (5\text{-}11)$$

The minus sign occurs because the closed contour $C + C_2$ is traversed in a

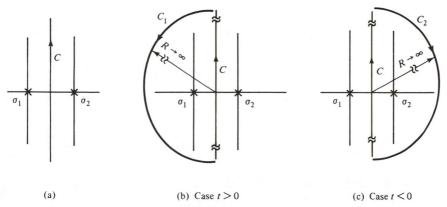

(a)   (b) Case $t > 0$ (c) Case $t < 0$

Figure 5-3 (a) The contour C for finding the inverse Laplace transform; (b) closing C with C_1 for $t > 0$; (c) closing C with C_2 for $t < 0$.

clockwise manner. Summarizing, we conclude:

If $F(s) = N(s)/D(s)$, $\sigma_1 < \sigma < \sigma_2$ where the order of $D(s)$ is higher than $N(s)$, then

For $t > 0$

$$f(t) = \Sigma [\text{residues of the poles of } F(s)e^{st} \text{ to the left of } \sigma]$$

For $t < 0$

$$f(t) = -\Sigma [\text{residues of the poles of } F(s)e^{st} \text{ to the right of } \sigma]$$

In addition, the uniqueness theorem for the Laplace transform is:

$$f(t) = \frac{1}{2\pi j} \int_{\sigma - j\infty}^{\sigma + j\infty} \left[\int_{-\infty}^{\infty} f(p)e^{-sp} \, dp \right] e^{st} \, ds \qquad (5\text{-}12)$$

which says that a time function and its Laplace transform constitute a unique pair.

We evaluate some previously considered inverses using residue theory.

EXAMPLE 5-8

Using residues, find the inverse Laplace transforms of the following functions:

(a) $F_1(s) = \dfrac{-1}{s + 2}$, $\sigma < -2$

(b) $F_2(s) = \dfrac{1}{s + 2}$, $\sigma > -2$

(c) $F_3(s) = \dfrac{3s + 2}{(s + 1)^2(s - 2)}$, $-1 < \sigma < 2$

(d) $F_4(s) = \dfrac{3s + 2}{(s + 1)^2(s - 2)}$, $\sigma < -1$

(e) $F_5(s) = \dfrac{3s + 2}{(s + 1)^2(s - 2)}$, $\sigma > 2$

(f) $F_6(s) = \dfrac{2s + 3}{(s + 4)(s^2 + 2s + 3)}$, $-4 < \sigma < -1$

Solution

(a) As shown in Figure 5-4(a) $C + C_1$ enclose no poles of $F(s)$, and therefore $f(t) = 0$, if $t > 0$.
For $t < 0$, $C + C_2$ encloses the pole at $s = -2$

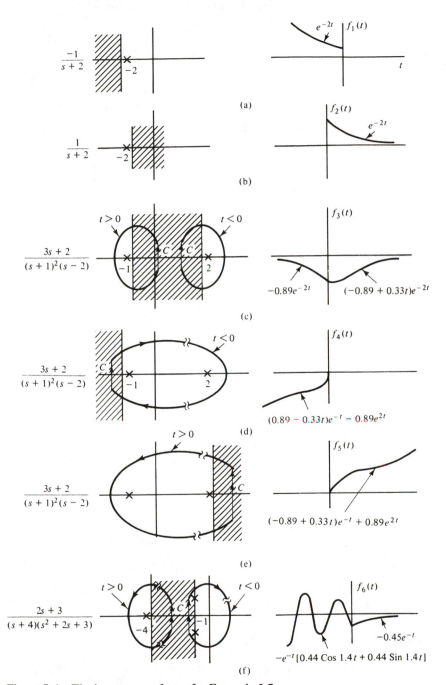

Figure 5-4 The inverse transforms for Example 5-7.

and thus $f_1(t) = -\left[\text{residue of pole at } s = -2 \text{ for } \dfrac{-1}{s+2} e^{st}\right]$

$$= -\left[(s+2)\dfrac{-1}{s+2} e^{st}\Big|_{s=-2}\right]$$

$$= e^{-2t}$$

Therefore $f_1(t) = e^{-2t}u(-t)$

(b) $f_2(t) = 0$, $t < 0$ as $C + C_2$ encloses no poles in Figure 5-4(b). For $t > 0$,

$$f_2(t) = \left[\text{residue of pole at } s = -2 \text{ for } \dfrac{1}{s+2} e^{st}\right]$$

Therefore $f_2(t) = e^{-2t}u(t)$

(c) $F_3(s) = \dfrac{3s+2}{(s+1)^2(s-2)}$, $-1 < \sigma < 2$

as shown in Figure 5-4(c).

For t > 0

$f_3(t) = [\text{residue of the second-order pole at } s = -1 \text{ for } F(s)e^{st}]$

$$= \dfrac{d}{ds}\left[\dfrac{3s+2}{s-2} e^{st}\right]_{s=-1}$$

$$= \dfrac{-1}{-3} te^{-t} + e^{st} \dfrac{(s-2)(3) - (3s+2)1}{(s-2)^2}\Big|_{s=-1}$$

$$= 0.33te^{-t} - 0.89e^{-t}$$

For t < 0

$$f_3(t) = -[\text{residue of pole at } s = 2 \text{ for } F(s)e^{st}]$$

$$= -\dfrac{3(2)+2}{(2+1)^2} e^{2t}$$

$$= -0.89e^{2t}$$

Therefore $f_3(t) = (-0.89 + 0.33t)e^{-t}u(t) - 0.89e^{2t}u(-t)$

This result agrees with Example 5-7(a) where the inverse was found by partial fractions.

(d) $F_4(s) = \dfrac{3s+2}{(s+1)^2(s-2)}$, $\sigma < -1$

as shown in Figure 5-4(d).

$$f_4(t) = 0 \text{ for } t > 0$$

since there are no poles to the left of σ.

For t < 0

$$f_4(t) = - \text{[sum of residues of poles of } F(s)e^{st}]$$

$$= (-0.33te^{-t} + 0.89e^{-t} - 0.89e^{2t})u(-t)$$

which agrees with Example 5-7(b).

(e) $F_5(s) = \dfrac{3s + 2}{(s + 1)^2(s - 2)}, \qquad \sigma > 2$

as shown in Figure 5-4(e).

$$f_5(t) = 0, \qquad t < 0$$

since there are no poles to the right of σ.

For t > 0

$$f_5(t) = \text{[sum of the residues of the poles of } F(s)e^{st}]$$

Therefore $\qquad f_5(t) = (0.33te^{-t} - 0.89e^{-t} + 0.89e^{2t})u(t)$

(f) $F_6(s) = \dfrac{2s + 3}{(s + 4)(s^2 + 2s + 3)}, \qquad -4 < \sigma < -1$

as shown in Figure 5-4(f).

$$= \dfrac{2s + 3}{(s + 4)(s + 1 + j1.4)(s + 1 - j1.4)}$$

For t > 0

$$f_6(t) = \text{[residue of the pole at } s = -4 \text{ for } F(s)e^{st}]$$

$$= -\frac{5}{11} e^{-4t}$$

$$= -0.45e^{-4t}$$

For t < 0

$$f_6(t) = - \text{[sum of the residues of the poles at } s = -1 - j1.4 \text{ and } -1 + j1.4]$$

$$= - \left[\frac{2(-1 - j1.4) + 3}{(3 - j1.4)(-j2.8)} e^{(-1-j1.4)t} + r_1^* e^{(-1+j1.4)t} \right]$$

where $\qquad\qquad r_1 = 0.22 + j0.22$

$$f_6(t) = - [(0.22 + j0.22)e^{(-1-j1.4)t} + (0.22 - j0.22)e^{(-1+j1.4)t}]$$

$$= - [e^{-t}(0.44 \, \text{Cos} \, 1.4t + 0.44 \, \text{Sin} \, 1.4t)]$$

Finally, as in Example 5-7(d) we obtain:

$$f_6(t) = -0.45e^{-4t}u(t) + [-0.44e^{-t}(\text{Cos} \, 1.4t + \text{Sin} \, 1.4t)]u(-t)$$

5-3-3 Complex Convolution

Complex convolution is an excellent illustration of the use of residues in transform theory. We now derive the Laplace transform for the product of two functions and illustrate complex convolution for some simple cases.

EXAMPLE 5-9
    Prove:

$$\mathcal{L}[x(t)y(t)] = \frac{1}{2\pi j} \int_{\substack{a-j\infty \\ C}}^{a+j\infty} Y(p)X(s-p)\, dp, \qquad \sigma_{s1} < \mathrm{Re}(s) < \sigma_{s2}$$

Pay particular attention to the relation between a and s.

Solution

Let $\qquad\qquad x(t) \leftrightarrow X(s), \qquad \sigma_{x1} < \sigma < \sigma_{x2}$

and $\qquad\qquad y(t) \leftrightarrow Y(s), \qquad \sigma_{y1} < \sigma < \sigma_{y2}$

The product function $x(t)y(t)$ must then have a region of convergence $\sigma_{x1} + \sigma_{y1} < \sigma < \sigma_{x2} + \sigma_{y2}$. This is so because $\int_0^\infty |x(t)|e^{-\sigma t}\, dt$ exists for $\sigma > \sigma_{x1}$ and $\int_0^\infty |y(t)|e^{-\sigma t}\, dt$ exists for the $\sigma > \sigma_{y1}$, and from the concept of the region of convergence then $\int_0^\infty |x(t)y(t)|e^{-\sigma t}\, dt$ exists for $\sigma > \sigma_{x1} + \sigma_{y1}$. Similarly, the behavior of $x(t)y(t)$ for $t < 0$ implies $\sigma < \sigma_{x2} + \sigma_{y2}$.

$$\mathcal{L}[x(t)y(t)] = \int_{-\infty}^{\infty} x(t)y(t)e^{-st}\, dt$$

$$= \int_{-\infty}^{\infty} x(t)\left[\frac{1}{2\pi j} \int_{\substack{\sigma_y - j\infty \\ C_1}}^{\sigma_y + j\infty} Y(p)\, e^{pt}\, dp \right] e^{-st}\, dt$$

where $\sigma_{y1} < \sigma_y < \sigma_{y2}$. Assuming it is permissible to interchange the order of integration, we have

$$\mathcal{L}[x(t)y(t)] = \frac{1}{2\pi j} \int_{\substack{\sigma_y - j\infty \\ C_1}}^{\sigma_y + j\infty} Y(p)\left[\int_{-\infty}^{\infty} x(t)e^{-(s-p)t}\, dt \right] dp$$

$$= \frac{1}{2\pi j} \int_{\substack{a - j\infty \\ C}}^{a+j\infty} Y(p)X(s-p)\, dp$$

$$= X(s) * Y(s)$$

We must be very careful with the allowable values of a. $X(s)*Y(s)$ exists for $\sigma_{x1} + \sigma_{y1} < \sigma < \sigma_{x2} + \sigma_{y2}$. In addition, to satisfy $Y(p)$ a must be such that $\sigma_{y1} < a < \sigma_{y2}$ and to satisfy $X(s-p)$ a must be such that $\sigma_{x1} < \mathrm{Re}(s) - a < \sigma_{x2}$ or $-\sigma_{x2} + \mathrm{Re}(s) < a < -\sigma_{x1} + \mathrm{Re}(s)$. To clarify any confusion about this, we now solve a few problems.

EXAMPLE 5-10

(a) Given:

$$x(t) = e^{2t} u(t) \quad \text{and} \quad y(t) = tu(t)$$

use complex convolution to find the Laplace transform of $x(t)y(t)$.

(b) Given:

$$x(t) = e^t u(-t) + u(t) \quad \text{and} \quad y(t) = e^{2t} u(-t)$$

use complex convolution to find the Laplace transform of $x(t)y(t)$.

Solution

(a) $$x(t) = e^{2t} u(t)$$

therefore $$X(s) = \frac{1}{s-2}, \quad \sigma > 2$$

$$y(t) = tu(t)$$

thus $$Y(s) = \frac{1}{s^2}, \quad \sigma > 0$$

$\overline{x(t)y(t)}$ will exist for $\sigma > 2$.

$$\overline{x(t)y(t)} = \frac{1}{2\pi j} \int_{\substack{a-j\infty \\ c}}^{a+j\infty} \frac{1}{p^2} \frac{1}{s-p-2} dp$$

where $\text{Re}(s) > 2$ and there is a first-order pole at $p = s - 2$. We note that $\text{Re}(s - p) > 2$, and therefore $\text{Re}(s) - a > 2$ or $a < \text{Re}(s) - 2$, and in addition, $a > 0$. Figure 5-5(a) shows a plot of the two poles at $p = 0$ and $p = s - 2$, and since $\text{Re}(s) > 2$, the pole at $s - 2$ is always to the right of the pole at $p = 0$. Using the inside–outside theorem, we may close C to the left or right.

Therefore $\overline{x(t)y(t)}$ = [residue of the pole at $p = 0$
(when we close the contour
to the left)]

$$= \frac{d}{dp} \left[\frac{1}{s-p-2} \right]_{p=0}$$

$$= -(s-p-2)^{-2}(-1)|_{p=0}$$

$$= \frac{1}{(s-2)^2}$$

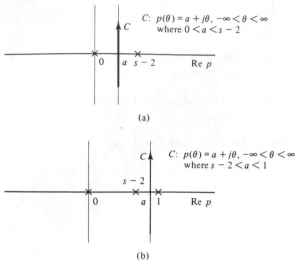

Figure 5-5 (a) Poles of $Y(p)$ and $X(s - p)$ for Example 5-9(a); (b) poles of $X(p)$ and $Y(s - p)$ for Example 5-9(b).

Also,
$$\overline{x(t)y(t)} = -[\text{residue of the pole at } p = s - 2$$
$$(\text{when we close the contour}$$
$$\text{to the right})]$$

$$= -\left[(p - (s - 2)) \frac{1}{p^2[-(p - s + 2)]} \right]\Bigg|_{p=s-2}$$

$$= \frac{1}{(s - 2)^2}$$

Therefore $\mathcal{L}[x(t)y(t)] = \dfrac{1}{(s - 2)^2}$, $\sigma > 2$

which can easily be verified directly.

(b) $x(t) = e^t u(-t) + u(t)$

and $X(s) = \dfrac{-1}{s - 1} + \dfrac{1}{s}$, $0 < \sigma < 1$

$$= \frac{-1}{s(s - 1)},   0 < \sigma < 1$$

$$y(t) = e^{2t} u(-t)$$

and $Y(s) = \dfrac{-1}{s - 2}$, $\sigma < 2$

Using complex convolution, we obtain:

$$\overline{x(t)y(t)} = \frac{1}{2\pi j} \int_{a-j\infty}^{a+j\infty} \frac{-1}{p(p - 1)} \frac{-1}{s - p - 2} dp$$

where the transform exists for $-\infty < \text{Re}(s) < 3$ and a for the path is such that $0 < a < 1$ and $\text{Re}(s - p) < 2$ or $\text{Re}(s) - 2 < a$ for any acceptable s. Figure 5-5(b) shows a plot of the poles of $-1/p(p - 1)(p - s + 2)$ where $(s - 2) < a < 1$ for any acceptable $\text{Re}(s)$. Again, we can find $x(t)y(t)$ in two ways by closing C to the left or right

$$\overline{x(t)y(t)} = [\text{residues of the poles at } p = 0 \text{ and } p = s - 2]$$

$$= \frac{-1}{-1(-s + 2)} + \frac{-1}{(s - 2)(s - 3)}$$

$$= \frac{-s + 3 - 1}{(s - 2)(s - 3)} = \frac{-1}{s - 3}$$

Also, $$\overline{x(t)y(t)} = -[\text{residue of the pole at } p = 1]$$

$$= -\left(\frac{-1}{-s + 3}\right)$$

$$= \frac{-1}{s - 3} \qquad \text{as before}$$

Therefore $$\overline{x(t)y(t)} = \frac{-1}{s - 3}, \qquad -\infty < \text{Re}(s) < 3$$

The integration is very tricky. We carefully note the pole at $p = s - 2$ is always to the left of the pole at $p = 1$.

Although using such tricky mathematics to find the Laplace transform of easy product functions may seem cumbersome, trying to understand the principle is worthwhile. Historically, complex convolution was instrumental in the development of the fast Fourier transform.

A very important application of complex convolution is to relate the Laplace transform of a continuous function $f(t)$ to the Laplace transform of $f^*(t) = \tau \sum_{-\infty}^{\infty} f(n\tau)\delta(t - n\tau)$. We will carry this out for causal functions and leave the general case as an exercise.

EXAMPLE 5-11

Given $f(t)$ has a Laplace transform $F(s)$ with $\sigma > \sigma_1$, use complex convolution to find the Laplace transform of:

$$f^*(t) = \tau \sum_{0}^{\infty} f(n\tau)\delta(t - n\tau)$$

Solution. Using the sifting property of delta functions $f(t)\delta(t - a) = f(a)\delta(t - a)$, we obtain:

$$f^*(t) = \sum_{0}^{\infty} \tau f(n\tau)\delta(t - n\tau)$$

$$= \tau f(t) \sum_{0}^{\infty} \delta(t - n\tau)$$

$$\mathcal{L}\left[\sum_{0}^{\infty} \delta(t - n\tau)\right] = 1 + e^{-\tau s} + e^{-2\tau s} + \cdots + e^{-n\tau s} + \cdots$$

$$= \frac{1}{1 - e^{-\tau s}}, \qquad \sigma > 0$$

Using complex convolution, we have:

$$\mathcal{L}\left[\tau f(t) \sum_{0}^{\infty} \delta(t - n\tau)\right] = \frac{1}{2\pi j} \int_{C} \frac{\tau F(p)}{1 - e^{-\tau(s-p)}} \, dp$$

If $C = \sigma + j\beta$, $-\infty < \beta < \infty$, then σ must be such that $\sigma > \sigma_1$, $\mathrm{Re}(s) - \sigma > 0$ for $\mathrm{Re}(s) > 0$.

EXAMPLE 5-12

Given

$$f(t) = 3e^{+4t}u(t)$$

and $f^*(t)$ is found by sampling $f(t)$ every 0.01 s and approximating it by:

$$f^*(t) = \sum_{n=0}^{\infty} 0.01 f\left(\frac{n}{100}\right) \delta\left(t - \frac{n}{100}\right)$$

find $F^*(s)$.

Solution

$$F(s) = \frac{3}{s + 4}, \qquad \sigma > -4$$

$$\mathcal{L}\left[\sum_{0}^{\infty} \delta\left(t - \frac{n}{100}\right)\right] = \frac{1}{1 - e^{-0.01s}}, \qquad \sigma > 0$$

and

$$F^*(s) = F(s) * \frac{0.01}{1 - e^{-0.01s}}$$

$$= \frac{1}{2\pi j} \int_{C} \frac{0.03}{p + 4} \frac{1}{1 - e^{-0.01(s-p)}} \, dp$$

where $C = \sigma + j\beta$ is such that $\sigma > -4$ and $\mathrm{Re}(s) - \sigma > 0$ or $\sigma < \mathrm{Re}(s)$ for any $\mathrm{Re}(s) > 0$. The path C is always to the right of $\mathrm{Re}(p) = -4$ and to the left of the poles due to $1 - e^{-0.01(s-p)}$. We should make a pole zero sketch similar to that of Figure 5-5(b).

Finding $F^*(s)$ by closing C to the left, we obtain:

$$F^*(s) = [\text{residue of the pole at } p = -4]$$

$$= \frac{0.03}{1 - e^{-0.01s}e^{-0.04}}$$

The form of $F^*(s)$ is somewhat unwieldy and from introductory complex variables we can see that $F^*(s)$ contains an infinite number of poles, all with the same real part. (Can you find them?) However, an important continuation of this problem is considered in Chapter 6 when the Z transform is found for the discrete function obtained by sampling $f(t)$ every τ seconds.

5-4 LINEAR SYSTEMS WITH RANDOM AND SIGNAL PLUS NOISE INPUTS

In Chapter 3 we found the output autocorrelation function and cross-correlation function of the input with the output when the input to a LTIC continuous system is a noise waveform with autocorrelation function $R_{xx}(\tau)$:

$$R_{yy}(\tau) = C_{hh}(\tau) * R_{xx}(\tau) \tag{5-13}$$

and

$$R_{xy}(\tau) = R_{xx}(\tau) \oplus h(\tau) \tag{5-14a}$$

or

$$R_{xy}(\tau) = h(\tau) * R_{xx}(\tau) \tag{5-14b}$$

These results are shown schematically in Figure 5-6(a). Let us denote $\mathcal{L}[R_{xx}(\tau)]$ by $S_{xx}(s)$, $\mathcal{L}[R_{xy}(\tau)]$ by $S_{xy}(s)$, $\mathcal{L}[R_{yx}(\tau)]$ by $S_{yx}(s)$, $\mathcal{L}[R_{yy}(\tau)]$ by $S_{yy}(s)$, and $\mathcal{L}[C_{hh}(\tau)]$ by $T(s)$. We call $S_{xx}(s)$ the power spectral density of $x(t)$, $S_{xy}(s)$ the cross-spectral density of $x(t)$ and $y(t)$, $S_{yx}(s)$ the cross-spectral density of $y(t)$ and $x(t)$, and $S_{yy}(s)$ the power spectral density of $y(t)$ and $T(s)$ the power transfer function. Technically, these names are more meaningful to physical interpretation when the Fourier transforms of $R_{xx}(\tau)$, $R_{xy}(\tau)$, $R_{yx}(\tau)$, and $C_{hh}(\tau)$ are used.

We now find expressions for these spectral quantities using the convolution and correlation theorems:

$$S_{yy}(s) = \mathcal{L}[h(t) \oplus h(t)] S_{xx}(s)$$

$$= [H(s)H(-s)] S_{xx}(s) \tag{5-15}$$

where

$$T(s) = H(s)H(-s)$$

$$S_{xy}(s) = \mathcal{L}[R_{xx}(\tau) \oplus h(\tau)]$$

$$= H(s)S_{xx}(-s) \tag{5-16}$$

or

$$S_{xy}(s) = \mathcal{L}[h(\tau) * R_{xx}(\tau)]$$

$$= H(s)S_{xx}(s) \tag{5-17}$$

These results are tabulated in Figure 5-6b. It can be shown that since $R_{xx}(\tau)$ is even, $S_{xx}(s)$ and $S_{xx}(-s)$ are equivalent. Before applying these formulas, we develop some symmetry properties for spectral functions.

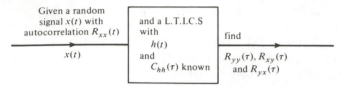

(a) Time-Domain Results

$$R_{yy}(\tau) = C_{hh}(\tau) * R_{xx}(\tau)$$

$$R_{xy}(\tau) = R_{xx}(\tau) \oplus h(\tau)$$

$$= h(\tau) * R_{xx}(\tau)$$

$$R_{yx}(\tau) = h(\tau) \oplus R_{xx}(\tau)$$

(b) Transform Results

$$L[h(t)] = H(s), \qquad L[C_{hh}(\tau)] = H(s)H(-s) = T(s), \qquad L[R_{xx}(\tau)] = S_{xx}(s)$$

$$S_{yy}(s) = S_{xx}(s)T(s)$$

$$S_{xy}(s) = H(s)S_{xx}(s) \quad \text{or} \quad H(s)S_{xx}(-s)$$

$$S_{yx}(s) = H(-s)S_{xx}(s) = S_{xy}(-s)$$

Figure 5-6 (a) The time-domain results for a system with a random input; (b) the transform results.

5-4-1 Properties of Spectral Functions

For $S_{xx}(s)$

It is easy to show $S_{xx}(s) = S_{xx}(-s)$. By definition:

$$S_{xx}(s) = \int_{-\infty}^{\infty} R_{xx}(\tau)e^{-s\tau}\, d\tau$$

Let $\tau = -p$

therefore
$$S_{xx}(s) = \int_{\infty}^{-\infty} R_{xx}(-p)e^{-s(-p)} - dp$$

$$= \int_{-\infty}^{\infty} R_{xx}(p)e^{sp}\, dp,$$

since
$$R_{xx}(p) = R_{xx}(-p)$$

and
$$S_{xx}(s) = S_{xx}(-s) \tag{5-18}$$

This implies that if $S_{xx}(s)$ contains a pole at $s = s_p$, it must also contain a pole at $s = -s_p$, and similarly, if $S_{xx}(s)$ contains a zero at $s = s_z$, it must also contain a zero at $s = -s_z$.

This is also clear from the time domain. If an even function contains $e^{\alpha t}u(t)$, then it must also contain $e^{-\alpha t}u(-t)$. The same is also true for $C_{hh}(\tau)$ where $C_{hh}(\tau) = h(\tau) \oplus h(\tau)$.

$$\mathcal{L}[h(\tau) \oplus h(\tau)] = H(s)H(-s)$$
$$= T(s)$$
$$T(s) = T(-s) \tag{5-19}$$

Since $S_{xx}(s)$ and $T(s)$ contain product terms such as $(s - s_p)(s + s_p) = (s^2 - s_p^2)$ in the denominator and $(s - s_z)(s + s_z) = (s^2 - s_z^2)$ in the numerator, then both the numerator and denominator will be real even polynomials in s.

$$S_{xx}(s) \text{ or } T(s) = \frac{b_m s^{2m} + b_{m-2}s^{2m-2} + \cdots + b_0}{a_n s^{2n} + a_{n-2}s^{2n-2} + \cdots + a_0} \tag{5-20}$$

For $S_{yx}(s)$

Since $R_{xy}(\tau) = R_{yx}(-\tau)$, it is easy to show that $S_{xy}(s) = S_{yx}(-s)$. By definition:

$$S_{xy}(s) = \int_{-\infty}^{\infty} R_{xy}(\tau)e^{-s\tau} d\tau$$

Let $p = -\tau$

therefore

$$S_{xy}(s) = \int_{\infty}^{-\infty} R_{xy}(-p)e^{sp} - dp$$
$$= \int_{-\infty}^{\infty} R_{xy}(-p)e^{sp} dp$$
$$= \int_{-\infty}^{\infty} R_{yx}(p)e^{sp} dp$$
$$= S_{yx}(-s)$$

and

$$S_{yx}(s) = H(-s)S_{xx}(s) \tag{5-21}$$

This implies that if $S_{xy}(s)$ has a pole at s_p or a zero at s_z, then $S_{yx}(s)$ has a pole at $-s_p$ or a zero at $-s_z$ and vice versa.

Summarizing the main properties of spectral functions, we have:

Property 1

$$S_{xx}(s) = S_{xx}(-s)$$

or

$$T(s) = T(-s)$$

where

$$T(s) = \mathcal{L}[C_{hh}(\tau)]$$

This implies a spectral function is the ratio of two even polynomials of s.

Property 2

$$S_{xy}(s) = S_{yx}(-s)$$

EXAMPLE 5-13

Given the impulse response of a system is:

$$h(t) = (3e^{-t} + 2te^{-2t})u(t)$$

Use the Laplace transform to find the power transfer function $T(s)$, and hence $C_{hh}(\tau)$.

Solution

$$h(t) = (3e^{-t} + 2te^{-2t})u(t)$$

$$H(s) = \frac{3}{s+1} + \frac{2}{(s+2)^2}$$

$$= \frac{3s^2 + 14s + 14}{(s+1)(s+2)^2}$$

$$= \frac{3(s+3.1)(s+1.5)}{(s+1)(s+2)^2}$$

$$L[h(t) \oplus h(t)] = T(s)$$

$$= \frac{3(s+3.1)(s+1.5)}{(s+1)(s+2)^2} \times \frac{3(-s+3.1)(-s+1.5)}{(-s+1)(-s+2)^2}$$

$$= \frac{9(+1)(s+3.1)(s-3.1)(s+1.5)(s-1.5)}{(-1)(s+1)(s-1)(-1)^2[(s+2)(s-2)]^2}$$

$$= \frac{-9(s+3.1)(s-3.1)(s+1.5)(s-1.5)}{(s+1)(s-1)[(s+2)(s-2)]^2},$$

$$-1 < \sigma < 1$$

As the ratio of two even polynomials this becomes:

$$T(s) = \frac{-9(s^2 - 9.6)(s^2 - 2.25)}{(s^2 - 1)(s^2 - 4)^2}$$

$$= \frac{-9[s^4 - 11.85s^2 + 21.6]}{s^6 - 9s^4 + 24s^2 - 16}, \quad -1 < \sigma < 1$$

and $\quad C_{hh}(\tau) = \frac{1}{2\pi j} \int_{\sigma-j\infty}^{\sigma+j\infty} \frac{-9(s^2 - 9.6)(s^2 - 2.25)}{(s-1)(s+1)(s-2)^2(s+2)^2} e^{s\tau}\, ds$

For $\tau > 0$

$$C_{hh}(\tau) = (\text{residue of pole at } -1) + (\text{residue of pole at } -2)$$

$$= \frac{-9(-8.6)(-1.25)}{(-2)(-3)^2(1)^2} e^{-\tau}\Big|_{s=-1}$$

$$+ \frac{d}{ds}\left[\frac{-9(s^4 - 11.85s^2 + 21.6)}{(s-1)(s+1)(s-2)^2} e^{s\tau} \right]\Big|_{s=-2}$$

$$= 5.4e^{-\tau} + \frac{-9(16 - 47.4 + 21.6)}{48}\tau e^{-2\tau}$$

$$+ e^{-2\tau} \frac{d}{ds}\left[\frac{-9(s^4 - 11.85s^2 + 21.6)}{s^4 - 4s^3 + 3s^2 + 4s - 4}\right]\Bigg|_{s=-2}$$

$$= 5.4e^{-\tau} + 1.8\tau e^{-2\tau} + 4.33e^{-2\tau}$$

Using the evenness of $C_{hh}(\tau)$, we get:

$$C_{hh}(\tau) = 5.4e^{-|\tau|} + 1.8\,|\tau|\,e^{-2|\tau|} + 4.33e^{-2|\tau|}$$

5-4-2 Deterministic Signal Plus Uncorrelated Zero-Mean Noise

Figure 5-7(a) shows a linear system with system function $H(s)$ and power transfer function $T(s) = H(s)H(-s)$. The input is $x(t) = f(t) + n(t)$, where $f(t)$ is a deterministic signal and $n(t)$ is zero-mean uncorrelated noise (i.e., $R_{fn}(\tau) = 0$) with autocorrelation function $R_{nn}(\tau)$. In Chapter 3 we found the deterministic output as:

$$g(t) = f(t) * h(t)$$

and the output noise autocorrelation function as:

$$R_{mm}(\tau) = R_{nn}(\tau) * C_{hh}(\tau)$$

and the cross-correlation of $n(t)$ with $m(t)$ as:

$$R_{nm}(\tau) = h(\tau) * R_{xx}(\tau)$$

Using the bilateral Laplace transform, we obtain:

$$G(s) = F(s)H(s) \tag{5-22}$$

$$S_{mm}(s) = S_{nn}(s)[H(s)H(-s)] \tag{5-23}$$

$$S_{nm}(s) = H(s)S_{nn}(s) \tag{5-24}$$

$$S_{mn}(s) = H(-s)S_{nn}(s) \tag{5-25}$$

These relations are summarized in Figure 5-7(b). We conclude this section by resolving Example 3-17 from Chapter 3 by using the bilateral Laplace transform.

EXAMPLE 5-14

Consider a linear system with impulse response $h(t) = 2e^{-3t}u(t)$ and with a deterministic input $f(t) = 3\,\mathrm{Cos}\,2t$ plus uncorrelated white noise with a mean square value of 100 whose autocorrelation function may be approximated by $R_{nn}(\tau) = 4\delta(\tau)$. Find the output signal, the output power spectral density $S_{mm}(s)$, and hence the output autocorrelation function and the input and output signal to noise ratios.

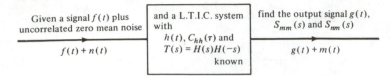

Given a signal $f(t)$ plus uncorrelated zero mean noise	and a L.T.I.C. system with $h(t)$, $C_{hh}(\tau)$ and $T(s) = H(s)H(-s)$ known	find the output signal $g(t)$, $S_{mm}(s)$ and $S_{nm}(s)$
$f(t) + n(t)$		$g(t) + m(t)$

(a) Time-Domain Results (Chapter 3)

$$g(t) = f(t)*h(t)$$

$$R_{mm}(\tau) = R_{nn}(\tau)*C_{hh}(\tau)$$

$$R_{nm}(\tau) = h(\tau)*R_{nn}(\tau)$$

(b) Transform Results

$$G(s) = H(s)F(s)$$

$$S_{mm}(s) = S_{nn}(s)T(s)$$

$$S_{nm}(s) = H(s)S_{nn}(s)$$

$$S_{mn}(s) = H(-s)S_{nn}(s)$$

Figure 5-7 (a) The time-domain results for a system with a signal plus uncorrelated noise input; (b) the transform results.

Solution

The Output Signal
As in Example 3-17, using phasors, we get:

$$g(t) = \text{Re}\left[\frac{2}{j2 + 3} \, 3\angle 0° \, e^{j2t}\right]$$

$$= 1.66 \, \text{Cos} \, (2t - 33°)$$

The Output Power Spectral Density

$$S_{nn}(s) = \overline{[4\delta(\tau)]} = 4$$

$$H(s) = \mathcal{L} \, [2e^{-3t}u(t)]$$

$$= \frac{2}{s + 3}$$

and the power transfer function is:

$$T(s) = H(s)H(-s)$$

$$= \frac{2}{s + 3}\frac{2}{-s + 3}$$

$$= \frac{-4}{s^2 - 9}; \qquad -3 < \sigma < 3$$

$$C_{hh}(\tau) = \frac{1}{2\pi j} \int_{\sigma - j\infty}^{\sigma + j\infty} \frac{-4}{s^2 - 9} e^{s\tau} \, ds$$

For $\tau > 0$

$$C_{hh}(\tau) = [\text{residue of the pole at } s = 3]$$

$$= \frac{-4}{-6} e^{-3\tau} u(\tau)$$

therefore

$$C_{hh}(\tau) = 0.67 e^{-3\tau} u(\tau) + 0.67 e^{3\tau} u(-\tau)$$

$$S_{mm}(s) = T(s) S_{nn}(s)$$

$$= \frac{-16}{s^2 - 9}, \qquad -3 < \sigma < 3$$

and

$$R_{mm}(\tau) = 2.67 e^{-3|\tau|}$$

As before, we can find:

$$\left. \frac{S}{N} \right|_{\text{input}} = \frac{\left(\dfrac{3}{\sqrt{2}}\right)^2}{100} = 0.045$$

and

$$\left. \frac{S}{N} \right|_{\text{output}} = \frac{\left(\dfrac{1.66}{\sqrt{2}}\right)^2}{R_{mm}(0)} = 0.53$$

This problem was just as easy in the time domain since the assumption of white noise made the calculation:

$$R_{mm}(\tau) = C_{hh}(\tau) * R_{nn}(\tau)$$

trivial. If $R_{nn}(\tau)$ is not so simple, the transform approach is easier, and in more difficult cases finding $R_{mm}(\tau)$ as the inverse of $S_{mm}(s)$ is available as a computer program.

SUMMARY

The two-sided Laplace transform was defined as $F(s) = \int_{-\infty}^{\infty} f(t) e^{-st} \, dt$, and if it exists it does so for a region of convergence $\sigma_1 < \text{Re}(s) < \sigma_2$. The behavior of $f(t)$ for positive time places the lower bound σ_1 on $\text{Re}(s)$ and the behavior of $f(t)$ for negative time places the upper bound of σ_2 on $\text{Re}(s)$. The experience of evaluating one-sided transforms may be utilized to find two-sided transforms. If $f(t) = f_1(t) u(t) + f_2(t) u(-t)$ and $F(s)$ exists, then $F(s) = F_1(s) + F_2(s)$ where $F_2(s)$ is the one-sided transform of $f_2(-t) u(t)$ with $s = -s$.

The most commonly occurring two-sided functions in linear system theory are correlation functions, whether $C_{hh}(\tau)$, the correlation of the impulse response of a system with itself, or the autocorrelation and cross-correlation functions of ergodic noise waveforms. When studying the main properties of two-sided transforms the reader should pay particular attention to the transform of correlation integrals:

$$\mathcal{L}\,[x(t) \oplus y(t)] = Y(s)X(-s)$$

and
$$\mathcal{L}\,[x(t) \oplus x(t)] = X(s)X(-s)$$

The time-domain results for a linear system with an ergodic random input whose autocorrelation function is $R_{xx}(\tau)$ are:

$$R_{yy}(\tau) = R_{xx}(\tau) * C_{hh}(\tau)$$

and
$$R_{xy}(\tau) = h(\tau) * R_{xx}(\tau)$$

In the transform domain these results are:

$$S_{yy}(s) = S_{xx}(s)\,T(s)$$

where
$$T(s) = H(s)\,H(-s)$$

and
$$S_{xy}(s) = H(s)S_{xx}(s)$$

where $S_{xx}(s)$ and $S_{yy}(s)$, the transforms of the autocorrelation functions, are called the power spectral densities. Using inverse transforms, we find that the correlation functions are:

$$R_{yy}(\tau) = \mathcal{L}^{-1}[S_{xx}(s)\,T(s)]$$

and
$$R_{xy}(\tau) = \mathcal{L}^{-1}[S_{xx}(s)\,H(s)]$$

Inverse transforms were evaluated in two ways. Any transform $F(s) = N(s)/D(s)$ where the order of $D(s)$ is at least one higher than $N(s)$ may be expanded in partial fractions and $f(t)$ is then found by table reference. For example, if $F(s) = 2/(s + 2) + 4/(s - 1)^2$; $-2 < \sigma < 1$, then $f(t) = 2e^{-2t}u(t) - 4te^{t}u(-t)$. Alternatively, the inverse may be found by the residue theory. From the definition of the inverse we have:

$$f(t) = \frac{1}{2\pi j} \int_{\substack{\sigma - j\infty \\ C}}^{\sigma + j\infty} F(s)e^{st}\,ds$$

By judiciously closing C to the left or right, we obtain:

When $t > 0$

$$f(t) = \Sigma[\text{sum of the residues of the poles of } F(s)e^{st} \text{ to the left of } \sigma]$$

When $t < 0$

$$f(t) = -\Sigma[\text{sum of the residues of the poles of } F(s)e^{st} \text{ to the right of } \sigma],$$

For example, the inverse transform of:

$$F(s) = \frac{2s^2 + 10}{(s + 2)(s - 1)^2}, \qquad -2 < \sigma < 1$$

is $\qquad f(t) = \dfrac{2(4) + 10}{9} e^{-2t} u(t) - \dfrac{d}{ds}\left[\dfrac{2s^2 + 10}{s + 2} e^{st}\right]_{s-1} u(-t)$

$\qquad\qquad = 2e^{-2t} u(t) - 4te^t u(-t)$

PROBLEMS

5-1. Find the two-sided Laplace transform and state the region of convergence for the following functions:

(a) $f_1(t) = (2 + 3t)u(t) - 3e^t u(-t)$ (b) $f_2(t) = (2 + 3t - 3e^t)u(t)$
(c) $f_3(t) = (2 + 3t - 3e^t)u(-t)$ (d) $f_4(t) = t^{10} e^{-2t} u(-t)$
(e) $f_5(t) = 6.2 \cos(4t - 40°)u(-t)$ (f) $f_6(t) = 2e^{-2|t|}$
(g) $(1 - |t|)[u(t + 1) - u(t + 1)]$

5-2. (a) Given:

$$f(t) = te^{-2t}u(t) - 4e^t u(-t)$$

and $\qquad\qquad g(t) = f(t - 4)$

Use the shifting theorem to find $G(s)$. How does the pole zero configuration and region of convergence of $G(s)$ compare to those of $F(s)$?

(b) Plot the following functions and find their two-sided Laplace transforms:

(i) $e^{-(t-3)}u(t-3)$ (ii) $2(t + 4)e^{-(t+4)}u(-t - 4)$
(iii) $3e^{3t}u(-t + 3)$

5-3. As quickly as possible find the inverse transforms of:

(a) $\dfrac{2}{s + 2},\qquad \sigma < -2$ (b) $\dfrac{2}{(s + 3)^2},\qquad \sigma < -3$

(c) $\dfrac{2s + 3}{s^2 + 4s + 3},\qquad \sigma < -3$

5-4. Evaluate the following inverse Laplace transforms using partial fractions and table reference. Plot the time function in each case:

(a) $F_1(s) = \dfrac{s^2}{(2s + 1)(s - 3)^2},\qquad -0.5 < \sigma < 3$

(b) $F_2(s) = \dfrac{s^2}{(2s + 1)(s - 3)},\qquad -0.5 < \sigma < 3$

(c) $F_3(s) = \dfrac{3s + 2}{s^2 - 9},\qquad \sigma < -3$

(d) $F_4(s) = \dfrac{2s^2 - 1}{s^3 + 3s + 2},\qquad$ where $f_4(t)$ is causal.

5-5. Repeat Problem 5-3 using the residue theory.

5-6. If possible, evaluate the following convolution and correlation integrals:

(a) $\delta(t - a) * \delta(t - b)$ (b) $\delta(t - a) \oplus \delta(t - b)$
(c) $\delta(t - b) \oplus \delta(t - a)$ (d) $2e^{-2t}u(t) * 2e^{-t}u(t)$
(e) $2e^{-2t}u(t) \oplus 2e^{-t}u(t)$ (f) $2e^{-2t}u(t) * 2e^{-t}u(-t)$
(g) $2e^{-2t}u(t) \oplus 2e^{-t}u(-t)$ (h) $2e^{-|t|} * 3e^{-2|t|}$
(i) $2e^{-|t|} \oplus 3e^{-2|t|}$ (j) $3e^{-2|t|} \oplus 2e^{-|t|}$

5-7. (a) If $x(t) \leftrightarrow X(s)$, $\sigma_{x1} < \sigma < \sigma_{x2}$
and $y(t) \leftrightarrow Y(s)$, $\sigma_{y1} < \sigma < \sigma_{y2}$
will the Laplace transform of $x(t)y(t)$ always exist?

(b) If $x(t) \leftrightarrow X(s)$, $-5 < \sigma < 3$
and $y(t) \leftrightarrow Y(s)$, $-1 < \sigma < 5$

and both denominator polynomials are of second order:
(1) for what range of $\text{Re}(s)$ does $\overline{x(t)y(t)}$ exist
(2) sketch the poles of $X(p)Y(s-p)$ and the contour of integration for

$$\overline{x(t)y(t)} = \frac{1}{2\pi j} \int_{\substack{\sigma - j\infty \\ C}}^{\sigma + j\infty} X(p)Y(s-p)\,dp$$

5-8. (a) Use the complex convolution to find the Laplace transform of $x(t)y(t)$

where $x(t) = 2te^{-t}u(-t) + e^{-4t}u(t)$

and $y(t) = e^{t}u(-t) + tu(t)$

(b) Check your answer by finding the transform directly.

5-9. Given:

$$\overline{x(t)} = X(s), \qquad \sigma_{x1} < \sigma < \sigma_{x2}$$

and $$\overline{y(t)} = Y(s), \qquad \sigma_{y1} < \sigma < \sigma_{y2}$$

(a) What are the conditions for $x(t)$ and $y(t)$ to be stable?
(b) List when the following are stable for $x(t)$ and $y(t)$ stable, and if they can be unstable give a specific example for $x(t)$ and $y(t)$:
(1) $x(t)y(t)$ (2) $x(t)*y(t)$ (3) $x(t) \oplus x(t)$
(4) $x(t) \oplus y(t)$ (5) $y(t) \oplus x(t)$
(c) Sketch pole diagrams.

5-10. (a) Find the output of a system with the system function:

$$H(s) = 3/(s+2), \qquad \sigma > -2$$

when the input is:

$$x(t) = 2u(-t) + te^{-t}u(t)$$

(b) Plot $y(t)$.

5-11. (a) Prove that the mean square value of a random process with power spectral density $S_{xx}(s)$ is:

$$\overline{x^2(t)} = \frac{1}{2\pi j} \int_{\sigma - j\infty}^{\sigma + j\infty} S_{xx}(s)\,ds, \qquad \sigma_1 < \sigma < \sigma_2$$

$$= \Sigma[\text{residues of the poles of } S_{xx}(s) \text{ to the left of } \sigma]$$

or $$= -\Sigma[\text{residues of the poles of } S_{xx}(s) \text{ to the right of } \sigma]$$

(b) For the continuous process:

$$S_{xx}(s) = \frac{-s^2 + 9}{s^4 - 5s^2 + 4}$$

find the mean square value of $x(t)$ using residue theory.

5-12. Given the input to a system with system function $H(s) = 2/(s+5)$ is essentially white noise with $S_{xx}(s) = 4$ and $\overline{x^2(t)} = 50$. Find:

(a) the output power spectral density and cross-power spectral density between the input and output.

(b) the output noise fluctuations $y^2(t)$ using:

$$\overline{y^2(t)} = R_{yy}(0) = \frac{1}{2\pi j} \int_{\sigma-j\infty}^{\sigma+j\infty} S_{yy}(s) \, ds$$

$$= \Sigma \,[\text{residues of the poles to the left of } \sigma]$$

$$= -\Sigma \,[\text{residues of the poles to the right of } \sigma].$$

5-13. If the input to the system of Problem 5-12 consists of the deterministic signal $f(t) = 20 \cos(50t - 30°)$ plus the same white noise with $S_{nn}(s) = 4$ and $\overline{n^2(t)} = 50$, find the input and output signal to noise ratios.

5-14 The power transfer function is defined as:

$$T(s) = H(s)H(-s)$$

(a) Find the power transfer function for the following systems and plot the pole zero diagram:

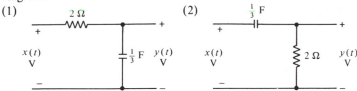

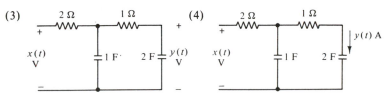

(b) If the input to each of the four systems of part (a) is assumed white noise with $S_{xx}(s) = 2$ and $\overline{x^2(t)} = 100$ use residue theory to find the mean squared fluctuations at the output $\overline{y^2(t)} = R_{yy}(0)$.

5-15. (a) If the input to system (1) of part (a) of the Problem 5-14 is random noise with $R_{xx}(\tau) = 50e^{-4|\tau|}$, find the power spectral density of the output noise $S_{yy}(s)$ and the cross-spectral density of the input and output noise $S_{xy}(s)$.

(b) give a pole zero plot for $S_{xy}(s)$, $S_{yx}(s)$, and $S_{yy}(s)$.

(c) Find the signal to noise ratio at the input and output if an input signal $f(t) = 2$ is added to the input noise.

5-16. A power spectral density $S_{xx}(s)$ or power transfer function may be written as:

$$S_{xx}(s) = G(s)G(-s) \quad \text{or} \quad T(s) = H(s)H(-s)$$

where $G(s)$ or $H(s)$ have their poles and zeros in the left half plane. Design a "shaping filter" $H(s)$ that transforms white noise with $S_{xx}(s) = 2$ to noise with a power spectral density $S_{yy}(s) = (-s^2 + 1)/(s^4 + 81)$.

5-17. In the statistical communication theory, a system is often designed assuming input white noise and a prefilter is then used to transform the actual power spectral density $S_{xx}(s)$ to white noise. This is the reverse of Problem 5-16. Find $H(s)$ to convert $S_{xx}(s) = (-s^2 + 1)/(s^4 + 81)$ to white noise $S_{yy}(s) = 1$.

5-18. Which of the following functions qualify as power spectral densities?

(a) $\dfrac{2}{s^2 - 8}$ (b) $\dfrac{-2}{s^2 - 8}$

(c) $\dfrac{s}{s^2 - 8}$ (d) $\dfrac{-s^2 + 1}{s^2 - 8}$

(e) $\dfrac{-4}{(s^2 - 9)^2}$ (f) $\dfrac{1 - s^2}{(s^2 - 9)^2}$

(g) $\dfrac{s^2 - 1}{(s^2 - 9)^2}$ (h) $\dfrac{1 - s^2}{s^4 - 8s^2 + 16}$

The One-sided Z Transform

INTRODUCTION

In Chapter 2 the time-domain analysis of LTIC discrete systems was treated. Linear difference equations of the type:

$$a_n y(n) + \cdots + a_{n-p} y(n - p) = f(n)$$

were solved classically and iteratively. It was seen that the homogeneous solution contained terms of the form $A_1 \alpha^n$ or $(A_1 + A_2 n)\alpha^n$, and so on. Similarly, when $f(n)$ was of the form $(A_0 + A_1 n + A_2 n^2)\alpha^n$ the forced response was readily found by logically assuming a solution and using substitution. A general LTIC discrete system is characterized by a difference equation:

$$a_n y(n) + a_{n-1} y(n - 1) + \cdots + a_{n-p} y(n - p)$$
$$= b_n x(n) + \cdots + b_{n-l} x(n - l)$$

where $x(n)$ and $y(n)$ denote the input and output, respectively. The pulse response $h(n)$ was defined as the output when the input $x(n) = \delta(n)$ and pulse responses were solved for classically, by assuming a homogeneous type solution. If the input to a LTIC discrete system is $x(n) = \Sigma_k x(k)\delta(n - k)$ then the zero-state output $y(n)$ was found as the convolution of the input and pulse response:

$$y(n) = x(n) * h(n)$$

$$= \sum_p x(p)h(n - p) \quad \text{or} \quad \sum_l h(l)x(n - l)$$

In Chapter 6 we consider the Z transform analysis of LTIC systems with causal inputs. The Z transform will be developed in an analogous manner to the Laplace transform of Chapter 4, using the following stages:

1. The definition is given and a number of transforms evaluated. From our discussion of Chapter 2 we focus on functions of the form $f(n) = (a_1 + a_2 n + \cdots)\alpha^n$.
2. The important theorems and properties of the Z transform are considered.
3. Inverse transforms are found by using partial fractions and table reference.
4. The Z transform is used to solve difference equations, to find pulse responses, and to determine the outputs of LTIC discrete systems for transformable inputs.

6-1 DEFINITION AND EVALUATION OF SOME TRANSFORMS

6-1-1 Definition

The **one-sided Z transform** of a *real* discrete function $f(n)$ is defined as:

$$F(z) = \sum_{n=0}^{\infty} f(n)z^{-n} \tag{6-1}$$

where $|z|$ is such that the series $\sum_0^\infty f(n)z^{-n}$ converges. This leads to an allowable "annulus" of convergence $|z| > p$. The definition is abstract but we will proceed, finding a few transforms by adhering to the definition and then coming back and commenting on its motivation. The following notations will be used interchangeably for the Z transform of $f(n)$: $F(z)$, $\overline{f(n)}$, and $Z[f(n)]$, whereas $f(n)$, $\overline{F(z)}$, and $Z^{-1}[F(z)]$ will be used for the inverse. In addition $f(n) \leftrightarrow F(z)$ indicates a Z transform pair.

A few series and results taken from calculus will constantly be used. An infinite geometric progression:

$$S = a + ar + ar^2 + \cdots + ar^n + \cdots$$

converges if $|r| < 1$ and its sum is $a/(1 - r)$. If r is complex, we obtain a real series plus j times another real series and both converge if $|r| < 1$. Some famous series, which carry over to the complex case and their regions of convergence are:

$$\frac{1}{1 + x} = 1 - x + x^2 - x^3 + x^4 - \cdots, \qquad = \sum_{n=0}^{\infty} (-x)^n, \qquad |x| < 1$$

$$\frac{1}{1 - x} = 1 + x + x^2 + x^3 + x^4 + \cdots, \qquad = \sum_{n=0}^{\infty} x^n, \qquad |x| < 1$$

$$e^x = 1 + x + \frac{1}{2}x^2 + \frac{1}{3!}x^3 + \frac{1}{4!}x^4 + \cdots, \qquad = \sum_{n=0}^{\infty} \frac{1}{n!}x^n, \qquad |x| < \infty$$

$$\text{Cos } x = 1 - \frac{1}{2}x^2 + \frac{1}{4!}x^4 + \cdots, \qquad = \sum_{n=0}^{\infty} (-1)^n \frac{x^{2n}}{(2n)!} \qquad |x| < \infty$$

$$\text{Sin } x = x - \frac{1}{3!}x^3 + \frac{1}{5!}x^5 + \cdots, \qquad = \sum_{n=0}^{\infty} (-1)^n \frac{x^{2n+1}}{(2n+1)!} \qquad |x| < \infty$$

EXAMPLE 6-1

Evaluate the Z transforms for the following functions:

(a) $f_1(n) = u(n)$ or   1*
(b) $f_2(n) = (0.5)^n u(n)$ or $(0.5)^n$
(c) $f_3(n) = (-3)^n u(n)$ or $(-3)^n$
(d) $f_4(n) = 2\delta(n) - 3\delta(n-2) + 4\delta(n-5)$
(e) $f_5(n) = 2 + 4(-3)^n$

Solution
(a) From the preceding definition we have:

$$F_1(z) = 1 + z^{-1} + z^{-2} + z^{-3} + \cdots$$

$$= \frac{1}{1 - z^{-1}}, \qquad \text{if } |z^{-1}| < 1$$

$$= \frac{z}{z - 1}, \qquad |z| > 1$$

Figure 6-1(a) shows $f_1(n)$ and $F_1(z)$ with its pole zero diagram.

(b) $f_2(n) = (0.5)^n$

Therefore $F_2(z) = 1 + 0.5z^{-1} + 0.5^2 z^{-2} + \cdots + 0.5^n z^{-n} + \cdots$

$$= \frac{1}{1 - 0.5z^{-1}}, \qquad |0.5z^{-1}| < 1$$

$$= \frac{z}{z - 0.5}, \qquad |z| > 0.5$$

$f_2(n)$ and $F_2(z)$ are shown in Figure 6-1(b).

(c) $f_3(n) = (-3)^n$

Therefore $F_3(z) = 1 + (-3)z^{-1} + (-3)^2 z^{-2} + (-3)^3 z^{-3} + \cdots$

$$= \frac{1}{1 - (-3z^{-1})}, \qquad |-3z^{-1}| < 1$$

*Because of the definition of the one-sided Z transform, the inclusion of $u(n)$ is immaterial: $f(n)$ and $f(n)u(n)$ have the same one-sided transform.

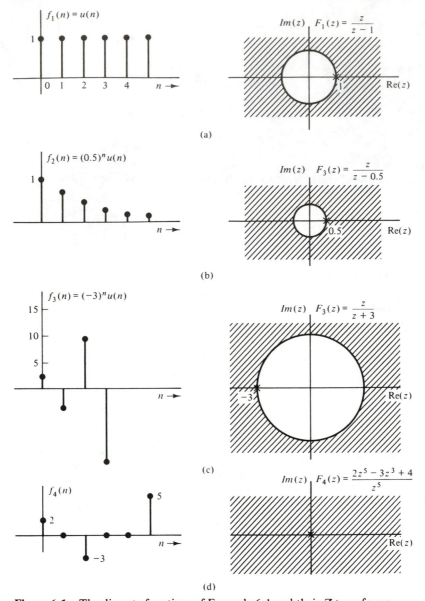

Figure 6-1 The discrete functions of Example 6-1 and their Z transforms.

$$= \frac{z}{z + 3}, \qquad |z| > 3$$

$f_3(n)$ and $F_3(z)$ are shown in Figure 6-1(c).

(d) $f_4(n) = 2\delta(n) - 3\delta(n - 2) + 4\delta(n - 5)$

Therefore $F_4(z) = 2 - 3z^{-2} + 4z^{-5}, \qquad$ for all z^{-1}

$$= \frac{2z^5 - 3z^3 + 4}{z^5}, \qquad |z| > 0$$

$f_4(n)$ and $F_4(z)$ are shown in Figure 6-1(d) and $F_4(z)$ has a pole at $z = 0$.

(e) $f_5(n) = 2 + 4(-3)^n$

Therefore
$$F_5(z) = \sum_{n=0}^{\infty} (2 + 4(-3)^n)z^{-n}$$

$$= \frac{2z}{z - 1} + \frac{4z}{z + 3}, \qquad |z| > 1 \cap |z| > 3$$

$$= \frac{6z^2 + 2z}{(z - 1)(z + 3)}, \qquad |z| > 3$$

Some facts are almost intuitively apparent from Example 6-1. If $F(z) = N(z)/D(z)$ then $F(z)$ converges $|z| > \rho$ where ρ is the distance from the origin to the furthest pole. Also, the order of $N(z)$ is at most the same as that of $D(z)$.

The definition may be extended to complex time functions where the Z transform is the transform of the real sequence plus j times the transform of the imaginary sequence. *Rarely* will we use complex time sequences.

6-1-2 Motivation for the Z Transform

In Chapter 2 the output of a LTIC discrete system was derived as:

$$y(n) = x(n)*h(n)$$

$$= \sum_{\text{all } p} x(p)h(n - p)$$

For $x(n)$ causal the summation extends from $p = 0$ to $p = n$. For example, if

$$\{x\} = \{x(0), x(1), x(2)\}$$

and
$$\{h\} = \{h(0), h(1), h(2)\}$$

then the output is:

$$\{x\}*\{h\} = \{x(0)h(0), x(1)h(0) + x(0)h(1), x(2)h(0) + x(1)h(1)$$
$$+ x(0)h(2), x(2)h(1) + x(1)h(2), x(2)h(2)\}$$

where
$$y(0) = x(0)h(0), \; y(1) = x(1)h(0) + x(0)h(1), \cdots, \quad \text{and}$$

$$y(4) = x(2)h(2)$$

The same result could be obtained for the coefficients of the resultant series if we multiply $[x(0) + x(1)z^{-1} + x(2)z^{-2}]$ by $[h(0) + h(1)z^{-1} + h(2)z^{-2}]$ to obtain:

$$x(0)h(0) + [x(1)h(0) + x(0)h(1)]z^{-1}$$

$$+ [x(2)h(0) + x(1)h(1) + x(0)h(2)]z^{-2} + \cdots + x(2)h(2)z^{-4}$$

Therefore multiplying series and carrying out convolution are identical. If we can obtain a quick way to multiply series, then we could utilize it to perform convolution quickly. For example, if we want to find:

$$u(n)*0.5^n u(n) = (1, 1, 1, \cdot \cdot \cdot)*(1, \tfrac{1}{2}, \tfrac{1}{4}, \tfrac{1}{8}, \cdot \cdot \cdot)$$

we can see this is equivalent to multiplying:

$$\left(\frac{1}{1 - z^{-1}}\right)\left(\frac{1}{1 - 0.5z^{-1}}\right) = \frac{z^2}{(z - 1)(z - 0.5)} = Y(z)$$

Instead of multiplying the two series longhand, there exists a means for expressing $Y(z)$ in a series. This is the Laurent series of complex variables. Although we discuss the Laurent series in Chapter 7, here we will use a rapid technique for finding the result by employing partial fractions and table reference. The reason for defining the Z transform to assign a series of negative powers of z to $x(n)$;

$$X(z) = x(0) + x(1)z^{-1} + x(2)z^{-2} + \cdot \cdot \cdot$$

is to achieve conformity with the Laplace transform. For all one-sided Z transforms the annulus of convergence is $|z| > \rho$ or z outside all the poles, which corresponds to the region of convergence $\sigma > \sigma_1$ or $\mathrm{Re}(s)$ to the right of all the poles for the Laplace transform.

6-1-3 An Expanded Set of Transforms

So far we have considered the transforms:

$$u(n) \longleftrightarrow \frac{1}{1 - z^{-1}} = \frac{z}{z - 1}$$

and

$$a^n \longleftrightarrow \frac{1}{1 - az^{-1}} = \frac{z}{z - a}$$

Indeed, the first is a special case of the second with $a = 1$. We now extend the transform for $a^n u(n)$ to find many others. Since:

$$\frac{z}{z - a} = 1 + az^{-1} + a^2 z^{-2} + a^3 z^{-3} + \cdot \cdot \cdot$$

$$= \sum_{n=0}^{\infty} a^n z^{-n} \tag{6-2}$$

we can differentiate both sides with respect to a to obtain:

$$\frac{-z(-1)}{(z - a)^2} = z^{-1} + 2az^{-2} + 3a^2 z^{-3} + \cdot \cdot \cdot + na^{n-1}z^{-n} + \cdot \cdot \cdot$$

$$= \sum_{n=0}^{\infty} na^{n-1}z^{-n} \quad \text{or} \quad \sum_{n=1}^{\infty} na^{n-1}z^{-n}$$

Therefore $\quad\dfrac{z}{(z-a)^2} = \displaystyle\sum_{n=0}^{\infty} na^{n-1}z^{-n}$ $\qquad\qquad$ (6-3)

and the sequence $(0, 1, 2a, 3a^2, \cdots , na^{n-1}, \cdots)$ or the function $f(n) = na^{n-1}$

has the Z transform $z/(z-a)^2$:

$$na^{n-1} \longleftrightarrow \dfrac{z}{(z-a)^2} \qquad\qquad (6\text{-}4)$$

Proceeding and differentiating both sides of expression (6-3) with respect to a, we obtain:

$$\dfrac{2z}{(z-a)^3} = \sum_{n=0}^{\infty} n(n-1)a^{n-2}z^{-n} \quad\text{or}\quad \sum_{n=2}^{\infty} n(n-1)a^{n-2}z^{-n} \qquad (6\text{-}5)$$

Since $n(n-1) = 0$ for $n = 0$ and 1, the lower limit on the summation may be 0, 1, or 2 as the sequence commences $0, 0, 2(1)z^{-2}$, and so on.

From Equation 6-5 we obtain the Z transform pair:

$$n(n-1)a^{n-2} \longleftrightarrow \dfrac{2z}{(z-a)^3} \qquad\qquad (6\text{-}6)$$

Now differentiating expression (6-6) with respect to a, we obtain from both sides:

$$\dfrac{(3 \times 2)z}{(z-a)^4} = \sum_{0}^{\infty} n(n-1)(n-2)a^{n-3}z^{-n} \qquad\qquad (6\text{-}7)$$

and proceeding inductively, we conclude:

$$\dfrac{(p-1)!z}{(z-a)^p} = \sum_{0}^{\infty} (n)(n-1) \times \cdots \times (n-p+2)a^{n-(p-1)}z^{-n} \qquad (6\text{-}8)$$

Equation 6-8 may compactly be written as:

$$\dfrac{(p-1)!z}{(z-a)^p} = \sum_{n=0}^{\infty} (n)_{p-1}a^{n-p+1}z^{-n} \qquad\qquad (6\text{-}9)$$

where $(n)_{p-1}$ is called "n truncated $p-1$" and is $(n)(n-1) \times \cdots \times (n-p+2)$. For example, $(7)_3 = 7 \times 6 \times 5$ and $(12)_5 = 12 \times 11 \times 10 \times 9 \times 8$. Equation 6-9 yields the very general and powerful Z transform pair:

$$\boxed{\; (n)_{p-1}a^{n-p+1} \longleftrightarrow \dfrac{(p-1)!z}{(z-a)^p} \;} \qquad \text{for any integer } 1 \le p < n \qquad (6\text{-}10)$$

Now we need to pause and apply some of the preceding derivations. We should be able inductively to proceed rapidly from:

$$\frac{z}{z-a} = \frac{1}{1-az^{-1}} = 1 + az^{-1} + a^2z^{-2} + \cdots$$

$$= \sum_{n=0}^{\infty} a^n z^{-n}$$

to Equation 6-10 and then deductively use Equation 6-10 to write expressions (6-2) and (6-3) or (6-4) and (6-5) and so on. How do we get Equation 6-3 from Equation (6-10)? We must have $p = 2$, $(p - 1)! = 1$, $(n)_1 = n$ and $a^{n-p+1} = a^{n-1}$. The reader is encouraged to concentrate on and become proficient with Equations 6-2 through 6-10.

We can now enumerate the following transforms involving higher-order poles in $F(z)$, all stemming from $a^n \leftrightarrow z/z - a$:

$$na^{n-1} \leftrightarrow \frac{z}{(z-a)^2}$$

$$n(n-1)a^{n-2} \leftrightarrow \frac{2z}{(z-a)^3}$$

$$n(n-1)(n-2)a^{n-3} \leftrightarrow \frac{3!z}{(z-a)^4}$$

and $\qquad (n)_p a^{n-p} \leftrightarrow \dfrac{p!z}{(z-a)^{p+1}}$ (This is Equation 6-10 slightly adjusted.)

for any $p \geq 0$. These transforms are plotted in Figure 6-2 for some general a where a can be any real number $-\infty < a < \infty$.

EXAMPLE 6-2
Find the Z transform of:

(a) $nu(n)$
(b) $n^2u(n)$
(c) $n^3u(n)$

Solution
(a) Since:

$$na^{n-1}u(n) \leftrightarrow \frac{z}{(z-a)^2}$$

then with $a = 1$ we obtain:

$$nu(n) \leftrightarrow \frac{z}{(z-1)^2}$$

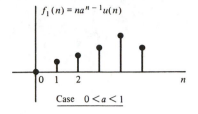

$f_1(n) = na^{n-1}u(n)$

Case $0 < a < 1$

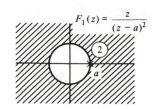

$F_1(z) = \dfrac{z}{(z-a)^2}$

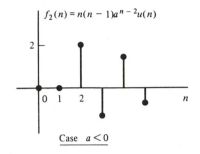

$f_2(n) = n(n-1)a^{n-2}u(n)$

Case $a < 0$

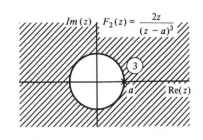

$F_2(z) = \dfrac{2z}{(z-a)^3}$

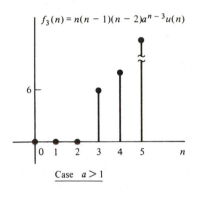

$f_3(n) = n(n-1)(n-2)a^{n-3}u(n)$

Case $a > 1$

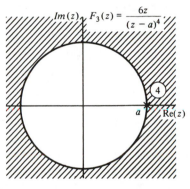

$F_3(z) = \dfrac{6z}{(z-a)^4}$

Figure 6-2 Some important Z transform pairs.

(b) From the transform pair:

$$n(n-1)a^{n-2}u(n) \leftrightarrow \frac{2z}{(z-a)^3}$$

with $a = 1$ we get:

$$Z[n(n-1)u(n)] = \frac{2z}{(z-1)^3}$$

$$n^2u(n) = n(n-1)u(n) + nu(n)$$

Therefore $\qquad Z[n^2u(n)] = \dfrac{2z}{(z-1)^3} + \dfrac{z}{(z-1)^2}$

$$= \dfrac{z^2 + z}{(z-1)^3}$$

(c) From the transform pair:

$$n(n-1)(n-2)a^{n-3}u(n) \leftrightarrow \dfrac{6z}{(z-a)^4}$$

with $a = 1$ we obtain:

$$Z[n(n-1)(n-2)u(n)] = \dfrac{6z}{(z-1)^4}$$

or $\qquad Z[(n^3 - 3n^2 + 2n)u(n)] = \dfrac{6z}{(z-1)^4}$

$$n^3u(n) = [n(n-1)(n-2)]u(n)$$

$$+ 3n^2u(n) - 2nu(n)$$

Therefore $\qquad Z[n^3u(n)] = \dfrac{6z}{(z-1)^4} + \dfrac{3z^2 + 3z}{(z-1)^3} - \dfrac{2z}{(z-1)^2}$

$$= \dfrac{z^3 + 4z^2 + z}{(z-1)^4}$$

Table 6-1 gives a list of Z transforms. It is remarkable that the first 12 transforms are all derivations of the basic transform pair $a^n \leftrightarrow z/(z-a)$. In addition, some extra transforms involving complex conjugate roots in $F(z)$ are included. In the text we confine our discussion to transforms with simple and higher-order real poles. Several more transforms will now be determined for typically occurring discrete functions.

EXAMPLE 6-3
Find the Z transforms of the following effectively using Table 6-1:

(a) $f_1(n) = 6n^2(0.8)^n$
(b) $f_2(n) = (2 + 3n)(-3)^n + (n^2 - 4)(-0.6)^{n-1}$
(c) $f_3(n) = (a_1 + a_2n + a_3n^2)\alpha^n$

Solution
(a) From Table 6-1 we have:

$$Z[6\,n(n-1)\,(0.8)^{n-2}] = \dfrac{12z}{(z-0.8)^3}$$

TABLE 6-1 A TABLE OF Z TRANSFORMS

$f(n)$	$F(z)$
a^n or $a^n u(n)$	$\dfrac{z}{z-a}$, $\|z\| > \|a\|$
$\delta(n)$	1, for all z
$u(n)$ or 1	$\dfrac{z}{z-1}$, $\|z\| > 1$
na^{n-1}	$\dfrac{z}{(z-a)^2}$, $\|z\| > \|a\|$
n	$\dfrac{z}{(z-1)^2}$, $\|z\| > 1$
$n(n-1)a^{n-2}$	$\dfrac{2z}{(z-a)^3}$, $\|z\| > \|a\|$
$n(n-1)$	$\dfrac{2z}{(z-1)^3}$, $\|z\| > 1$
n^2	$\dfrac{z^2+z}{(z-1)^3}$, $\|z\| > 1$
$n(n-1)(n-2)a^{n-3}$	$\dfrac{6z}{(z-a)^4}$, $\|z\| > \|a\|$
$n(n-1)(n-2)$	$\dfrac{6z}{(z-1)^4}$, $\|z\| > 1$
n^3	$\dfrac{*}{(z-1)^4}$, $\|z\| > 1$
$(n)_{p-1}a^{n-p+1}$	$\dfrac{(p-1)!\,z}{(z-a)^p}$, $\|z\| > \|a\|$
e^{jnw}	$*$,
$\text{Cos } nw$	$\dfrac{z^2-z\,\text{Cos}w}{(z-1\angle w)(z-1\angle -w)}$
$\text{Sin } nw$	$\dfrac{z\,\text{Sin}w}{(z-1\angle w)(z-1\angle -w)}$

*The reader should fill in the two blanks

We now write:

$$f_1(n) = 6n^2(0.8)^n = 6[n\,(n-1) + n]0.8^n$$

$$= 6n(n-1)0.8^2\,0.8^{n-2} + 6n(0.8)(0.8)^{n-1}$$

Therefore $$F_1(z) = \frac{12(0.8)^2 z}{(z-0.8)^3} + \frac{6(0.8)z}{(z-0.8)^2}$$

The result may also be given as:

$$F_1(z) = \frac{4.8z^2 + 3.8z}{(z-0.8)^3}$$

(b) In order to utilize our basic transforms, we write:

$$f_2(n) = (2 + 3n)(-3)^n + (n^2 - 4)(-0.6)^{n-1}$$

$$f_2(n) = 2(-3)^n + 3(-3)n(-3)^{n-1}$$

$$+ n(n - 1)(-0.6)(-0.6)^{n-2}$$

$$+ n(-0.6)^{n-1} - 4(-0.6)^{-1}(-0.6)^n$$

Therefore $$F_2(z) = \frac{2z}{(z + 3)} + \frac{-9z}{(z + 3)^2} + \frac{2(-0.6)z}{(z + 0.6)^3} + \frac{z}{(z + 0.6)^2}$$

$$+ \frac{6.7z}{z + 0.6}$$

$$= \frac{N(z)}{(z + 3)^2 (z + 0.6)^3}$$

where the reader may find $N(z)$.

(c) $f_3(n) = (a_1 + a_2 n + a_3 n^2) \alpha^n$

$$= a_1\alpha^n + a_2\alpha n\alpha^{n-1} + a_3 n(n - 1)\alpha^2\alpha^{n-2} + a_3 n\alpha\alpha^{n-1}$$

$$= \frac{a_1 z}{z - \alpha} + \frac{(a_2\alpha + a_3\alpha)z}{(z - \alpha)^2} + \frac{2a_3\alpha^2 z}{(z - \alpha)^3}$$

$$= \frac{N(z)}{(z - \alpha)^3}$$

where the reader may find $N(z)$. It is worth noting that both numerator and denominator polynomials are of order 3, and $N(z)$ does not contain a constant term.

6-1-4 Some Comments About Z Transforms

If we examine closely the Z transforms found so far, we see that certain characteristics are evident. For example, we determined the following transform pairs:

$$1 \leftrightarrow \frac{z}{z - 1} \tag{1}$$

$$a^n \leftrightarrow \frac{z}{z - a} \tag{2}$$

$$n^2 \leftrightarrow \frac{z^2 + z}{(z - 1)^3} \tag{3}$$

$$n(n - 1)a^{n-2} \leftrightarrow \frac{2z}{(z - a)^3} \tag{4}$$

$$(2 + 3n) \leftrightarrow \frac{2z^2 + z}{(z - 1)^2} \tag{5}$$

We note that for transforms (1), (2), and (5) the order of the numerator is the same as that of the denominator. Also, for each of our transforms the numerator does not contain a constant term. Writing a Z transform longhand, we have that:

$$F(z) = f(0) + f(1)z^{-1} + \cdots$$

indicates the order of the numerator and denominator are the same if $f(0) \neq 0$. For example, in transform (3) since $n^2 = 0$ when $n = 0$ the order of the numerator is one less than the denominator and in transform (4) since $n(n-1)a^n$ equals zero for $n = 0$ and $n = 1$, the order of the numerator is two less than the order of the denominator.

6-2 IMPORTANT THEOREMS OF THE Z TRANSFORM

Table 6-2 lists some of the more important Z transform theorems. As in the case of the Laplace transform, these theorems are used repeatedly in applications and must form part of a basic vocabulary. We now prove and illustrate a number of them by using examples.

EXAMPLE 6-3
   (a) Prove the transform-scaling theorem.
   (b) Using the transform pair:

$$u(n) \leftrightarrow \frac{z}{z-1}$$

verify the transform of $(-2)^n u(n)$ with this theorem.

TABLE 6-2   SOME ONE-SIDED Z TRANSFORM THEOREMS

Given $f(n) \leftrightarrow F(z)$,	$\|z\| > \rho_1$,	$g(n) \leftrightarrow G(z)$,	$\|z\| > \rho_2$
Function	Transform		Theorem's name
$a^n f(n)$	$F\left(\dfrac{z}{a}\right)$		Transform-scaling
$f(n-1)u(n-1)$	$z^{-1}F(z)$		Shifting for a causal function
$f(n-k)u(n-k)$	$z^{-k}F(z)$		
$f(n-1)$	$z^{-1}F(z) + f(-1)$		
$f(n-2)$	$z^{-2}F(z) + z^{-1}f(-1) + f(-2)$		Shifting with initial conditions
$f(n-k)$	$z^{-k}F(z) + \sum\limits_{p-1}^{k} z^{-k+p}f(-p)$		
$f(n)u(n) * g(n)u(n)$	$F(z)G(z)$		Convolution
$= \left[\sum\limits_{0}^{n} f(k)g(n-k)\right] \cdot u(n)$			
$f(0) = \lim\limits_{z \to \infty} F(z)$			Initial value

Solution

(a) $Z[f(n)] = F(z)$

$$= f(0) + f(1)z^{-1} + f(2)z^{-2} + \cdots$$

$$Z[a^n f(n)] = f(0) + af(1)z^{-1} + a^2 f(2)z^{-2} + \cdots$$

$$= f(0) + f(1)\left(\frac{z}{a}\right)^{-1} + f(2)\left(\frac{z}{a}\right)^{-2} + \cdots$$

$$= F\left(\frac{z}{a}\right)$$

(b) $u(n) \longleftrightarrow \dfrac{z}{z-1}$

therefore $(-2)^n u(n) \longleftrightarrow \dfrac{z/-2}{(z/-2)-1} = \dfrac{z}{z+2}$ as expected.

EXAMPLE 6-4

(a) Prove the shifting theorem for causal functions.
(b) Use this theorem to find the inverse transform of $1/(z - 1)$ and $1/z^4(z + 2)$.
(c) Find the transform of $(n - 2)u(n - 3)$.

Solution

(a) $Z[f(n)] = f(0) + f(1)z^{-1} + f(2)z^{-2} + \cdots$

$$Z[f(n - 1)u(n - 1)] = 0 + f(0)z^{-1} + f(1)z^{-2} + \cdots$$

$$= z^{-1}F(z)$$

We now find $Z[f(n - k)u(n - k)]$ a little more neatly.

$$Z[f(n - k)u(n - k)] = \sum_{p=0}^{\infty} [f(p - k)u(p - k)]z^{-p}$$

$$= \sum_{p=k}^{\infty} z^{-k}[f(p - k)z^{-p+k}]$$

$$= z^{-k}F(z) \qquad\qquad (6\text{-}11)$$

Short-hand proofs are very compact but should not be done mechanically. We must always visualize the series that is being represented.

(b) (i) $\dfrac{1}{z-1} = z^{-1}\left(\dfrac{z}{z-1}\right)$

therefore $\left[\dfrac{1}{z-1}\right] = u(n-1)$

(ii) $\dfrac{1}{z^4(z+2)} = z^{-5}\left(\dfrac{z}{z+2}\right)$

thus $\left[\dfrac{1}{z^4(z+2)}\right] = (-2)^{n-5}u(n-5)$

(c) $(n-2)u(n-3) = (n-3)u(n-3) + u(n-3)$

therefore $Z[(n-2)u(n-3)] = z^{-3}\dfrac{z}{(z-1)^2} + z^{-3}\dfrac{z}{z-1}$

$$= \dfrac{z^{-2} + z^{-2}(z-1)}{(z-1)^2}$$

$$= \dfrac{1}{z(z-1)^2}$$

EXAMPLE 6-5

(a) Inductively prove the shifting theorem incorporating initial conditions.
(b) Set up the solution of the difference equation:

$$y(n) + 0.6y(n-1) = 2^n, \qquad \text{given } y(-1) = 3$$

Solution

(a) $Z[f(n-1)] = f(-1) + f(0)z^{-1} + f(1)z^{-2}$

$$+ \cdots + f(n)z^{-n-1} + \cdots$$

$$= f(-1) + z^{-1}[f(0) + f(1)z^{-1} + \cdots]$$

$$= f(-1) + z^{-1}F(z) \tag{6-12}$$

Now proceeding inductively, we get:

$$Z[f(n-2)] = Z[g(n-1)]$$

where $g(n) = f(n-1)$

Therefore $Z[f(n-2)] = g(-1) + z^{-1}[f(-1) + z^{-1}F(z)]$

$$= z^{-2}F(z) + z^{-1}f(-1) + f(-2) \tag{6-13}$$

and $Z[f(n-k)] = z^{-k}F(z) + z^{-k+1}f(-1)$

$$+ \cdots + f(-k) \tag{6-14}$$

or more compactly:

$$Z[f(n-k)] = z^{-k}F(z) + \sum_{p-1}^{k} f(-p)z^{-k+p} \tag{6-15}$$

(b) Consider:

$$y(n) + 0.6y(n-1) = 2^n, \qquad \text{given } y(-1) = 3$$

Taking the Z transform, we obtain:

$$Y(z) + 0.6(z^{-1}Y(z) + 3) = \frac{z}{z - 2}$$

This becomes

$$Y(z) = \frac{z^2 - 1.8z(z - 2)}{(z - 2)(z + 0.6)}$$

and $y(n) = Z^{-1}[Y(z)]$

Inverse Z transforms will be treated in the next section.

EXAMPLE 6-6

(a) Prove the convolution theorem for causal functions.

(b) Use the convolution theorem to find the transform of:

$$2^n u(n) * (n - 1)(\tfrac{1}{3})^{n-1} u(n - 1)$$

Solution

(a) By definiton, for two causal functions:

$$f(n) * g(n) = \sum_{k=0}^{\infty} f(k)g(n - k)$$

the Z transform is:

$$Z[f(n) * g(n)] = \sum_{n=0}^{\infty} \left[\sum_{k=0}^{\infty} f(k)g(n - k)\right] z^{-n}$$

Assuming it is permissible to interchange the order of summation, we get:

$$\text{LHS} = \sum_{k=0}^{\infty} f(k)\left[\sum_{n=0}^{\infty} g(n - k)z^{-n}\right]$$

Let $n - k = p$ and this becomes:

$$\text{LHS} = \sum_{k=0}^{\infty} f(k)\left[\sum_{p=-k}^{\infty} g(p)z^{-p}\right] z^{-k}$$

$$= F(z)G(z)$$

$g(p) = 0$, $-k < p < 0$, since $g(n)$ is assumed causal. We should be careful and slow in appreciating this definition. For example, because the functions are assumed causal, we could have started with the definition for convolution as $\sum_{k=0}^{n} f(k)g(n - k)$ since $g(n - k) = 0, k > n$. Be convinced the same answer is still obtained. What is the region or annulus of convergence? Apart from doing it mathematically, is it reasonable to assume the order of summation may be interchanged?

(b) $Z\left[2^n u(n) * (n-1)\left(\dfrac{1}{3}\right)^{n-1} u(n-1)\right] = \dfrac{z}{z-2}\overline{\left[\dfrac{1}{3}(n-1)\left(\dfrac{1}{3}\right)^{n-2} u(n-1)\right]}$

$$= \dfrac{0.33z}{(z-2)(z-0.33)^2} \text{ (with work)}$$

Therefore $\quad y(n) = Z^{-1}\left[\dfrac{0.33z}{(z-2)(z-0.33)^2}\right]$

6-3 THE INVERSE Z TRANSFORM

The definition of the inverse Z transform is:

$$f(n) = \frac{1}{2\pi j}\oint_C F(z)z^{n-1}\,dz \tag{6-16}$$

where C includes all the poles. On evaluation this becomes:

$$f(n) = 0, \quad \text{if } n < 0$$

$$= \Sigma\,[\text{residues of the poles of } F(z)z^{n-1}], \quad n \geq 0 \tag{6-17}$$

The uniqueness theorem for the Z transform is:

$$f(n) = \frac{1}{2\pi j}\oint_C\left[\sum_{p=0}^{\infty} f(p)z^{-p}\right]z^{n-1}\,dz \tag{6-18}$$

These formulas which involve residue theory will be utilized in Chapter 7 and are not pursued in this chapter. Here we find inverse transforms by representing $F(z)$ in a form where it may be found by table reference and by using the theorems.

The good news is that all the required partial fraction theory from Chapter 4 carries over. We now solve an extensive problem and then make some general observations.

EXAMPLE 6-7
Find the inverse Z transforms of the following functions and plot $f(n)$ versus n:

(a) $\dfrac{z}{z+0.6}$

(b) $\dfrac{z+2}{z+0.6}$

(c) $\dfrac{z^2+4z}{(z+0.5)(z-1)}$

(d) $\dfrac{4}{z^3(z + 0.2)}$

(e) $\dfrac{z^2 + 2z}{(z + 1)(z - 0.5)^2}$

Solution

(a) By inspection:

$$\frac{z}{z + 0.6} \leftrightarrow (-0.6)^n u(n)$$

and $F(z)$ and $f(n)$ are shown in Figure 6-3(a).

(b) $$\frac{z + 2}{z + 0.6} = \frac{z}{z + 0.6} + \frac{2}{z + 0.6}$$

therefore $\dfrac{z + 2}{z + 0.6} \leftrightarrow (-0.6)^n u(n) + 2(-0.6)^{n-1} u(n - 1)$ (1)

$$= \delta(n) + (-0.6)^n(1 - 3.3)u(n - 1)$$
$$= \delta(n) - 2.3(-0.6)^n u(n - 1) \qquad (2)$$

$F(z)$ and $f(n)$ are shown in Figure 6-3(b). Form (2) is somewhat simpler than (1) for plotting.

(c) Now

$$F(z) = \frac{z^2 + 4z}{(z + 0.5)(z - 1)}$$

In order to find $f(n)$ we need to obtain:

$$F(z) = \frac{A_1 z}{z + 0.5} + \frac{A_2 z}{z - 1}$$

Since we cannot use partial fractions except when the order of $D(z)$ is higher than $N(z)$, we will express:

$$\frac{F(z)}{z} = \frac{z + 4}{(z + 0.5)(z - 1)}$$

in partial fraction form:

$$\frac{z + 4}{(z + 0.5)(z - 1)} = \frac{-2.33}{z + 0.5} + \frac{3.33}{z - 1}$$

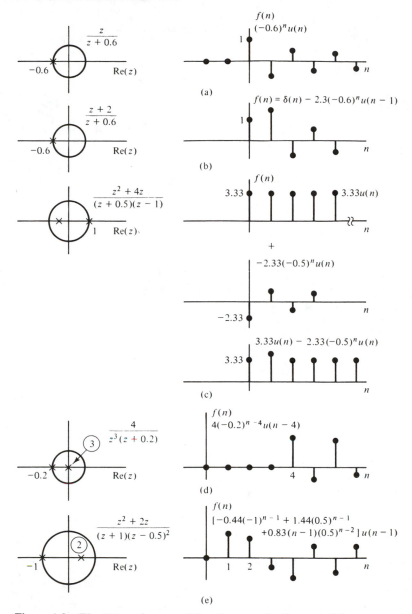

Figure 6-3 The Z transforms and their inverses for Example 6-7.

Therefore

$$F(z) = \frac{-2.33z}{z + 0.5} + \frac{3.33z}{z - 1}$$

and

$$f(n) = -2.33(-0.5)^n u(n) + 3.33u(n)$$

$F(z)$ and $f(n)$ are shown in Figure 6-3(c) and the reader should later use the initial value theorem to check $f(0)$.

(d)
$$F(z) = \frac{4}{z^3(z + 0.2)}$$

$$= z^{-4}\frac{4z}{z + 0.2}$$

Therefore $f(n) = 4(-0.2)^{n-4}u(n - 4)$

The term z^3 in the denominator indicates a shift and not to use partial fractions. $F(z)$ and $f(n)$ are shown in Figure 6-3(d).

(e)
$$F(z) = \frac{z^2 + 2z}{(z + 1)(z - 0.5)^2}$$

$$\frac{z^2 + 2z}{(z + 1)(z - 0.5)^2} = \frac{A_1}{z + 1} + \frac{A_2}{z - 0.5} + \frac{A_3}{(z - 0.5)^2}$$

$$= \frac{-0.44}{z + 1} + \frac{A_2}{z - 0.5} + \frac{0.83}{(z - 0.5)^2}$$

If $z = 0$:

$$0 = -0.44 - 2A_2 + 3.32$$

therefore $A_2 = 1.44$

and $f(n) = -0.44(-1)^{n-1}u(n - 1)$

$$+ 1.44(0.5)^{n-1}u(n - 1)$$

$$+ 0.83(n - 1)(0.5)^{n-2}u(n - 1)$$

$F(z)$ and $f(n)$ are shown in Figure 6-3(e).

    We now summarize finding inverse transforms for the case of poles on the real axis. This is the most common situation and complex conjugate poles may be studied when encountered.

If $$F(z) = \frac{N(z)}{D(z)} = \frac{b_m z^m + b_{m-1}z^{m-1} + \cdots + b_1 z}{a_n z^n + a_{n-1}z^{n-1} + \cdots + a_0}$$

then there are two important cases:

Case m = n

If $m = n$, we expand $F(z)/z$ in partial fractions (when $b_0 = 0$) to obtain:

$$\frac{F(z)}{z} = \frac{A_1}{(z - \alpha_1)} + \frac{A_2}{(z - \alpha_1)^2} + \cdots$$

$$+ \frac{B_1}{(z - \alpha_2)} + \frac{C_1}{(z - \alpha_3)} + \cdots$$

and $$F(z) = \frac{A_1 z}{(z - \alpha_1)} + \frac{A_2 z}{(z - \alpha_1)^2} + \cdots$$

Corresponding to different order poles, we obtain:

$$\frac{A_1 z}{(z - \alpha_1)} \leftrightarrow A_1 (\alpha_1)^n u(n)$$

$$\frac{A_2 z}{(z - \alpha_1)^2} \leftrightarrow A_2 n (\alpha_1)^{n-1} u(n)$$

and
$$\frac{A_p z}{(z - \alpha_1)^{p+1}} \leftrightarrow \frac{A_p}{p!} (n)_p a^{n-p} u(n)$$

where $(n)_{p-1} = n \times (n - 1) \times \cdots \times (n - p + 2)$

Case $m < n$

There is only a slight adjustment in this case.

If $$F(z) = \frac{A_1}{(z - \alpha_1)} + \frac{A_2}{(z - \alpha_1)^2} + \cdots + \frac{B_1}{(z - \alpha_2)} + \cdots$$

then corresponding to the different order poles, we obtain:

$$\frac{A_1}{z - \alpha_1} \leftrightarrow A_1 (\alpha_1)^{n-1} u(n - 1)$$

$$\frac{A_2}{(z - \alpha_1)^2} \leftrightarrow A_2 (n - 1)(\alpha_1)^{n-2} u(n - 1)$$

and
$$\frac{A_p}{(z - \alpha_1)^p} \leftrightarrow \frac{A_p}{(p - 1)!} (n - 1)_{p-1} (\alpha_1)^{n-p} u(n - 1)$$

If $n - m = k$, then the first nonzero value of $f(n)$ is $f(k)$. We should note that:

$$f(n) = n(n - 1)(\alpha_1)^{n-2} u(n) = n(n - 1)(\alpha_1)^{n-2} u(n - 2)$$

since we have $f(0) = f(1) = 0$.

Drill Set: Inverse Z Transforms

1. Evaluate the inverse Z transforms:

(a) $\dfrac{3z^2 + 2z}{(z - 0.8)^2}$

(b) $\dfrac{3z^2 + 2z}{(z - 0.8)^3}$

(c) $\dfrac{3z^2 + 2z}{(z + 2)(z - 1)^2}$

2. (a) Prove the initial value theorem:

$$f(0) = \lim_{z \to \infty} F(z)$$

(b) Extend the theorem to find an expression for $f(1), f(2)$, and $f(p)$ in terms of $F(z)$ as $z \rightarrow \infty$.

6-4 APPLICATIONS OF THE Z TRANSFORM

In this section we consider two important applications of the one-sided Z transform. These are:

1. the solution of linear difference equations with constant coefficients and the determination of pulse responses
2. analysis of LTIC discrete systems

6-4-1 The Solution of Difference Equations

The procedure which is very simple will be illustrated by giving a few examples.

EXAMPLE 6-8
   Solve:

$$y(n) - y(n - 1) + 0.25y(n - 2) = 2^n$$

$$\text{given } y(-1) = 2, \qquad y(-2) = 1$$

Solution.   Taking the Z transform of both sides of the equation, we get:

$$Y(z) - [z^{-1}Y(z) + 2]$$

$$+ 0.25[z^{-2}Y(z) + 2z^{-1} + 1] = \frac{z}{z - 2}$$

therefore

$$(1 - z^{-1} + 0.25z^{-2})Y(z) = \frac{z}{z - 2} + 2 - 0.5z^{-1} - 0.25$$

$$Y(z) = \frac{z^3 + (1.75z^2 - 0.5z)(z - 2)}{(z - 2)(z^2 - z + 0.25)}$$

Thus

$$\frac{Y(z)}{z} = \frac{z^2 + (1.75z - 0.5)(z - 2)}{(z - 2)(z - 0.5)^2}$$

$$= \frac{A_1}{z - 2} + \frac{A_2}{z - 0.5} + \frac{A_3}{(z - 0.5)^2}$$

$$= \frac{1.78}{z - 2} + \frac{A_2}{z - 0.5} + \frac{0.21}{(z - 0.5)^2}$$

if $z = 0$:

$$-2 = -0.89 - 2A_2 + 0.84$$

and $A_2 = 0.98$:

$$y(n) = 1.78(2)^n u(n) + 0.98(0.5)^n u(n) + 0.21n(0.5)^{n-1} u(n-1)$$

This solution satisfies the difference equation for n, $-\infty < n < \infty$.

EXAMPLE 6-9

Find the pulse response $h(n)$ for the system governed by:

$$y(n) + 0.6y(n-1) = 3x(n) + x(n-1)$$

Solution.   This problem was solved in the time domain in Example 2.17(b). If $x(n) = \delta(n)$, the equation is:

$$y(n) + 0.6y(n-1) = 3\delta(n) + \delta(n-1)$$

Taking the Z transform with $y(-1) = 0$, we get:

$$Y(z) = \frac{3 + z^{-1}}{1 + 0.6z^{-1}}$$

$$= \frac{3z + 1}{z + 0.6}$$

$$= 3 - \frac{0.8}{z + 0.6}$$

therefore $h(n) = 3\delta(n) - 0.8(-0.6)^{n-1} u(n-1)$

This is identical to the solution in Example 2-17(b) after a slight manipulation.

6-4-2 System Analysis

The System Function H(z).   Given a linear time-invariant causal system (LTIC) is governed by:

$$a_n y(n) + a_{n-1} y(n-1) + \cdots + a_{n-p} y(n-p)$$
$$= b_n x(n) + \cdots + b_{n-m} x(n-m) \qquad (6-19)$$

we define the system function $H(z)$ as:

$$H(z) \triangleq \frac{Y(z)}{X(z)}$$

which is the Z transform of the output divided by the Z transform of the input with $y(n)$ equal to zero for n negative. Applying the definition to 6-19, we obtain:

$$(a_n + a_{n-1}z^{-1} + \cdots + a_{n-p}z^{-p})\,Y(z)$$

$$= (b_n + b_{n-1}z^{-1} + \cdots + b_{n-m}z^{-m})\,X(z)$$

and
$$H(z) = \frac{b_n + b_{n-1}z^{-1} + \cdots + b_{n-m}z^{-m}}{a_n + a_{n-1}z^{-1} + \cdots + a_{n-p}z^{-p}}$$

$$= \frac{b_n z^p + b_{n-1}z^{p-1} + \cdots + b_{n-m}z^{p-m}}{a_n z^p + a_{n-1}z^{p-1} + \cdots + a_{n-p}} \qquad (6\text{-}20)$$

This is identical to the function obtained in Chapter 2 when the system function was defined as the forced response to z^n divided by z^n.

The Zero-State Output
Consider a system as in Figure 6-4. The Z transform of the zero-state output is:

$$Y(z) = H(z)X(z)$$

and
$$y(n) = Z^{-1}[H(z)X(z)] \qquad (6\text{-}21)$$

$y(n)$ may be found by taking partial fractions and using table reference.

H(z) and h(n)
If $x(n) = \delta(n)$, then the response is:

$$h(n) = Z^{-1}[H(z)(1)]$$

$$= Z^{-1}[H(z)] \qquad (6\text{-}22)$$

The pulse response $h(n)$ and the system function $H(z)$ form probably the most famous Z transform pair:

$$h(n) \leftrightarrow H(z) \qquad (6\text{-}23)$$

Figure 6-4 summarizes the solution of LTIC systems using transform and time domain analysis.

EXAMPLE 6-10
Consider the system described by:

$$y(n) - 0.5y(n-1) = x(n)$$

Find:

(a) the system function $H(z)$
(b) the pulse response $h(n)$
(c) the zero-state response when $x(n) = u(n)$
(d) the complete response when $x(n) = u(n)$ and $y(-1) = 2$

$x(n) \rightarrow$

$$a_n y(n) + a_n y(n-1) + \cdots a_{n-p} y(n-p) = b_n x(n) + \cdots + b_{n-m} x(n-m)$$

$\rightarrow y(n)$

z Domain

$$H(z) \triangleq \frac{Y(z)}{X(z)} = \frac{b_n z^p + b_{n-1} z^{p-1} + \cdots + b_{n-m} z^{p-m}}{a_n z^p + a_{n-1} z^{p-1} + \cdots + a_{n-p}}$$

$$Y(z) = H(z)X(z)$$

$$y_{zs}(n) = Z^{-1}[H(z)X(z)]$$

Time Domain

$$H(\alpha) = \left.\frac{y_{f0}(n)}{x(n)}\right|_{x(n)=\alpha^n} = \frac{b_n + b_{n-1}\alpha^{-1} + \cdots + b_{n-m}\alpha^{-m}}{a_n + a_{n-1}\alpha^{-1} + \cdots + a_{n-p}\alpha^{-p}}$$

$$H(z) = \left.\frac{y_{f0}(n)}{x(n)}\right|_{x(n)=z^n} = \frac{b_n z^p + b_{n-1} z^{p-1} + \cdots + b_{n-m} z^{p-m}}{a_n z^p + a_{n-1} z^{p-1} + \cdots + a_{n-p}}$$

from $a_n h(n) + \cdots + a_{n-p} h(n-p) = b_n \delta(n) + \cdots + b_{n-m} \alpha(n-m)$, $h(n)$ may be found

and $$y_{zs}(n) = x(n) * h(n)$$

Figure 6-4 Transform and time-domain results for a LTIC discrete system.

Solution

(a) $$H(z) = \frac{1}{1 - 0.5z^{-1}} = \frac{z}{z - 0.5}$$

(b) $h(n) = (0.5)^n u(n)$

(c)
$$Y(z) = \frac{z^2}{(z-0.5)(z-1)}$$

$$\frac{Y(z)}{z} = \frac{z}{(z-0.5)(z-1)} = \frac{2}{z-1} - \frac{1}{z-0.5}$$

and $$y(n) = [2 - (0.5)^n]u(n)$$

(d) Now the homogeneous response is different in $y(n)$:

$$y(n) = 2u(n) + A(0.5)^n u(n)$$

If $y(-1) = 2$ then from the system difference equation:

$$y(0) = 0.5(2) + 1$$

$$= 2$$

and $2 = 2 + A$

therefore $A = 0$

and $y(n) = 2u(n)$

In this case the transient term is zero.

6-4-3 Analog to Digital Conversion

In this section we consider replacing a continuous system with system function $H(s)$ by a discrete system with system function $H_d(z)$. Figure 6-5(a) and (b) shows the continuous system and its discrete model. From Chapter 2:

$$y(n\tau) \approx \tau \sum_{p=0}^{n} x(p\tau) h(n\tau - p\tau) \tag{6-24}$$

and the approximation is good if $x(t)$ and $h(t)$ change little over any range τ. If we define as our discrete input:

$$x(n) = x(t)\big|_{t=n\tau}$$

and the discrete pulse response:

$$h_d(n) = \tau h(n\tau)$$

then $y(n) = x(n) * h_d(n)$

$$= \sum_{0}^{n} x(p) h_d(n - p) \tag{6-25}$$

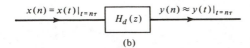

$x(t)$ → $\boxed{H(s)}$ → $y(t)$

(a)

$x(n) = x(t)|_{t=n\tau}$ → $\boxed{H_d(z)}$ → $y(n) \approx y(t)|_{t=n\tau}$

(b)

Figure 6-5 (a) A continuous system with $H(s)$; (b) the discrete system $H_d(z)$ with $h_d(n) = \tau h(t)\big|_{t=n\tau}$.

is identical to the result in Equation 6-24. We therefore require for our discrete system function:

$$H_d(z) = Z[h_d(n)]$$

$$= \tau[h(0) + h(\tau)z^{-1} + h(2\tau)z^{-2} + \cdots]$$

$$= \tau \sum_{p=0}^{\infty} h(p\tau)z^{-p} \tag{6-26}$$

We now explore $H_d(z)$ for systems where the roots of $H(s)$ are real.

First-Order System
Given:

$$H(s) = \frac{c_1}{s + \alpha_1} \quad \text{or} \quad h(t) = c_1 e^{-\alpha_1 t} u(t)$$

then Equation 6-26 becomes:

$$H_d(z) = c_1 \tau [1 + e^{-\alpha_1 \tau} z^{-1} + e^{-\alpha_1 2\tau} z^{-2} + \cdots]$$

$$= \frac{c_1 \tau}{1 - e^{-\alpha_1 \tau} z^{-1}}$$

$$= \frac{c_1 \tau z}{z - e^{-\alpha_1 \tau}} \tag{6-27}$$

Second-Order Overdamped System
If:

$$H(s) = \frac{c_1}{s + \alpha_1} + \frac{c_2}{s + \alpha_2}$$

then

$$H_d(z) = \frac{c_1 \tau z}{z - e^{-\alpha_1 \tau}} + \frac{c_2 \tau z}{z - e^{-\alpha_2 \tau}} \tag{6-28}$$

In general, for an nth order system with distinct real roots in the denominator of $H(s)$:

$$H(s) = \sum_{i=1}^{n} \frac{c_i}{s + \alpha_i}$$

and

$$H_d(z) = \sum_{i=1}^{n} \frac{c_i \tau z}{z - e^{-\alpha_i \tau}} \tag{6-29}$$

The investigation of $H_d(z)$ can be extended for the case of repeated real roots and complex conjugate roots and some of this is required in the chapter problems. A few simple examples will now be solved.

EXAMPLE 6-11

Given a continuous system with system function $H(s) = 2/(s + 3)$ and input $x(t) = 5u(t)$, replace $H(s)$ by a discrete system and find $y(n) \approx y(t)$ at $t = n\tau$.

Solution

$$h(t) = 2e^{-3t}u(t)$$

The time constant for $h(t)$ is $\tau_1 = 0.33$ s. Let us choose $\tau = 0.01$. Since $x(t)$ is constant it does not affect the choice of τ.

$$H_d(z) = \frac{2(0.01)z}{z - e^{-0.03}}$$

and since $x(n) = 5$:

$$Y(z) = \frac{0.1z^2}{(z - 1)(z - 0.97)}$$

$$\frac{Y(z)}{z} = \frac{A_1}{z - 1} + \frac{A_2}{z - 0.97}$$

$$= \frac{0.1/0.03}{z - 1} + \frac{0.1(0.97)/-0.03}{z - 0.97}$$

$$Y(z) = \frac{3.33z}{z - 1} + \frac{-3.23z}{z - 0.97}$$

$$y(n) = [3.33 - 3.23(0.97)^n]u(n)$$

The reader should find it instructive to solve for $y(t)$ for the continuous system and to compare the results.

EXAMPLE 6-12

Replace the differential equation:

$$y''(t) + 5y'(t) + 4y(t) = 2x(t) + x'(t)$$

by a difference equation.

Solution

$$H(s) = \frac{s + 2}{(s + 4)(s + 1)}$$

$$= \frac{0.67}{s + 4} + \frac{0.33}{s + 1}$$

$$H_d(z) = \frac{0.67\tau z}{z - e^{-4\tau}} + \frac{0.33\tau z}{z - e^{-\tau}}$$

$$= \frac{\tau z^2 - 0.67\tau e^{-\tau}z - 0.33\tau e^{-4\tau}z}{z^2 - (e^{-4\tau} + e^{-\tau})z + e^{-5\tau}}$$

The equivalent difference equation is:

$$y(n) - (e^{-4\tau} + e^{-\tau})y(n-1) + e^{-5\tau}y(n-2)$$
$$= \tau[x(n) - (0.67e^{-\tau} + 0.33e^{-4\tau})x(n-1)]$$

For any input $x(t)$, τ must be chosen small compared to $\tau_1 = 0.25$ the lesser time constant of the system and such that $x(t)$ changes negligibly over any range τ.

6-4-4 General Relation of the Laplace to Z Transform*

The general conversion of the Laplace transform of $y(t)$, $Y(s)$ to $Y_d(z)$, the Z transform of $y(t)$ sampled every τ seconds where $y(n) = y(n\tau)$ is very important. Let us consider the following notation:

1. $y(t)$ and $Y(s)$ for a causal function and its Laplace transform
2. $y^*(t) = \Sigma_0^\infty y(n\tau)\delta(t - n\tau)$ and $Y^*(s)$, for a string of delta functions weighted by $y(n\tau)$ and its Laplace transform $Y^*(s)$
3. $y(n) = y(n\tau)$ and its Z transform $Y_d(z)$

Conversion

Now

$$y^*(t) = \sum_0^\infty y(n\tau)\delta(t - n\tau)$$

$$= y(t)\sum_0^\infty \delta(t - n\tau)$$

Therefore
$$Y^*(s) = Y(s) * \frac{1}{1 - e^{-\tau s}} \qquad (6\text{-}30)$$

and using complex convolution, we obtain:

$$Y^*(s) = \frac{1}{2\pi j}\int_{a-j\infty}^{a+j\infty}{}_C\, Y(p)\frac{1}{1 - e^{-\tau(s-p)}}\,dp$$

$$= \frac{1}{2\pi j}\oint_{C_2}\frac{Y(p)}{1 - e^{-\tau(s-p)}}\,dp$$

where C_2 is the closed contour obtained when C is closed to the left. Since $1/(1 - e^{-\tau s})$ does not have any poles inside C_2 if all the poles of $Y(p)$ are in the left half plane, then

$$Y^*(s) = \sum_{\substack{\text{poles} \\ \text{of} \\ Y(p)}}\left[\text{residues of }\frac{Y(p)}{1 - e^{-\tau(s-p)}}\right] \qquad (6\text{-}31)$$

Since:

$$y^*(t) = \sum_0^\infty y(n\tau)\delta(t - n\tau)$$

then

$$Y^*(s) = \sum_0^\infty y(n\tau)e^{-n\tau s}$$

and

$$Y_d(z) = \sum_0^\infty y(n\tau)z^{-n}$$

$$= \sum_{-0}^\infty y(n)z^{-n}$$

therefore

$$Y_d(z) = Y^*(s) \qquad \text{with } e^{\tau s} = z$$

and

$$Y_d(z) = \sum_{\substack{\text{poles} \\ \text{of} \\ Y(p)}} \left[\text{residues of } \frac{Y(p)}{1 - e^{\tau p}z^{-1}} \right] \qquad (6\text{-}32)$$

EXAMPLE 6-13

Given the system function $H(s) = 1/(s + 1)$, find by using Equation 6-31 the discrete system function $H_d(z)$ for handling discrete inputs with a sampling interval τ.

Solution

$$H(z) = \sum_{\substack{\text{pole} \\ \text{at } p = -1}} \text{residues of } \frac{1/(p + 1)}{1 - e^{\tau p}z^{-1}}$$

$$= \text{residue of pole at } p = -1 \text{ in} \left[\frac{z}{(p + 1)(z - e^{\tau p})} \right]$$

$$= \frac{z}{z - e^{-\tau}}$$

6-4-5 Transform Proofs of Time Relations

Many of the properties involving operations on discrete functions may be proved very simply by using Z transforms.

EXAMPLE 6-14

Prove the following by using Z transforms and assuming all the functions are causal:

(a) $f(n)*\delta(n - k) = f(n - k)$
(b) $f(n)*[g(n)*h(n)] = [f(n)*g(n)]*h(n)$

Solution

(a)
$$\overline{f(n)*\delta(n - k)} = F(z)z^{-k}$$

Therefore
$$f(n)*\delta(n - k) = f(n - k)$$

(b) $\overline{f(n)*[g(n)*h(n)]} = F(z)[G(z)H(z)] = F(z)G(z)H(z)$

$\overline{[f(n)*g(n)]*h(n)} = [F(z)G(z)]H(z)$

$= \text{LHS}$

SUMMARY

Chapter 6 covered the one-sided Z transform. This is the basic transform as far as discrete systems, digital communications, digital control theory, or any application involving analog to digital conversion is concerned.

The one-sided Z transform was defined and appreciated by forming a vocabulary of transforms all essentially stemming from $a^n u(n)$. This is not surprising because, just as the Laplace transform was a decomposition of a function into exponentials, the Z transform is a decomposition of functions into exponents. The main theorems of the Z transform, that is, the z-scaling, shifting, and convolution theorems were introduced and proved. In particular, the shifting theorems were interesting and they are the counterpart of the derivative theorems of the Laplace transform. Incorporating $f(-1), f(-2)$, and so on into the shifting theorem allows for the solution of systems with nonzero initial states.

Inverse transforms were evaluated by using partial fractions and the theorems. If the order of the numerator and denominator are equal, then partial fractions are found for $F(z)/z$ and $f(0)$ is always nonzero. If the order of the denominator exceeds the order of the numerator by k, then $f(n)$ is zero for $n < k$ and commences with a nonzero value for $f(k)$. The section on the inverse Z transform was simpler than for the one-sided Laplace transform since the case of complex conjugate poles was not investigated.

Finally, some of the main applications of Z transforms were discussed. These were: solving difference equations incorporating initial conditions, finding pulse responses, the transform analysis of linear time-invariant causal systems with discrete deterministic inputs and converting continuous to discrete systems. We concluded with carrying out proofs of the discrete time-domain relations by means of Z transforms.

Chapters 4 and 6 represent the transform treatment of Chapter 2 on the analysis of LTIC continuous and discrete systems.

PROBLEMS

6-1. Find the Z transform of:

(a) $f_1(n) = 3(-0.6)^{n-3}$

(b) $f_2(n) = 3n(-0.6)^{n-1}$

(c) $f_3(n) = 3(n-2)(-0.6)^{n-2}$

(d) $f_4(n) = \sum_{k=0}^{\infty}(0.5)^k \delta(n-2k)$

(e) $f_5(n) = \sum_{k=1}^{\infty} 2k\delta(n-2k+1)$

(f) $f_6(n) = (-2+n^2)(3)^n$

(g) $f_7(n) = 2^{n-5}u(n-3)$

(h) $f_8(n) = (4n-2)\delta(n-4)$

(i) $f_9(n) = (2-3n+4n^2)(-1)^n$

(j) $f_{10}(n) = (-0.8)^n u(n-3)$

(k) $f_{11}(n) = (n-2)0.7^n u(n-3)$

6-2. Find the inverse Z transform $f(n)$ and plot it for $0 \le n \le 4$.

(a) $\dfrac{z + 2}{z + 0.6}$ (b) $\dfrac{z^2 + z}{(z + 1)(z - 0.6)}$

(c) $\dfrac{4z}{(z - 2)^3}$ (d) $\dfrac{28z + 2}{z^3}$

(e) $\dfrac{3}{z^2(z + 1)^2}$ (f) $\dfrac{z^3 + 2z}{(z + 1)^2(z + 0.8)}$

6-3. Convolve using the Z transform:

(a) $2^n u(n) * (0.6)^n u(n)$ (b) $n(0.5)^n u(n) * nu(n)$

(c) $(0.5)^n u(n - 2) * (-0.7)^n u(n)$

6-4. For a system governed by:

$$y(n) + 0.6\, y(n - 1) = 2x(n)$$

Find:

(a) the system function $H(z)$
(b) the pulse response as the inverse of $H(z)$
(c) the unit step response $r(n)$ when $x(n) = u(n)$
(d) check that $r(n) = \sum_{k=0}^{n} h(k)$
(e) the zero-state response when $x(n) = 0.5^n u(n)$
(f) the complete response when $x(n) = 0.5^n u(n)$ and $y(-1) = 2$

6-5. Repeat all the questions of Problem 6-4 for the system governed by:

$$y(n) - 3y(n - 1) + 2y(n - 2) = 4x(n)$$

In part (f) use $y(-1) = 2,$ $y(-2) = -1$.

6-6. Find a difference equation whose solution is:

$$y(n) = (2 + n)(-1)^n + 6$$

where $y_{fo}(n) = 6$ and $y_{homo}(n) = (2 + n)(-1)^n$.

6-7. (a) Given the pulse response of a system is:

$$h(n) = 2(-0.6)^n u(n)$$

find the unit step response $r(n)$ when $x(n) = u(n)$.
(b) Given:

$$r(n) = [3(-0.6)^n + 2^n - 4]u(n)$$

find the pulse response $h(n)$.
(c) Show the relation between $r(n)$ and $h(n)$.

6-8. Given the following continuous system functions, find $H_d(z)$ for an equivalent discrete system:

(a) $H(s) = \dfrac{c_1}{s^2 + \beta^2}$

(b) $H(s) = \dfrac{c_1}{(s + \alpha)^2}$

(c) $H(s) = \dfrac{c_1}{(s + \alpha)^2 + \beta^2}$

6-9. (a) Using $\tau = 0.01$, replace the following differential equations by difference equations and solve for $y(n) = y(t)$ with $t = (0.01)n$, where $y(t)$ is the zero-state response.

   (1) $y'(t) + 3y(t) = t$

   (2) $y''(t) + 4y'(t) + 4y(t) = 6$

 (b) Check $y(n)$ for an appropriately large value of n against $y_{f0}(t)$.

 (c) Repeat part (a) by solving recursively for $y(n)$ for a few values.

 (d) Do part (a)(1) for $0 < n < 100$ using a computer program.

 (e) How good was the choice of $\tau = 0.01$?

6-10. Consider approximately finding the area of a causal function $x(t)$ where $x(0)$, $x(\tau), \ldots, x(N\tau)$ are known. Letting $y_A(p)$ be the area from $t = 0$ to $t = (N + 1)\tau$ we find:

$$y_A(1) = \tau x(0)$$

$$y_A(2) = \tau[x(0) + x(1)]$$

and

$$y_A(p) = \tau \sum_{i=0}^{p-1} x(i)$$

 (a) Find a difference equation relating $y_A(n)$ and $x(n)$.

 (b) If $x(t) = t$, $0 < t < 4$ and is zero otherwise, find $y_A(n)$ using $\tau = 0.1$ and plot $y_A(n)$ for $0 \le n \le 10$.

6-11. Determine α and β in the following system if its unit pulse response is:

$$h(n) = 0.33\delta(n) + 7\delta(n - 1) - \tfrac{22}{3}(-3)^n u(n)$$

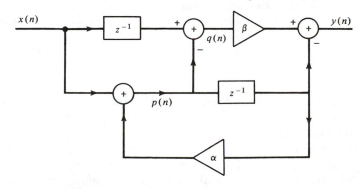

6-12. (a) For the system shown, find the system function

$$H(z) = \frac{Y(z)}{X(z)}$$

 (b) For what value of K is the system stable?

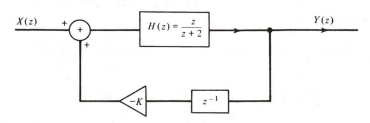

The Two-sided *Z* Transform

INTRODUCTION

Chapter 7 discusses the two-sided *Z* transform whose main application is to solve LTIC discrete systems with random or signal plus random inputs. This chapter is the transform analysis of the discrete material of Chapter 3.

The material is traversed by using what is now our standard treatment of any transform. The different stages are:

1. The transform is defined and a number of transforms are evaluated. We will utilize all our knowledge of one-sided transforms to help evaluate two-sided ones.
2. The properties and theorems are given and attention is focused on the transform of convolution and correlation summations.
3. The inverse transform is treated using previously mastered partial fraction techniques combined with table reference and also by using Laurent series plus residue theory from complex variables.
4. LTIC systems are solved with random or signal plus random inputs.

7-1 THE DEFINITION AND EVALUATION OF SOME TRANSFORMS

The two-sided or bilateral *Z* transform of a real discrete function $f(n)$ is defined as:

$$F_B(z) \triangleq \sum_{n=-\infty}^{\infty} f(n)z^{-n} \qquad (7\text{-}1)$$

If $F_B(z)$ exists it will do so for all complex z in an annulus $\rho_1 < |z| < \rho_2$. Normally, the subscript "B" is omitted and from the context it will be clear whether the one- or two-sided Z transform is being used. Alternate notations for $F(z)$ are $Z[f(n)]$, and $\overline{f(n)}$. $f(n) \leftrightarrow F(z)$ is used to indicate the transform pair.

A number of transforms will now be evaluated and the relationship between $f(n)$ for positive and negative n to the annulus of convergence explored.

EXAMPLE 7-1

Find the two-sided Z transforms of the following functions and state the annulus of convergence:

(a) $f_1(n) = (-0.5)^n u(n)$
(b) $f_2(n) = (-0.5)^n u(-n)$
(c) $f_3(n) = 3(0.5)^n u(n) + 3^n u(-n)$
(d) $f_4(n) = 3(0.5)^n u(n) + 3^n u(n)$
(e) $f_5(n) = 3(0.5)^n u(-n) + 3^n u(-n)$
(f) $f_6(n) = A_1(\alpha)^n u(n) + A_2(\beta)^n u(-n)$ in general for all α and β.

Solution

(a)
$$f_1(n) = (-0.5)^n u(n)$$

Therefore
$$F_1(z) = \frac{z}{z + 0.5}, \qquad |z| > 0.5$$

$f_1(n)$ and $F_1(z)$ are shown in Figure 7-1(a), and since $f(n)$ is a causal function, the one- and two-sided transforms are identical.

(b)
$$f_2(n) = (-0.5)^n u(-n)$$

therefore
$$F_2(z) = 1 - 2z + 4z^2 - 8z^3 + \cdots$$

$$= \frac{1}{1 + 2z}, \qquad 2|z| < 1$$

$$= \frac{0.5}{z + 0.5}, \qquad |z| < 0.5$$

$f_2(n)$ and $F_2(z)$ are plotted in Figure 7-1(b). It is easier to compare the transforms of $(-0.5)^n u(n)$ and $(-0.5)^n u(-n)$ when we write them as $1/(1 + 0.5z^{-1})$ and $1/[1 + (0.5z^{-1})^{-1}]$ and we may predict in general that:

$$a^n u(n) \leftrightarrow \frac{1}{1 - az^{-1}} = \frac{z}{z - a}, \qquad |z| > |a|$$

and
$$a^n u(-n) \leftrightarrow \frac{1}{1 - z/a} = \frac{-a}{z - a}, \qquad |z| < |a|$$

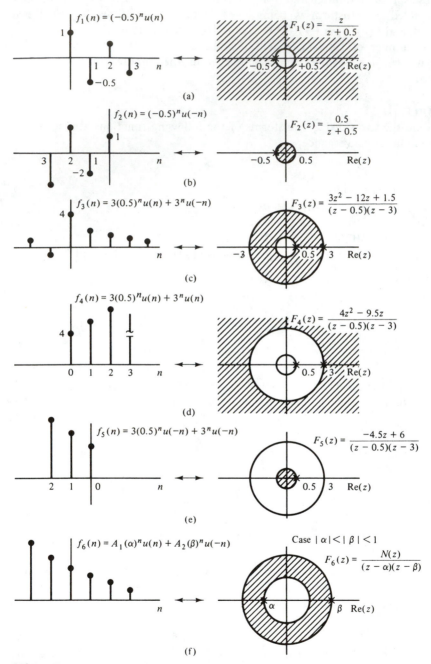

Figure 7-1 The discrete time functions of Example 7-1 and their two-sided Z transforms.

(c)
$$f_3(n) = 3(0.5)^n u(n) + 3^n u(-n)$$

Therefore $\quad F_3(z) = \dfrac{3z}{z - 0.5} + \dfrac{-3}{z - 3},\qquad |z| > 0.5 \cap |z| < 3$

$$= \dfrac{3z^2 - 12z + 1.5}{(z - 0.5)(z - 3)},\qquad 0.5 < |z| < 3$$

$f_3(n)$ and $F_3(z)$ are shown in Figure 7-1(c).

(d)
$$f_4(n) = 3(0.5)^n u(n) + 3^n u(n)$$

therefore $\quad F_4(z) = \dfrac{3z}{z - 0.5} + \dfrac{z}{z - 3},\qquad |z| > 0.5 \cap |z| > 3$

$$= \dfrac{4z^2 - 9.5z}{(z - 0.5)(z - 3)},\qquad |z| > 3$$

$f_4(n)$ and $F_4(z)$ are plotted in Figure 7-1(d), and since $f(n)$ is zero for $n < 0$, the annulus of convergence is outside all the poles.

(e)
$$f_5(n) = 3(0.5)^n u(-n) + 3^n u(-n)$$

$$F_5(z) = \dfrac{-1.5}{z - 0.5} + \dfrac{-3}{z - 3},\qquad |z| < 0.5 \cap |z| < 3$$

$$= \dfrac{-4.5z + 6}{(z - 0.5)(z - 3)},\qquad |z| < 0.5$$

$f_5(n)$ and $F_5(z)$ are shown in Figure 5-1(e), and since the function is zero for $n > 0$ the annulus of convergence is inside all the poles.

(f)
$$f_6(n) = A_1(\alpha)^n u(n) + A_2(\beta)^n u(-n)$$

therefore $\quad F_6(z) = \dfrac{A_1 z}{z - \alpha} - \dfrac{A_2 \beta}{z - \beta},\qquad |z| > |\alpha| \cap |z| < |\beta|$

The Z transform will exist for all α, β such that $|\alpha| < |\beta|$. This general situation is demonstrated in Figure 7-1(f).

Reflection on Example 7-1 indicates that the behavior of $f(n)$ for $n < 0$ places an upper bound on $|z|$ and that the behavior of $f(n)$ for $n \geq 0$ places a lower bound on $|z|$. If $f(n)$ for both positive and negative n consists of products of exponents and polynomials of n (e.g., $f(n) = [(2^n + 3n(0.5)^n)u(n) + (2n + 3)3^n u(-n)]$, then if $F(z)$ exists it will be the ratio of two equal order polynomials of z (if $f(0) \neq 0$).

The evaluation of Z transforms for any function of the form:

$$f(n) = f_1(n) u(n) + f_2(n) u(-n) \tag{7-2}$$

is straightforward for functions for which the one-sided Z transforms of $f_1(n) u(n)$ and $f_2(-n) u(n)$ are known:

$$Z[f(n)] = \sum_0^\infty f_1(n) z^{-n} + \sum_{-\infty}^0 f_2(n) z^{-n}$$

$$= (f_1(0) + f_1(1)z^{-1} + \cdots)$$
$$+ (f_2(0) + f_2(-1)z + f_2(-2)z^2 + \cdots)$$
$$= Z[f_1(n)u(n)] + Z[f_2(-n)u(n)]|_{z=z^{-1}} \qquad (7\text{-}3)$$

The clear insightful understanding of:

$$\boxed{Z[f_2(n)u(-n)] = Z[f_2(-n)u(n)]|_{z=z^{-1}} \qquad (7\text{-}4)}$$

for $|z^{-1}| > \rho$ or $|z| < \rho^{-1}$ is very important.

We now find some two-sided Z transforms using a table of one-sided transforms and Equation 7-3.

EXAMPLE 7-2

Find the Z transform of the following functions using Equation 7-3.

(a) $f_1(n) = a^n u(-n)$
(b) $f_2(n) = (-0.5)^n u(n) + (3 + n)(-3)^n u(-n)$

Solution

(a)
$$Z[a^n u(-n)] = Z[a^{-n} u(n)]|_{z=z^{-1}}$$

$$= \left.\frac{z}{z - a^{-1}}\right|_{z=z^{-1}}$$

$$= \frac{z^{-1}}{z^{-1} - a^{-1}}$$

$$= \frac{-a}{z - a}, \qquad |z^{-1}| > a^{-1}$$

$$= \frac{-a}{z - a}, \qquad |z| < a$$

This agrees with our result from Example 7-1(b) when $a = -0.5$.

(b) Using Equation 7-3, we obtain:

$$Z[(-0.5)^n u(n) + (3 + n)(-3)^n u(-n)] = \frac{z}{z + 0.5} + Z[(3 - n)(-3)^{-n} u(n)]|_{z=z^{-1}}$$

$$= \frac{z}{z + 0.5} + Z\left[3\left(-\frac{1}{3}\right)^n u(n)\right.$$

$$\left.-\left(-\frac{1}{3}\right)\left(-\frac{1}{3}\right)^{n-1} n\, u(n)\right]\Bigg|_{z=z^{-1}}$$

$$= \frac{z}{z + 0.5} + \frac{3z^{-1}}{z^{-1} + \frac{1}{3}} + \frac{1}{3}\frac{z^{-1}}{(z^{-1} + \frac{1}{3})^2}$$

$$= \frac{z}{z + 0.5} + \frac{9}{z + 3} + \frac{3z}{(z + 3)^2},$$

$$|z| > 0.5 \cap |z^{-1}| > \frac{1}{3}$$

$$= \frac{z^3 + 40.5z^2 + 24z}{(z + 0.5)(z + 3)^2}, \, 0.5 < |z| < 3$$

Finally, Table 7-1 gives a short list of two-sided Z transforms.

7-2 IMPORTANT THEOREMS OF BILATERAL Z TRANSFORMS

Table 7-2 lists some important theorems for two-sided Z transforms. Since the main application of two-sided Z transforms is to solve LTIC discrete systems with random or signal plus random inputs the convolution and correlation theorems are of the utmost importance and will now be proved and demonstrated.

EXAMPLE 7-3

Prove the convolution theorem and comment on the region of convergence.

TABLE 7-1 A TABLE OF TWO-SIDED Z TRANSFORMS

$f(n)$	$F(z) = \sum_{-\infty}^{\infty} f(n)z^{-n}$	Region of convergence
$a^n u(n)$	$\dfrac{z}{z - a}$	$|z| > |a|$
$na^{n-1}u(n)$	$\dfrac{z}{(z - a)^2}$	$|z| > |a|$
$n(n - 1)a^{n-2}u(n)$	$\dfrac{2z}{(z - a)^3}$	$|z| > |a|$
$a^n u(-n)$	$\dfrac{-a}{z - a}$	$|z| < |a|$
$na^{n+1}u(-n)$	$\dfrac{-za^2}{(z - a)^2}$	$|z| < |a|$
$n(n + 1)a^{n+2}u(-n)$	$\dfrac{-2z^2a^3}{(z - a)^3}$	$|z| < |a|$
$f_1(n)u(n) + f_2(n)u(-n)$	$Z[f_1(n)u(n)] + Z[f_2(-n)u(n)]\|_{z \to z^{-1}}$	$\rho_1 < |z| < \rho_2$
$f_2(n)u(-n)$	$Z[f_2(-n)u(n)]\|_{z \to z^{-1}}$	$|z^{-1}| > \rho_2' \cup |z| < \rho_2 = \dfrac{1}{\rho_2'}$

TABLE 7-2   TWO-SIDED Z TRANSFORM THEOREMS

Theorem	Time function	Two-sided Z transform	Region of convergence
	$f(n)$	$F(z)$	$\rho_{11} < \|z\| < \rho_{12}$
	$g(n)$	$G(z)$	$\rho_{21} < \|z\| < \rho_{22}$
Linearity	$af(n) + bg(n)$	$aF(z) + bG(z)$	$\max(\rho_{11}, \rho_{21}) < \|z\| < \min(\rho_{12}, \rho_{22})$
Shifting	$f(n - k)$	$z^{-k}F(z)$	?
Convolution	$f(n)*g(n)$ $= \sum_{-\infty}^{\infty} f(p)g(n - p)$	$F(z)G(z)$	$\max(\rho_{11}, \rho_{21}) < \|z\| < \min(\rho_{12}, \rho_{22})$
Correlation	$f(n) \oplus g(n)$ $= \sum_{-\infty}^{\infty} f(p)g(p + n)$	$G(z)F(z^{-1})$	$\max(\rho_{21}, \rho_{12}^{-1}) < \|z\| < \min(\rho_{22}, \rho_{11}^{-1})$
	$g(n) \oplus f(n)$	$F(z)G(z^{-1})$	?
	$f(n) \oplus f(n)$	$F(z)F(z^{-1})$	$\max(\rho_{11}, \rho_{12}^{-1}) < \|z\| < \min(\rho_{12}, \rho_{11}^{-1})$

Note: The reader should fill in the two blanks marked by the question mark.

Solution.   The convolution theorem states that if:

$$f(n) \leftrightarrow F(z), \qquad \rho_{f_1} < |z| < \rho_{f_2}$$

and

$$g(n) \leftrightarrow G(z), \qquad \rho_{g_1} < |z| < \rho_{g_2}$$

then

$$f(n)*g(n) \leftrightarrow F(z)G(z), \qquad \rho_1 < |z| < \rho_2$$

where ρ_1 and ρ_2 will be found. To prove this, we have:

$$f(n)*g(n) = \sum_{-\infty}^{\infty} f(p)g(n - p)$$

therefore

$$Z[f(n)*g(n)] = \sum_{n=-\infty}^{\infty}\left[\sum_{p=-\infty}^{\infty} f(p)g(n - p)\right] z^{-n}$$

Interchanging the order of summation, we obtain

$$Z[f(n)*g(n)] = \sum_{p=-\infty}^{\infty} f(p)\left(\sum_{n=-\infty}^{\infty} g(n - p)z^{-n}\right)$$

and letting $n - p = l$, we get:

$$Z[f(n)*g(n)] = \sum_{p=-\infty}^{\infty} f(p) \sum_{l=-\infty}^{\infty} g(l)z^{-p-l}$$

$$= \sum_{p} f(p)z^{-p} \sum_{l} g(l)z^{-l}$$

Therefore $Z[f(n)*g(n)] = F(z)G(z),$ for

$$(\rho_{f_1} < |z| < \rho_{f_2}) \cap (\rho_{g_1} < |z| < \rho_{g_2})$$

or

$$\max(\rho_{f_1}, \rho_{g_1}) < |z| < \min(\rho_{f_2}, \rho_{g_2}) \qquad (7\text{-}5)$$

EXAMPLE 7-4

Find the Z transforms of the following and denote the region of convergence if the transform exists:

(a) $f_1(n) = (0.5)^n u(n) * (-0.6)^n u(-n)$
(b) $f_2(n) = u(n) * (0.5)^n u(-n)$
(c) $f_3(n) = (0.5)^{|n|} * u(n)$

Solution

(a) $F_1(z) = \dfrac{z}{z - 0.5} \dfrac{0.6}{z + 0.6}$, $|z| > 0.5 \cap |z| < 0.6$

$\qquad = \dfrac{0.6z}{(z - 0.5)(z + 0.6)}$, $0.5 < |z| < 0.6$

(b) $F_2(z) = \dfrac{z}{z - 1} \dfrac{-0.5}{z - 0.5}$, $|z| > 1 \cap |z| < 0.5 = \emptyset$

therefore $F_2(z)$ does not exist.

(c) $(0.5)^{|n|} = 0.5^n u(n) + 0.5^{-n} u(-n) - \delta(n)$

We note the term $-\delta(n)$ is necessary since $(0.5)^n u(n)$ and $(0.5)^{-n} u(-n)$ each contribute a value of magnitude one at $n = 0$. So we must subtract $\delta(n)$.

$$Z[0.5^n u(n) + (0.5)^{-n} u(-n) - \delta(n)] = \frac{z}{z - 0.5} + \frac{-2}{z - 2} - 1,$$

$$|z| > 0.5 \cap |z| < 2$$

$$= \frac{z^2 - 4z + 1}{(z - 0.5)(z - 2)}$$

$$- \frac{z^2 - 2.5z + 1}{(z - 0.5)(z - 2)}$$

$$= \frac{-1.5z}{(z - 0.5)(z - 2)},$$

$$0.5 < |z| < 2$$

Therefore $Z[(0.5)^{|n|} * u(n)] = \dfrac{-1.5z}{(z - 0.5)(z - 2)} \dfrac{z}{z - 1}$,

$$1 < |z| < 2$$

$$= \frac{-1.5z^2}{(z - 0.5)(z - 1)(z - 2)},$$

$$1 < |z| < 2$$

We now have two ways to evaluate discrete convolution summations:

1. directly from $\Sigma_p f(p)g(n-p)$ or
2. as the inverse transform of $F(z)G(z)$

EXAMPLE 7-5

Prove the correlation theorems:

(a) $x(n) \oplus y(n) \leftrightarrow Y(z)X(z^{-1})$
(b) $x(n) \oplus x(n) \leftrightarrow X(z)X(z^{-1})$

and discuss the regions of convergence.

Solution

(a) Now by definition:

$$Z\left[\sum_{k-\infty}^{\infty} y(k)x(k-n)\right] = \sum_{n=-\infty}^{\infty}\left[\sum_{k=-\infty}^{\infty} y(k)x(k-n)\right]z^{-n}$$

Interchanging the order of summation and using the substitution of variable $k - n = p$, we obtain:

$$Z[x(n) \oplus y(n)] = \sum_{k=-\infty}^{\infty} y(k)\left[\sum_{n=-\infty}^{\infty} x(k-n)z^{-n}\right]$$

$$= \sum_{k=-\infty}^{\infty} y(k) \sum_{p=-\infty}^{\infty} x(p)z^{p-k}$$

$$= \sum_{k} y(k)z^{-k} \sum_{p} x(p)z^{p}$$

$$= Y(z)X(z^{-1}), \qquad \text{for}$$

$$(\rho_{y_1} < |z| < \rho_{y_2}) \cap (\rho_{x_1} < |z^{-1}| < \rho_{x_2})$$

The annulus $(\rho_{x_1} < z^{-1} < \rho_{x_2})$ is equivalent to $\rho_{x_2}^{-1} < |z| < \rho_{x_1}^{-1}$.

Therefore $Z[x(n) \oplus y(n)] = Y(z)X(z^{-1}), \qquad \text{for}$

$$(\rho_{y_1} < |z| < \rho_{y_2}) \cap (\rho_{x_2}^{-1} < |z| < \rho_{x_1}^{-1})$$

or $Z[x(n) \oplus y(n)] = Y(z)X(z^{-1}),$

$$\max(\rho_{y_1}, \rho_{x_2}^{-1}) < |z|$$

$$< \min(\rho_{y_2}, \rho_{x_1}^{-1}) \qquad (7\text{-}6)$$

(b) The Z transform of an autocorrelation function is a special case of Equation 7-6:

$$Z[x(n) \oplus x(n)] = X(z)X(z^{-1}),$$

$$(\rho_{x_1} < |z| < \rho_{x_2}) \cap (\rho_{x_2}^{-1} < |z| < \rho_{x_1}^{-1}) \qquad (7\text{-}7)$$

The region of convergence becomes, max $(\rho_{x_1}, \rho_{x_2}^{-1}) < |z| < $ min $(\rho_{x_2}, \rho_{x_1}^{-1})$ and the transform exists if this annulus exists.

EXAMPLE 7-6

Find the Z transforms of the following correlation summations:

(a) $(-0.5)^n u(n) \oplus (-0.5)^n u(n)$
(b) $(0.5)^n u(n) \oplus 3^n u(-n)$
(c) $(0.5)^{|n|} \oplus u(n)$

Solution

(a) $(-0.5)^n u(n) \longleftrightarrow \dfrac{z}{z + 0.5}$, $|z| > 0.5$

therefore $\dfrac{Z[(-0.5)^n u(n)}{\oplus (-0.5)^n u(n)]} = \dfrac{z}{z + 0.5} \dfrac{z^{-1}}{z^{-1} + 0.5}$

$$= \frac{z}{z + 0.5} \frac{2}{z + 2},$$

$$0.5 < |z| < 2$$

$$= \frac{2z}{(z + 0.5)(z + 2)},$$

$$0.5 < |z| < 2$$

(b) $\dfrac{Z[(0.5)^n u(n)}{\oplus 3^n u(-n)]} = \dfrac{-3}{z - 3} \dfrac{z^{-1}}{z^{-1} - 0.5}$

$$= \frac{+6}{(z - 3)(z - 2)}, \quad \max(0, 0) < |z| < \min(3, 2)$$

$$= \frac{+6}{(z - 3)(z - 2)}, \quad 0 < |z| < 2$$

(c) $(0.5)^{|n|} = (0.5)^n u(n) + (0.5)^{-n} u(-n) - \delta(n)$

$$= (0.5)^n u(n) + 2^n u(-n) - \delta(n)$$

Therefore $Z[(0.5)^{|n|}] = \dfrac{z}{z - 0.5} + \dfrac{-2}{z - 2} - 1$

$$= \frac{-1.5z}{(z - 0.5)(z - 2)}, \quad 0.5 < |z| < 2$$

as in Example 7-4.

Therefore $Z[(0.5)^{|n|} \oplus u(n)] = \dfrac{z}{z - 1} \dfrac{-1.5z^{-1}}{(z^{-1} - 0.5)(z^{-1} - 2)}$

$$= \dfrac{z}{z - 1} \dfrac{-1.5z}{0.5(z - 2)^2(z - 0.5)}$$

$$= \dfrac{-1.5z^2}{(z - 0.5)(z - 1)(z - 2)},$$

$$|z| > 1 \cap [0.5 < |z| < 2]$$

$$= \dfrac{-1.5z^2}{(z - 0.5)(z - 1)(z - 2)},$$

$$1 < |z| < 2$$

If we need to find $0.5^{|n|} \oplus u(n)$, we now have two approaches:

1. evaluate $\Sigma_k \, y(k)x(k - n)$ or
2. find $Z^{-1} [-1.5z^2/(z - 0.5)(z - 1)(z - 2)]$, where Z^{-1} indicates the inverse Z transform.

7-3 THE INVERSE TWO-SIDED Z TRANSFORM

In this section two techniques for finding inverse two-sided transforms will be discussed:

1. the use of partial fraction expansions plus table reference
2. the classical evaluation using the theory of Laurent series and residue theory

7-3-1 Inverse Transforms Using Partial Fractions

Given:

$$F(z) = \frac{N(z)}{D(z)} = \frac{b_m z^m + b_{m-1}z^{m-1} + \cdots + b_0}{a_n z^n + a_{n-1}z^{n-1} + \cdots + a_0}, \qquad \rho_1 < |z| < \rho_2$$

where the order of $N(z)$ is at most the same as that of $D(z)$ (is this common for Z transforms?) we can expand $F(z)$ or $z^{-1}F(z)$ into partial fractions and from a table of one-sided transforms plus the fact that $Z[f_2(n)u(-n)] = Z[f_2(-n)u(n)] \,|_{z=z^{-1}}$ call off $f(n)$. A number of inverse transforms will now be evaluated.

EXAMPLE 7-7

Find the inverse Z transforms of the following functions using partial fractions:

(a) $F_1(z) = \dfrac{z^3 + 2z^2 + 2z}{(z + 1)^2(z - 2)}, \qquad 1 < |z| < 2$

(b) $F_2(z) = \dfrac{z^3 + 2z^2 + 2z}{(z + 1)^2(z - 2)}, \qquad |z| > 2$

(c) $F_3(z) = \dfrac{z^3 + 2z^2 + 2z}{(z + 1)^2(z - 2)}, \qquad |z| < 1$

Solution

(a) Since the order of the numerator and denominator are the same, we express $F(z)/z$ in partial fractions:

$$\frac{F(z)}{z} = \frac{z^2 + 2z + 2}{(z + 1)^2(z - 2)}, \qquad 1 < |z| < 2$$

$$= \frac{A_1}{z + 1} + \frac{A_2}{(z + 1)^2} + \frac{A_3}{z - 2}$$

$$A_2 = \frac{1 - 2 + 2}{-3} = -0.33,$$

$$A_3 = \frac{4 + 4 + 2}{9} = 1.11$$

and

$$A_1 = \left[\frac{d}{dz} \frac{z^2 + 2z + 2}{z - 2}\right]_{z=-1}$$

$$= \frac{-3(0) - 1(1)}{9} = -0.11$$

Therefore

$$F(z) = \frac{-0.11z}{z + 1} + \frac{-0.33z}{(z + 1)^2} + \frac{1.11z}{z - 2},$$

$$1 < |z| < 2$$

From our experience we know that the pole at $z = -1$ contributes to $f(n)$ for $n > 0$ and the pole at $z = 2$ contributes to $f(n)$ for $n < 0$

Therefore $f_1(n) = -0.11(-1)^n u(n) - 0.33n(-1)^{n-1}u(n)$

$$- 1.11(2)^n u(-n - 1)$$

The inverse of $1.11z/(z - 2), |z| < 2$ requires some thought.

$$Z^{-1}\left[\frac{1.11}{z - 2}\right] = -1.11\left(\frac{1}{2}\right)(2)^n u(-n) = g(n)$$

Therefore the inverse of $Z[1.11z/(z - 2)]$ is $g(n + 1) = -1.11$ $(\frac{1}{2})(2^{n+1}u(-n - 1)) = -1.11(2)^n u(-n - 1)$, as was written in the expression for $f_1(n)$.

(b) $F_2(z) = \dfrac{z^3 + 2z^2 + 2z}{(z + 1)^2(z - 2)}, \qquad |z| > 2$

Since all the poles are inside $|z| = 2$, then $f(n)$ is zero for $n < 0$

and $f_2(n) = -0.11(-1)^n u(n) - 0.33n(-1)^{n-1} u(n) + 1.11(2)^n u(n)$

(c) $F_3(z) = \dfrac{z^3 + 2z^2 + 2z}{(z + 1)^2(z - 2)}$, $|z| < 1$

Since all the poles are outside $|z| = 1$, then $f_3(n)$ is zero for $n > 0$.

$$F_3(z) = \frac{-0.11z}{z + 1} + \frac{-0.33z}{(z + 1)^2} + \frac{1.11z}{z - 2}, |z| < 1$$

$$f_3(n) = 0.11(-1)^n u(-n - 1) + \frac{-0.33z}{(z + 1)^2} - 1.11(2)^n u(-n - 1)$$

We must now discuss the inverse of $-0.33/(z + 1)^2$. In general:

$$a^n u(-n) \longleftrightarrow \frac{-a}{z - a}$$

Therefore $na^{n-1} u(-n) \longleftrightarrow \dfrac{(z - a)(-1) + a(-1)}{(z - a)^2}$

$$= \frac{-z}{(z - a)^2}$$

Using this relation, we have:

$$\frac{-0.33z}{(z + 1)^2} \longleftrightarrow 0.33n(-1)^{n-1} u(-n - 1)$$

and $f_3(n) = [0.11(-1)^n + 0.33n(-1)^{n-1} - 1.11(2)^n] u(-n - 1)$

Figure 7-2 parts (a) to (c) show $F(z)$ and its corresponding discrete time function for this problem.

From Example 7-6 it can be seen that all the work required to evaluate inverse two-sided Z transforms by partial fractions was already mastered for the one-sided case. The poles inside $|z|$ where $\rho_1 < |z| < \rho_2$ determine $f(n)$ for $n \geq 0$, whereas the poles outside $|z|$ determine $f(n)$ for $n < 0$. If $\alpha_1 \leq \rho_1$, then a term $A_1 z/(z - \alpha_1)$ contributes $(\alpha_1)^n u(n)$, whereas $A_1/(z - \alpha_1)$ contributes $(\alpha_1)^{n-1} u(n - 1)$. If $\alpha_1 \geq \rho_2$, then a term $A_2/(z - \alpha_2)$ contributes $A_2(\alpha_2)^n u(-n)$, whereas $-A_2 z/(z - \alpha_2)$ contributes $A_2(\alpha_2)^{n+1} u(-n - 1)$.

7-3-2 Inverse Two-sided Z Transforms Using Residues

The Appendix on complex variables summarizes the theory of Laurent series. If the function $F(z) = N(z)/D(z)$ is expanded in a Laurent series in the region $\rho_1 < |z| < \rho_2$ which represents an annulus between two consecutive poles, then:

$$F(z) = \sum_{n=-\infty}^{\infty} A_n z^n,   \rho_1 < |z| < \rho_2 \tag{7-8}$$

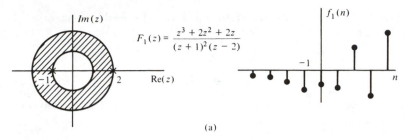

(a)

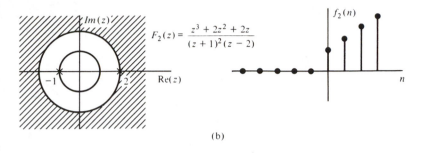

(b)

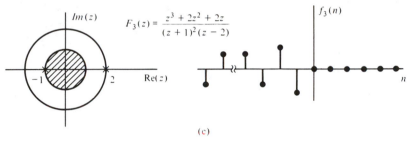

(c)

Figure 7-2 The Z transforms and their inverses for Example 7-7.

where the coefficients are given by:

$$A_n = \frac{1}{2\pi j} \oint_C \frac{F(z)}{z^{n+1}} \, dz \tag{7-9}$$

where C is defined by $z(\theta) = \rho e^{j\theta}$, $0 < \theta \leq 2\pi$ with $\rho_1 < \rho < \rho_2$. Further, if the order of $N(z)$ is at most the order of $D(z)$, then the inside–outside theorem yields:

For n > 0

$$A_n = -\Sigma \left[\text{residues of the poles of } \frac{F(z)}{z^{n+1}} \text{ outside } C\right] \tag{7-10}$$

For n ≤ 0

$$A_n = \Sigma \left[\text{residues of the poles of } \frac{F(z)}{z^{n+1}} \text{ inside } C\right] \tag{7-11}$$

As was seen in the Appendix the use of the inside–outside theorem allows us to avoid finding the residue of a higher-order pole at $z = 0$ for $n > 0$. We now must carefully adjust this theory of the Laurent series to evaluate inverse Z transforms. By definition:

$$Z[f(n)] = \cdots f(-n)z^n + \cdots + f(-1)z + f(0)$$
$$+ f(1)z^{-1} + f(2)z^{-2} + \cdots, \qquad \rho_1 < |z| < \rho_2$$

The Z transform is a Laurent series expansion where the Laurent coefficients are related to the discrete time values by:

$$f(-n) = A_n$$

or
$$f(n) = A_{-n}$$

Therefore, given:

$$Z[f(n)] = F(z), \qquad \rho_1 < |z| < \rho_2$$

we have:

$$f(n) = A_{-n} = \frac{1}{2\pi j} \oint \frac{F(z)}{z^{-n+1}} \, dz$$

$$= \frac{1}{2\pi j} \oint z^{n-1} F(z) \, dz \qquad\qquad (7\text{-}12)$$

and from Equations 7-10 and 7-11 we obtain:

For n ≥ 0

$$f(n) = \frac{1}{2\pi j} \oint_C z^{n-1} F(z) \, dz$$

$$= \Sigma \text{ [residues of the poles of } z^{n-1} F(z) \text{ inside } C] \qquad (7\text{-}13)$$

For n < 0

$$f(n) = \frac{1}{2\pi j} \oint_C z^{n-1} F(z) \, dz$$

$$= -\Sigma \text{ [residues of the poles of } z^{n-1} F(z) \text{ outside } C] \qquad (7\text{-}14)$$

since $z^{n-1} = 1/z^{|n|+1}$ causes the order of the denominator to be more than one higher than the numerator and the inside–outside theorem may be used.

Summarizing, we conclude:

If

$$Z[f(n)] = F(z) = \frac{N(z)}{D(z)}, \qquad \rho_1 < |z| < \rho_2$$

and the order of $N(z)$ is at most equal to the order of $D(z)$, then the inverse transform:

$$f(n) = A_n = \frac{1}{2\pi j} \oint_C z^{n-1} F(z) \, dz$$

is:

for n ≥ 0

$$f(n) = \Sigma \text{ [residues of the poles of } z^{n-1} F(z) \text{ inside } C]$$

for n < 0

$$f(n) = -\Sigma \text{ [residues of the poles of } z^{n-1} F(z) \text{ outside } C]$$

where C is defined by $z(\theta) = \rho e^{j\theta}$, $\rho_1 < \rho < \rho_2$.

We now find some inverse two-sided Z transforms using residue theory.

EXAMPLE 7-8

Find the inverse Z transforms of the following functions using residue theory:

(a) $F_1(z) = \dfrac{z^3 + 2z^2 + 2z}{(z + 1)^2(z - 2)}$, $1 < |z| < 2$

(b) $F_2(z) = \dfrac{z^3 + 2z^2 + 2z}{(z + 1)^2(z - 2)}$, $|z| > 2$

(c) $F_3(z) = \dfrac{z^3 + 2z^2 + 2z}{(z + 1)^2(z - 2)}$, $|z| < 1$

Solution. We are now finding by residue theory the inverse transforms of the same functions whose inverses were found by partial fraction theory in Example 7-7.

(a) Figure 7-2(a) showed a pole zero diagram for $F_1(z)$.

Therefore $f_1(n) = \dfrac{1}{2\pi j} \oint_C z^{n-1} \dfrac{z^3 + 2z^2 + 2z}{(z + 1)^2(z - 2)} dz$

For n ≥ 0

$f_1(n) = $ [residue of the second-order pole at $z = -1$]

$$= \frac{d}{dz}\left[\frac{z^n(z^2 + 2z + 2)}{(z - 2)}\right]_{z=-1}$$

$$= \frac{[(z - 2)[nz^{n-1}(z^2 + 2z + 2) + z^n(2z + 2)] - z^n(z^2 + 2z + 2)(1)]}{(z - 2)^2}\Bigg|_{z=-1}$$

$$= \frac{1}{9}\{-3n(-1)^{n-1}(1) + (-1)^n(0)] - (-1)^n(1)\}$$

$$= -0.33n(-1)^{n-1} - 0.11(-1)^n$$

We note when $n = 0$ the pole at $z = 0$ has a zero residue:

For $n < 0$

$$f_1(n) = - \text{ [residue of the pole at } z = 2]$$

$$= - \left. \frac{z^n(z^2 + 2z + 2)}{(z + 1)^2} \right|_{z=2}$$

$$= -2^n \left(\frac{10}{9} \right)$$

$$= -1.11(2)^n$$

Summarizing the inverse transform yields:

$$f_1(n) = [-0.11(-1)^n - 0.33n(-1)^{n-1}]u(n) - 1.11(2)^n u(-n - 1)$$

This result agrees with part (a) of Example 7-7 and was shown in Figure 7-2(a).

(b) $f_2(n) = \dfrac{1}{2\pi j} \displaystyle\oint_{|z|=\rho} z^n \dfrac{z^2 + 2z + 2}{(z + 1)^2(z - 2)} \, dz, \qquad \rho > 2$

For $n \geq 0$

$$f_2(n) = \Sigma \text{ [residues of the poles at } z = -1 \text{ and } z = +2]$$

$$= [-0.11(-1)^n - 0.33n(-1)^{n-1} + 1.11(2)^n]$$

For $n < 0$

Since there are no poles of $F(z)z^{n-1}$ outside C, then:

$$f_2(n) = 0$$

Finally, the inverse transform is:

$$f_2(n) = [-0.11(-1)^n - 0.33n(-1)^{n-1} + 1.11(2)^n]u(n)$$

This agrees with Example 7-7(b), which is shown in Figure 7-2(b).

(c) $f_3(n) = \dfrac{1}{2\pi j} \displaystyle\oint_{|z|=\rho} z^n \dfrac{z^2 + 2z + 2}{(z + 1)^2(z - 2)} \, dz, \qquad \rho < 1$

For $n \geq 0$

Since there are no poles inside C then $f_3(n) = 0$

For $n < 0$

$$f_3(n) = -\Sigma \text{ [residues of the poles at } z = -1 \text{ and } z = 2]$$

$$= [0.11(-1)^n + 0.33n(-1)^{n-1} - 1.11(2)^n]$$

The inverse transform is:

$$f_3(n) = [0.11(-1)^n + 0.33n(-1)^{n-1} - 1.11(2)^n]u(-n-1)$$

This agrees with Example 7-7(c) and is plotted in Figure 7-2(c).

7-3-3 Complex Convolution

In Chapter 5 we discussed complex convolution when finding the Laplace transform of the product of continuous functions. We now consider complex convolution for the product of discrete functions.

EXAMPLE 7-9
Prove:

$$K(z) = Z[f(n)g(n)] = \frac{1}{2\pi j} \oint_C F(p)G\left(\frac{z}{p}\right)p^{-1} \, dp$$

$$= F(z)*G(z)$$

given: $f(n) \leftrightarrow F(z)$, $\rho_{f1} < \rho < \rho_{f2}$

and $g(n) \leftrightarrow G(z)$, $\rho_{g1} < \rho < \rho_{g2}$

Pay particular attention to the restrictions on C and the region of convergence for $K(z)$.

Solution. Before starting our proof, we note that if $F(z)$ converges $\rho_{f1} < \rho < \rho_{f2}$ and $G(z)$ converges $\rho_{g1} < \rho < \rho_{g2}$, then $k(n) = f(n)g(n)$ must have a Z transform that converges $\rho_{f1}\rho_{g1} < \rho < \rho_{f2}\rho_{g2}$ [think carefully about this]. By definition:

$$K(z) = \sum_{n=-\infty}^{\infty} f(n)g(n)z^{-n}$$

$$= \sum_{-\infty}^{\infty} g(n)\left[\frac{1}{2\pi j} \oint_{C_1} F(p)p^{n-1} \, dp\right]z^{-n}$$

Now assuming it is permissible to interchange the order of summation and integration, we obtain:

$$K(z) = \frac{1}{2\pi j} \oint_{C_1} F(p)\left[\sum_{n=-\infty}^{\infty} g(n)\left(\frac{z}{p}\right)^{-n} p^{-1}\right] dp$$

$$= \frac{1}{2\pi j} \oint_C F(p)G\left(\frac{z}{p}\right)p^{-1} \, dp$$

$$= F(z)*G(z)$$

Now we must carefully discuss $C = \rho_k e^{j\phi}$. First, ρ_k must satisfy $\rho_{f1} < \rho_k < \rho_{f2}$. Also for any z such that $\rho_{f1}\rho_{g1} < |z| < \rho_{f2}\rho_{g2}$, we must have $\rho_{g1} < |z|$

$\div \rho_k < \rho_{g2}$. The solution of an actual problem will make us appreciate these restrictions.

EXAMPLE 7-10

Consider finding $Z[f(n)g(n)]$ where $f(n) = 2^n u(-n) + u(n)$ and $g(n) = u(-n) + 0.5^n u(n)$ by complex convolution.

(a) Find the annulus of convergence for which $F(z)*G(z)$ exists.
(b) Sketch a pole zero diagram showing the poles of $F(p)G(z/p)p^{-1}$ and indicate where C is constrained in the p plane.
(c) Evaluate $K(z)$.

Solution

(a)
$$F(z) = Z[2^n u(-n) + u(n)]$$

$$= \frac{-2}{z-2} + \frac{z}{z-1}, \qquad 1 < |z| < 2$$

$$G(z) = Z[u(-n) + 0.5^n u(n)]$$

$$= \frac{-1}{z-1} + \frac{z}{z-0.5}, \qquad 0.5 < |z| < 1$$

$K(z) = F(z)*G(z)$ will exist for $0.5 < |z| < 2$ as is easily seen by finding $f(n)g(n)$.

(b) With some work:

$$K(z) = \frac{1}{2\pi j} \oint_C \frac{p^2 - 4p + 2}{(p-2)(p-1)} \frac{p^2 - 4pz + 2z^2}{(p-z)(p-2z)p} dp$$

We note $K(z)$ has poles at $p = 0, 1, z, 2,$ and $2z$, and we must satisfy two conditions for ρ_k in C defined by $\rho_k e^{j\phi}, 0 < \phi \le 2\pi$; first, $1 < \rho_k < 2$, and second, $|z| < \rho_k < 2|z|$ where $0.5 < |z| < 2$ from part (a). This requires the pole at $p = z$ is always inside the pole at $p = 2$ and the pole at $p = 2z$ is always outside the pole at $p = 1$. Therefore the contour $\rho_k e^{j\phi}$ always has the poles at $p = 0, p = 1,$ and $p = z$ inside it and the poles at $p = 2z$ and $p = 2$ outside. The pole zero diagram is shown for different cases in Figure 7-3. These moving poles at $p = z$ and $p = 2z$ in the p plane are tricky to visualize.

(c) The direct evaluation of:

$$K(z) = \frac{1}{2\pi j} \oint_C \frac{(p^2 - 4p + 2)(p^2 - 4pz + 2z^2)}{p(p-1)(p-z)(p-2)(p-2z)} dp$$

$$= \sum [\text{residues at } p = 0, 1, \text{ and } z]$$

is very messy and so we will handle it in simple parts.

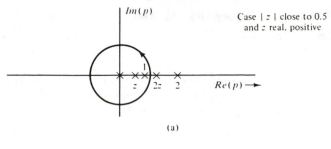

(a)

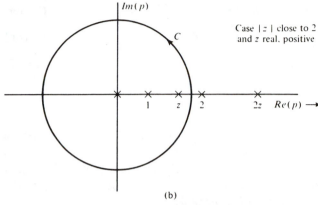

(b)

Figure 7-3   Pole zero diagrams for $F(p)G(z/p)p^{-1}$ for Example 7-10.

$$K(z) = \frac{1}{2\pi j} \oint_C \left(\frac{-2}{p-2} - \frac{p}{p-1} \right) \left(\frac{p}{p-z} - \frac{2z}{p-2z} \right) \frac{1}{p} \, dp$$

$$= \frac{1}{2\pi j} \oint_C \left[\frac{-2}{(p-2)(p-z)} + \frac{p}{(p-1)(p-z)} \right.$$

$$\left. - \frac{2z}{(p-1)(p-2z)} + \frac{4z}{(p-2)(p-2z)p} \right] dp$$

$$= \frac{-2}{z-2} + 1 + \frac{z}{z-0.5} + 1$$

$$= \frac{3z^2 - 9z + 3}{(z-0.5)(z-2)}, \qquad 0.5 < |z| < 2$$

As a check we find $Z[f(n)g(n)]$ directly.

$$Z[2^n u(-n) + u(n)][u(-n) + 0.5^n u(n)]$$

$$= Z[2^n u(-n) + 2\delta(n) + 0.5^n u(n)]$$

therefore $K(z) = \dfrac{-2}{z-2} + 2 + \dfrac{z}{z-0.5},$ $|z| < 2 \cap |z| > 0.5$

This checks with our previous result. It is important when finding $f(n)g(n)$ to note that $u(n)u(-n) = \delta(n)$ and not 0.

7-4 LINEAR SYSTEMS WITH RANDOM AND SIGNAL PLUS NOISE INPUTS

In Chapter 3 we found that the output autocorrelation function and the cross-correlation of the input with the output when the input to a LTIC discrete system is a noise waveform with autocorrelation function $R_{xx}(n)$ were:

$$R_{yy}(n) = C_{hh}(n)*R_{xx}(n) \tag{7-15}$$

$$R_{xy}(n) = h(n)*R_{xx}(n) \tag{7-16a}$$

or

$$= R_{xx}(n) \oplus h(n) \tag{7-16b}$$

and

$$R_{yx}(n) = R_{xy}(-n) \tag{7-17}$$

$$= h(n) \oplus R_{xx}(n) \tag{7-18}$$

These results are shown schematically in Table 7-3(a).

Let us denote $Z[R_{xx}(n)]$ by $S_{xx}(z)$, $Z[R_{xy}(n)]$ by $S_{xy}(z)$, $Z[R_{yx}(n)]$ by $S_{yx}(z)$, $Z[R_{yy}(n)]$ by $S_{yy}(z)$, and $Z[C_{hh}(n) = h(n) \oplus h(n)]$ by $T(z)$. We call $S_{xx}(z)$ the power spectral density of $x(n)$, $S_{xy}(z)$ the cross-spectral density of $x(n)$ with $y(n)$, $S_{yx}(z)$ the cross-spectral density of $y(n)$ with $x(n)$, $S_{yy}(z)$ the power spectral density of $y(n)$, and $T(z)$ the power transfer function.

Using the convolution and correlation theorems, we find that Equations

TABLE 7-3

(a)   (b)

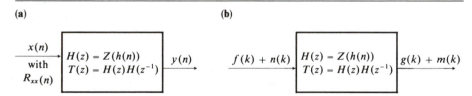

	Case $x(n)$ random	Case $f(k)$ deterministic, $n(k)$ zero-mean random and uncorrelated
Time-Domain Results from Chapter 3	$R_{yy}(n) = C_{hh}(n)*R_{xx}(n)$ $R_{xy}(n) = h(n)*R_{xx}(n)$ $R_{yx}(n) = R_{xy}(-n)$	$g(k) = f(k)*h(k)$ $R_{mm}(k) = C_{hh}(k)*R_{nn}(k)$ $R_{nm}(k) = h(k)*R_{nn}(k)$ $R_{mn}(k) = R_{nm}(-k)$
Transform Results	$S_{yy}(z) = T(z)S_{xx}(z)$ $S_{xy}(z) = H(z)S_{xx}(z)$ $S_{yx}(z) = H(z^{-1})S_{xx}(z)$	$G(z) = F(z)H(z)$ $S_{mm}(z) = T(z)S_{nn}(z)$ $S_{nm}(z) = H(z)S_{nn}(z)$ $S_{mn}(z) = H(z^{-1})S_{nn}(z)$
Properties	$S_{nn}(z) = S_{nn}(z^{-1})$ $T(z) = T(z^{-1})$ $S_{nm}(z) = S_{mn}(z^{-1})$	

7-15 through 7-18 become:

$$S_{yy}(z) = [H(z)H(z^{-1})]S_{xx}(z) \qquad (7\text{-}19)$$

where

$$T(z) = H(z)H(z^{-1})$$

and

$$S_{xy}(z) = S_{xx}(z)H(z) \qquad (7\text{-}20)$$

Similarly,

$$S_{yx}(z) = S_{xx}(z)H(z^{-1}) \qquad (7\text{-}21)$$

These results are tabulated in Figure 7-3(a). Before applying these formulas, we will comment on the symmetry properties of spectral functions.

7-4-1 Properties of Spectral Functions

The properties of spectral functions for continuous functions were developed in detail in Chapter 5. The proofs involving the spectral functions for discrete waveforms are almost identical to those for continuous waveforms except we use summations instead of integrals. Table 7-3 lists many of the main properties for discrete waveforms and a few of them will be demonstrated.

$S_{xx}(z), S_{yy}(z), T(z)$

Power spectral and power transfer functions have the same properties since they are the Z transforms of correlation functions.

EXAMPLE 7-11

Show that:

(a) $S_{xx}(z) = S_{xx}(z^{-1})$
(b) $S_{xy}(z) = S_{yx}(z^{-1})$

Solution

(a)
$$S_{xx}(z) = \sum_{-\infty}^{\infty} R_{xx}(n)z^{-n}$$

Let $p = -n$

and
$$S_{xx}(z) = \sum_{-\infty}^{\infty} R_{xx}(-p)z^{p}$$

$$= \sum_{-\infty}^{\infty} R_{xx}(p)z^{p},$$

(since $R_{xx}(p)$ is even)

Therefore
$$S_{xx}(z) = S_{xx}(z^{-1}) \qquad (7\text{-}22)$$

Since $S_{xx}(z) = S_{xx}(z^{-1})$ we note that if $z - a$ is in the numerator or denominator, we must also have the term $(z^{-1} - a)$ or $(z - 1/a)$

present. Any power spectral density function or power transfer function

$$T(z) = Z[C_{hh}(n)]$$

has this property.

(b)
$$S_{xy}(z) = \sum_{-\infty}^{\infty} R_{xy}(n)z^{-n}$$

$$\text{Let } p = -n$$

Therefore
$$S_{xy}(z) = \sum_{-\infty}^{\infty} R_{xy}(-p)z^{p}$$

$$= \sum_{-\infty}^{\infty} R_{yx}(p)z^{p}$$

$$(\text{since } R_{yx}(\tau) = R_{xy}(-\tau)$$

and
$$S_{xy}(z) = S_{yx}(z^{-1}) \tag{7-23}$$

EXAMPLE 7-12

Given the pulse response of a system is:

$$h(n) = [(-0.6)^n + (0.5)^n]u(n)$$

use the Z transform to find the power transfer function and hence $C_{hh}(n)$.

Solution

$$h(n) = [(-0.6)^n + (0.5)^n]u(n)$$

therefore
$$H(z) = \frac{z}{z + 0.6} + \frac{z}{z - 0.5}, \qquad |z| > 0.6$$

$$= \frac{2z^2 + 0.1z}{(z + 0.6)(z - 0.5)}$$

$$T(z) = H(z)H(z^{-1})$$

$$= \frac{2z^2 + 0.1z}{(z + 0.6)(z - 0.5)} \frac{2z^{-2} + 0.1z^{-1}}{(z^{-1} + 0.6)(z^{-1} - 0.5)}$$

$$= \frac{2z^2 + 0.1z}{(z + 0.6)(z - 0.5)} \frac{2 + 0.1z}{(1 + 0.6z)(1 - 0.5z)}$$

$$= \frac{2z(z + 0.05)0.1(z + 20)}{0.6(z + 0.6)(z + 1.7)(z - 0.5)(-0.5)(z - 2)}$$

$$= \frac{-0.67z(z + 0.05)(z + 20)}{(z + 0.6)(z + 1.7)(z - 0.5)(z - 2)}, \qquad \text{for}$$

$$0.6 < |z| < 1.7$$

The correlation of $h(n)$ with itself may now be found using the inverse transform:

$$C_{hh}(n) = \frac{1}{2\pi j} \oint_C \frac{-0.67z(z + 0.05)(z + 20)}{(z + 0.6)(z + 1.7)(z - 0.5)(z - 2)} z^{n-1} dz$$

For $n \geq 0$

$$C_{hh}(n) = \sum [\text{residues of the poles at } z = -0.6 \text{ and } 0.5]$$

$$= \frac{-0.67(-0.55)(19.4)}{1.1(-1.1)(-2.6)}(-0.6)^n + \frac{-0.67(0.55)(20.5)}{1.1(2.2)-1.5)}(0.5)^n$$

$$= 2.27(-0.6)^n + 2.08(0.5)^n$$

For $n < 0$

We can now find $C_{hh}(n)$ as minus the residues of the poles at $z = -1.7$ and $z = 2$, or using the fact $R_{xx}(n) = R_{xx}(-n)$, we have:

$n < 0$

$$R_{xx}(n) = 2.27(-1.7)^n + 2.08(2)^n$$

$$= 2.27(-0.6)^{-n} + 2.08(0.5)^{-n}$$

7-4-2 Deterministic Signal Plus Uncorrelated Zero-Mean Noise

Table 7-3 shows a linear system with system function $H(z)$ and power transfer function $T(z) = H(z)H(z^{-1})$. The input is $x(k) = f(k) + n(k)$ where $f(k)$ is deterministic and $n(k)$ is a zero-mean uncorrelated noise waveform $[R_{fn}(k) = 0]$ with autocorrelation function $R_{nn}(k)$. In Chapter 3 we found the deterministic output as:

$$g(k) = f(k)*h(k) \tag{7-24}$$

and the output noise autocorrelation as:

$$R_{mm}(k) = C_{hh}(k)*R_{nn}(k)$$

and the cross-correlation of the input and noise as:

$$R_{nm}(k) = h(k)*R_{nn}(k)$$

and

$$R_{mn}(k) = R_{nm}(-k)$$

Using the Z transform, we obtain:

$$G(z) = F(z)H(z) \tag{7-25}$$

and as previously demonstrated:

$$S_{mm}(z) = Z(C_{hh}(n))S_{nn}(z)$$

$$= T(z)S_{nn}(z)$$

where
$$T(z) = H(z)H(z^{-1})$$
$$S_{nm}(z) = H(z)S_{nn}(z)$$
and
$$S_{mn}(z) = H(z^{-1})S_{nn}(z)$$

These results are summarized in Figure 7-3(b).

EXAMPLE 7-13

Consider a linear system with pulse response $h(n) = (0.6)^n u(n)$ has as its input $x(k) = u(k) + n(k)$ where $n(k)$ is an ergodic noise waveform with $R_{nn}(k) = 2\delta(n)$. Find the output signal for $k \gg 0$, the output noise power spectral density $S_{mm}(z)$, the output autocorrelation function, and the input and output signal to noise ratios for $k \gg 0$.

Solution

The Output Signal

$$H(z) = \frac{z}{z - 0.6}, \qquad |z| > 0.6$$

$$Y(z) = \frac{z}{z - 0.6}\frac{z}{z - 1}$$

$$\frac{Y(z)}{z} = \frac{z}{(z - 0.6)(z - 1)}$$

$$= \frac{-1.5}{z - 0.6} + \frac{2.5}{z - 1}$$

therefore
$$Y(z) = \frac{-1.5z}{z - 0.6} + \frac{2.5z}{z - 1}$$

and
$$y(n) = -1.5(0.6)^n u(n) + 2.5u(n)$$

and for $n \gg 0$, $y(n) = 2.5$.

The Output Noise Power Spectral Density

$$S_{mm}(z) = S_{nn}(z)T(z)$$

where $S_{nn}(z) = 2$ and $T(z) = H(z)H(z^{-1})$.

$$T(z) = \frac{z}{z - 0.6}\frac{z^{-1}}{z^{-1} - 0.6}$$

$$= \frac{z}{(z - 0.6)(-0.6)(z - 1.7)}$$

$$= \frac{-1.7z}{(z - 0.6)(z - 1.7)}$$

therefore
$$S_{mm}(z) = \frac{-3.4z}{(z - 0.6)(z - 1.7)}, \qquad 0.6 \le |z| < 1.7$$

$$R_{mm}(n) = \frac{1}{2\pi j}\oint z^n \frac{-3.4}{(z - 0.6)(z - 1.7)}\, dz$$

For n > 0

$$R_{mm}(n) = [\text{residue of the pole at } z = 0.6]$$

$$= \frac{-3.4}{-1.1}(0.6)^n$$

$$= 3.1(0.6)^n$$

and by symmetry:

$$R_{mm}(n) = 3.1(0.6)^n u(n) + 3.1(1.7)^n u[-n - 1]$$

$$= 3.1(0.6)^{|n|}$$

Signal to Noise Ratios

At the input:

$$\frac{S}{N} = \frac{1}{R_{nn}(0)}$$

$$= 0.5$$

whereas at the output:

$$\frac{S}{N} = \frac{2.5^2}{3.1}$$

$$= 2.01$$

SUMMARY

The two-sided Z transform was defined as $F(z) = \Sigma_{n=-\infty}^{\infty} f(n)z^{-n}$ and if it exists it does so in an annulus $\rho_1 < |z| < \rho_2$. The behavior of $f(n)$ for $n < 0$ places the upper bound ρ_2 on $|z|$ and the behavior of $f(n)$ for $n > 0$ places the lower bound ρ_1 on $|z|$. The previously mastered material on the one-sided Z transform was utilized to facilitate the evaluation of two-sided transforms. If $f(n) = f_1(n)u(n) + f_2(n)u(-n)$, then $F(z) = F_1(z) + F_2(z)$, $\rho_1 < |z| < \rho_2$ where $F_1(z)$ is the one-sided Z transform of $f_1(n)$ and $F_2(z)$ is the one-sided Z transform of $f_2(-n)$ with z replaced by z^{-1}.

The most commonly occurring noncausal time functions are auto- and cross-correlation functions whether associated with ergodic noise waveforms or the correlation of the impulse response $h(n)$ with itself. In Chapter 3 $R_{xx}(n)$, $R_{xy}(n)$, $R_{yx}(n)$, and $C_{hh}(n)$ were defined and studied. Here their Z transforms, the spectral functions, $S_{xx}(z)$, $S_{xy}(z)$, $S_{yx}(z)$, and $T(z)$ were studied. Using the famous transform pairs, we obtain:

$$Z[x(n)*y(n)] = X(z)Y(z)$$

$$Z[x(n) \oplus y(n)] = Y(z)X(z^{-1})$$

and $$Z[y(n) \oplus x(n)] = X(z)Y(z^{-1})$$

The time-domain results for a linear discrete system with pulse response $h(n)$ whose input is an ergodic noise waveform with autocorrelation function $R_{xx}(n)$ were:

$$R_{yy}(n) = R_{xx}(n)*C_{hh}(n)$$

and $$R_{xy}(n) = h(n)*R_{xx}(n)$$

Using the Z transform, we find:

$$S_{yy}(z) = S_{xx}(z)T(z)$$

and $$S_{xy}(z) = H(z)S_{xx}(z)$$

$S_{yy}(z)$ and $S_{xx}(z)$ are power spectral densities, $S_{xy}(z)$ and $S_{yx}(z)$ are cross-spectral densities and $T(z) = Z[C_{hh}(n)] = H(z)H(z^{-1})$ is the power transfer function.

Inverse transforms were evaluated by either the use of partial fractions or residues. If $F(z) = F_1(z) + F_2(z)$, where the poles of $F_1(z)$ are inside $|z| = \rho_1$ and the poles of $F_2(z)$ are outside $|z| = \rho_2$, then the inverse is found as $f_1(n)u(n) + f_2(n)u(-n)$ by table reference.

Using residues the inverse transform $f(n)$ is defined as:

$$f(n) = \frac{1}{2\pi j} \oint_C z^{n-1} F(z)\, dz$$

and $f(n)$ is found as:

For $n \geq 0$

$$f(n) = \sum [\text{residues of the poles of } F(z)z^{n-1} \text{ inside } |z| = \rho]$$

For $n < 0$

$$f(n) = -\sum [\text{residues of the poles of } F(z)z^{n-1} \text{ outside } |z| = \rho]$$

PROBLEMS

7-1. Evaluate the two-sided Z transforms of the following functions:

(a) $f_1(n) = 3\delta(n + 2) - \delta(n - 1)$ (b) $f_2(n) = \sum_{k=0}^{\infty} a^{2k}\delta(n + 2k)$

(c) $f_3(n) = 2$ (d) $f_4(n) = (3n - 1)u(-n) + 3^n u(n)$

(e) $f_5(n) = (3n - 1)u(-n - 1)$ (f) $f_6(n) = 2^n u(-n - 1)$
$\quad\quad\quad + 3^{n-1}u(n - 1)$ $\quad\quad\quad + 3n(-1)^n u(n)$

(g) $f_7(n) = (3n^2 - 2n + 2)(0.5)^{-n}u(-n) + (3n - 2)(0.5)^n u(n)$

(h) Without any work, what is the denominator polynomial and region of convergence of the Z transform of $(an^2 - b)(-2)^n u(n) + (cn + d)(0.7)^n u(-n - 1)$?

7-2. Given:

$$x(n) \leftrightarrow X(z), \qquad \rho_{11} < |z| < \rho_{12}$$
$$y(n) \leftrightarrow Y(z), \qquad \rho_{21} < |z| < \rho_{22}$$
$$w(n) \leftrightarrow W(z), \qquad \rho_{31} < |z| < \rho_{32}$$

Find the Z transform and its region of convergence for:
(a) $[x(n) \oplus y(n)] * z(n)$ (b) $[x(n) * y(n)] \oplus z(n)$
(c) $x(n) \oplus [y(n) * z(n)]$

7-3. (a) Prove whether or not:

$$[x(n) \oplus y(n)] * z(n) = x(n) \oplus [y(n) * z(n)]$$

(b) Under what conditions of evenness or oddness for $x(n)$ or $y(n)$ is:
(1) $x(n) \oplus y(n) = y(n) \oplus x(n)$
(2) $x(n) \oplus y(n) = x(n) * y(n)$
(3) $y(n) \oplus x(n) = x(n) * y(n)$

7-4. Evaluate the inverse transform of and plot $f(n)$ versus n for:

(a) $\dfrac{2z^3 + 3z^2 + 1}{z^2}$, for all z (b) $\dfrac{2z^2}{(z + 1)^2}$, $|z| < 1$

(c) $\dfrac{2z^2}{(z + 1)^2}$, $|z| > 1$ (d) $\dfrac{3z^3 + 2z^2 + z}{(z + 3)^2(z - 2)}$, $|z| < 2$

(e) $\dfrac{3z^3 + 2z^2 + z}{(z + 3)^2(z - 2)}$, $2 < |z| < 3$ (f) $\dfrac{3z^3 + 2z^2 + z}{(z + 3)^2(z - 2)}$, $|z| > 3$

(g) $\dfrac{z^6}{(z + 3)^2(z - 2)}$, $2 < |z| < 3$

7-5. If possible evaluate:
(a) $3^n u(-n) \oplus 2^n u(-n)$ (b) $3^n u(-n) * 2^n u(-n)$
(c) $2(0.6)^{|n|} \oplus 2(0.6)^{|n|}$ (d) $2(0.6)^{|n|} * 2(0.6)^{|n|}$
(e) $(3 + n)(-0.5)^n u(n) \oplus (2 + n)u(n)$ (f) $(3 + n)(-0.5)^n u(n) * (2 + n)u(n)$

7-6. Given a linear system with pulse response $h(n) = (-0.8)^n u(n)$ has as its input $x(n) = 4u(-n - 1) + (0.6)^n u(n)$. Find the output $y(n)$.

7-7. If $x(n) \leftrightarrow X(z)$, $0.2 < |z| < 2$

and $y(n) \leftrightarrow Y(z)$, $0.8 < |z| < 3$

show:

$$Z[x(n)y(n)] = X(z) * Y(z)$$
$$= \frac{1}{2\pi j} \oint_C X(p) Y\left(\frac{z}{p}\right) p^{-1} dp$$

Carefully explain for what annulus, $\rho_1 < |z| < \rho_2$, $X(z) * Y(z)$ exists and plot the poles and C on the p plane.

7-8. Use complex convolution to find the Z transform of $x(n)y(n)$, where:

(a) $x(n) = (-2)^n u(n)$ and $y(n) = nu(n)$

(b) $x(n) = 2^n u(-n-1) + u(n)$

$y(n) = (-0.6)^n u(-n-1) + (0.5)^n u(n)$

7-9. Given

$$\overline{x(n)} = X(z), \qquad \rho_{x1} < |z| = \rho < \rho_{x2}$$

and

$$\overline{y(n)} = Y(z), \qquad \rho_{y1} < |z| = \rho < \rho_{y2}$$

(a) What are the conditions for $x(n)$ and $y(n)$ to be stable?

(b) List when the following are stable, for $x(n)$ and $y(n)$ stable; and if they can be unstable, give a specific example for $x(n)$ and $y(n)$:

(1) $x(n)y(n)$ (2) $x(n)*y(n)$ (3) $x(n) \oplus x(n)$

(4) $x(n) \oplus y(n)$ (5) $y(n) \oplus x(n)$

Sketch pole diagrams for each case.

7-10. Given a linear system with pulse response $h(n) = (-0.8)^n u(n)$ has as its input a deterministic signal $f(n) = (-1)^n u(n)$ plus zero-mean independent white noise $n(k)$ with autocorrelation function $R_{nn}(k) = 6\delta(k)$:

(a) Find the output signal $g(n)$ and spectral densities $S_{mm}(z)$, $S_{nm}(z)$, and $S_{mn}(z)$.

(b) Do the spectral functions possess their expected properties?

(c) Find the output mean square fluctuations $m^2(n)$ using residue theory.

7-11. The power transfer function is defined as:

$$T(z) = H(z)H(z^{-1})$$

(a) Find the power transfer functions for the following systems and plot their pole zero diagram:

(1)

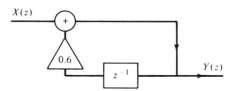

(2)

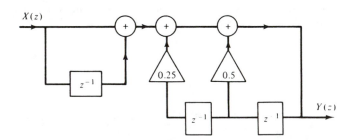

(b) If the input to each of the systems of part (a) is assumed white noise with $S_{xx}(z) = 2$, use residue theory to find the mean-squared output fluctuations

$$\overline{y^2(n)} = R_{yy}(0) = \frac{1}{2\pi j} \oint S_{yy}(z)z^{-1} dz$$

at the output.

7-12. **(a).** If the input to system (1) of the previous problem has $R_{xx}(n) = 10(0.6)^{|n|}$, find the power spectral density of the output noise and the cross-spectral density of the input and output noise $S_{xy}(z)$.

(b). Give a pole zero plot for $S_{xy}(z)$, $S_{yx}(z)$, and $S_{yy}(z)$.

(c). Find the signal to noise ratio at the input and output for $n \gg 0$ if an input signal $f(n) = 6u(n)$ is added to the input noise.

7-13. A power spectral density $S_{xx}(z)$ or power transfer function $T(z)$ may be written as:

$$S_{xx}(z) = G(z)G(z^{-1}) \quad \text{or} \quad T(z) = H(z)H(z^{-1})$$

where $G(z)$ and $H(z)$ have their poles or zeros inside the unit circle $z = 1$.

    Design a "shaping filter" that transforms white noise with $S_{xx}(z) = 1$ to noise with a power spectral density:

$$S_{yy}(z) = \frac{-1.5z}{(z - 0.5)(z - 2)}, \qquad 0.5 < |z| < 2$$

7-14. Which of the following functions qualify as power spectral densities?

(a) $\dfrac{-0.5z}{z^2 + 2.5z + 1}$ **(b)** $\dfrac{0.5z}{z^2 + 2.5z + 1}$

(c) $\dfrac{0.5}{z^2 + 2.5z + 1}$ **(d)** $\dfrac{-z}{z^2 - 16}$

(e) $\dfrac{2z}{(z + 2)^2(z + 0.5)^2}$

Chapter 8

The Fourier Transform

INTRODUCTION

This chapter will develop the Fourier transform and discuss a number of its properties and applications. The material is treated in a manner slightly different from previous transform chapters: instead of starting with the Fourier transform definition, Fourier series analysis will provide our point of departure. The conceptually and intuitively appealing ideas of Fourier series analysis are probably familiar to most readers. However, we begin from basics in Section 8-1 and then briefly consider the generalized Fourier series in Section 8-2. The Fourier transform is developed in Section 8-3 and early emphasis will be placed on deriving a number of Fourier transform pairs, that is, the time function $f(t)$ and its corresponding transform $F(j\omega)$. In Section 8-4 a number of Fourier transform properties will be developed, including the time-shift property and the convolution property. The last section in this chapter examines a number of Fourier transform applications, including filters, modulation, and multiplexing. For the most part, the signals considered in this chapter will be continuous. Chapter 9 deals with the Fourier analysis of discrete signals.

Jean Baptiste Joseph Fourier (1768–1830) was a French mathematician and physicist who did extensive study of heat conduction. He developed what is now called the **Fourier series analysis** to be applied to the solution of partial differential equations that arose out of his heat conduction studies. There was some heated controversy surrounding his publications, however, because he was not able to prove in a general fashion that his infinite series of sines and cosines actually converged to the function they were supposed to represent. No one could prove it at first. It took about a hundred years and the invention of the Lebesgue integral to do the job. For an overview of Fourier's life and times see Oppenheim,

Willsky, and Young, pp. 162–168. One rather strange thing in Fourier's life might be explained by the time he spent in Egypt and his interests in Egyptology, as well as his intense involvement in heat studies: He believed dry desert heat to be ideal for health and lived the latter part of his life wrapped like a mummy in overheated rooms. Genius is permitted its eccentricities!

To motivate this Fourier analysis project, imagine that we have an $f(t)$ sampled to yield $f(n)$. Then, recalling Equation 1-15, we can write:

$$f(n) = \sum_{k=-\infty}^{\infty} f(k)\delta(k - n) \tag{8-1}$$

Thus any $f(t)$ can be approximated by a sum of unit pulse functions. The $f(k)$ values are constants that represent samples of the original $f(t)$. If $f(n)$ is the input to a linear system with unit pulse response $h(n)$, then the output of that system—as we saw in Chapter 2—was $g(n)$, where:

$$g(n) = f(n)*h(n) \tag{8-2}$$

The response of a linear system to a signal represented by unit pulses requires convolution. Determining the input signal representation is usually easy compared to performing the convolution required in Equation 8-2. Not only is convolution difficult to perform, but also the convolution operation calls for the unit pulse response function which may be difficult to determine.

Different kinds of representations of signals, however, might permit simpler response calculations. We know, for example, from sinusoidal steady-state circuit analysis that representing signals as sinusoids is computationally attractive. If the input to an *RLC* circuit is a sinusoid, then the output sinusoid of such a circuit has the same frequency as the input and differs only in amplitude and phase. Representing signals as sinusoids or as sums of sinusoids has certain advantages over representing signals as sums of unit pulses. These signals or functions "in terms of which" a given function is to be represented are called **basis functions.** We consider the employment of basis functions in Section 8-2. Note now only that unit pulse functions and sinusoids are particularly useful basis functions. Sinusoids or complex exponentials will be the type of basis function that is most commonly encountered throughout the rest of this Fourier transform chapter.

8-1 THE TRIGONOMETRIC FOURIER SERIES

Fourier's genius developed the insight that any periodic $f(t)$ can be expressed as a sum of sinusoids. A periodic function, $f(t)$, is one such that $f(t + nT) = f(t)$ for all integers n. T is the period.

Then
$$f(t) = a_0 + \sum_{n=1}^{\infty} a_n \cos n\omega_0 t + b_n \sin n\omega_0 t \tag{8-3}$$

where
$$a_n = \frac{2}{T} \int_T f(\xi) \cos n\omega_0 \xi \, d\xi \tag{8-4}$$

$$b_n = \frac{2}{T} \int_T f(\xi)\,Sin\;n\omega_0\xi\;d\xi \qquad\qquad (8\text{-}5)$$

for $n = 1, 2, \ldots$ The a_0 term is the dc component or average value of $f(t)$:

$$a_0 = \frac{1}{T} \int_T f(\xi)\,d\xi \qquad\qquad (8\text{-}6)$$

In these equations the integration symbol with "T" subscript implies that we can integrate over *any* period. Also, the ω_0 term which is called the fundamental angular frequency is $\omega_0 = 2\pi/T$. Note that all the sinusoids appearing in Equation 8-3 have frequencies that are integer multiples of the fundamental. This Fourier series representation in Equation 8-3 is known as the **trigonometric Fourier series.**

EXAMPLE 8-1

Determine the trigonometric Fourier series expansion of $f(t)$ in Figure 8-1.

Solution

$T = 2$, $\omega_0 = \pi$, and $f(t) = t$ in the region $0 \le t \le 1$.

$$a_0 = \frac{1}{2} \int_0^1 \xi\,d\xi = \frac{1}{2}\frac{\xi^2}{2}\bigg|_0^1 = \frac{1}{4}$$

$f = \frac{1}{2}$

$w_0 = 2\pi f$

$\quad = 2\pi \frac{1}{2}$

$$a_n = \int_0^1 \xi\,Cos\;n\pi\xi\;d\xi$$

$$= \frac{(-1)^n - 1}{n^2\pi^2}$$

$w_0 = \pi$

and

$$b_n = \int_0^1 \xi\,Sin\;n\pi\xi\;d\xi$$

$$= \frac{1}{n\pi}(-1)^{n+1}$$

Therefore

$$f(t) = \frac{1}{4} + \sum_{n=1}^{\infty}\left[\frac{(-1)^n - 1}{(n\pi)^2}\,Cos\;n\pi t + \frac{(-1)^{n+1}}{n\pi}\,Sin\;n\pi t\right]$$

EXAMPLE 8-2

Determine the trigonometric Fourier series expansion of $f(t)$ in Figure 8-2.

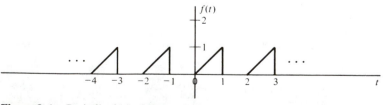

Figure 8-1 Periodic $f(t)$ of Example 8-1.

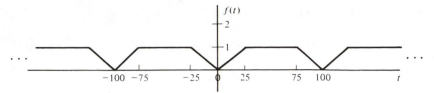

Figure 8-2 Periodic $f(t)$ of Example 8-2.

Solution

$T = 100$, $\omega_0 = \pi/50$. By inspection $a_0 = \frac{3}{4}$.

$$b_n = \frac{1}{50} \int_0^{25} \frac{\xi}{25} \operatorname{Sin} \frac{n\pi\xi}{50} \, d\xi + \frac{1}{50} \int_{25}^{75} (1) \operatorname{Sin} \frac{n\pi\xi}{50} \, d\xi$$

$$+ \frac{1}{50} \int_{75}^{100} \left(4 - \frac{\xi}{25} \right) \operatorname{Sin} \frac{n\pi\xi}{50} \, d\xi$$

$$= \frac{1}{50} \left[\int_0^{25} \frac{\xi}{25} \operatorname{Sin} \frac{n\pi\xi}{50} \, d\xi - \int_{75}^{100} \frac{\xi}{25} \operatorname{Sin} \frac{n\pi\xi}{50} \, d\xi \right] + \frac{1}{50} \int_{25}^{75} \operatorname{Sin} \frac{n\pi\xi}{50} \, d\xi$$

$$+ \frac{4}{50} \int_{75}^{100} \operatorname{Sin} \frac{n\pi\xi}{50} \, d\xi = 0$$

and from an equation similar to the equation for b_n, we get:

$$a_n = \frac{4(\operatorname{Cos} n\pi/2 - 1)}{(n\pi)^2}$$

Therefore $$f(t) = \frac{3}{4} - \frac{4}{\pi^2} \left(\operatorname{Cos} \frac{\pi}{50}t + \frac{1}{2} \operatorname{Cos} \frac{2\pi}{50}t + \frac{1}{9} \operatorname{Cos} \frac{3\pi}{50}t + \cdots \right)$$

Strictly speaking, the functions $f(t)$ that we are representing must be well behaved in order that the series expressing them will converge. This means that the periodic $f(t)$ of interest needs to satisfy what are called the **Dirichlet conditions:** $f(t)$ must have at most a finite number of maxima and minima and finite discontinuities in one period, and $f(t)$ must be absolutely integrable over one period. Absolute integrability means that:

$$\int_{-T/2}^{T/2} |f(t)| \, dt < \infty \tag{8-7}$$

If these conditions are satisfied, then the Fourier series representation of $f(t)$ converges to the actual $f(t)$. The Dirichlet conditions are sufficient conditions; that is, it is not necessarily true that if the series converges then the conditions are satisfied. Fortunately, most engineering applications employ functions that do satisfy the Dirichlet conditions.

Now, if a Fourier series representation of a periodic signal is obtained and this signal is used as an input to a linear system, then what is the forced output of the system? In order to deal with this situation most effectively, we need to express our periodic signals in a way that combines the sine and cosine terms in the original expansion into a single term with a phase shift. We can write a cosinusoidal Fourier series as:

$$f(t) = a_0 + \sum_{n=1}^{\infty} \tilde{c}_n \cos(n\omega_0 t + \theta_n) \qquad (8\text{-}8)$$

where

$$\tilde{c}_n = \sqrt{a_n^2 + b_n^2} \quad \text{and} \quad \theta_n = -\text{Tan}^{-1} b_n/a_n \qquad (8\text{-}9)$$

Now let $x(t)$ be a periodic input to a system that has a transfer function $H(j\omega)$, and let $y(t)$ be the system output.

Then

$$x(t) = a_0 + \sum_{n=1}^{\infty} \tilde{c}_n \cos(n\omega_0 t + \theta_n) \qquad (8\text{-}10)$$

$$H(jn\omega_0) = H(j\omega)\mid_{\omega - n\omega_0} = \mid H(jn\omega_0)\mid \angle \arg H(jn\omega_0) \qquad (8\text{-}11)$$

and

$$y(t) = a_0 H(0)$$

$$+ \sum_{n=1}^{\infty} \tilde{c}_n \mid H(jn\omega_0)\mid \cos(n\omega_0 t + \theta_n + \arg H(jn\omega_0)) \qquad (8\text{-}12)$$

EXAMPLE 8-3

Consider the response $y(t)$ of the system $H(j\omega) = j\omega/(j\omega + 2)$ when the input $x(t)$ is a periodic signal of period $T = 4$

and

$$x(t) = 0, \qquad -2 \le t \le -1$$

$$= \cos\frac{\pi}{2}t, \qquad -1 \le t \le 1$$

$$= 0, \qquad 1 \le t \le 2$$

This is actually a half-wave rectified cosine function, which appears as in Figure 8-3. Determine the dc term, the first harmonic (fundamental), and the second harmonic in the response $y(t)$.

Solution

$$x(t) = \frac{1}{\pi} + \frac{1}{2}\cos\frac{\pi t}{2} - \sum_{n=1}^{\infty} \frac{2(-1)^n}{\pi(4n-1)}\cos n\pi t$$

follows from a straightforward (but very tedious) application of Equations 8-4 through 8-6. Now $\omega_0 = 2\pi/T = \pi/2$. The a_0 or dc term in $x(t)$ is $1/\pi$. The first harmonic term in $x(t)$ is $\frac{1}{2}\cos \pi t/2$. The second harmonic term in

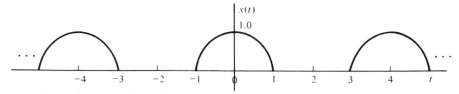

Figure 8-3   Periodic $x(t)$ of Example 8-3.

$x(t)$ is the first term in the summation: $-2(-1)/\pi(4 - 1)$ Cos $\pi t =$ $(2/3\pi)$ Cos πt. Now from the system transfer function:

$$H(jn\omega_0) = H\left(jn\frac{\pi}{2}\right) = \frac{jn\pi/2}{jn\pi/2 + 2}$$

$$= \frac{n\pi/2}{\sqrt{(n\pi/2)^2 + 4}} \angle\left(90° - \text{Tan}^{-1}\frac{n\pi}{4}\right)$$

Therefore $H(0) = 0,$ $H\left(j\frac{\pi}{2}\right) = \frac{\pi}{\sqrt{\pi^2 + 16}} \angle\left(90° - \text{Tan}^{-1}\frac{\pi}{4}\right)$

$$= 0.618 \angle 51.85°$$

$$H(j\pi) = \frac{\pi}{\sqrt{\pi^2 + 4}} \angle\left(90° - \text{Tan}^{-1}\frac{\pi}{2}\right) = 0.844 \angle 32.48°$$

and in the output, we get:

$$dc = a_0 H(0) = 0$$

$$\text{first harmonic} = 0.618(0.5) \text{ Cos}\left(\frac{\pi t}{2} + 51.85°\right) = 0.309 \text{ Cos}\left(\frac{\pi t}{2} + 51.85°\right)$$

$$\text{second harmonic} = 0.844\left(\frac{2}{3\pi}\right) \text{Cos}(\pi t + 32.48°) = 0.179 \text{ Cos}(\pi t + 32.48°)$$

The concepts of evenness and oddness are useful in the Fourier series theory. An even function $f_e(t)$ is one such that:

$$f_e(-t) = f_e(t) \tag{8-13}$$

An odd function $f_0(t)$ is one such that:

$$f_0(-t) = -f_0(t) \tag{8-14}$$

An interesting fact is that *any* $f(t)$ can be written:

$$f(t) = f_e(t) + f_0(t) \tag{8-15}$$

where $$f_e(t) = \frac{f(t) + f(-t)}{2} \tag{8-16}$$

and $$f_0(t) = \frac{f(t) - f(-t)}{2} \tag{8-17}$$

EXAMPLE 8-4

Determine and plot $f_e(t)$ and $f_0(t)$ if $f(t) = u(t)$.

Solution

$$f_e(t) = \frac{u(t) + u(-t)}{2} = \frac{1}{2}, \qquad \text{for all } t, \text{ except } t = 0$$

$$f_0(t) = \frac{u(t) - u(-t)}{2} = -\frac{1}{2}, \qquad \text{for } t < 0$$

$$= \frac{1}{2}, \qquad \text{for } t > 0$$

The only problem here is the value of these functions at $t = 0$. Since $u(t)$ is defined to be 1.0 for $t \geq 0$, then $u(-t) = 1$ for $t \leq 0$.

Therefore $f_e(0) = 1$ and $f_0(0) = 0$

The functions $f_e(t)$ and $f_0(t)$ are plotted in Figure 8-4.

The evenness and oddness of certain functions can be used to simplify the calculations for a_n and b_n required in the trigonometric Fourier series. Given some $f(t)$, if this $f(t)$ is *odd*,

then $a_n = 0,$ for $n = 0, 1, 2, \cdots$ (8-18)

$$b_n = \frac{4}{T}\int_{T/2} f(\xi) \text{ Sin } n\omega_0\xi \, d\xi, \qquad \text{for } n = 1, 2, \cdots$$

If some given $f(t)$ is *even*,

then $a_0 = \frac{2}{T}\int_{T/2} f(\xi) \, d\xi$ (8-19)

$$a_n = \frac{4}{T}\int_{T/2} f(\xi) \text{ Cos } n\omega_0\xi \, d\xi, \qquad \text{for } n = 1, 2, \cdots$$

$$b_n = 0, \qquad \text{for } n = 1, 2, \cdots$$

These results follow from the fact that $\int_T = 2\int_{T/2}$ if the integrand is *even* and $\int_T = 0$ if the integrand is *odd*.

EXAMPLE 8-5

Determine the trigonometric Fourier series expansion for the $f(t)$ given in Figure 8-5.

Solution

$$T = 3, \omega_0 = \frac{2\pi}{T} = \frac{2}{3}\pi$$

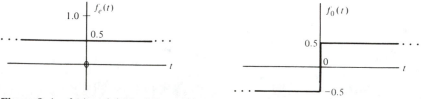

Figure 8-4 $f_e(t)$ and $f_0(t)$ of Example 8-4.

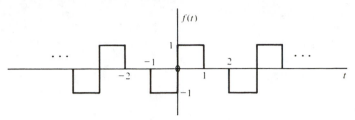

Figure 8-5 Periodic $f(t)$ of Example 8-5.

By inspection $f(t)$ is odd.

Therefore $\qquad a_n = 0,\ b_n = \dfrac{4}{3}\displaystyle\int_0^1 (1)\ \text{Sin}\ n\dfrac{2\pi}{3}\xi\ d\xi$

$$= -\dfrac{4}{3}\left(\dfrac{3}{2\pi n}\right)\left[\text{Cos}\ \dfrac{2\pi n}{3} - 1\right]$$

thus $\qquad f(t) = \dfrac{2}{\pi}\displaystyle\sum_{n=1}^{\infty}\dfrac{1}{n}\left(1 - \text{Cos}\ \dfrac{2\pi n}{3}\right)\text{Sin}\ \dfrac{2\pi n}{3}t$

8-2 GENERALIZED FOURIER SERIES

The trigonometric Fourier series represents a periodic function as an infinite series of sinusoids. In a more general sense, any function can be expressed as an infinite series of other functions or can at least be approximated by a finite series of other functions. Call these "other functions" **basis functions:** $\phi_0(t)$, $\phi_1(t)$, $\phi_2(t)$, ... Assume we can approximate $f(t)$ over a certain range $t_1 \le t \le t_2$ with the function $\hat{f}(t)$:

$$\hat{f}(t) = \sum_{i=0}^{N} \alpha_i\phi_i(t) \qquad (8\text{-}20)$$

where N may be infinity. Generally speaking, the more terms we take, the closer $\hat{f}(t)$ will be to $f(t)$. We will limit the possible spread of the basis functions by demanding that they have certain properties that will result in elegant formulations. Insist that the basis functions be **orthogonal** over the range $t_1 \le t \le t_2$; that is:

$$\int_{t_1}^{t_2} \phi_i(t)\phi_j^*(t)\ dt = 0, \qquad i \ne j$$

$$= \lambda_i, \qquad i = j \qquad (8\text{-}21)$$

The asterisk (*) notation indicates complex conjugation and must be employed if the basis functions are complex functions of time. There is a special case of Equation 8-21 where $\lambda_i = 1$ for all i. In this case the basis functions are said to be **orthonormal.** Referring to Equation 8-20, assuming the ϕ_i's are known, the problem is to determine the proper α_i values. To do so, multiply both sides of

Equation 8-20 by $\phi_j^*(t)$ and integrate over $[t_1, t_2]$ to get:

$$\int_{t_1}^{t_2} \hat{f}(t)\phi_j^*(t)\,dt = \sum_{i=0}^{N} \alpha_i \int_{t_1}^{t_2} \phi_i(t)\phi_j^*(t)\,dt \qquad (8\text{-}22)$$

But from Equation 8-21, the integral on the right is zero unless $i = j$. Thus only the term for $i = j$ in the summation will remain and when $i = j$ the integral on the right becomes λ_j. We get:

$$\int_{t_1}^{t_2} \hat{f}(t)\phi_j^*(t)\,dt = \alpha_j\lambda_j \qquad (8\text{-}23)$$

Unfortunately, this integral requires us to use $\hat{f}(t)$ which is what we are trying to determine. But $\hat{f}(t)$ is supposed to be "close" to $f(t)$. Substituting $f(t)$ for $\hat{f}(t)$, we obtain what are called the **generalized Fourier series coefficients** (changing the index j to i).

$$\alpha_i = \frac{1}{\lambda_i} \int_{t_1}^{t_2} f(t)\phi_i^*(t)\,dt \qquad (8\text{-}24)$$

These α_i values substituted back into Equation 8-20 give $\hat{f}(t)$ as the best approximation to $f(t)$, best in the sense of what is called the "minimum mean square error." The function $\hat{f}(t)$ is supposed to be a good approximation to $f(t)$. The mean square error (MSE) is a measure of how good "good" is:

$$\text{MSE} = \frac{1}{t_2 - t_1} \int_{t_1}^{t_2} \left| f(t) - \sum_{i=0}^{N} \alpha_i\phi_i(t) \right|^2 dt \qquad (8\text{-}25)$$

Intuitively, as N gets larger, in most cases the MSE gets smaller. If $\lim_{N\to\infty} \hat{f}(t) = f(t)$, then the basis functions are said to be **complete.** In this case the MSE $\to 0$ and by equating Equation 8-25 to zero we can derive what is called **Parseval's relation:**

$$\int_{t_1}^{t_2} f^2(t)\,dt = \sum_{i=0}^{\infty} |\alpha_i|^2 \lambda_i \qquad (8\text{-}26)$$

EXAMPLE 8-6

Demonstrate Parseval's relation.

Solution. Carry out Equation 8-25:

$$\text{MSE} = \frac{1}{t_2 - t_1} \int_{t_1}^{t_2} \left[f^2(t) - f(t)\sum_{i=0}^{N} \alpha_i\phi_i(t) \right.$$

$$\left. - f(t)\sum_{i=0}^{N} \alpha_i^*\phi_i^*(t) + \sum_{j=0}^{N} \alpha_j\phi_j(t)\sum_{i=0}^{N} \alpha_i^*\phi_i^*(t) \right] dt$$

since for complex numbers $|z|^2 = zz^*$. From Equation 8-21, we get the last term:

$$\frac{1}{t_2 - t_1} \sum_{i=0}^{N} \lambda_i |\alpha_i|^2$$

But the two middle terms become:

$$\frac{-2}{t_2 - t_1} \sum_{i=0}^{N} |\alpha_i|^2 \lambda_i$$

from Equation 8-24. This assumes that λ_i and $f(t)$ take on only real values, whereas ϕ_i and α_i may be complex.

Hence
$$\text{MSE} = \frac{1}{t_2 - t_1} \int_{t_1}^{t_2} f^2(t)\, dt - \frac{1}{t_2 - t_1} \sum |\alpha_i|^2 \lambda_i$$

and if MSE $\rightarrow 0$,

then
$$\int_{t_1}^{t_2} f^2(t)\, dt \rightarrow \sum_{i=0}^{N} |\alpha_i|^2 \lambda_i$$

which becomes, as $N \rightarrow \infty$:

$$\int_{t_1}^{t_2} f^2(t)\, dt = \sum_{i=0}^{\infty} |\alpha_i|^2 \lambda_i.$$

Parseval's relation is an equation relating energy in the actual signal to energy contained in the signal's representation. If we can make the calculation on the left side of the equation— call it E—then it is often desired to compare E, not to the infinite sum on the right side, but to the sum of a finite number of these terms. For example, we might want to take enough terms in the summation such that the right side is at least 95% of E. Or we can calculate $(E - \sum |\alpha_i|^2 \lambda_i)/E$ as a relative energy error and try to minimize this term. In the real world we typically approximate a given $f(t)$ by as few basis functions as possible. The level of accuracy needed in a given problem is usually determined by the overall problem context.

EXAMPLE 8-7

Assume we are given the basis functions ϕ_1, ϕ_2, and ϕ_3 of Figure 8-6. Approximate:

$$f(t) = t, \qquad 0 \leq t \leq 1$$
$$= 0, \qquad \text{otherwise}$$

in terms of ϕ_1, ϕ_2, and ϕ_3 as $\hat{f}(t) = \alpha_1 \phi_1 + \alpha_2 \phi_2 + \alpha_3 \phi_3$ and determine the relative energy error.

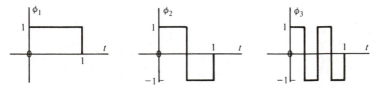

Figure 8-6   Basis functions for Example 8-7.

Solution. We need to check our basis functions using Equation 8-21. In total, we have six integrals to compute. Since $\int_0^1 \phi_i\phi_i \, dt = 1$ for $i = 1, 2, 3$, and $\int_0^1 \phi_i\phi_j \, dt = 0$ for $i \neq j$, we can conclude that the basis functions are orthonormal and $\lambda_1 = \lambda_2 = \lambda_3 = 1$. From Equation 8-24 we have:

$$\alpha_1 = \int_0^1 t \, dt = 0.5,$$

$$\alpha_2 = \int_0^{0.5} t \, dt - \int_{0.5}^1 t \, dt = -0.25$$

and

$$\alpha_3 = \int_0^{0.25} t \, dt - \int_{0.25}^{0.5} t \, dt + \int_{0.5}^{0.75} t \, dt$$
$$- \int_{0.75}^1 t \, dt = -0.125$$

Therefore $\hat{f}(t) = 0.5 \, \phi_1 - 0.25 \, \phi_2 - 0.125 \, \phi_3$

A comparison between $f(t)$ and $\hat{f}(t)$ is shown in Figure 8-7.

Also, $$E = \int_0^1 f^2 \, dt = \int_0^1 t^2 \, dt = \frac{1}{3} t^3 \,|_0^1 = 0.333$$

and $$\Sigma \, \alpha_i^2\lambda_i = (0.5)^2 + (0.25)^2 + (0.125)^2 = 0.328$$

Therefore $$\frac{E - \Sigma \, \alpha_i^2\lambda_i}{E} = \frac{0.333 - 0.328}{0.333} = 0.0146 \rightarrow 1.46\%$$

This small error implies that $\hat{f}(t)$ very closely approximates $f(t)$.

Now in these generalized Fourier Series representations we have assumed that a finite time duration is of interest: $t_1 \leq t \leq t_2$. If we are dealing with periodic functions, then that duration can be considered to be one period of the periodic function. Let us apply the generalized Fourier series ideas to the development of the complex exponential Fourier series. Assume we are given $f(t)$ which is periodic with period T. Let $\phi_n(t)$ be a *complete* set of basis functions: $\phi_n(t) = e^{jn\omega_0 t}$, $n = 0, \pm 1, \pm 2, \ldots$, where $\omega_0 = 2\pi/T$.

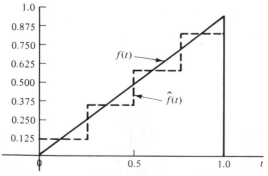

Figure 8-7 Comparison of $f(t)$ and $\hat{f}(t)$ for Example 8-7.

Then
$$f(t) = \hat{f}(t) = \sum_{n=-\infty}^{\infty} \alpha_n \phi_n(t) \tag{8-27}$$

The one-sided summation used in Equation 8-20 in this case becomes a two-sided summation. This is permissible because we have an infinite number of terms and the indexing is arbitrary. For purposes of symmetry, the two-sided format is convenient for the complex exponential Fourier series. Note that n instead of i is the index in Equation 8-27. This is because the basis functions here are complex and contain j terms and we want to avoid mixing i and j terms. Mathematicians use i where engineers typically use j. Also, the basis functions employed in this representation are orthogonal over $0 < t < T$ with $\lambda_n = T$ for all n.

EXAMPLE 8-8

Prove that the exponential Fourier series basis functions are orthogonal with $\lambda_n = T$ for all n over $0 \leq t \leq T$.

Solution

$$\int_{t_1}^{t_2} \phi_n(t) \phi_m{}^*(t)\, dt = \int_{t_0}^{t_0+T} e^{jn\omega_0 t}\, e^{-jm\omega_0 t}\, dt$$

$$= \frac{1}{j\omega_0(n-m)} [e^{j\omega_0(n-m)(t_0+T)} - e^{j\omega_0(n-m)t_0}]$$

but $n - m = k$ an integer and $e^{j\omega_0 k t_0} e^{j\omega_0 k T} - e^{j\omega_0 k t_0} = e^{j\omega_0 k t_0}(e^{j\omega_0 k T} - 1)$ and $\omega_0 k T = 2\pi k$ and $e^{j2\pi k} = 1$. Thus the term in square brackets $= 0$. Only when $n = m$ will things be otherwise. When $n = m$ we get:

$$\int_{t_1}^{t_2} \phi_n(t) \phi_m^*(t)\, dt = \int_{t_0}^{t_0+T} dt = T$$

Therefore
$$\int_{t_1}^{t_2} \phi_n(t) \phi_m^*(t)\, dt = T, \qquad \text{if } n = m$$

$$= 0, \qquad \text{otherwise}$$

and the exponential Fourier series has orthogonal basis functions with $\lambda_n = T$ for all n.

Now any periodic $f(t)$ can be written

$$f(t) = \sum_{n=-\infty}^{\infty} \alpha_n \phi_n(t) = \sum_{n=-\infty}^{\infty} \alpha_n e^{jn\omega_0 t} \tag{8-28}$$

In order to determine α_n, we use Equation 8-24:

$$\alpha_n = \frac{1}{T} \int_T f(t) e^{-jn\omega_0 t}\, dt \tag{8-29}$$

These are called **complex exponential Fourier series coefficients** and usually are written as c_n instead of α_n in order to distinguish this particular Fourier series.

The complex exponential Fourier series is closely related to the trigonometric Fourier series. The relationship between these representations can be made explicit by applying the Euler identities. We can write the complex exponential in Equation 8-28 as $\text{Cos } nw_0 t + j \text{ Sin } nw_0 t$. Then comparing Equation 8-28 to the representation in Equation 8-3, we can deduce the following:

$$a_n = c_n + c_{-n}$$

$$b_n = j(c_n - c_{-n})$$

$$c_n = \frac{a_n - jb_n}{2} \quad \text{and} \quad c_{-n} = \frac{a_n + jb_n}{2}, \quad \text{for } n > 0$$

$$c_0 = a_0 \tag{8-30}$$

EXAMPLE 8-9

Determine the exponential Fourier series representation of the periodic $f(t)$ which $= e^t$, $0 \leq t \leq 1$, and which has $T = 1$. This $f(t)$ is sketched in Figure 8-8.

Solution

$$w_0 = 2\pi, \quad c_n = \frac{1}{1} \int_0^1 e^t e^{-jnw_0 t}\, dt = \frac{1}{1 - jnw_0} (e^{1-jnw_0} - 1)$$

But
$$e^{-jnw_0} = e^{-jn2\pi} = 1$$

and
$$e^{1-jnw_0} = e^1 = 2.718$$

Therefore
$$c_n = \frac{1.718}{1 - jn2\pi} \quad \text{and} \quad f(t) = 1.718 \sum_{n=-\infty}^{\infty} \frac{e^{jn2\pi t}}{1 - jn2\pi}$$

$$= 1.718 \sum_{n=-\infty}^{\infty} \frac{e^{jnw_0 t}}{1 - jnw_0}$$

Often, the relationship between $f(t)$ and c_n, such as the relationship between $f(t)$ and its Laplace transform or between $f(n)$ and its Z transform, is indicated by the double arrow notation: $f(t) \leftrightarrow c_n$. Note that c_n in the last example was a complex number. We can write:

$$c_n = |c_n| \angle \theta_n = |c(nw_0)| \angle \theta(nw_0) \tag{8-31}$$

If we plot $|c(nw_0)|$ versus nw_0 and $\theta(nw_0)$ versus nw_0, we have what is called the

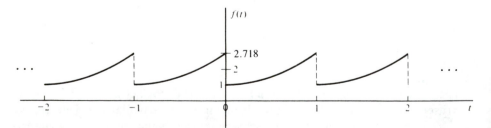

Figure 8-8 Periodic $f(t)$ of Example 8-9.

complex Fourier spectrum. Generally, these plots are points, discrete numbers at discrete values of $n\omega_0$. We can make the plots more dramatic by dropping the points to the $n\omega_0$ axis to form **line spectra,** typified by the plots in Figure 8-9 which represent the line spectra of the previous example. These lines indicate the spectral content of the signal. We normally plot the magnitude and phase line spectra as functions of $n\omega_0$ instead of just n because later, in the development of the Fourier transform, we will have ω as our independent variable and ω comes directly from $n\omega_0$.

Observation of Figure 8-9 reveals an interesting result: The magnitude spectrum is an *even* function of $n\omega_0$ and the phase spectrum is an *odd* function of $n\omega_0$. This is true for the $f(t)$ of Example 8-9, but is it always the case? To determine what must be the case for c_n to have an even magnitude spectrum and an odd phase spectrum, we take the complex conjugate of Equation 8-29, with α_n replaced by c_n:

$$c_n^* = \frac{1}{T} \int_T \{f(t)e^{-jn\omega_0 t}\, dt\}*$$

$$= \frac{1}{T} \int_T \{f(t)e^{-jn\omega_0 t}\}* \, dt$$

$$= \frac{1}{T} \int_T f^*(t)e^{jn\omega_0 t}\, dt \qquad (8\text{-}32)$$

Now from Equation 8-31

$$c_n^* = \{|c(n\omega_0)|\, \angle\theta(n\omega_0)\}* = \{|c(n\omega_0)|\, e^{j\theta(n\omega_0)}\}*$$

$$= |c(n\omega_0)|\, e^{-j\theta(n\omega_0)} \qquad (8\text{-}33)$$

Assume that the magnitude is an even function of $n\omega_0$,

then $$|c(-n\omega_0)| = |c(n\omega_0)|$$

Assume that the phase is an odd function of $n\omega_0$

then $$\theta(-n\omega_0) = -\theta(n\omega_0)$$

Therefore $$c_n^* = |c(-n\omega_0)|\, e^{j\theta(-n\omega_0)} = c_{-n} \qquad (8\text{-}34)$$

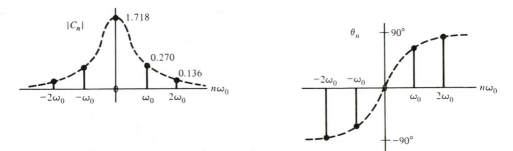

Figure 8-9 The magnitude and phase of c_n from Example 8-9.

But from the integral equation, Equation 8-29, we get:

$$c_{-n} = \frac{1}{T} \int_T f(t) e^{jn\omega_0 t} \, dt \qquad (8\text{-}35)$$

and if this equals

$$c_n^* = \frac{1}{T} \int_T f^*(t) e^{jn\omega_0 t} \, dt$$

then $f^*(t) = f(t)$; that is, the time function whose Fourier series we are interested in must be a *real* function. This is the condition under which the magnitude and phase of c_n are respectively even and odd functions of $n\omega_0$.

Also, note that since, for real $f(t)$, $\theta(0) = 0$, c_0 is a real number. This should make intuitive sense because from Equation 8-30 we have $c_0 = a_0$. This is just the average or dc value of the given time function $f(t)$.

Now, as we have seen, the trigonometric and the exponential Fourier series are closely related. The trigonometric series very clearly displays the given periodic $f(t)$ as a dc term plus a sum of sinusoids. The exponential Fourier series, on the other hand, is a compact expression. That is its appeal, plus the fact that it leads very nicely into the Fourier transform which is considered in the next section. A point of confusion concerning the exponential Fourier series is often expressed in the question: How can a *real* $f(t)$ be represented by a summation of basis functions that are *complex*? The answer is that although the basis functions are complex, they appear in complex conjugate pairs that reduce to real sines and cosines.

The generalized Fourier series methods permit us to represent a given signal in terms of other signals that may be easier to handle. For periodic signals, the trigonometric, cosinusoidal, and exponential Fourier series methods provide useful representations that reveal the spectral content of the given signal. Intuitively, the $f(t)$ in Figure 8-8, for instance, is composed of a dc term plus a number of sinusoids. These functions can be generated from the c_n plots of Figure 8-9. Fourier methods applied to periodic signals, then, provide representations *and* reveal spectral content. The generalized Fourier series methods applied to nonperiodic signals, on the other hand, are used typically to provide alternative representations for a given signal. They are seldom concerned with spectral content. To reveal the spectral content of nonperiodic signals, we use the methods of Fourier transform analysis. In fact, in the next section we develop the magnitude and phase of the Fourier transform to show the spectral content of nonperiodic signals, just as the c_n terms stand out in the complex exponential Fourier series to represent the spectral content of periodic signals.

Before turning to the Fourier transform, let us consider one more Fourier series example.

EXAMPLE 8-10

Determine the complex exponential Fourier series coefficients for the $f(t)$ represented in Figure 8-10. Then consider the effect of shifting $f(t)$ $d/2$ units to the right.

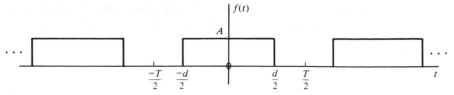

Figure 8-10 $f(t)$ for Example 8-10.

Solution

$$c_n = \frac{1}{T} \int_T f(t) e^{-jn\omega_0 t} \, dt$$

which becomes:

$$c_n = \frac{1}{T} \int_{-d/2}^{d/2} A e^{-jn\omega_0 t} \, dt = \frac{A}{T} \frac{1}{(-jn\omega_0)} \left(e^{-jn\omega_0(d/2)} - e^{jn\omega_0(d/2)} \right)$$

$$= \frac{2A}{Tn\omega_0} \mathrm{Sin}\left(n\omega_0 \frac{d}{2} \right) = \frac{Ad}{T} \frac{\mathrm{Sin}\,(n\omega_0(d/2))}{(n\omega_0 d/2)}$$

Now recall from Chapter 1 that $\mathrm{Sinc}\,(t) = \mathrm{Sin}\,\pi t / \pi t$.

Therefore
$$c_n = \frac{Ad}{T} \mathrm{Sinc}\left(\frac{n\omega_0 d}{2\pi} \right) = \frac{Ad}{T} \mathrm{Sinc}\left(\frac{nd}{T} \right)$$

For purposes of illustration we plot $|c_n|$ versus $n\omega_0$ in Figure 8-11 in the case where $A = 10$, $T = 10$, and $d = 2$. Note that the envelope of this curve is the familiar $\mathrm{Sinc}\,(x)$ pattern. Now if we shift $f(t)$ to the right by $d/2$ units, we can write:

$$c_n = \frac{1}{T} \int_0^d A e^{-jn\omega_0 t} \, dt = \frac{A}{T} \frac{1}{(-jn\omega_0)} \left(e^{-jn\omega_0 d} - 1 \right)$$

$$= \frac{A}{T} \frac{e^{-jn\omega_0(d/2)}}{(-jn\omega_0)} \left(e^{-jn\omega_0(d/2)} - e^{jn\omega_0(d/2)} \right) = e^{-jn\omega_0(d/2)} \frac{Ad}{T} \mathrm{Sinc}\left(\frac{n\omega_0 d}{2\pi} \right)$$

which is the same as c_n for the unshifted function except for the phase term $\angle -n\omega_0 d/2$. Therefore the magnitude of this new c_n will be the same as before and only the phase will be changed. This result, in fact, is very

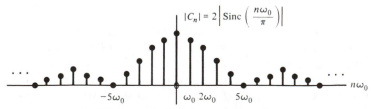

Figure 8-11   Plot of $|c_n|$ versus $n\omega_0$ for Example 8-10.

general and can be stated as follows: if $f(t) \leftrightarrow c_n$, then $f(t - t_0) \leftrightarrow e^{-jn\omega_0 t_0} c_n$. To prove it, let $f(t - t_0) \leftrightarrow \hat{c}_n$.

thus

$$\hat{c}_n = \frac{1}{T} \int_0^T f(t - t_0) e^{-jn\omega_0 t} dt$$

Let $\lambda = t - t_0$, $d\lambda = dt$.

then

$$\hat{c}_n = \frac{1}{T} \int_{-t_0}^{T-t_0} f(\lambda) e^{-jn\omega_0(\lambda + t_0)} d\lambda = e^{-jn\omega_0 t_0} c_n$$

Drill Set: Fourier Series

1. Prove that $\{\phi_n\} = \{\text{Sin } nt, \text{Cos } nt, n = 1, 2, \ldots\}$ constitute an orthogonal set of basis functions over the range $0 < t < 2\pi$.
2. $\phi_1(t) = \frac{1}{3} t$ and $\phi_2(t) = d_1 t^2 + d_2$ are known to be a pair of orthonormal basis functions on the interval $0 < t < t_1$. Find d_1, d_2, and t_1. Determine α_1 and α_2 where $\hat{f}(t) = \alpha_1 \phi_1 + \alpha_2 \phi_2$ and the function we want $\hat{f}(t)$ to approximate is the pulse $f(t) = u(t) - u(t - t_1)$.
3. Expand $f(t) = \text{Sin}^2 2\pi t \text{ Cos } \pi t$ into a trigonometric Fourier series and into a complex exponential Fourier series. Plot c_n.
4. Determine and sketch the even and odd components of

$$f(t) = e^{-t} \text{ Cos } t, \qquad t > 0$$

$$= 0, \qquad t < 0$$

5. Consider the periodic impulse train $f(t) = \sum_{n=-\infty}^{\infty} \delta(t + nT)$. Determine and plot the exponential Fourier series coefficients. How is c_n in this case unlike c_n terms in previous examples?
6. Let

$$f(t) = 10, \qquad 0 \le t \le 1$$

$$= 0, \qquad 1 \le t \le 2$$

be a periodic function with period $T = 2$. Assume $f(t) \approx k_1 + k_2 \text{ Sin } \omega_0 t + k_3 \text{ Sin } 3\omega_0 t$. Determine k_1, k_2, and k_3.

8-3 THE FOURIER TRANSFORM

Assume we have $f(t)$ in the range $-d/2 < t < d/2$. Outside this range $f(t) = 0$. Now the complex exponential Fourier series of Equation 8-28 can be used to describe $f(t)$ within the given range, but outside this range Equation 8-28 would not describe $f(t)$—since its true value is zero—but would instead describe a periodic extension of $f(t)$ in the range $-d/2 < t < d/2$. Assume that this periodic extension has a period T and that $d < T$. As an example, a glance at the periodic

$f(t)$ represented in Figure 8-10 might be helpful. As T gets larger and larger, the given $f(t)$ is more and more accurately represented by the right-hand side of Equation 8-28. As $T \to \infty$, Equation 8-28, in theory, exactly represents the given $f(t)$ which is nonzero for $-d/2 < t < d/2$ and equal to zero for t outside this range. For example, the $f(t)$ in Figure 8-10 would have only the middle pulse remaining as $T \to \infty$. This new function, instead of being considered a periodic function with infinite period, will be considered a nonperiodic function.

In order to formalize this development, let $\alpha_n = c_n$ in Equation 8-29 and plug it into Equation 8-28 to get:

$$f(t) = \sum_{n=-\infty}^{\infty} \frac{1}{T} \int_T f(t)e^{-jn\omega_0 t}\, dt\, e^{jn\omega_0 t} \tag{8-36}$$

Now the spacing between harmonics, as illustrated in Figure 8-9, is just ω_0. But $\omega_0 = 2\pi/T$. In the limit as $T \to \infty$, ω_0 becomes infinitesimal; call it $d\omega$. Also, $n\omega_0$ becomes a continuous variable; call it ω. In addition, the summation becomes an integral. We can summarize the changes made in Equation 8-36 as $T \to \infty$:

$$\int_T \to \int_{-\infty}^{\infty}$$

$$\omega_0 \to d\omega$$

$$n\omega_0 \to \omega$$

$$\sum_{-\infty}^{\infty} \to \int_{-\infty}^{\infty}$$

$$\frac{1}{T} = \frac{\omega_0}{2\pi} \to \frac{d\omega}{2\pi} \tag{8-37}$$

Incorporating these changes in Equation 8-36, we obtain:

$$f(t) = \frac{1}{2\pi} \int_{-\infty}^{\infty} \left[\int_{-\infty}^{\infty} f(t)e^{-j\omega t}\, dt \right] e^{j\omega t}\, d\omega \tag{8-38}$$

The term in the brackets is called the **Fourier transform of** $f(t)$ and is indicated by $F(j\omega)$.

Thus
$$F(j\omega) = \mathrm{FT}\{f(t)\} = \int_{-\infty}^{\infty} f(t)e^{-j\omega t}\, dt \tag{8-39}$$

Then the inverse Fourier transform is written

$$f(t) = \mathrm{IFT}\{F(j\omega)\} = \frac{1}{2\pi} \int_{-\infty}^{\infty} F(j\omega)e^{j\omega t}\, d\omega \tag{8-40}$$

Equations 8-39 and 8-40 constitute what are called Fourier transform pairs and can be represented, like other transform pairs, as follows:

$$f(t) \leftrightarrow F(j\omega)$$

For the most part, corresponding to $f(t)$ there is a unique $F(j\omega)$ and corresponding to $F(j\omega)$ there is a unique $f(t)$. To get one from the other, we use an integral operator. The Fourier transform transforms a time-domain function into the frequency domain. The inverse Fourier transform transforms a frequency-domain function into the time domain. The frequency domain is indicated by ω, the radian frequency, which has units of radians/second. The more "natural" frequency domain is indicated by f, which has units of hertz. Of course, $f = \omega/2\pi$. Some texts will use the notation $F(f)$ or $F(\omega)$ to indicate the Fourier transform. These are particularly popular in texts whose focus is communication theory. For our purposes, however, $F(j\omega)$ will be a more useful notation because of our concern with the Laplace transform, $F(s)$. Often these transforms will be identical if we let $s = j\omega$ in the Laplace transform.

EXAMPLE 8-11

Determine the Fourier transform of the $f(t)$ indicated in Figure 8-10 when $T \rightarrow \infty$.

Solution

$$F(j\omega) = \int_{-\infty}^{\infty} f(t)e^{-j\omega t}\, dt = \int_{-d/2}^{d/2} Ae^{-j\omega t}\, dt$$

$$= \frac{A}{-j\omega}\{e^{-j\omega d/2} - e^{j\omega d/2}\} = \frac{2A}{\omega}\, \text{Sin}\left(\frac{\omega d}{2}\right) = Ad\, \text{Sinc}\left(\frac{\omega d}{2\pi}\right)$$

Letting $A = 10$ and $d = 2$, we plot $F(j\omega)$ in Figure 8-12 as a familiar Sinc (x) curve.

Now in view of the results of this example, we cannot help noticing that there is a striking resemblance between this $F(j\omega)$ and the c_n from the previous example. From Example 8-10 we had $c_n = (Ad/T)\, \text{Sinc}\,(n\omega_0 d/2\pi)$. Letting $n\omega_0 = \omega$ and multiplying c_n by T, we obtain the result of Example 8-11; that is, $F(j\omega) = Ad\, \text{Sinc}\,(\pi d/2\pi)$. This is a very general result. Imagine we have c_n for a periodic $f(t)$. Let

$$\tilde{f}(t) = f(t), \qquad \frac{-T}{2} \le t \le \frac{T}{2}$$

and assume $\tilde{f}(t)$ is zero outside this range.

Let $F(j\omega)$ be the Fourier transform of $\tilde{f}(t)$. Then c_n and $F(j\omega)$ are related as follows:

$$F(j\omega) = Tc_n\Big|_{\substack{n\omega_0 \rightarrow \omega \\ T \rightarrow \infty}} \qquad\qquad (8\text{-}41)$$

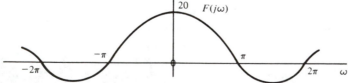

Figure 8-12   Plot of $F(j\omega)$ versus ω for Example 8-11.

and
$$c_n = \frac{1}{T} F(j\omega) \big|_{\omega \to n\omega_0} \tag{8-42}$$

This procedure can make the determination of the Fourier transform a trivial matter. However, it presupposes the existence of the corresponding complex exponential Fourier series coefficients. If these are not available, then we must revert to the defining equation, Equation 8-39, or look up the result in a table of transform pairs. We present a table subsequently.

What does this Fourier transform do? Why use it? What does it mean? Like the Fourier series coefficients, the Fourier transform reveals the spectral content of a signal. It will not normally indicate that some $f(t)$ contains specific frequencies, say, at ω_1, ω_2, and ω_3, but rather, it shows a range of frequencies, say, $\omega_1 < \omega < \omega_2$, over which $f(t)$ contains significant spectral content. If within this range $F(j\omega)$ hits a very narrow peak, say, $\omega \approx \omega_x$, this often indicates the presence of a sinusoid of that specific frequency. This sinusoid might be buried in noise to form $f(t)$ as some data record of "signal plus noise." To ferret signals out of given data records, numerous techniques have been developed from what is called spectral estimation theory. Such studies are beyond our current scope. However, the basics of the Fourier transform are essential to this area and to many fields of sophisticated research in engineering and science.

The Fourier transform is generally a complex function of frequency. We can write:

$$F(j\omega) = |F(j\omega)| \angle \text{arg } F(j\omega) \tag{8-43}$$

A plot of the amplitude spectrum is typically all we need to have a good idea of the spectral content of a given signal. But in order to return from the frequency domain to find $f(t)$, we need both magnitude and phase of $F(j\omega)$. To obtain $f(t)$, given an analytical expression for $F(j\omega)$, we do not normally use the inverse Fourier transform equation, Equation 8-40. Usually, as with inverse Laplace transforms, we would try to break up a given $F(j\omega)$ into terms that are readily inverse transformable, for example, by observation of simple terms that might appear in a table of transform pairs.

EXAMPLE 8-12

Determine the Fourier transform of the following functions:

(a) $f(t) = e^t [u(t) - u(t - 1)]$
(b) $f(t) = e^{5t} u(-t) + e^{-t} u(t)$
(c) $f(t) = \Pi(0.5(t - 2))$
(d) $f(t) = t e^{-t} u(t)$
(e) $f(t) = \delta(t - t_0)$
(f) $f(t) = u(t + 1) - 2u(t) + u(t - 1)$
(g) $f(t) = t u(t)$

Solution

(a) $F(j\omega) = \int_0^1 e^t e^{-j\omega t} \, dt = \frac{1}{1 - j\omega} (e^{1-j\omega} - 1)$

Now from Equation 8-42

$$c_n = \frac{1}{T} F(j\omega)|_{\omega \to n\omega_0} = \frac{1}{T}\left(\frac{1}{1 - jn\omega_0}\right)(e^{1-jn\omega_0} - 1)$$

but $T = 1$ and $\omega_0 = 2\pi/T = 2\pi$ and $e^{-jn\omega_0} = e^{-jn2\pi} = 1$

Therefore $e^{1-jn\omega_0} = (e)(1) = 2.718$

Thus $$c_n = \frac{1.718}{1 - jn2\pi}$$

These c_n values are the exponential Fourier series coefficients for the periodic signal $f(t) = e^t, 0 \le t \le 1$ which has a period $T = 1$.

(b) $F(j\omega) = \displaystyle\int_{-\infty}^0 e^{5t}e^{-j\omega t}\, dt + \int_0^\infty e^{-t}e^{-j\omega t}\, dt$

$$= \frac{1}{5 - j\omega}(1 - 0) + \frac{1}{-1 - j\omega}(0 - 1)$$

$$= \frac{1}{5 - j\omega} + \frac{1}{1 + j\omega} = \frac{6}{(5 - j\omega)(1 + j\omega)}$$

(c) Recall from Chapter 1 that

$$\sqcap(t) = 1, \qquad \text{for } -0.5 \le t \le 0.5$$
$$= 0, \qquad \text{otherwise}$$

Therefore $\sqcap(0.5(t - 2)) = 1, \qquad \text{for } 1 \le t \le 3$
$$= 0, \qquad \text{otherwise}$$

and $F(j\omega) = \displaystyle\int_1^3 1 e^{-j\omega t}\, dt = \frac{-1}{j\omega}(e^{-3j\omega} - e^{-j\omega})$

$$= \frac{e^{-2j\omega}(e^{j\omega} - e^{-j\omega})}{2j(\omega)}(2) = \frac{2e^{-2j\omega}}{\omega} \operatorname{Sin}\omega$$

(d) $F(j\omega) = \displaystyle\int_0^\infty te^{-t}e^{-j\omega t}\, dt = \frac{e^{at}}{a^2}(at - 1)|_0^\infty, \qquad a = (-1 - j\omega)$

$$= 0 - \frac{1}{a^2}(-1) = \frac{1}{a^2} = \frac{1}{(1 + j\omega)^2}$$

(e) $F(j\omega) = \displaystyle\int_{-\infty}^\infty \delta(t - t_0)e^{-j\omega t}\, dt = e^{-j\omega t_0}$

(from the properties of the impulse function)

(f) $F(j\omega) = \displaystyle\int_{-1}^0 e^{-j\omega t}\, dt + \int_0^1 (-1)e^{-j\omega t}\, dt$

$$= \frac{-1}{j\omega}\{1 - e^{j\omega}\} + \frac{1}{j\omega}\{e^{-j\omega} - 1\}$$

$$= \frac{-2}{j\omega} + \frac{1}{j\omega}(e^{j\omega} + e^{-j\omega}) = \frac{1}{j\omega}\{-2 + 2\,\text{Cos}\,\omega\} = \frac{-4\,\text{Sin}^2(\omega/2)}{j\omega}$$

$$= j\omega\left(\frac{\text{Sin}\,(\omega/2)}{\omega/2}\right)^2$$

(g) $F(j\omega) = \int_0^\infty te^{-j\omega t}\,dt = \frac{e^{at}}{a^2}(at-1)\Big|_0^\infty,\qquad a = -j\omega$

$$= \frac{e^{-j\omega\infty}(-j\omega\infty - 1)}{-\omega^2} - \frac{(-1)}{-\omega^2}$$

which is undefined. Thus the Fourier transform of $tu(t)$ does not exist.

The last part of Example 8-12 illustrates that the existence of the Fourier transform, like the Fourier series, is contingent on certain conditions. In order for $f(t)$ to have a Fourier transform, it is sufficient that $f(t)$ have a finite number of maxima, minima, and finite discontinuities in any finite interval. Most functions of interest to engineers will satisfy these restrictions. Another sufficient condition—which is often problematic—is that $f(t)$ be absolutely integrable:

$$\int_{-\infty}^{\infty}|f(t)|\,dt < \infty \qquad\qquad (8\text{-}44)$$

These sufficiency conditions for the existence of the Fourier transform, as was the case with the Fourier series, are called Dirichlet conditions. A typical signal encountered in engineering work, like a burst signal from a radar, will satisfy the integrability condition because such a signal starts and stops at finite time points and always has a finite value. However, some simple functions, like $u(t)$, do not satisfy the condition of absolute integrability. Still, by indirect procedures and assuming the existence of $\delta(\omega)$ in the frequency domain, Fourier transforms for such functions can be developed. Functions like the unit step are called *power signals* and are distinguished from *energy signals*.

Energy Signals   These are functions $f(t)$ such that $\int_{-\infty}^{\infty} f^2(t)\,dt < \infty$. The integral of a function squared is often taken as a measure of the energy contained in the signal. **Energy signals,** then, are functions that represent finite energy phenomena.

Power Signals   These are functions $f(t)$ such that $\lim_{\tau\to\infty} 1/\tau \int_{-\tau/2}^{\tau/2} f^2(t)\,dt < \infty$. Typical examples of these are periodic signals, dc wave forms, and the unit step function. **Power signals** will have infinite energy but will have finite power, whereas energy signals will have finite energy but will have zero power.

Now, in general, an energy signal will die out as $t \to \pm\infty$. The functions considered in Example 8-12 were all energy signals, except for the last function. Signals with finite energy also satisfy the Dirichlet condition of absolute integrability. Their Fourier transforms can be directly computed. The signal $f(t) = tu(t)$ from Example 8-12(g) is neither an energy signal nor a power signal.

Compute $\lim_{\tau \to \infty} 1/\tau \int_0^{\tau/2} t^2 \, dt = \lim_{\tau \to \infty} 1/\tau \frac{1}{3} (\tau^3/8) = \infty$ which is not finite. If we permit frequency-domain impulse functions, then any signal that is either a power or an energy signal will have a Fourier transform and any signal that is neither an energy nor a power signal will not have a Fourier transform. The unit step function is not an energy signal but it is a power signal and does have a Fourier transform.

EXAMPLE 8-13

Show that $u(t)$ is a power signal and determine its Fourier transform.

Solution. Compute

$$\lim_{\tau \to \infty} \frac{1}{\tau} \int_0^{\tau/2} (1) \, dt = \lim_{\tau \to \infty} \frac{1}{\tau} \left(\frac{\tau}{2} \right) = \frac{1}{2}$$

which is finite. Therefore $u(t)$ is a power signal. Now if we decompose $u(t)$ into its even and odd components, we can write:

$$u(t) = \tfrac{1}{2} f_1(t) + \tfrac{1}{2} f_2(t), \qquad \text{where } f_1(t) = 1 \text{ for } -\infty < t < \infty \text{ is even}$$

and
$$f_2(t) = -1, \qquad \text{for } t < 0$$
$$= 0, \qquad \text{for } t = 0$$
$$= 1, \qquad \text{for } t > 0$$

which is an odd function. This $f_2(t)$ function is sometimes called the **signum function:** $f_2(t) = \text{sgn}(t)$. Taking the Fourier transform, we obtain:

$$\text{FT}\{u(t)\} = \tfrac{1}{2}[\text{FT}\{f_1(t)\} + \text{FT}\{f_2(t)\}]$$

To get the Fourier transforms of $f_1(t)$ and $f_2(t)$, we represent these time functions as limiting processes:

$$f_1(t) = \lim_{a \to 0} e^{at}, \qquad t \le 0$$
$$= \lim_{a \to 0} e^{-at}, \qquad t \ge 0$$

and
$$f_2(t) = \lim_{a \to 0} -e^{at}, \qquad t < 0$$
$$= 0, \qquad t = 0$$
$$= \lim_{a \to 0} e^{-at}, \qquad t > 0$$

Then
$$F_1(j\omega) = \int_{-\infty}^{\infty} f_1(t) e^{-j\omega t} \, dt$$
$$= \lim_{a \to 0} \int_{-\infty}^{0} e^{at} e^{-j\omega t} \, dt + \lim_{a \to 0} \int_{0}^{\infty} e^{-at} e^{-j\omega t} \, dt$$
$$= \lim_{a \to 0} \left[\frac{1}{a - j\omega} (e^0 - e^{-\infty}) + \frac{1}{-a - j\omega} (e^{-\infty} - e^0) \right]$$
$$= \lim_{a \to 0} \left[\frac{1}{a - j\omega} + \frac{1}{a + j\omega} \right] = \lim_{a \to 0} \left[\frac{2a}{a^2 + \omega^2} \right]$$

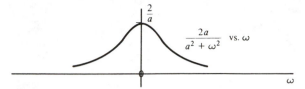

Figure 8-13 Plot of $2a/(a^2 + \omega^2)$ versus ω for Example 8-13.

For a positive finite value of a, if we plot the term in brackets versus ω, we get a function like the one in Figure 8-13. The area under this curve, from a table of definite integrals, is 2π, independent of the value of a. As a gets smaller and smaller, since the peak is $2/a$ at $\omega = 0$, the curve gets sharper and sharper with $2/a \rightarrow \infty$ as $a \rightarrow 0$. Since the area remains fixed, we end up with an impulse of weight 2π centered at the origin in the ω domain.

$$F_1(j\omega) = 2\pi\delta(\omega)$$

In words, the Fourier transform of a constant is an impulse in the frequency domain.

Now

$$F_2(j\omega) = \int_{-\infty}^{\infty} f_2(t)e^{-j\omega t} \, dt$$

$$= \lim_{a \to 0} \int_{-\infty}^{0} - e^{at}e^{-j\omega t} \, dt + \lim_{a \to 0} \int_{0}^{\infty} e^{-at}e^{-j\omega t} \, dt$$

$$= \lim_{a \to 0} \left[-\frac{1}{a - j\omega} (e^0 - e^{-\infty}) + \frac{1}{-a - j\omega}(e^{-\infty} - e^0) \right]$$

$$= \lim_{a \to 0} \left[\frac{-1}{a - j\omega} + \frac{1}{a + j\omega} \right]$$

$$= \lim_{a \to 0} \frac{-2j\omega}{a^2 + \omega^2} = \frac{-2j}{\omega} = \frac{2}{j\omega}$$

The Fourier transform of the signum function is $2/j\omega$.

Therefore
$$FT\{u(t)\} = \frac{1}{2}\left\{2\pi\delta(\omega) + \frac{2}{j\omega}\right\}$$

or
$$u(t) \leftrightarrow \pi\delta(\omega) + \frac{1}{j\omega}$$

The Fourier transform exists for many other power signals. Some of these are easily determined by employing some of the properties of the Fourier transform. Properties of the Fourier transform are the topic of the next section. Before turning to that material, note the summary of Fourier transform pairs presented in Table 8-1. Many of these could be worked out as additional exercises. We will do number 17 as a final example in this section.

TABLE 8-1   FOURIER TRANSFORM TABLE

$f(t)$	$\longleftrightarrow$	$F(j\omega)$
1. $\delta(t)$		1
2. 1		$2\pi\delta(\omega)$
3. $u(t)$		$\pi\delta(\omega) + \dfrac{1}{j\omega}$
4. $e^{-\alpha t}u(t)$		$1/(j\omega + \alpha), \qquad \alpha > 0$
5. $t^n e^{-\alpha t}u(t)$		$n!/(\alpha + j\omega)^{n+1}, \qquad \alpha > 0$
6. $\lvert t \rvert$		$\dfrac{-2}{\omega^2}$
7. $\operatorname{Sin} \omega_0 t$		$j\pi[\delta(\omega + \omega_0) - \delta(\omega - \omega_0)]$
8. $\operatorname{Cos} \omega_0 t$		$\pi[\delta(\omega - \omega_0) + \delta(\omega + \omega_0)]$
9. $\dfrac{\operatorname{Sin} \omega_0 t}{\pi t}$		$\begin{cases} 1, & \lvert \omega \rvert < \omega_0 \\ 0, & \lvert \omega \rvert > \omega_0 \end{cases}$
10. $\begin{cases} 1, & \lvert t \rvert < T \\ 0, & \lvert t \rvert > T \end{cases}$		$\dfrac{2 \operatorname{Sin} \omega T}{\omega}$
11. $e^{j\omega_0 t}$		$2\pi\delta(\omega - \omega_0)$
12. $\delta(t - t_0)$		$e^{-j\omega t_0}$
13. $e^{-\alpha t} \operatorname{Cos} \omega_0 t\, u(t)$		$\dfrac{\alpha + j\omega}{(\alpha + j\omega)^2 + \omega_0^2}$
14. $e^{-\alpha t} \operatorname{Sin} \omega_0 t\, u(t)$		$\dfrac{\omega_0}{(\alpha + j\omega)^2 + \omega_0^2}$
15. $e^{-\alpha^2 t^2}$		$\dfrac{\sqrt{\pi}}{\alpha} e^{-\omega^2/4\alpha^2}$
16. $e^{-\alpha \lvert t \rvert}, \qquad \alpha > 0$		$\dfrac{2\alpha}{\alpha^2 + \omega^2}$
17. $\operatorname{Cos} \omega_0 t[u(t + T) - u(t - T)]$		$\left[\dfrac{\operatorname{Sin}(\omega - \omega_0)T}{(\omega - \omega_0)} + \dfrac{\operatorname{Sin}(\omega + \omega_0)T}{(\omega + \omega_0)}\right]$
18. $\begin{cases} A\left[1 - \dfrac{\lvert t \rvert}{T}\right], & \lvert t \rvert < T \\ 0, & \lvert t \rvert > T \end{cases}$		$AT\left[\dfrac{\operatorname{Sin} \omega T/2}{\omega T/2}\right]^2$

EXAMPLE 8-14

Determine the Fourier transform for:

$$f(t) = \operatorname{Cos} \omega_0 t\, [u(t + T) - u(t - T)]$$

Solution

$$F(j\omega) = \int_{-\infty}^{\infty} \left(\frac{e^{j\omega_0 t} + e^{-j\omega_0 t}}{2}\right) e^{-j\omega t} \{u(t + T) - u(t - T)\}\, dt$$

$$= \frac{1}{2} \int_{-T}^{T} \left(e^{jt(\omega_0 - \omega)} + e^{-jt(\omega_0 + \omega)} \right) dt$$

$$= \frac{1}{2} \left\{ \frac{1}{j(\omega_0 - \omega)} \left[e^{jT(\omega_0 - \omega)} - e^{-jT(\omega_0 - \omega)} \right] \right.$$

$$\left. + \frac{1}{-j(\omega_0 + \omega)} \left[e^{-jT(\omega_0 + \omega)} - e^{jT(\omega_0 + \omega)} \right] \right\}$$

This can be written as:

$$F(j\omega) = \frac{\text{Sin}(\omega - \omega_0)T}{\omega - \omega_0} + \frac{\text{Sin}(\omega_0 + \omega)T}{\omega_0 + \omega}$$

which is the result presented in Table 8-1. However, note that:

$$\omega_0 = \frac{2\pi}{T}$$

Therefore $\qquad \omega_0 T = 2\pi \quad$ and $\quad e^{j\omega_0 T} = e^{-j\omega_0 T} = 1$

Thus $\qquad F(j\omega) = \frac{1}{2} \left\{ \frac{1}{j(\omega_0 - \omega)} \left[e^{-j\omega T} - e^{j\omega T} \right] \right.$

$$\left. - \frac{1}{j(\omega_0 + \omega)} \left[e^{-j\omega T} - e^{j\omega T} \right] \right\}$$

$$= \left(\frac{e^{-j\omega T} - e^{j\omega T}}{2j} \right) \left(\frac{1}{\omega_0 - \omega} - \frac{1}{\omega_0 + \omega} \right)$$

$$= -\text{Sin}\,\omega T \left(\frac{\omega_0 + \omega - \omega_0 + \omega}{\omega_0^2 - \omega^2} \right)$$

or $\qquad F(j\omega) = \frac{2\omega \sin \omega T}{\omega^2 - \omega_0^2}$

which is a simplified version.

8-4 FOURIER TRANSFORM PROPERTIES

In Table 8-2 we list some of the more important Fourier transform properties. These properties are labor-saving devices that enable us to determine Fourier transforms or inverse Fourier transforms with a minimum of effort. Employing these properties not only saves work but often provides significant insights into complicated problems. We now prove and demonstrate the use of a number of these properties.

EXAMPLE 8-15

Prove the evenness and oddness property.

Solution. First assume $f(t)$ is even.

Then $\qquad F(j\omega) = \int_{-\infty}^{\infty} f(t) e^{-j\omega t}\, dt = \int_{-\infty}^{\infty} f(t)(\text{Cos }\omega t - j\,\text{Sin }\omega t)\, dt$

Now $f(t)$ Cos ωt is an even function and $f(t)$ Sin ωt is odd. Therefore integrating from $-\infty$ to $+\infty$, we get:

$$F(j\omega) = 2\int_{0}^{\infty} f(t)\,\text{Cos }\omega t\, dt + 0$$

and $F(j\omega)$ is even.

Now assume $f(t)$ is odd.

Then $\qquad F(j\omega) = \int_{-\infty}^{\infty} f(t)(\text{Cos }\omega t - j\,\text{Sin }\omega t)\, dt$

$$= 0 - j2\int_{0}^{\infty} f(t)\,\text{Sin }\omega t\, dt$$

and $F(j\omega)$ is odd.

EXAMPLE 8-16.

Prove the time shift property, then use it to determine the Fourier transform of:

$$f(t) = \sqcap(t) = u(t + \tfrac{1}{2}) - u(t - \tfrac{1}{2})$$

Solution.   We know:

$$\text{FT}\{f(t)\} = \int_{-\infty}^{\infty} f(t) e^{-j\omega t}\, dt$$

Then $\qquad\qquad \text{FT}\{f(t - t_0)\} = \int_{-\infty}^{\infty} f(t - t_0) e^{-j\omega t}\, dt$

Let $\qquad\qquad\qquad\qquad t - t_0 = \lambda, \qquad dt = d\lambda$

and $\qquad \int_{-\infty}^{\infty} f(t - t_0) e^{-j\omega t}\, dt = \int_{-\infty}^{\infty} f(\lambda) e^{-j\omega(\lambda + t_0)}\, d\lambda$

$$= e^{-j\omega t_0}\int_{-\infty}^{\infty} f(\lambda) e^{-j\omega \lambda}\, d\lambda$$

Therefore $\qquad\qquad e^{-j\omega t_0} F(j\omega) \leftrightarrow f(t - t_0)$

Now we know:

$$u(t) \leftrightarrow \pi\delta(\omega) + \frac{1}{j\omega}$$

Thus $\qquad u\!\left(t + \frac{1}{2}\right) \leftrightarrow e^{j\omega 1/2}\left\{\pi\delta(\omega) + \frac{1}{j\omega}\right\},$

$$u\!\left(t - \frac{1}{2}\right) \leftrightarrow e^{-j\omega 1/2}\left\{\pi\delta(\omega) + \frac{1}{j\omega}\right\}$$

and $\qquad \sqcap(t) \leftrightarrow \left(e^{j\omega 1/2}\pi\delta(\omega) + \frac{e^{j\omega 1/2}}{j\omega}\right)$

$$- \left(e^{-j\omega 1/2}\pi\delta(\omega) + \frac{e^{-j\omega 1/2}}{j\omega}\right)$$

TABLE 8-2 FOURIER TRANSFORM PROPERTIES

Given $f(t) \leftrightarrow F(j\omega)$, $\quad g(t) \leftrightarrow G(j\omega)$

Function	Transform	Property
1. $\alpha f(t) + \beta g(t)$	$\alpha F(j\omega) + \beta G(j\omega)$	Linearity
2. $f(t)$ even	$F(j\omega) = 2 \int_0^\infty f(t) \cos \omega t \, dt$	Evenness and
$\quad f(t)$ odd	$F(j\omega) = -j2 \int_0^\infty f(t) \sin \omega t \, dt$	Oddness
3. $f(t - t_0)$	$e^{-j\omega t_0} F(j\omega)$	Time shift
4. $f(\alpha t)$	$\dfrac{1}{|\alpha|} F\left(j \dfrac{\omega}{\alpha}\right)$	Time scale
5. $F(jt)$	$2\pi f(-\omega)$	Duality
6. $f(t) * g(t)$	$F(j\omega) G(j\omega)$	Time convolution
7. $f(t) g(t)$	$\dfrac{1}{2\pi} F(j\omega) * G(j\omega)$	Frequency convolution
8. $\dfrac{d^n}{dt^n} f(t)$	$(j\omega)^n F(j\omega)$	Time differentiation
9. $\int_{-\infty}^t f(\lambda) \, d\lambda$	$\dfrac{1}{j\omega} F(j\omega) + \pi F(0)\delta(\omega)$	Integration
10. $t^n f(t)$	$\dfrac{(j)^n d^n}{d\omega^n} F(j\omega)$	Frequency differentiation
11. $e^{j\omega_0 t} f(t)$	$F(j[\omega - \omega_0])$	Modulation (frequency shift)
12. $\int_{-\infty}^\infty f(\lambda - t) g(\lambda) \, d\lambda$	$F(-j\omega) G(j\omega)$	Correlation

but
$$e^{j\omega 1/2} \pi\delta(\omega) = e^{-j\omega 1/2} \pi\delta(\omega) = \pi\delta(\omega)$$

Therefore
$$\text{FT}\{\sqcap(t)\} = \frac{1}{j\omega}(e^{j\omega 1/2} - e^{-j\omega 1/2})$$

$$= \frac{2}{\omega} \sin \omega \frac{1}{2}$$

which agrees with the result of number 10 from Table 8-1 with $T = \frac{1}{2}$.

EXAMPLE 8-17

The time scale property, which is sometimes known as the reciprocal spreading property, indicates that an expansion in the time domain results in a contraction in the frequency domain and vice versa. Demonstrate this result by determining and plotting $f(10t)$ and $f(\frac{1}{2}t)$ where $f(t)$ is the triangular function shown in Figure 8-14.

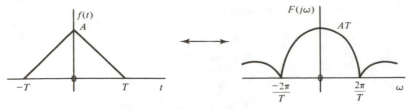

Figure 8-14   Fourier transform pair for Example 8-17.

Solution.   Plots of $f(10t)$ and $f(\frac{1}{2}t)$ appear in Figure 8-15. From Property 4, $f(10t) \leftrightarrow \frac{1}{10} F(j(\omega/10))$ and $f(\frac{1}{2}t) \leftrightarrow 2F(j2\omega)$. Plots of $\frac{1}{10} F(j\omega/10)$ and $2F(j2\omega)$ appear in Figure 8-16. Note that if the time function is contracted, then the transform will be expanded. If the time function is expanded, then its transform will be contracted.

EXAMPLE 8-18
(a) Prove the duality property.
(b) Use it to determine the Fourier transform of $f(t) = 10/(t^2 + 1)$.
(c) Use it to determine the Fourier transform of $f(t) = \text{Sin } t/t$.

Solution

(a) $f(t) = \dfrac{1}{2\pi} \displaystyle\int_{-\infty}^{\infty} F(j\omega) e^{j\omega t} \, d\omega$

Changing the dummy variable ω to x, we obtain:

$$2\pi f(t) = \int_{-\infty}^{\infty} F(jx) e^{jxt} \, dx$$

Now replace t by $-\omega$ to yield:

$$2\pi f(-\omega) = \int_{-\infty}^{\infty} F(jx) e^{-jx\omega} \, dx$$

Now on the right-hand side change the dummy variable x to t which gives:

$$2\pi f(-\omega) = \int_{-\infty}^{\infty} F(jt) e^{-j\omega t} \, dt = \text{FT}\{F(jt)\}$$

Therefore $F(jt) \leftrightarrow 2\pi f(-\omega)$

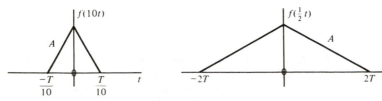

Figure 8-15   Plots of $f(10t)$ and $f(\frac{1}{2}t)$ for Example 8-17.

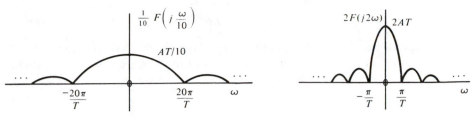

Figure 8-16 Plots of $\frac{1}{10}F(j\frac{\omega}{10})$ and $2F(j2\omega)$ for Example 8-17.

(b) We know that:

$$e^{-|t|} \leftrightarrow \frac{2}{\omega^2 + 1}$$

where $f(t) = e^{-|t|}$ and $F(j\omega) = \dfrac{2}{\omega^2 + 1}$

therefore $F(jt) = \dfrac{2}{t^2 + 1} \leftrightarrow 2\pi f(-\omega) = 2\pi e^{-|-\omega|} = 2\pi e^{-|\omega|}$

Thus $\dfrac{10}{t^2 + 1} \leftrightarrow 10\pi e^{-|\omega|}$

(c) We know that:

$$f(t) = u(t + T) - u(t - T) \leftrightarrow F(j\omega) = \frac{2 \text{ Sin } \omega T}{\omega}$$

Therefore $F(jt) = \dfrac{2 \text{ Sin } tT}{t} \leftrightarrow 2\pi f(-\omega)$

Since $f(-t) = f(t)$, $f(-\omega) = u(\omega + T) - u(\omega - T)$

If $T = 1$:

$$\frac{2 \text{ Sin } t}{t} \leftrightarrow 2\pi[u(\omega + 1) - u(\omega - 1)]$$

Thus $\dfrac{\text{Sin } t}{t} \leftrightarrow \pi[u(\omega + 1) - u(\omega - 1)]$

Note that this result checks with number 9 in Table 8-1.

As mentioned earlier, we use $F(j\omega)$ instead of $F(\omega)$ or $F(f)$ for the Fourier transform. However, the $F(f)$ notation provides an interesting symmetry when employed in the duality property. Let:

$$F(f) = \int_{-\infty}^{\infty} f(t)e^{-j2\pi ft} \, dt \tag{8-45}$$

which is just $F(j\omega)$ with $\omega = 2\pi f$.

$$f(t) = \int_{-\infty}^{\infty} F(f)e^{j2\pi ft}\, df \tag{8-46}$$

Note the absence of the 2π term in the inverse Fourier transform. Using the $F(f)$ notation, we find that the 2π term is also absent in the statement of the duality property: If $f(t) \leftrightarrow F(f)$, then $F(t) \leftrightarrow f(-f)$. From the previous example, for instance:

$$e^{-|t|} \leftrightarrow \frac{2}{\omega^2 + 1} \quad \text{and} \quad \frac{1}{t^2 + 1} \leftrightarrow \pi e^{-|\omega|}$$

But $F(f) = F(j\omega)|_{\omega = 2\pi f}$.

Thus

$$\frac{2}{(2\pi f)^2 + 1} \leftrightarrow e^{-|t|} \quad \text{and} \quad \frac{2}{(2\pi t)^2 + 1} \leftrightarrow e^{-|f|}$$

Both versions of the duality property then will yield similar results. The differences lie primarily in scaling.

EXAMPLE 8-19

Prove the time convolution property and use it to determine the system input when the system has an impulse response

$$h(t) = e^{-10t}u(t)$$

and the system output is:

$$y(t) = (e^{-5t} - e^{-15t})u(t)$$

Solution

$$FT\{f(t)*g(t)\} = \int_{-\infty}^{\infty} f(t)*g(t)e^{-j\omega t}\, dt$$

$$= \int_{-\infty}^{\infty} \int_{-\infty}^{\infty} f(\lambda)g(t - \lambda)e^{-j\omega t}\, dt\, d\lambda$$

Let $t - \lambda = v$, then $dt = dv$

and

$$f(t)*g(t) \leftrightarrow \int_{-\infty}^{\infty} \int_{-\infty}^{\infty} f(\lambda)g(v)e^{-j\omega v}e^{-j\omega \lambda}\, d\lambda\, dv$$

$$= \int_{-\infty}^{\infty} f(\lambda)e^{-j\omega \lambda}\, d\lambda \int_{-\infty}^{\infty} g(v)e^{-j\omega v}\, dv$$

$$= F(j\omega)G(j\omega)$$

Therefore $f(t)*g(t) \leftrightarrow F(j\omega)G(j\omega)$

Now we know that

$$y(t) = h(t)*x(t)$$

The time convolution property indicates that the Fourier transform of the output is:

$$Y(j\omega) = H(j\omega)X(j\omega)$$

The system function:

$$H(j\omega) = \frac{1}{j\omega + 10} \quad \text{and} \quad Y(j\omega) = \frac{1}{j\omega + 5} - \frac{1}{j\omega + 15}$$

$$Y(j\omega) = \frac{10}{(j\omega + 5)(j\omega + 15)}$$

Then

$$X(j\omega) = \frac{10/(j\omega + 5)(j\omega + 15)}{1/(j\omega + 10)} = \frac{10(j\omega + 10)}{(j\omega + 5)(j\omega + 15)}$$

Next, using partial fraction expansion, we can write:

$$X(j\omega) = \frac{A}{j\omega + 5} + \frac{B}{j\omega + 15} = \frac{5}{j\omega + 5} + \frac{5}{j\omega + 15}$$

Therefore

$$x(t) = (5e^{-5t} + 5e^{-15t})u(t)$$

EXAMPLE 8-20

Use the frequency convolution property to verify number 13 in Table 8-1.

Solution

$$f_1(t) = e^{-at} \cos \omega_0 t\, u(t)$$

Let:

$$f(t) = e^{-at}u(t) \quad \text{and} \quad g(t) = \cos \omega_0 t$$

$$F(j\omega) \leftrightarrow \frac{1}{j\omega + \alpha} \quad \text{and} \quad G(j\omega) \leftrightarrow \pi[\delta(\omega - \omega_0) + \delta(\omega + \omega_0)]$$

Then

$$F_1(j\omega) = \frac{1}{2\pi} F(j\omega) * G(j\omega) = \frac{1}{2\pi} \int_{-\infty}^{\infty} F(j\lambda)G(j[\omega - \lambda])\, d\lambda$$

$$= \frac{\pi}{2\pi} \int_{-\infty}^{\infty} \left(\frac{1}{j\lambda + \alpha}\right)(\delta(\omega - \lambda - \omega_0) + \delta(\omega - \lambda + \omega_0))\, d\lambda$$

$$= \frac{1}{2}\left\{\frac{1}{j(\omega - \omega_0) + \alpha} + \frac{1}{j(\omega + \omega_0) + \alpha}\right\}$$

$$= \frac{j\omega + \alpha}{[j(\omega - \omega_0) + \alpha][j(\omega + \omega_0) + \alpha]}$$

$$F_1(j\omega) = \frac{j\omega + \alpha}{(\alpha + j\omega)^2 + \omega_0^2}$$

EXAMPLE 8-21

Prove the time differentiation property and use it to determine the Fourier transform of the $f(t)$ in Figure 8-17.

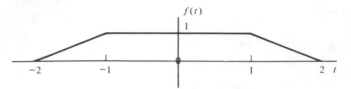

Figure 8-17   Time function used in Example 8-21.

Solution.   The inverse Fourier transform is:

$$f(t) = \frac{1}{2\pi} \int_{-\infty}^{\infty} F(j\omega)e^{j\omega t}\, d\omega$$

Then

$$\frac{df(t)}{dt} = \frac{d}{dt}\left\{\frac{1}{2\pi} \int_{-\infty}^{\infty} F(j\omega)e^{j\omega t}d\omega\right\} = \int_{-\infty}^{\infty} \frac{1}{2\pi} j\omega F(j\omega)e^{j\omega t}\, d\omega$$

Therefore $\dot{f}(t) \leftrightarrow j\omega F(j\omega)$

Likewise, $\ddot{f}(t) \leftrightarrow (j\omega)^2 F(j\omega)$

and in general:

$$\frac{d^n f(t)}{dt^n} \leftrightarrow (j\omega)^n F(j\omega)$$

Now if we differentiate $f(t)$, then differentiate again, we obtain the plots indicated in Figure 8-18. The second derivative consists of four impulses. Impulses have very simple transforms:

$$\delta(t - t_0) \leftrightarrow e^{-j\omega t_0}$$

Therefore:

$$\ddot{f}(t) \leftrightarrow e^{+2j\omega} - e^{j\omega} - e^{-j\omega} + e^{-2j\omega} = 2\,\text{Cos}\,2\omega - 2\,\text{Cos}\,\omega$$

But this is $(j\omega)^2 F(j\omega)$.

Thus $F(j\omega) = \dfrac{2\,\text{Cos}\,2\omega - 2\,\text{Cos}\,\omega}{(j\omega)^2}$

or $F(j\omega) = \dfrac{2\,\text{Cos}\,\omega - 2\,\text{Cos}\,2\omega}{\omega^2}$

This procedure is often useful: (1) Given $f(t)$, (2) differentiate $f(t)$ enough

Figure 8-18   First and second derivatives of $f(t)$.

times to yield only impulses or their derivatives, (3) transform, and (4) divide by $(j\omega)^k$ where k is the number of derivatives performed.

EXAMPLE 8-22

Use the frequency differentiation property to determine the Fourier transform of the following:

(a) $f_1(t) = te^{-5t}u(t)$
(b) $f_2(t) = te^{-t^2}$
(c) $f_3(t) = te^{-|t|}$
(d) $f_4(t) = t^2e^tu(-t)$
(e) $f_s(t) = tu(t)$

Solution

(a) Let:

$$tf(t) = te^{-5t}u(t).$$

$$f(t) = e^{-5t}u(t) \leftrightarrow \frac{1}{5 + j\omega}$$

Then
$$tf(t) \leftrightarrow j\frac{d}{d\omega}\left(\frac{1}{5 + j\omega}\right) = j(-1)(5 + j\omega)^{-2}(j)$$

Therefore
$$te^{-5t}u(t) \leftrightarrow \frac{1}{(j\omega + 5)^2}$$

(b) Let:

$$te^{-t^2} = tf(t) \leftrightarrow j\frac{d}{d\omega}F(j\omega)$$

where
$$F(j\omega) = \sqrt{\pi}e^{-\omega^2/4}$$

$$\frac{d}{dx}e^u = e^u\frac{du}{dx} \rightarrow \frac{d}{d\omega}e^{-\omega^2/4}$$

$$= e^{-\omega^2/4}\frac{d}{d\omega}\left(-\frac{1}{4}\omega^2\right)$$

$$= e^{-\omega^2/4}\left(-\frac{1}{2}\omega\right)$$

Thus
$$te^{-t^2} \leftrightarrow -j\frac{\omega}{2}\sqrt{\pi}e^{-\omega^2/4}$$

(c)
$$e^{-|t|} \leftrightarrow \frac{2}{1 + \omega^2}$$

and
$$\frac{d}{d\omega}2(1 + \omega^2)^{-1} = -2(1 + \omega^2)^{-2}2\omega$$

Therefore $\qquad te^{-|t|} \leftrightarrow \dfrac{-4j\omega}{(1 + \omega^2)^2}$

(d) $\qquad\qquad FT\{e^t u(-t)\} = \displaystyle\int_{-\infty}^{0} e^t e^{-j\omega t}\, dt$

$$= \frac{1}{1 - j\omega}(1 - 0) = \frac{1}{1 - j\omega}$$

Then $\qquad \dfrac{d}{d\omega}(1 - j\omega)^{-1} = -(1 - j\omega)^{-2}(-j) = j(1 - j\omega)^{-2}$

$$\frac{d^2}{d\omega^2}(1 - j\omega)^{-1} = -2j(1 - j\omega)^{-3}(-j) = \frac{-2}{(1 - j\omega)^3}$$

and $\qquad\qquad t^2 f(t) \leftrightarrow (j)^2 \dfrac{d^2}{d\omega^2} F(j\omega) = \dfrac{2}{(1 - j\omega)^3}$

(e) Let:

$$f(t) = u(t) \leftrightarrow \pi\delta(\omega) + \frac{1}{j\omega} = F(j\omega)$$

$$\frac{d}{d\omega} F(j\omega) = \pi\dot{\delta}(\omega) - \frac{1}{j\omega^2}$$

Then $\qquad\qquad tu(t) \leftrightarrow j\dfrac{d}{d\omega} F(j\omega) = j\pi\dot{\delta}(\omega) - \dfrac{1}{\omega^2}$

Note that Example 8-22(e) presents a dilemma. Part (g) in Example 8-12 asked for the Fourier transform directly. The conclusion there was that FT $\{tu(t)\}$ does not exist. The reason was that $tu(t)$ was neither an energy signal, whose Fourier transforms are not problematic, nor a power signal, whose Fourier transforms are not problematic as long as we allow $\delta(\omega)$ functions to exist in the frequency domain. Note the result for Example 8-22(e). $F(j\omega)$ contains a unit-doublet. From generalized function theory, which is beyond the scope of this book, it can be shown that if $\dot{f}(t)$ is a power signal *and* we let $\dot{\delta}(\omega)$ exist in the frequency domain, then $F(j\omega)$ can be developed. Likewise, if $\ddot{f}(t)$ is a power signal *and* we let $\ddot{\delta}(\omega)$ exist, then $F(j\omega)$ can be developed, and so on. Due to the abstract nature of these issues, we will not consider them further. We turn instead to an examination of the frequency shift or modulation property. This property proves to be very useful in a number of different areas of communication theory.

EXAMPLE 8-23

Use the modulation property to determine the Fourier transform of the following functions:

(a) $f_1(t) = f(t) \cos \omega_c t$

(b) $f_2(t) = f(t) \sin \omega_c t$

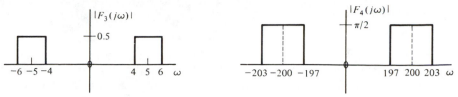

Figure 8-19 Plot of magnitudes of F_3 and F_4 from Example 8-23.

(c) $f_3(t) = \dfrac{\text{Sin } t \text{ Sin } 5t}{\pi t}$

(d) $f_4(t) = \dfrac{\text{Sin } 3t \text{ Cos } 200t}{t}$

(e) plot $|F_3(j\omega)|$ and $|F_4(j\omega)|$

Solution

(a)
$$f_1(t) = f(t) \text{ Cos } \omega_c t = \frac{f(t)}{2} \{e^{j\omega_c t} + e^{-j\omega_c t}\}$$

$$\longleftrightarrow \frac{F(j[\omega - \omega_c]) + F(j[\omega + \omega_c])}{2}$$

(b)
$$f_2(t) = f(t) \text{ Sin } \omega_c t = \frac{f(t)}{2j} \{e^{j\omega_c t} - e^{-j\omega_c t}\}$$

$$\longleftrightarrow \frac{F(j[\omega - \omega_c]) - F(j[\omega + \omega_c])}{2j}$$

(c)
$$f_3(t) = \frac{\text{Sin } t}{\pi t} \frac{e^{5jt} - e^{-5jt}}{2j} \quad \text{but} \quad \frac{\text{Sin } t}{\pi t} \longleftrightarrow \Pi\left(\frac{\omega}{2}\right)$$

Therefore
$$F_3(j\omega) = \frac{\Pi((\omega - \omega_c)/2) - \Pi((\omega + \omega_c)/2)}{2j}$$

$$= \frac{\Pi((\omega - 5)/2) - \Pi((\omega + 5)/2)}{2j}$$

(d)
$$f_4(t) = \frac{\text{Sin } 3t}{t} \left\{\frac{e^{200jt} + e^{-200jt}}{2}\right\} \quad \text{but} \quad \frac{\text{Sin } 3t}{t} \longleftrightarrow \pi\Pi\left(\frac{\omega}{6}\right)$$

Thus $F_4(j\omega) = \dfrac{\pi}{2}\Pi\left(\dfrac{\omega - 200}{6}\right) + \dfrac{\pi}{2}\Pi\left(\dfrac{\omega + 200}{6}\right)$

(e) Plots of $|F_3(j\omega)|$ and $|F_4(j\omega)|$ appear in Figure 8-19.

EXAMPLE 8-24

Use the modulation property to determine the Fourier transform of an arbitrary periodic $f(t)$ which is represented as a complex exponential Fourier series.

Solution. Let:

$$f(t) = \sum_{n=-\infty}^{\infty} c_n e^{jn\omega_0 t}$$

Each of the terms in the summation is a constant multiplied by a complex exponential. The Fourier transform of a constant is an impulse; that is:

$$FT\{c_n\} = 2\pi c_n \delta(\omega)$$

Therefore the Fourier transform of $f(t)$ is a summation of impulses, each of which is shifted in frequency due to the complex exponential terms; that is

$$FT\{f(t)\} = 2\pi \sum_{n=-\infty}^{\infty} c_n \delta(\omega - n\omega_0)$$

The Fourier transform of a Fourier series consists of a sequence of impulses. Each impulse is weighted by $2\pi c_n$ and all impulses are separated from each other by ω_0. Although the term ω_0 is similar to the period of the transform, the Fourier transform is not a periodic function. Even though the impulses are all separated by the same amount, their weights are all different. The best way to understand the relationship between the Fourier series and the Fourier transform is to imagine that the line spectra in the Fourier series are replaced by infinite lines or impulses in the Fourier transform. Each Fourier transform impulse is weighted with the corresponding complex exponential Fourier series coefficient c_n (times 2π).

EXAMPLE 8-25

Demonstrate the correlation property.

Solution. The property states that the Fourier transform of;

$$\int_{-\infty}^{\infty} f(\lambda - t)g(\lambda)\, d\lambda \quad \text{is the product} \quad F(-j\omega)G(j\omega)$$

This of course is very similar to the convolution property. The integral expression is written $f(t) \oplus g(t)$ analogous to the convolution notation. In the integral, if we let $\lambda - t = p$, then the integral becomes:

$$\int_{-\infty}^{\infty} f(p)g(\lambda)\, d\lambda$$

The Fourier transform of this integral then is:

$$\int_{-\infty}^{\infty} \int_{-\infty}^{\infty} f(p)g(\lambda)e^{-j\omega t}\, dt\, d\lambda$$

but $dt = -dp$, so we obtain:

$$\int_{-\infty}^{\infty} \int_{+\infty}^{-\infty} f(p)g(\lambda)e^{-j\omega(\lambda-p)}(-dp)\, d\lambda$$

The minus sign with dp reverses the limits on the second integral and we can write:

$$\int_{-\infty}^{\infty} f(p)e^{j\omega p}\left[\int_{-\infty}^{\infty} g(\lambda)e^{-j\omega\lambda}\, d\lambda\right] dp = F(-j\omega)G(j\omega)$$

Now before concluding this section on properties of the Fourier transform, we consider Parseval's theorem. In the Fourier series discussion we discussed what was called Parseval's relation. This equation related the energy contained in a finite time interval of a function to the Fourier series coefficients of that function. Parseval's theorem is similar. Consider a real energy signal $f(t)$ with the Fourier transform $F(j\omega)$. Let the energy contained in $f(t)$ be:

$$\mathcal{E} = \int_{-\infty}^{\infty} f^2(t)\, dt \tag{8-47}$$

Since $f(t) = (1/2\pi) \int_{-\infty}^{\infty} F(j\omega) e^{j\omega t}\, d\omega$ is the inverse Fourier transform of $F(j\omega)$, we can write:

$$\mathcal{E} = \int_{-\infty}^{\infty} f(t) \left[\frac{1}{2\pi} \int_{-\infty}^{\infty} F(j\omega) e^{j\omega t}\, d\omega \right] dt \tag{8-48}$$

which can be further expressed as:

$$\mathcal{E} = \frac{1}{2\pi} \int_{-\infty}^{\infty} F(j\omega) \left[\int_{-\infty}^{\infty} f(t) e^{j\omega t}\, dt \right] d\omega \tag{8-49}$$

But note that the term in parentheses here is just $F(-j\omega)$, and since we are assuming that $f(t)$ is a real function of time, we know that $F(-j\omega) = F^*(j\omega)$. Also, we know for any complex function that $FF^* = |F|^2$. Thus we have:

$$\mathcal{E} = \frac{1}{2\pi} \int_{-\infty}^{\infty} F(j\omega) F(-j\omega)\, d\omega = \frac{1}{2\pi} \int_{-\infty}^{\infty} |F(j\omega)|^2\, d\omega \tag{8-50}$$

Relating time- and frequency-domain integrals, we can write Parseval's theorem:

$$\int_{-\infty}^{\infty} f^2(t)\, dt = \frac{1}{2\pi} \int_{-\infty}^{\infty} |F(j\omega)|^2\, d\omega \tag{8-51}$$

The term $|F|^2$ is called the **energy spectral density** and indicates a distribution of energy over a spectral band. For instance, if $F(j\omega)$ is fairly constant over a small band $\Delta\omega = \omega_2 - \omega_1$, then the energy contained in that band is approximately $|F|^2 \Delta\omega/2\pi$. This result can be obtained from Equation 8-51 if we let F be constant and integrate from ω_1 to ω_2 instead of from $-\infty$ to $+\infty$. Then we have $\mathcal{E} = |F|^2 \Delta\omega/2\pi$ as the energy contained in the spectral band $\omega_1 \le \omega \le \omega_2$. We can write $|F|^2 = 2\pi\mathcal{E}/\Delta\omega$. Dividing $\mathcal{E}$ by $\Delta\omega$ gives a kind of energy density: We have an amount of energy per $\Delta\omega$. This is the motivation for calling the term $|F|^2$ the energy spectral density. Note that when we talk about continuous energy spectral densities or continuous Fourier transforms, the energy over a band of frequencies—never the energy contained in a single frequency—is of interest.

With regard to linear systems, the Parseval theorem can be useful. If $f(t) \leftrightarrow F(j\omega)$ is the input, $g(t) \leftrightarrow G(j\omega)$ is the output, and $H(j\omega)$ is the system transfer function, then the output energy spectral density is:

$$|G(j\omega)|^2 = |F(j\omega)|^2 |H(j\omega)|^2 \tag{8-52}$$

The term $|H(j\omega)|^2$ is called the **energy transfer function.** It relates the input energy spectral density to the output energy spectral density. Because of the magnitude-squared nature of these terms, the output and the input energy spectral densities are both independent of any phase variations that might be present.

Another use for Parseval's theorem is in what is called **energy localization.** Assume for some given $f(t)$ that the left-hand side of Equation 8-51 can be computed. This yields the total energy contained in the signal. Now on the right-hand side of Equation 8-51, note first that $|F|^2$ is an even function of ω. Thus we can write:

$$\int_{-\infty}^{\infty} f^2(t)\, dt = \frac{1}{\pi} \int_0^{\infty} |F(j\omega)|^2\, d\omega \qquad (8\text{-}53)$$

Often the energy spectrum $|F|^2$ will be concentrated over a finite band of frequencies. A typical question in this area is to determine such a frequency band within which a certain percentage of the total energy will be localized.

EXAMPLE 8-26

Determine a frequency band $(0, \omega_c)$ over which one half the energy in $f(t) = e^{-t}u(t)$ will be localized.

Solution. The energy in $f(t)$ is:

$$\mathscr{E} = \int_{-\infty}^{\infty} f^2(t)\, dt = \int_0^{\infty} e^{-2t}\, dt = 0.5$$

Now, from Equation 8-53, we can write:

$$\frac{1}{2}(0.5) = \frac{1}{\pi} \int_0^{\omega_c} |F|^2\, d\omega$$

equating one half the energy to the integral with finite upper limit. We know that:

$$F(j\omega) = \frac{1}{1 + j\omega}$$

Thus
$$|F|^2 = \frac{1}{1 + \omega^2}$$

and
$$0.25 = \frac{1}{\pi} \int_0^{\omega_c} \frac{1}{1 + \omega^2}\, d\omega = \frac{1}{\pi} \{\mathrm{Tan}^{-1} \omega |_0^{\omega_c}\}$$

or
$$0.25\pi = \mathrm{Tan}^{-1} \omega_c - \mathrm{Tan}^{-1} 0 = \mathrm{Tan}^{-1} \omega_c$$

therefore
$$\omega_c = \mathrm{Tan}(\pi/4) = 1 \text{ rad/s}$$

The discussion on Parseval's theorem provides a transition between the properties and the applications of the Fourier transform. The result postulated in Parseval's theorem employs the idea of signal energy and follows directly from the definitions of the Fourier transform and the inverse Fourier transform. Using Parseval's theorem in the energy localization problem introduces Fourier transform applications. Applications of the Fourier transform span a wide variety of disciplines. Some of these applications will be dealt with in Section 8.6.

At this point, we pause in order to consolidate our results. We studied the Fourier series and from it developed the Fourier transform. A number of properties of the Fourier transform were considered, not only as an aid to obtain Fourier transform functions, but also as a means to gain deeper insights into the essence of the Fourier transform. Even further appreciation can be obtained by comparing the Fourier transform to the Laplace transform, which has already been discussed in Chapters 4 and 5. A basic understanding of the Laplace transform is presupposed. The next short section deals with the relationship between the Fourier and Laplace transforms.

8-5 THE FOURIER TRANSFORM AND THE LAPLACE TRANSFORM: A COMPARISON

From a cursory glance at the two transforms we might conclude that $F(j\omega)$ is just $F(s)$ with s replaced by $j\omega$. This, however, is not always the case. It is so if $f(t) = 0, t < 0$, and $\int_0^\infty |f(t)|\, dt < \infty$; that is, if $f(t)$ is absolutely integrable.

EXAMPLE 8-27
Determine $F(j\omega)$ from $F(s)$ for:

(a) $f_1(t) = e^{-10t}u(t)$
(b) $f_2(t) = e^{-t}\operatorname{Cos} 10tu(t)$
(c) $f_3(t) = u(t) - u(t - 10)$

Solution

(a)
$$F_1(s) = \frac{1}{s + 10}$$

Since $f_1(t)$ is zero for $t < 0$ and $f_1(t)$ is absolutely integrable:

$$F_1(j\omega) = \frac{1}{10 + j\omega}$$

(b)
$$F_2(s) = \frac{s + 1}{(s + 1)^2 + 100} \quad \cdot \quad F_2(j\omega) = \frac{j\omega + 1}{(j\omega + 1)^2 + 100}$$

(c)
$$F_3(s) = \frac{1}{s} - \frac{1}{s}e^{-10s} \quad \cdot \quad F_3(j\omega) = \frac{1}{j\omega} - \frac{1}{j\omega}e^{-10j\omega}$$

Now in this case $u(t) \leftrightarrow F(j\omega) = \pi\delta(\omega) + 1/j\omega$ and the shifted $u(t)$ has the transform:

$$u(t - 10) \leftrightarrow e^{-j\omega 10} F(j\omega) = e^{-j\omega 10}\left(\pi\delta(\omega) + \frac{1}{j\omega}\right)$$

$$= e^0 \pi\delta(\omega) + e^{-10j\omega} \frac{1}{j\omega}$$

$$= \pi\delta(\omega) + \frac{e^{-10j\omega}}{j\omega}$$

Therefore

$$F_3(j\omega) = \left(\pi\delta(\omega) + \frac{1}{j\omega}\right)$$

$$- \left(\pi\delta(\omega) + \frac{e^{-10j\omega}}{j\omega}\right)$$

$$= \frac{1}{j\omega} - \frac{1}{j\omega} e^{-10j\omega}$$

which checks with the preceding result.

Under the two constraints of $f(t) = 0$, $t < 0$, and $f(t)$ being absolutely integrable, we can do the reverse and get $F(s)$ from $F(j\omega)$ by letting $F(s) = F(j\omega)|_{\omega - s/j}$. If we employ the two-sided Laplace transform, we can relax the constraint that $f(t)$ be a causal signal; that is $f(t) = 0$, $t < 0$. The interesting cases, however, are those in which simple substitution does not work. This occurs when absolute integrability does not hold. To handle these cases, assume $f(t)$ is causal so we need only consider the one-sided Laplace transform. If this is the case, then the Laplace transform is more inclusive than the Fourier transform; it exists for a wider class of functions. To put it differently, the existence of $F(j\omega)$ implies the existence of $F(s)$, but the existence of $F(s)$ does not necessarily imply the existence of $F(j\omega)$. Let us examine the issue of absolute integrability by distinguishing the various possible regions of convergence of a given Laplace transform function. These regions can be considered in terms of the s plane pole locations of $F(s)$.

Region of Convergence I   If $F(s)$ has all poles in the LHP (Left Hand Plane), then $f(t)$ is absolutely integrable

and

$$F(j\omega) = F(s)|_{s = j\omega} \tag{8-54}$$

and

$$F(s) = F(j\omega)|_{\omega = s/j} \tag{8-55}$$

Region of Convergence II   If $F(s)$ has any nonrepeated poles on the $j\omega$ axis (with possibly other poles in the LHP), then $f(t)$ is not absolutely integrable but it is a power signal. The Fourier transform of these signals contains impulses in the frequency domain. We can then write:

$$F(j\omega) = F(s)|_{s=j\omega} + \pi \sum_i k_i \delta(\omega - \omega_i) \tag{8-56}$$

The k_i terms are the residues at the poles on the $j\omega$ axis: $s = j\omega_i$. The reverse is easier. Given $F(j\omega)$, simply let $\omega = s/j$ and zero out all impulses $\delta(\omega - \omega_i)$ in order to get $F(s)$ from $F(j\omega)$. The case of repeated poles on the $j\omega$ axis is more difficult because $F(j\omega)$ contains $\dot{\delta}$, $\ddot{\delta}$, and so on, terms. To get $F(j\omega)$ in these cases, we obtain $f(t)$ from $F(s)$, then work with $f(t)$ instead of $F(s)$. To get $F(s)$ from $F(j\omega)$ is also easy: Simply let $\omega = s/j$ and zero out all δ, $\dot{\delta}$, $\ddot{\delta}$, and so on, terms.

Region of Convergence III   If $F(s)$ has any poles in the RHP (Right Hand Plane), then $F(j\omega)$ does not exist.

EXAMPLE 8-28
(a) Given $F(s) = 10/s(s + 10)$, determine $F(j\omega)$.
(b) Given $F(j\omega) = 10/\omega(1 - \omega^2)(10j - \omega) + \pi\delta(\omega) - 5\pi/101$ $(10 + j)\delta(\omega + 1) - 5\pi/101 (10 - j)\delta(\omega - 1)$, determine $F(s)$.
(c) Given $F(s) = 10/s^2(s + 1)$, determine $F(j\omega)$.

Solution
(a) Write:

$$F(s) = \frac{A}{s} + \frac{B}{s + 10} = \frac{1}{s} - \frac{1}{s + 10} = \frac{10}{s(s + 10)}$$

therefore from Equation 8-56:

$$F(j\omega) = \frac{10}{j\omega(j\omega + 10)} + \pi\delta(\omega)$$

(b) To obtain $F(s)$, zero out the $\delta(\omega)$, $\delta(\omega + 1)$, and $\delta(\omega - 1)$ terms, then let $\omega = s/j$:

Thus $\quad F(s) = \dfrac{10}{s/j(1 - (s/j)^2)(10j - s/j)} = \dfrac{10}{(10s + s^2)(1 + s^2)}$

$$F(s) = \frac{10}{s(s + 10)(s^2 + 1)}$$

(c) Since $F(s)$ has repeated poles on the $j\omega$ axis, we get $f(t)$ first, because Equation 8-56 is not directly applicable in this case. Write

$$F(s) = \frac{A}{s} + \frac{B}{s^2} + \frac{C}{s + 1} = \frac{10}{s + 1} + \frac{10}{s^2} - \frac{10}{s}$$

$$f(t) = 10e^{-t}u(t) + 10tu(t) - 10u(t)$$

We saw that $tu(t) \leftrightarrow j\pi\dot{\delta}(\omega) - 1/\omega^2$ from Example 8-22(e).

Thus $FT\{f(t)\} = F(j\omega)$

$$= \frac{10}{j\omega + 1} + 10j\pi\dot\delta(\omega) - \frac{10}{\omega^2} - 10\pi\delta(\omega) - \frac{10}{j\omega}$$

These considerations cover most of the possible relations between the Fourier transform and the Laplace transform. The fundamental idea here is:

$$LT\{f(t)\} = FT\{e^{-\sigma t}f(t)u(t)\} \qquad (8\text{-}57)$$

which is true as long as the Laplace transform of $f(t)$ exists. Working with s terms instead of $j\omega$ terms usually results in simpler algebraic manipulations. In addition, the Laplace transform readily applies to systems with initial conditions, whereas the Fourier transform does not. Although as a rule of thumb working with $F(s)$ is preferred, facility in switching from $F(s)$ to $F(j\omega)$ is essential, especially in the case where frequency response or spectral analysis is at issue. We stress the importance of Equation 8-56. A final example follows.

EXAMPLE 8-29

If $f(t) = e^{+10t}u(t)$, we know that $F(s) = 1/(s - 10)$ is the Laplace transform that exists as long as $\mathrm{Re}(s) = \sigma > 10$. To what might $F(j\omega)$ correspond?

Solution.   As we saw previously, if $F(s)$ has RHP poles, $F(j\omega)$ does not exist. Thus the given $f(t)$ does not have a Fourier transform. However, if $g(t) = -e^{10t}u(-t)$, then $G(j\omega) = 1/(j\omega - 10)$ is the Fourier transform of $g(t)$. Note also that $G(j\omega) = F(s)|_{s=j\omega} = F(j\omega)$. So $F(j\omega)$ is not the Fourier transform of the causal time function $f(t)$, but rather, the Fourier transform of the noncausal time function $g(t)$.

Drill Set: Fourier Transforms

1. Use the duality property to determine the Fourier transform of $(\mathrm{Sin}\ t/t)^2$.
2. Use the frequency differentiation property to determine the Fourier transform of $t^3 e^{-5t}u(t)$.
3. Determine the Fourier transform of the periodic signal

$$F(t) = \begin{cases} 1, & 0 < t < 1 \text{ which has a period } T = 3 \\ 2, & 1 < t < 2 \\ 0, & 2 < t < 3 \end{cases}$$

4. Prove that the Fourier transform of the correlation of the $x(t)$ with itself is equal to the magnitude of the Fourier transform squared.
5. If $F(j\omega)$ takes the following forms, determine the corresponding $f(t)$:
   (a) $(j\omega + 1)/(j\omega + 2)(j\omega + 3)$
   (b) $(1 - \omega^2 + j\omega)/(j\omega + 2)(2 - \omega^2 + j\omega)$
   (c) $(j\omega + 5)/(j\omega + 10)(j\omega + 20)^2$
   (d) $j\omega/(j\omega + 1)(j\omega + 2)(j\omega + 3)$

8-6 APPLICATIONS OF FOURIER THEORY

The Fourier transform and its digital counterpart, the discrete Fourier transform, which will be studied in Chapter 9, are widely employed in control systems, communication systems, and signal processing. Many of the signals considered, in particular, in the signal processing area, are in the form of data collected from radar tracks or measurements from various kinds of electromechanical devices. These signals invariably are noisy. Fourier analysis is used with these signals in an attempt to reveal their spectral content. This processing typically employs numerical techniques and the discrete Fourier transform. In addition to processing signals, Fourier analysis is used considerably in the control systems and the communication systems areas. Multiplexing and modulation of communication systems rely heavily on Fourier theory. The design of filters that are employed in control and communication systems is another field in which the Fourier theory is indispensable. Filters can be classified as either analog or digital. Digital filters usually employ the Z transform and analog filters the Fourier transform. As a first example of the application of Fourier theory we briefly consider analog filters.

8-6-1 Filters

One of the most basic applications of the Fourier transform employs the convolution property. Since the Fourier transform of $f(t)*g(t)$ is $F(j\omega)G(j\omega)$, we have an alternative to performing the convolution operation. If $f(t) = h(t)$, the impulse response, then $F(j\omega) = H(j\omega)$, the system function or filter transfer function.

Let $g(t)$ be the system input $x(t)$ and $f*g = h*x = y(t)$, the system output. Then $Y(j\omega) = H(j\omega)X(j\omega)$. To get the output $y(t)$, we need only take the inverse Fourier transform of $Y(j\omega)$. This is another version of the basic input–output relation of the linear systems theory and is useful in comparing real and ideal filters. Consider an ideal low-pass (LP) filter with transfer function $H(j\omega)$ as follows:

$$H(j\omega) = 1, \qquad -\omega_0 \le \omega \le \omega_0,$$

$$= 0, \qquad \text{otherwise} \qquad (8\text{-}58)$$

$H(j\omega)$ is plotted in Figure 8-20.

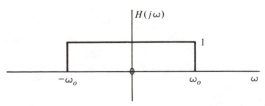

Figure 8-20   Ideal low-pass filter transfer function.

From entry number 9 in Table 8-1, the inverse Fourier transform of this $H(j\omega)$ is $h(t)$

where
$$h(t) = \frac{\text{Sin } \omega_0 t}{\pi t} = \frac{\omega_0}{\pi} \text{Sinc}\left(\frac{\omega_0}{\pi} t\right) \qquad (8\text{-}59)$$

This impulse reponse is noncausal. Noncausal systems are not physically realizable; that is, they cannot be built. This is why we say $H(j\omega)$ represents an ideal filter. Although not realizable, we can use $H(j\omega)$ as a standard against which real filters can be compared.

Often the step response instead of the impulse response is used in filter comparisons. We know that $y(t) = x(t)*h(t)$. If the input $x(t)$ is the unit step $u(t)$, then we can call the output $y(t)$, the unit step response $w(t)$. We have:

$$w(t) = u(t)*h(t)$$

$$= \int_{-\infty}^{\infty} h(\lambda)u(t - \lambda) \, d\lambda$$

$$= \int_{-\infty}^{t} h(\lambda) \, d\lambda \qquad (8\text{-}60)$$

and for the ideal low-pass filter

$$w(t) = \int_{-\infty}^{t} \frac{\text{Sin } \omega_0 \lambda}{\pi \lambda} \, d\lambda \qquad (8\text{-}61)$$

which is an integral of a Sinc function. In these cases $h(t)$ and $w(t)$ appear as in Figure 8-21. The ripple and overshoot in $w(t)$ on either side of the step discontinuity at $t = 0$ are known as the **Gibbs phenomenon,** named for Josiah Gibbs, a mathematical physicist who studied finite Fourier series approximation theory around 1900.

Now the occurrence for $t < 0$ of the Gibbs phenomenon in $w(t)$ is another indication of the noncausal nature of the ideal low-pass filter. The unit step input is applied at $t = 0$ but the output $w(t)$ has nonzero values for $t < 0$. The noncausal ideal low-pass filter is not physically realizable. To improve the filter somewhat, imagine $w(t)$ to be time shifted to the right by t_0 units. We get a plot of $w(t - t_0)$ which appears in Figure 8-22. This corresponds to a phase shift in $H(j\omega)$ to obtain:

$$\overline{H}(j\omega) = 1e^{-j\omega t_0}, \qquad -\omega_0 \leq \omega \leq \omega_0$$

$$= 0 \qquad \text{otherwise} \qquad (8\text{-}62)$$

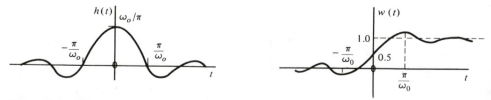

Figure 8-21   Plots of $h(t)$ and $w(t)$ for the ideal LP filter.

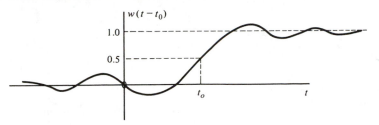

Figure 8-22   Plot of $w(t - t_0)$ for the ideal LP filter.

We can compare these ideal filter characteristics to those of a real low-pass filter, for example, the RC circuit presented in Figure 8-23. We can write:

$$H_{RC}(j\omega) = \frac{1}{1 + j\omega RC} = \frac{1}{1 + j(\omega/\omega_0)} = \frac{1}{\sqrt{1 + \omega^2/\omega_0^2}} e^{-j\mathrm{Tan}^{-1}\omega/\omega_0} \qquad (8\text{-}63)$$

where $\omega_0 = 1/RC$ is the 3 db or half power or cutoff frequency. ω_0 is the bandwidth of the low-pass filter.

This filter has the step response:

$$w(t) = (1 - e^{-\omega_0 t})u(t) \qquad (8\text{-}64)$$

The step response is plotted in Figure 8-24.

The ideal low-pass (LP) filter has a phase $\bar{\theta}(\omega) = \omega t_0$. From this relation we can define the time delay as follows:

$$t_0 = -\frac{d\bar{\theta}}{d\omega}(\omega)\Big|_{\omega=0} \qquad (8\text{-}65)$$

Now the phase of the real LP filter is $\theta(\omega) = -\mathrm{Tan}^{-1}\omega/\omega_0$. Applying the time delay definition to the real filter, we get:

$$t_0 = -\frac{d\theta}{d\omega}(\omega) = \frac{d}{d\omega}\left(\mathrm{Tan}^{-1}\frac{\omega}{\omega_0}\right) = \frac{1}{1 + (\omega/\omega_0)^2}\frac{1}{\omega_0} \qquad (8\text{-}66)$$

which when evaluated at $\omega = 0$ becomes:

$$t_0 = \frac{1}{\omega_0} \qquad (8\text{-}67)$$

Note that the product of the time delay and the filter bandwidth is constant: $t_0\omega_0 = 1$. This reciprocal relationship is important in Fourier transform theory and follows from the time-scaling property of the Fourier transform. If we desire a

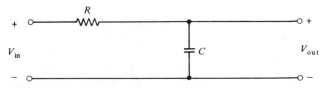

Figure 8-23 RC circuit as low-pass filter.

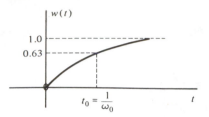

Figure 8-24   Step response of RC circuit.

fast response, for instance, a very small value for the t_0 indicated in Figure 8-24, then we must have a very large value for the bandwidth. Consider the rise time, t_r. Although there are a number of different ways to define t_r, we define it as the time from 10 to 90% of the final value.

Thus $\qquad\qquad\qquad\qquad 0.1 = 1 - e^{-\omega_0 t_1} \quad$ and $\quad 0.9 = 1 - e^{-\omega_0 t_2}$

and $\qquad\qquad\qquad\qquad\qquad t_r = t_2 - t_1$

$$e^{-\omega_0 t_1} = 0.9$$

$$e^{-\omega_0 t_2} = 0.1$$

$$e^{-\omega_0(t_2 - t_1)} = e^{-\omega_0 t_r} = \tfrac{1}{9}$$

$$\omega_0 t_r = \ln 9 = 2.2$$

Again the reciprocal relationship appears. If we want a very short rise time, for instance, we must have a filter with a very wide bandwidth.

   If we consider the real and ideal LP filter representations in the context of the Paley-Wiener criterion, some interesting results follow. Although the proof of this criterion is beyond our scope, the employment of it is fairly straightforward and it can serve as a useful test for causality. The Paley-Wiener criterion can be applied to any Fourier transform to indicate whether or not it corresponds to a causal time function. Let $H(j\omega) = A(\omega) \angle \theta(\omega)$ be the transfer function for a given filter. Then for $H(j\omega) \leftrightarrow h(t)$, if $h(t)$ is causal, the following inequality must hold:

$$\int_{-\infty}^{\infty} \frac{|\ln A(\omega)|}{1 + \omega^2}\, d\omega < \infty \qquad\qquad (8\text{-}68)$$

This criterion is obviously not satisfied by the ideal LP filter since $A(\omega) = 0$ for $|\omega| > \omega_0$ and the natural log of zero is infinity. For the RC LP filter, Equation 8.68 is satisfied since $A(\omega) \rightarrow 0$ and $\ln A(\omega) \rightarrow \infty$ only as $\omega \rightarrow \pm\infty$.

   The simple RC LP filter of Figure 8-23 can be improved by considering more complex circuitry. To improve the filter means to get it closer to the characteristics of the ideal filter. In Figure 8-25 we plot the frequency response magnitude curves for $\overline{H}$ and H_{RC} for positive frequencies. For the ideal LP filter, the spectrum from $0 \le \omega \le \omega_0$ is called the *pass-band* and from $\omega_0 \le \omega < \infty$ is called the *stop-band*. For the real LP filter there is a region around ω_0 called the *transition-band* which can specify the attenuation desired by a certain frequency beyond ω_0. More complex circuitry could result in a LP filter that more nearly approximates the curve for $|\overline{H}|$. These circuits typically employ operational

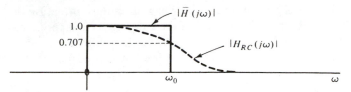

Figure 8-25 Real and ideal LP magnitude frequency response.

amplifiers and various passive circuit elements, usually resistors and capacitors. In practice, inductors are seldom used because of their size and weight.

EXAMPLE 8-30

The circuit shown in Figure 8-26 represents an LP filter that is an improvement over the filter of Figure 8-23. It has a sharper frequency response magnitude curve. Let $R = 1\ \Omega$ and determine C such that the filter has a half power point at 1 kHz.

Solution. At the node labeled V we can write the node equations:

$$\frac{V_{in} - V}{R} = \frac{V}{1/j\omega C} + \frac{V - V_{out}}{R}, \qquad \frac{V - V_{out}}{R} = \frac{V_{out}}{1/j\omega C}$$

then
$$V_{in} = (2 + j\omega C)\, V - V_{out}, \qquad V = V_{out} + j\omega C V_{out}$$
$$= [(2 + j\omega C)(1 + j\omega C) - 1]\, V_{out}$$

or
$$\frac{V_{out}}{V_{in}} = \frac{1}{1 + 3\, j\omega C + (j\omega C)^2}$$

and
$$\left| \frac{V_{out}}{V_{in}} \right| = \frac{1}{\sqrt{(1 - \omega^2 C^2)^2 + 9\omega^2 C^2}}$$

Setting the magnitude to $1/\sqrt{2}$, we can write:

$$2 = (1 - \omega^2 C^2)^2 + 9\omega^2 C^2 = 1 + 7\omega^2 C^2 + \omega^4 C^4$$

Let $x = \omega^2 C^2$ and write:

$$x^2 + 7x - 1 = 0 \quad \text{or} \quad x = 0.14,$$

taking only the positive value. Now $\omega = 2\pi f = 2\pi 10^3$.

Thus
$$C^2 = \frac{0.14}{4\pi^2 10^6} \quad \text{or} \quad C = 59.5\,\mu F$$

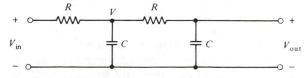

Figure 8-26 Low-pass filter for Example 8-30.

In the practical design of analog filters today engineers most often use the Butterworth, Bessel, or Chebyshev filters. Each of these designs is an approximation to the ideal LP filter. Each is "optimal," or best, but in a different sense. A detailed consideration of these filters is beyond our present scope. We mention only a few general characteristics:

1. The Butterworth filter is noted for having a response in the pass-band that is optimum in that it is as flat as possible for a given filter order. (**Filter order** is just the degree of the polynomial in the filter transfer function. The order is generally the number of energy storage devices required.)
2. The Bessel filter has a phase response that is as linear as possible for a given filter order.
3. The Chebyshev filter maintains a specified amplitude response in a given range of the pass-band. Although it has ripples in the pass-band, it is monotonic in the stop-band and yields maximum attenuation for a specified filter order. The Chebyshev and Bessel filter are so named because of their characterization in terms of Chebyshev polynomials and Bessel functions.

So far, the discussion on filters has centered around the LP filter. By employing some simple transformations, however, we can convert a given low-pass filter into a high-pass, band-pass, or notch filter. Assume we have a normalized LP filter described by $H(j\omega)$, where $\omega = 1$ is the cutoff frequency and $|H(j0)| = 1$. We can normalize the frequency by replacing ω by ω/ω_0. Replace ω by $-1/\omega$ and we get a high-pass (HP) filter. Replace ω by $(\omega^2 - \omega_1\omega_2)/\omega(\omega_2 - \omega_1)$ where ω_1 and ω_2 are the lower and upper cutoff frequencies and we get a band-pass filter with bandwidth $= \omega_2 - \omega_1$. Finally, replace ω by $\omega(\omega_2 - \omega_1)/(\omega_1\omega_2 - \omega^2)$ and we obtain a notch filter, where ω_1 and ω_2 are the cutoff frequencies. The network synthesis problem of determining the proper RLC component values or proper op-amp configuration for a given transfer function is a more difficult task and will not be pursued here. We merely note in concluding this discussion on filters that the Fourier transform does play an important role in both the analysis and the synthesis of filters and that filters of a variety of types are being used more and more in technical devices of all kinds.

8-6-2 Amplitude Modulation

Consider an application of Fourier analysis to amplitude modulation (AM). Although there are many ways to indicate the AM modulated signal, we will employ the form:

$$f(t) = (1 + ms(t)) \cos \omega_c t \qquad (8\text{-}69)$$

This represents what is called double-sideband amplitude modulation. Assume the audio signal is $s(t)$. Also, assume it is bounded in magnitude by 1.0 and its highest frequency is $\omega_s \ll \omega_c$, where ω_c is called the carrier frequency. Let m, the

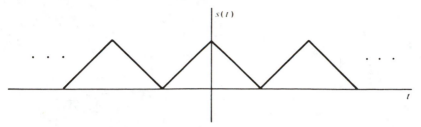

Figure 8-27   Triangular pulse train.

index of modulation, range from 0 to 1. Now the term $1 + ms(t)$ varies slowly compared to Cos $\omega_c t$, which means that we can view $1 + ms(t)$ as an envelope to Cos $\omega_c t$. The term $1 + ms(t)$ functions as the amplitude of the Cos $\omega_c t$ sinusoid. As $s(t)$ varies, this sinusoid then has an amplitude that varies with time: The amplitude is said to be modulated by the variations in $s(t)$. For instance, if $s(t)$ is a triangular pulse train as in Figure 8-27, then $f(t)$ might appear as in Figure 8-28. To consider the spectrum of $f(t)$ in an explicit fashion, let $s(t) = $ Cos $\omega_s t$, a very simple audio signal, but sufficient for purposes of illustration:

$$f(t) = (1 + m \text{ Cos } \omega_s t) \text{ Cos } \omega_c t$$

$$= \text{Cos } \omega_c t + \frac{m}{2} \text{ Cos } (\omega_c + \omega_s)t + \frac{m}{2} \text{ Cos } (\omega_c - \omega_s)t \qquad (8\text{-}70)$$

Expressing these cosines as complex exponentials, we can write the following:

$$f(t) = \tfrac{1}{2}e^{j\omega_c t} + \tfrac{1}{2}e^{-j\omega_c t} + \frac{m}{4} e^{j(\omega_c + \omega_s)t}$$

$$+ \frac{m}{4} e^{-j(\omega_c + \omega_s)t} + \frac{m}{4} e^{j(\omega_c - \omega_s)t} + \frac{m}{4} e^{-j(\omega_c - \omega_s)t} \qquad (8\text{-}71)$$

This time function has the complex Fourier line spectrum indicated by Figure 8-29. Note that all these frequencies are relatively "high" frequencies, which are essential for long-distance transmission. After this signal is transmitted it is necessary to recover $s(t)$, the signal of interest. This is called **demodulation.** There are many schemes available to do this, the simplest of which probably is

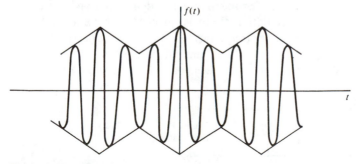

Figure 8-28   Plot of a typical A.M. signal.

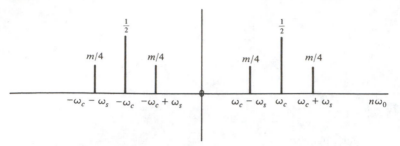

Figure 8-29   Spectrum of an A.M. signal.

the detector or envelope demodulator, a circuit that is shown in Figure 8-30. Referring to $f(t)$ in Figure 8-28, we find that if this signal is input to the circuit of Figure 8-30, the rectifier will pass the positive portion of $f(t)$ and the RC filter with $RC = 1/\omega_s$, will transmit only the envelope of this positive portion. The output will then be a reasonable facsimile of $s(t)$.

Now let $s(t)$ be generalized somewhat to be a time signal that is band-limited—instead of just a single sinusoid. Such an $s(t)$ will have a Fourier transform $S(j\omega)$ perhaps like the one in Figure 8-31. Band-limitedness in reality is a fiction because it implies an $s(t)$ that is of infinite duration. Often, however, a signal will display a spectrum that is negligibly small outside a certain band. These signals can be approximated by a band-limited spectrum, which is useful at least for the purpose of illustration. Consider the Fourier transform of the $f(t)$ of Equation 8-69.

$$F(j\omega) = \pi\delta(\omega - \omega_c) + \pi\delta(\omega + \omega_c) + \frac{m}{2}S(j(\omega - \omega_c))$$

$$+ \frac{m}{2}S(j(\omega + \omega_c)) \qquad (8\text{-}72)$$

This equation indicates that $S(j\omega)$ is shifted to yield a spectrum like the one in Figure 8-32. Again, once the modulated signal is transmitted, there is the need to demodulate. Again, a circuit similar to the circuit of Figure 8-30 will process the received $f(t)$ signal and yield a good approximation to $s(t)$ at the circuit output.

So far, all this AM discussion has focused on what is called **asynchronous** amplitude modulation. This refers primarily to the demodulation side of the system. In asynchronous demodulation there is no need to have available an oscillator synchronized to the carrier frequency ω_c. We merely send $f(t)$ through a circuit like that in Figure 8-30. Why is this advantageous? Employing an oscillator at the receiver synchronized to ω_c is a costly venture as well as a tricky

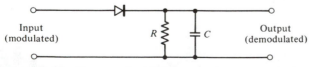

Figure 8-30   Detector demodulator.

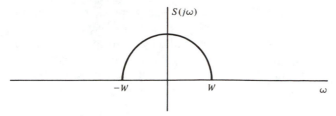

Figure 8-31   Transform of a band-limited signal.

stabilization problem. For accurate signal recovery, the receiver oscillator must be set and maintained at the exact frequency and phase of the oscillator at the transmitter. This is called **synchronous** amplitude modulation and although it is practically more difficult to implement than an asynchronous system, it is conceptually appealing because of its simplicity. We will present its fundamentals which will then be useful in a discussion of *multiplexing*.

In synchronous AM the transmitted signal is typically expressed as:

$$f(t) = s(t) \operatorname{Cos} \omega_c t \qquad (8\text{-}73)$$

Again, assume $\omega_s \ll \omega_c$, where ω_s is the highest frequency in $s(t)$. We illustrate this with the band-limited signal whose spectrum is represented in Figure 8-31.

Then
$$F(j\omega) = \frac{S(j[\omega + \omega_c]) + S(j[\omega - \omega_c])}{2} \qquad (8\text{-}74)$$

A plot of this would be similar to the one in Figure 8-32 except for the impulses at $\omega = \pm\omega_c$. Transmission of $f(t)$ proceeds just like in the asynchronous case. The major difference occurs at the demodulation receiver side of the operation. To demodulate the received $f(t)$, we multiply by Cos $\omega_c t$ and then perform a LP filtering operation. The frequency of this cosine must be exactly the same as (synchronized to) the frequency of the cosine on the transmitting side of the operation. This synchronization problem, keep in mind, is the difficult part. Here is the way it works:

$$f(t) \operatorname{Cos} \omega_c t = s(t) \operatorname{Cos}^2 \omega_c t = s(t) \left\{ \frac{1 + \operatorname{Cos} 2\omega_c t}{2} \right\} \qquad (8\text{-}75)$$

and this has the Fourier transform:

$$\mathrm{FT}\{f(t) \operatorname{Cos} \omega_c t\} = \frac{S(j\omega)}{2} + \frac{S(j[\omega + 2\omega_c]) + S(j[\omega - 2\omega_c])}{4} \qquad (8\text{-}76)$$

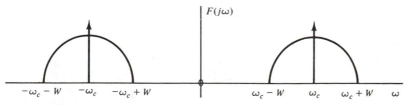

Figure 8-32   Transform of a modulated signal.

This spectrum appears as in Figure 8-33. Note that the shape of the part of the spectrum centered about the origin is exactly the same as the shape of the spectrum in Figure 8-31. This implies that we need only perform a LP filtering operation to recover the original $s(t)$. An ideal LP filter with cutoff frequency $\geq$ W will suffice. Deviations from ideality of course will produce some distortion in the received signal. This is where trade-offs become essential.

8-6-3 Multiplexing

Multiplexing is a technique of simultaneous transmission of a number of different signals over a single channel.

We consider here a type of multiplexing called frequency-division-multiplexing (FDM). In this process each input spectrum is assigned a distinct frequency band. The modulation and demodulation in an FDM system are both based on frequency translation. This is very similar to the synchronous AM discussed previously. Although large numbers of signals can be handled simultaneously, we examine only two of them. This will illustrate the procedure without too much clutter.

Imagine that we have $s_1(t)$ and $s_2(t)$ as information signals. Assume each is band-limited with Fourier transforms like those in Figure 8-34. If we modulate $s_2(t)$ with $\text{Cos } \omega_2 t$ and $s_1(t)$ with $\text{Cos } \omega_1 t$, we can form the signal $x(t)$:

$$x(t) = s_1(t) \text{ Cos } \omega_1 t + s_2(t) \text{ Cos } \omega_2 t \qquad (8\text{-}77)$$

then $$X(j\omega) = \frac{S_1(j[\omega + \omega_1]) + S_1(j[\omega - \omega_1])}{2}$$

$$+ \frac{S_2(j[\omega + \omega_2]) + S_2(j[\omega - \omega_2])}{2} \qquad (8\text{-}78)$$

which has a spectrum similar to that in Figure 8-35. It is important here to make sure that the spectra do not overlap; that is, $\omega_1 + W_1 < \omega_2 - W_2$. The signal $x(t)$ is then transmitted and received and must be demodulated to recover $s_1(t)$ and $s_2(t)$. In view of $X(j\omega)$ in Figure 8-35 to capture the spectrum of $s_1(t)$, we would need a band-pass filter centered around ω_1 and for $s_2(t)$ a band-pass filter centered around ω_2. Then, with the output of each of these band-pass filters, we would proceed as in the case of synchronous AM demodulation: Multiply by $\text{Cos } \omega_1 t$ (or $\text{Cos } \omega_2 t$) and pass through LP filters. The overall system of

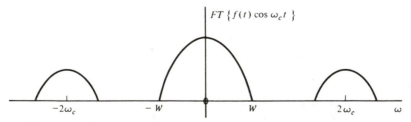

Figure 8-33   Spectrum of the synchronous demodulator.

Figure 8-34 Transform of S_1 and S_2 for the FDM system.

modulation, multiplexing, transmission, demultiplexing, and demodulation is illustrated in Figure 8-36.

Of course, the FDM system presented here is idealistic. As in the synchronous AM system, we need ideal LP filters, one with a cutoff frequency $\geq W_1$, the other with the cutoff $\geq W_2$. The ω_1 and ω_2 frequencies on the demodulation side must be perfect matches of those on the modulation side. The band-pass filters must also be ideal. The band-limitedness assumption is very troublesome in FDM systems. Since no real signals are ever band-limited, the $X(j\omega)$ spectrum will display overlap between the distinct frequency bands. This overlap results in a phenomenon called *aliasing,* which we will consider in greater detail in Chapter 9. Although many simplifying assumptions are involved in this discussion, these are the rudiments of frequency-division-multiplexing. Extension from two to n signals follows in a straightforward manner.

8-6-4 Frequency Modulation

Fourier methods can also be usefully applied to other forms of modulation. *Angle modulation* consists of two basic types: phase modulation (PM) and frequency modulation (FM). The preceding discussion on AM started with the typical carrier signal $f(t) = A \cos \omega_c t$, where $A = 1 + ms(t)$ in the asynchronous case and $A = s(t)$ in the synchronous case represented the time varying sinusondal amplitude. Now let:

$$f(t) = A \cos \theta(t) = A \cos (\omega_c t + \theta_c) \qquad (8\text{-}79)$$

where θ_c is the phase and ω_c is the frequency of the carrier. Assume A is constant. In PM the phase is modulated so that:

$$\theta_c = \theta_c(t) = \theta_0 + k_1 \mu_1(t) \qquad (8\text{-}80)$$

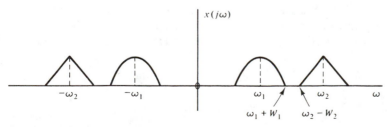

Figure 8-35 Spectrum of $x(t)$ for the FDM system.

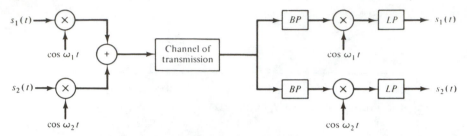

Figure 8-36   Frequency-division-multiplexing (FDM) system.

In FM the frequency is modulated. This is done by letting:

$$\theta(t) = \omega_c t + k_2 \int_0^t \mu_2(t)\, dt \tag{8-81}$$

Then the instantaneous frequency $\omega_i(t)$ is:

$$\omega_i(t) = \frac{d\theta(t)}{dt} = \omega_c + k_2\mu_2(t) \tag{8-82}$$

Since the mathematics of general angle modulation can get very complex, consider here only some basic FM ideas. Also, since even basic FM analysis is difficult, we deal with the special case where:

$$\mu_2(t) = \text{Cos } \omega_0 t \tag{8-83}$$

In spite of the simplicity of this assumption, it should give some good insight into the concept of frequency modulation. From Equation 8-81 we obtain:

$$\theta(t) = \omega_c t + \frac{k_2}{\omega_0} \text{Sin } \omega_0 t = \omega_c t + k \text{ Sin } \omega_0 t \tag{8-84}$$

Then
$$f(t) = A \text{ Cos } (\omega_c t + k \text{ Sin } \omega_0 t)$$

$$= A \text{ Re } \{e^{j(\omega_c t + k\text{Sin}\omega_0 t)}\}$$

$$= A \text{ Re } \{e^{j\omega_c k} e^{jk\text{Sin}\omega_0 t}\} \tag{8-85}$$

The second complex exponential can be expressed in terms of the Bessel functions:

$$e^{jk\text{Sin}\omega_0 t} = \sum_{n=-\infty}^{\infty} J_n(k)e^{jn\omega_0 t} \tag{8-86}$$

Bessel functions, $J_n(k)$, arise as solutions of certain kinds of differential equations. They are tabulated functions, available in any extensive tables of mathematical formulas. The Fourier transform of $f(t)$ can now be written:

$$F(j\omega) = A \text{ Re }\left\{ \sum_{n=-\infty}^{\infty} J_n(k)\delta(\omega - \omega_c - n\omega_0) \right\} 2\pi \tag{8-87}$$

and since the right-hand side is purely real, we can write:

$$F(j\omega) = A \sum_{n=-\infty}^{\infty} J_n(k)\delta(\omega - \omega_c - n\omega_0) \cdot 2\pi \tag{8-88}$$

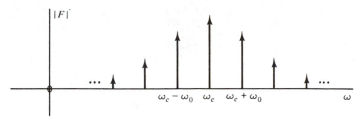

Figure 8-37 Spectrum for an FM signal.

Then the FM spectrum consists of an infinite number of impulses centered around $\omega = \omega_c$ and would appear as in Figure 8-37. Each impulse is weighted with a value obtained from Bessel functions. Such signals are called wide-band FM signals since there are now an infinite number of sidebands. The weights of the impulses representing the sidebands, however, become negligibly small after a very short excursion on either side of ω_c.

These are just the very basic ideas associated with frequency modulation. We have generated a signal that is of sufficiently high frequency to be transmitted. Then, of course, it will need to be received and demodulated. FM demodulation is a more difficult task than AM demodulation. We will leave it for a more involved study in the area of communication systems.

8-6-5 The Sampling Theorem

As a final application of Fourier theory we consider a theorem that has had an important impact on very large sectors of the technological world, especially the digital areas. The theorem discussed in this continuous time Fourier chapter can provide the introduction into the next chapter which deals with discrete time Fourier analysis. Very often discrete signals are obtained from continuous signals via a process of sampling.

There is a famous theorem called the **Shannon sampling theorem** which delimits the sampling process. It says that if $f(t) \leftrightarrow F(j\omega)$ is band-limited by $-\omega_B < \omega < \omega_B$, then $f(t)$ is recoverable from its samples $f(nT)$, $n = 0, \pm 1, \ldots$ if:

$$\omega_0 > 2\omega_B \tag{8-89}$$

where $\omega_0 = 2\pi/T$ is the sampling frequency. To illustrate what is involved here, we can employ Fourier transform ideas. Let $f(t)s(t) = g(t)$, the sampled function. Assume that $f(t)$ is continuous and that $s(t)$ is a sampling function

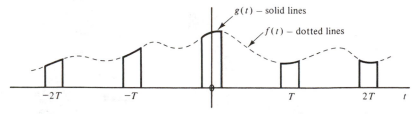

Figure 8-38 Output of a sampler.

consisting of a train of narrow pulses of amplitude 1, separated by T time units. Then $g(t)$ might appear as in Figure 8-38. Since $f(t)$ is band-limited, in order to illustrate the sampling theorem, we assume $F(j\omega)$ is as in Figure 8-39. Now since $s(t)$ is periodic, we know that it has a complex exponential Fourier series representation:

$$s(t) = \sum_{n=-\infty}^{\infty} c_n e^{jn\omega_0 t} \tag{8-90}$$

and
$$S(j\omega) = \sum_{n=-\infty}^{\infty} 2\pi c_n \delta(\omega - n\omega_0) \tag{8-91}$$

From the frequency domain convolution property we know that:

$$g(t) = f(t)s(t) \leftrightarrow G(j\omega) = \frac{F(j\omega) * S(j\omega)}{2\pi}$$

Therefore
$$G(j\omega) = \frac{1}{2\pi} \int_{-\infty}^{\infty} F(j\beta)S(j(\omega - \beta))d\beta \tag{8-92}$$

Since S from Equation 8-91 contains impulses, Equation 8-92 is easily integratable. We obtain:

$$G(j\omega) = \sum_{n=-\infty}^{\infty} c_n F(j(\omega - n\omega_0)) \tag{8-93}$$

illustrated in Figure 8-40. To recover the exact shape of $F(j\omega)$ we only need to pass $g(t)$ through an ideal LP filter whose cutoff frequency is ω_B. Note, however, that we are assuming $\omega_0 > 2\omega_B$ exactly as the sampling theorem requires. If $\omega_0 < 2\omega_B$, we get overlap in the spectrum of $G(j\omega)$. Then $f(t)$ is not recoverable. This critical frequency, ω_0, which is the sampling frequency, is also sometimes called the **Nyquist frequency.**

To conclude this section we mention an application of the sampling theorem to communication systems. Modern data communication and telephone systems often employ a technique called **time-division-multiplexing (TDM)**. In this process a large number of samples of different signals are transmitted over the same channel. Upon reception, complex synchronization equipment is needed to separate the various signals. Then from the samples the original signals can be recovered. This requires, however, a Nyquist frequency not just of ω_0, but if we multiplex, say, N signals, we will need a Nyquist frequency of $N\omega_0$. Further aspects of sampling will be considered in the next chapter when we deal with the discrete Fourier transform.

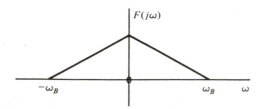

Figure 8-39 Spectrum of $f(t)$.

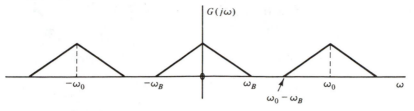

Figure 8-40 Spectrum of $g(t)$, the sampler output.

SUMMARY

This chapter considered the Fourier analysis of continuous time functions $f(t)$. The spectral content indicating the frequencies of significance contained in $f(t)$ played a large role in this chapter. For periodic $f(t)$'s the spectral content is revealed via the Fourier series analysis. We discussed the trigonometric, generalized, and exponential Fourier series. From the exponential Fourier series we developed the Fourier transform to deal with nonperiodic signals.

In the Fourier transform development a number of illustrative examples were worked out, then many properties were considered. Most of these properties were based on the defining integrals:

$$F(j\omega) = \int_{-\infty}^{\infty} f(t)e^{-j\omega t}\, dt$$

$$f(t) = \frac{1}{2\pi} \int_{-\infty}^{\infty} F(j\omega)e^{j\omega t}\, d\omega$$

Following the properties of the Fourier transform, a brief section was presented in which the Fourier and Laplace transforms were compared.

A few applications of the Fourier transform were considered in the last section. Among these applications were some brief discussions of analog filter design and amplitude and frequency modulation. Also, we discussed some multiplexing ideas: frequency- and time-division-multiplexing. Then we concluded with an introduction to the sampling theorem. Further applications of the Fourier transform will be dealt with in the next chapter when we study the Fourier analysis of discrete time signals.

PROBLEMS

8-1. Determine the trigonometric Fourier series expansion of the periodic signal $f(t)$ where $f(t) = t^2$ for $0 < t < 2\pi$ and $T = 2\pi$.

8-2. Determine the exponential Fourier series expansion of the periodic signal $f(t)$ where $f(t) = e^{-t}$ for $0 < t < 2$ and $T = 2$.

8-3. Consider the following set of basis functions, orthonormal over $0 \le t < \infty$.

$$\phi_1(t) = \sqrt{2}e^{-t}, \qquad t \ge 0$$

$$= 0, \qquad t < 0$$

and
$$\phi_2(t) = 6e^{-2t} - 4e^{-t}, \qquad t \geq 0$$
$$= 0, \qquad t < 0$$

Let
$$f(t) = 2e^{-3t}, \qquad t \geq 0$$
$$= 0, \qquad t < 0$$

Determine $\hat{f}(t) = c_1\phi_1(t) + c_2\phi_2(t)$ and examine the accuracy of this approximation using Parseval's relation.

8-4. A certain system has a system function:
$$H(j\omega) = \frac{1}{j\omega + 10}$$

Assume that the input $x(t)$ is periodic with period 2

and
$$x(t) = \begin{cases} 10, & 0 < t < 1 \\ 5, & 1 < t < 2 \end{cases}$$

Determine $y(t)$, the system output.

8-5. Consider
$$f(t) = f(t + 2)$$
$$= 1 - |t|, \qquad \text{for } -1 < t < 1$$

(a) Sketch $f(t)$, $f'(t)$, $f''(t)$. Do not forget the singularity functions that may occur.

(b) Let
$$f''(t) = \sum_{n=-\infty}^{\infty} C_n e^{jn\omega_0 t}$$

Determine the Fourier series coefficients C_n.

(c) Find the relation between the Fourier series coefficients for $f(t)$, $f'(t)$, and $f''(t)$.

(d) Express $f(t)$ in an exponential and trigonometric Fourier series.

(e) Find the Fourier transform of one period of $f(t)$, $f'(t)$, and $f''(t)$.

(f) Evaluate the exponential Fourier coefficients from the transform of a single period of $f(t)$, $f'(t)$, and $f''(t)$. Compare these with the results of parts (b) and (c).

8-6. Given the transform pair:
$$e^{-\pi t^2} \leftrightarrow e^{-\omega^2/4\pi}$$

Evaluate:

(a) $\displaystyle\int_0^\infty e^{-t^2}\, dt$

(b) $\displaystyle\int_0^\infty t^2 e^{-t^2}\, dt$

8-7. Find the Fourier transform and plot $|F(j\omega)|$ versus ω for the following
(a) $f(t) = e^{-2t}u(t) - e^{2t}u(-t)$

(b) $f(t) = \dfrac{\text{Sin } 2\pi t}{2\pi t}(4 + \text{Cos } 20t)$

(c) $f(t) = \dfrac{\text{Sin } \pi t}{\pi t}\,\text{Cos}^2\, 50t$

8-8. Consider the following periodic $f(t)$:

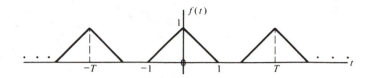

For this $f(t)$,

$$c_n = \frac{\text{Sin}^2(n\pi/T)}{n^2\pi^2/T}$$

(a) Assume T is very large so that $\text{Sin }(n\pi/T) \approx n\pi/T$. Determine T such that $c_5 = 1/100$.

(b) Now let $T = 3$ and determine the energy in $f(t)$ and the energy in

$$\hat{f}(t) = \sum_{n=-2}^{2} c_n e^{jn\omega_0 t}$$

Compare results.

8-9. Consider a periodic

$$f(t) = A, \qquad 0 < t < \frac{T}{2}$$

$$= 4, \qquad \frac{T}{2} < t < T$$

T is the period.

(a) Determine A if $c_1 = 2/\pi \,\underline{|90°}$

(b) Determine A if $c_1 = 2/\pi \,\underline{|-90°}$

(c) Determine c_2 if $A = 10$.

8-10. A certain periodic $f(t)$ has the Fourier series coefficients: $c_0 = 5$; $c_1 = 2 + 2j$; $c_{-1} = 2 + 2j$; all other c's $= 0$. We can write

$$f(t) = f_1(t) + jf_2(t),$$

a complex time function. Determine $f_1(t)$ and $f_2(t)$.

8-11. Let

$$f(t) \leftrightarrow F(j\omega) = R(\omega) + jX(\omega)$$

Assume $f(t) = 0$,   for $t < 0$. Let:

$$X(\omega) = e^{-\omega}$$

and

$$R(\omega) = \int_0^\infty g(t)\,\text{Cos }\omega t\, dt$$

Determine $g(t)$. [This problem is nontrivial. *Hint:* Solve for $R(\omega)$ as a function of $X(\omega)$.]

8-12. Determine $G(j\omega)$, where $g(t) = d^2/dt^2\,(f(t)\,\text{Cos}\,2t)$ and $g(t) \leftrightarrow G(j\omega)$:

$$f(t) \leftrightarrow F(j\omega) = \frac{5}{5 + j\omega}$$

8-13. Let $f(t)$ be two impulses:

$$f(t) = A\delta(t - t_1) + B\delta(t - t_2)$$

Assume:

$$F(j\omega) = e^{-j\omega}\,\text{Cos}\,\omega$$

Determine suitable values for A, B, t_1, and t_2.

8-14. Let:

$$f(t) = u(t) - u(t - 1) + u(t - 2) - u(t - 3)$$

(a) Determine $F(j\omega)$.
(b) Let:

$$g(t) = f(t)$$

Determine $G(j\omega)$.

8-15. Consider the following *RLC* circuit.

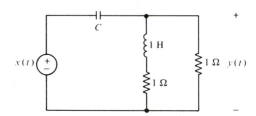

(a) Determine $H(j\omega) = Y(j\omega)/X(j\omega)$ as a function of C.
(b) Determine C such that $|H(j\omega)| = 1$ at $\omega = \sqrt{2}r/s$.
(c) Determine the phase of $H(j\omega)$ at $\omega = \sqrt{2}r/s$ and $C = 1F$.

8-16. Let $f_1(t) = \text{Cos}\,t$ and let $f_2(t)$ be as follows:

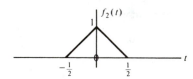

Determine the Fourier transform of:

$$f(t) = f_1(t)f_2(t)$$

8-17. Let:

$$F(j\omega) = \frac{1 + j\omega}{8 + 6j\omega + (j\omega)^2} \leftrightarrow f(t)$$

and $$\ddot{g}(t) + 2\dot{g}(t) + g(t) = f(t)$$

Determine $G(j\omega)$.

8-18. A certain system:

$$H(j\omega) = \frac{100}{(j\omega)^2 + 0.2j\omega + 100}$$

has an input:

$$x(t) = \frac{1}{2} + \frac{2}{\pi} \, \text{Sin} \, \frac{10}{3}t + \frac{2}{3\pi} \, \text{Sin} \, 10t + \frac{2}{5\pi} \, \text{Sin} \, \frac{50}{3}t + \cdots$$

Assume that the output is

$$y(t) \approx k \sin(\omega_0 t + \theta)$$

Determine k, ω_0, and θ.

8-19. Let:

$$f(t) = 3u(t) - 3u(t - 2) + 3u(t - 3) - 3u(t - 4)$$

Determine the energy spectral density for $f(t)$

and $$\int_{-\infty}^{\infty} F(j\omega)F(-j\omega) \, d\omega$$

8-20. For a certain $f(t)$:

$$F(j\omega) = \frac{1}{2 + j\omega}$$

(a) Determine $\int_{-\infty}^{\infty} f(t) \, dt$.

(b) Let:

$$F_1(j\omega) \longleftrightarrow f_1(t) = \int_{-\infty}^{t} f(\lambda) \, d\lambda$$

Determine $F_1(j\omega)$.

8-21. Let:

$$f(t) = u(t + 1) - u(t - 1)$$

(a) Show that $f(t)$ can be expressed as follows:

$$f(t) = k \int_{0}^{\infty} \frac{\text{Sin} \, \omega}{\omega} \, \text{Cos} \, \omega t \, d\omega$$

(b) Determine k.

8-22. Let:

$$f(t) = \frac{\text{Sin}^2 \, t}{t^2}$$

Using Parseval's equation, determine $\int_{-\infty}^{\infty} f(t)^2 \, dt$.

8-23. Determine the autocorrelation function $R(\tau)$ for:

$$f(t) = Au\left(t + \frac{a}{2}\right) - Au\left(t - \frac{a}{2}\right)$$

$$R(\tau) \triangleq \int_{-\infty}^{\infty} f(\lambda - \tau) f(\lambda) \, d\lambda$$

(This is just the correlation of a function with itself.)

8-24. Show that the inverse Fourier transform of $E(\omega) = F(j\omega) F(-j\omega)$ is the autocorrelation function of $f(t)$. Show that the autocorrelation function is an even function of its argument.

8-25. Let:

$$F(j\omega) = \frac{10 + 2j\omega}{(5 + 3j\omega)(2 + j\omega)}$$

Determine the Fourier transform of:
(a) $f(t) \operatorname{Sin} \omega_0 t$ (b) $f(t/5)$
(c) $t^3 f(t)$ (d) $d^3/dt^3 f(t)$

8-26. Assume $f(t) \leftrightarrow F(j\omega)$. Show that:

$$|F(j\omega)| \leq \frac{1}{|\omega|^n} \int_{-\infty}^{\infty} \left| \frac{d^n f}{dt^n} \right| dt, \qquad \text{for } n = 0, 1, 2, \cdots$$

where $|\omega|^0 = 1$ and $|d^0 f/dt^0| = |f|$.

8-27.

$$F_1(j\omega) \leftrightarrow f_1(t)$$

and

$$F_2(j\omega) \leftrightarrow f_2(t)$$

Let:

$$F_1(j\omega) = \frac{1}{(5 + j\omega)(2 + j\omega)}$$

Let:

$$F_2(j\omega) = F_1(j(\omega + \omega_0)) + F_1(j(\omega - \omega_0))$$

Determine $f_2(t)$.

8-28. Determine and sketch the autocorrelation function for $f(t)$

where

$$f(t) = 0, \qquad t < -\frac{1}{2}$$

$$= -10, \qquad -\frac{1}{2} < t < 0$$

$$= 10, \qquad 0 < t < \frac{1}{2}$$

$$= 0, \qquad t > \frac{1}{2}$$

Determine the autocorrelation function of $\bar{f}(t)$. $\bar{f}(t)$ is the periodic version of $f(t)$. Assume $T = 2$.

8-29. The power spectral density of a continuous random process $x(t)$ is defined as:

$$S_{xx}(\omega) = \int_{-\infty}^{\infty} R_{xx}(\tau)e^{-j\omega\tau} \, d\tau$$

and the cross-spectral densities for processes $x(t)$ and $y(t)$ as:

$$S_{xy}^{'}(\omega) = \int_{-\infty}^{\infty} R_{xy}(\tau)e^{-j\omega\tau} \, d\tau$$

and

$$S_{yx}(\omega) = \int_{-\infty}^{\infty} R_{yx}(\tau)e^{-j\omega\tau} \, d\tau$$

Using the properties of correlation functions, show:

(a) $S_{xx}(\omega) = S_{xx}(-\omega)$

(b) $P_{av} = x^2(t) = \int_{-\infty}^{\infty} S_{xx}(\omega) \, df$

(c) $S_{xy}(\omega) = S_{yx}^{*}(\omega)$

(d) $S_{xx}(\omega_0) \, \Delta f$ is the total average power from a narrow bandpass filter of bandwidth Δf about f_0. Demonstrate that all these properties hold for a process with autocorrelation function $R_{xx}(\tau) = e^{-2|\tau|}$ and cross-correlation function $R_{xy}(\tau) = 3e^{-|\tau-4|}$.

8-30. Given the input to a linear system with system function $H(j\omega)$ is zero-mean noise with a power spectral density $S_{xx}(\omega)$, transform the time domain results to show:

$$S_{yy}(\omega) = |H(\omega)|^2 S_{xx}(\omega)$$

$$S_{xy}(\omega) = S_{xx}(\omega)H(\omega)$$

and

$$S_{yx}(\omega) = S_{xx}(\omega)H(-\omega)$$

8-31. Given a linear system with system function $H(\omega) = 1/2 + j\omega$ has as its input white noise with actual mean square fluctuations $\overline{x^2}(t) = 20$ and power spectral density $S_{xx}(\omega) = 2$, find

(a) $S_{yy}(\omega)$, $S_{xy}(\omega)$, and $\overline{y^2}(t) = \int_{-\infty}^{\infty} S_{yy}(\omega) \, df$

(b) Do the spectral quantities have the desired properties?

8-32. Consider the frequency-division-multiplixing system:

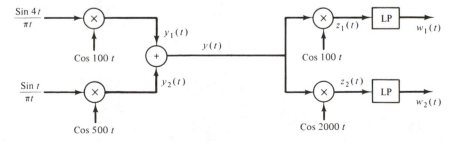

The blocks labeled LP are ideal low-pass filters. The top filter has a bandwidth $-4 \leq \omega \leq 4$ and the bottom filter has a bandwidth $-1 \leq \omega \leq 1$. Determine and plot

$Y_1(j\omega)$, $Y_2(j\omega)$, $Z_1(j\omega)$, $Z_2(j\omega)$, $w_1(t)$, and $w_2(t)$. $Y(j\omega) \leftrightarrow y(t)$ and $Z(j\omega) \leftrightarrow z(t)$.

8-33. Consider the following system:

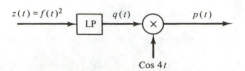

The block labeled LP is an ideal low-pass filter with bandwidth $-2 \leq \omega \leq 2$. The time function $f(t)$ has a Fourier transform $F(j\omega) = 10\{u(\omega + 2) - u(\omega - 2)\}$. Determine and plot $Z(j\omega)$, $Q(j\omega)$, $P(j\omega)$, and $q(t)$.

$$Z(j\omega) \leftrightarrow z(t), \qquad Q(j\omega) \leftrightarrow q(t), \qquad P(j\omega) \leftrightarrow p(t)$$

8-34. A low-pass filter has the system function:

$$H(j\omega) = \frac{10(10 + j\omega)}{(5 + j\omega)(20 + j\omega)}$$

Determine the cutoff frequency ω_0. Then normalize the frequency by replacing ω by ω/ω_0. Convert this normalized low-pass filter into a band-pass filter where $\omega_2 = 50$ and $\omega_1 = 40$. ω_2 and ω_1 are respectively the upper and lower cutoff frequencies. Plot the magnitude of this band-pass filter.

Chapter 9

The Discrete Fourier Transform and the Fast Fourier Transform

INTRODUCTION

In Section 8-6 we illustrated the essentials of the Fourier analysis by considering a number of applications. Most of these applications were from the communications area. Another area of engineering science that is becoming increasingly important is that of signal processing. Within this field, the digital or discrete Fourier transform is beginning to play a large role. Real signals, like radar tracks, which are often processed with the Fourier transform in order to reveal their spectral content, are typically measured at discrete points in time, resulting in discrete time signals, $f(n)$. These discrete or discretized time signals call for some kind of discrete Fourier transform (DFT).

Thus the need for a DFT arises from discrete signals. From a slightly different point of view, let us recall the definitions:

$$F(j\omega) = \int_{-\infty}^{\infty} f(t)e^{-j\omega t}\,dt \tag{9-1}$$

and

$$f(t) = \frac{1}{2\pi}\int_{-\infty}^{\infty} F(j\omega)e^{j\omega t}\,d\omega \tag{9-2}$$

The numerical computation of these integrals using digital computer processing requires that we take the continuous signals $f(t)$ and $F(j\omega)$ and discretize them. Also, we replace the integrals by finite summations. These manipulations lead directly to a discrete Fourier transform and an inverse discrete Fourier transform (IDFT). After a discussion of the DFT and the IDFT, we consider the problems of aliasing and leakage and the technique of windowing, all of which are relevant to the DFT. Then we investigate some of the DFT properties, after which we examine some efficient ways to compute the DFT and the IDFT.

349

The fast Fourier transform (FFT) is an efficient way to perform the computations called for in the DFT. Its efficiency results from minimizing the number of DFT mathematical operations. This is accomplished by taking advantage of certain periodicities that appear in the DFT. In this chapter we will focus on the development of two basic FFT algorithms: the decimation in time algorithm and the decimation in frequency algorithm.

This chapter, then, explores the mathematical basis of the DFT and the FFT. Although research into the theory and applications of the DFT and the FFT has expanded considerably since the early 1970s, there is still much important research to be done in these areas and interested readers are encouraged to consult the literature.

9-1 THE DISCRETE FOURIER TRANSFORM

Given $f(n)$, how do we determine its Fourier transform? The last chapter dealt only with continuous functions, $f(t)$. In order to arrive at a discrete Fourier transform, we follow a path that takes off from the theory of Fourier series (not transforms) and employs the duality property. Recall from Chapter 8 that if $f(t)$ is periodic

then
$$c_n = \frac{1}{T} \int_T f(t) e^{-jn\omega_0 t} \, dt \tag{9-3}$$

and
$$f(t) = \sum_{n=-\infty}^{\infty} c_n e^{jn\omega_0 t} \tag{9-4}$$

This is the complex exponential Fourier series representation and the Fourier series coefficients. This well-known pair can be represented as:

$$f(t) \leftrightarrow c_n \tag{9-5}$$

Corresponding to a periodic time function, we have a sequence of points in a discrete frequency domain that are the discrete exponential Fourier series coefficients.

Now if we start with a discrete function $f(n) = f(nT)$, then we have a discrete sequence of time points that are analogous to the discrete Fourier series coefficients. Duality ideas suggest that corresponding to the discrete time function $f(n)$ we would have a continuous frequency transform that is periodic in the frequency domain. This proves to be precisely the case.

Consider the transform pair:

$$f(t) \leftrightarrow F(j\omega) \tag{9-6}$$

To illustrate the development here, assume that we have an $f(t)$ and an $F(j\omega)$ as in Figure 9-1(a). The time function $f(t)$ is assumed to be time-limited: It is zero outside the range $-\alpha \le t \le \alpha$. The transform $F(j\omega)$ is assumed to be band-limited: It is zero outside the range $-\beta \le \omega \le \beta$. Such assumptions are fictitious, of course, because any time-limited $f(t)$ like the triangle in Figure

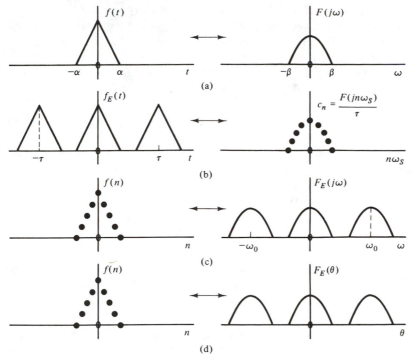

Figure 9-1 (a) Fourier transform pair; (b) periodic time function and discrete frequency function; (c) periodic frequency function and discrete time function; (d) same as (c) except frequency is normalized.

9-1(a) will always have a Fourier transform that is not band-limited. On the other hand, if $F(j\omega)$ is band-limited as in Figure 9-1(a), then the corresponding time function $f(t)$ will never be time-limited. Time-limiting and band-limiting are mutually exclusive phenomena. Although proving this in general is rather difficult, a glance at some of the famous Fourier transform pairs of the last chapter should be convincing. Note, for example, that the square pulse that is time-limited has a Sinc transform that is not band-limited. The consequences of the "useful fiction" to be employed in this development will be considered later.

Assume at first that $f(t)$ is not periodic. Now sample $F(j\omega)$ at $\omega = 0, \pm \omega_s,$ $\pm 2\omega_s, \ldots$. This yields $F(jn\omega_s)$, a discretized frequency domain function. Using this function, we can construct the frequency-domain points c_n which are actually the complex exponential Fourier series coefficients:

$$c_n = \frac{F(jn\omega_s)}{\tau} \tag{9-7}$$

Corresponding to these Fourier series coefficients we have $f_E(t)$, the periodic extension of the original $f(t)$, periodic with period τ. The functions $f_E(t)$ and c_n would appear as in Figure 9-1(b).

Next we employ duality and reverse the roles of $f_E(t)$ and c_n, the time-domain and frequency-domain functions. Instead of sampling $F(j\omega)$, we sample $f(t)$ at $t = 0, \pm T, \pm 2T, \ldots$. This yields $f(n) = f(nT)$, a discrete sequence of points that in the time domain is analogous to a discrete frequency-domain sequence of Fourier series coefficients. Employing the concept of duality, we correspond to this sequence the periodic extension of $F(j\omega)$, call it $F_E(j\omega)$. The functions $f(n)$ and $F_E(j\omega)$ would appear as in Figure 9-1(c). Assume $F_E(j\omega)$ is periodic with period ω_0 and we can write:

$$F_E(j\omega) = \sum_{n=-\infty}^{\infty} \tilde{c}_n e^{-jn\omega T} \qquad (9\text{-}8)$$

where $T = 2\pi/\omega_0$. Note that Equation 9-4 has a positive sign in its complex exponential. The negative sign in Equation 9-8 is due to the fact that an argument reversal occurs in applying the duality property.

Now let $\theta = \omega T$ and write:

$$F_E(\theta) = \sum_{n=-\infty}^{\infty} \tilde{c}_n e^{-jn\theta} \qquad (9\text{-}9)$$

The Fourier series coefficients appearing in this equation are the sampled values of $f(t)$, that is:

$$\tilde{c}_n = f(nT) = f(n) \qquad (9\text{-}10)$$

and
$$f(n) = \frac{1}{\omega_0} \int_{-\omega_0/2}^{\omega_0/2} F_E(j\omega) e^{jn\omega T} \, d\omega \qquad (9\text{-}11)$$

Since $\omega = \theta/T$, we can write $d\omega = d\theta/T$.

Then
$$f(n) = \frac{1}{2\pi} \int_{-\pi}^{\pi} F_E(\theta) e^{jn\theta} \, d\theta \qquad (9\text{-}12)$$

and
$$F_E(\theta) = \sum_{n=-\infty}^{\infty} f(n) e^{-jn\theta} \qquad (9\text{-}13)$$

Remember, $F_E(\theta)$ is periodic: $F_E(\theta + 2\pi) = F_E(\theta)$. Also, the independent variable is not ω but θ, where θ can be thought of as a normalized variable. For this reason, $F_E(\theta)$ is sometimes referred to as the "coordinate normalized Fourier transform." The functions $f(n)$ and $F_E(\theta)$ would appear as in Figure 9-1(d). Equations 9-12 and 9-13 constitute a discrete time Fourier transform pair. Some sufficiently simple discrete time functions $f(n)$ can yield a closed form solution for $F_E(\theta)$.

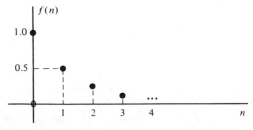

Figure 9-2   Sketch of $f(n)$ from Example 9-1.

EXAMPLE 9-1

Determine $F_E(\theta)$, the discrete time Fourier transform, for $f(n) = (\frac{1}{2})^n u(n)$. This function is sketched in Figure 9-2.

Solution

$$F_E(\theta) = \sum_{n=0}^{\infty} \left(\frac{1}{2}\right)^n e^{-jn\theta} = 1 + \frac{1}{2}e^{-j\theta} + \frac{1}{4}e^{-2j\theta} + \cdots$$

$$= \frac{1}{1 - \frac{1}{2}e^{-j\theta}} = \frac{1}{1 - \frac{1}{2}\cos\theta + \frac{1}{2}j\sin\theta}$$

$$F_E(\theta) = |F_E(\theta)| \angle \arg F_E(\theta)$$

where $\quad |F_E(\theta)| = \dfrac{1}{\sqrt{1.25 - \cos\theta}} \quad$ and

$$\arg F_E(\theta) = -\text{Tan}^{-1}\left(\frac{\sin\theta}{2 - \cos\theta}\right)$$

The magnitude and phase are plotted separately in Figure 9-3. Both functions are periodic with period 2π.

It is interesting to note that in Equation 9-13 if we let $e^{j\theta} = z$, we get

$$F_E(\theta)|_{e^{j\theta}=z} = F_E(z) = \sum_{n=-\infty}^{\infty} f(n)z^{-n} \qquad (9\text{-}14)$$

which is none other than the two-sided Z transform. Thus all of the two-sided Z transform theory applies to the discrete time Fourier transform. To determine the discrete time Fourier transform for some $f(n)$, find the two-sided Z transform of $f(n)$ and simply let $z = e^{j\theta}$.

EXAMPLE 9-2

Determine $F_E(\theta)$ for $f(n) = (\frac{1}{3})^n \cos(n\pi) u(n)$ using the Z transform theory.

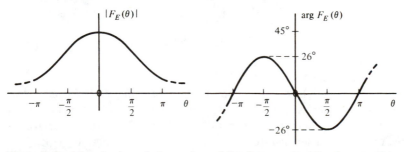

Figure 9-3 Magnitude and phase plots of the discrete time Fourier transform of $f(n) = (\frac{1}{2})^n u(n)$.

Solution. $f(n)$ has a Z transform

$$F(z) = \frac{1 - (\frac{1}{3} \text{Cos } \pi)z^{-1}}{1 - \frac{2}{3} \text{Cos } \pi \, z^{-1} + \frac{1}{9} z^{-2}}$$

$$= \frac{1 + \frac{1}{3} z^{-1}}{1 + \frac{2}{3} z^{-1} + \frac{1}{9} z^{-2}} = \frac{z(z + \frac{1}{3})}{z^2 + \frac{2}{3} z + \frac{1}{9}} = \frac{z}{z + \frac{1}{3}}$$

Then $\quad F_E(\theta) = F(z)|_{z=e^{j\theta}} = \dfrac{e^{j\theta}}{e^{j\theta} + \frac{1}{3}} = \dfrac{\text{Cos } \theta + j \text{ Sin } \theta}{\text{Cos } \theta + \frac{1}{3} + j \text{ Sin } \theta}$

whose magnitude is $3/\sqrt{10 + 6 \text{ Cos } \theta}$ and whose phase is $+\theta - \text{Tan}^{-1}$ (Sin $\theta/(\frac{1}{3} + \text{Cos } \theta))$.

Using the Z transform is fine for simple signals whose Z transforms are readily available. Unfortunately, the time functions processed by Fourier methods are typically very complicated sequences of points that are often representing some information signal hidden in noise. Closed form solutions for $F_E(\theta)$ or $F(z)$ for realistic signals are rather uncommon. Actual applications call for numerical techniques. Thus even though the Z transform ideas theoretically fit nicely into the discrete Fourier transform development, in practice, we usually resort to numerical computation to determine Fourier transforms. This indicates that we must limit our summations—as in Equation 9-13—to finite summations containing, for example, N points. Let N be an even number. Although not essential, it keeps the development simple. Let us change the infinite summation in Equation 9-13 to a finite sum from $n = 0$ to $n = N - 1$. Of course, the more points there are in the summation or the larger N is, the closer the finite sum approximates $F_E(\theta)$. To distinguish the finite sum from $F_E(\theta)$, we call it $F(\theta)$. Then we can write:

$$F(\theta) = \sum_{n=0}^{N-1} f(n)e^{-jn\theta} \tag{9-15}$$

Often, in real signals, most of the energy is confined to a finite time duration. This provides some justification for using F instead of F_E. However, Equation 9-15 is not without problems. Note that θ is still continuous. For computer computations, $F(\theta)$ and $f(n)$ need to be functions of discrete variables. We must choose values of θ for which the computation is to be performed. The variable n is already discrete. Let us sample the periodic and continuous $F(\theta)$ at N equally spaced points over one period of the transform: $\theta = 2\pi k/N$, where $k = 0, 1, \ldots, N - 1$. Then, instead of Equation 9-15, we can write $F(\theta)$ as $F(k)$

$$F(k) = \sum_{n=0}^{N-1} f(n)e^{-j(2\pi/N)kn} \tag{9-16}$$

This is the discrete Fourier transform (DFT).

So far, then, what have we accomplished? Starting with $F(j\omega)$ as the continuous Fourier transform of the continuous time function $f(t)$, we have—through a series of mathematical operations and transformations—arrived at $F(k)$ as the discrete Fourier transform of the discrete time function $f(n)$.

From the discrete time Fourier transform $F_E(\theta)$, we can return to the time domain to retrieve $f(n)$ by employing Equation 9-12. The $f(n)$ thus obtained would appear as the $f(n)$ in Figure 9-1(d). Note that this $f(n)$ is nonperiodic. Real-world Fourier calculations, however, generally employ $F(k)$, not $F_E(\theta)$. Thus Equation 9-12 for $f(n)$ will not do. Given the N samples of $F(k)$, nevertheless, the means to retrieve the time signal $f(n)$ do exist. The $f(n)$ that results from $F(k)$ turns out to be a periodic function. It is, in fact, a periodic extension of the $f(n)$ from Figure 9-1(d). Therefore the $F(k)$ and its corresponding $f(n)$ are both discrete periodic functions. They appear as in Figure 9-4. To determine $f(n)$ from $F(k)$, let $W = e^{-j(2\pi/N)}$.

Then
$$F(k) = \sum_{n=0}^{N-1} f(n) W^{nk}, \qquad k = 0, 1, \ldots, N - 1 \tag{9-17}$$

This equation can be expanded as follows:

$$F(0) = f(0) + f(1) + \cdots + f(N - 1)$$
$$F(1) = f(0) W^0 + f(1) W^1 + \cdots + f(N - 1) W^{N-1}$$
$$F(2) = f(0) W^0 + f(1) W^2 + \cdots + f(N - 1) W^{2(N-1)}$$
$$\vdots$$
$$F(N - 1) = f(0) W^0 + f(1) W^{N-1} + \cdots + f(N - 1) W^{(N-1)(N-1)}$$

$$
\begin{bmatrix} F(0) \\ F(1) \\ F(2) \\ \vdots \\ F(N-1) \end{bmatrix}
=
\begin{bmatrix}
W^0 & W^0 & W^0 & \cdots & W^{0(N-1)} \\
W^0 & W^1 & W^2 & \cdots & W^{1(N-1)} \\
W^0 & W^2 & W^4 & \cdots & W^{2(N-1)} \\
\vdots & \vdots & \vdots & & \vdots \\
W^0 & W^{N-1} & W^{2(N-1)} & & W^{(N-1)(N-1)}
\end{bmatrix}
\begin{bmatrix} f(0) \\ f(1) \\ f(2) \\ \vdots \\ f(N-1) \end{bmatrix}
\tag{9-18}
$$

These can be written as:

$$\bar{F} = \mathcal{W} \bar{f} \tag{9-19}$$

$\bar{F}$ is the N-dimensional vector on the left-hand side of Equation 9-18. $\bar{f}$ is the N-dimensional vector on the right-hand side of Equation 9-18. $\mathcal{W}$ is the $N \times N$ matrix that is multiplied by $\bar{f}$. From Equation 9-19 we get:

$$\bar{f} = \mathcal{W}^{-1} \bar{F} \tag{9-20}$$

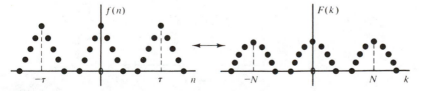

Figure 9-4   Discrete and periodic time and frequency functions.

We assume that the inverse exists and avoid those cases where the $\mathcal{W}$ matrix is singular.

EXAMPLE 9-3

Determine $\mathcal{W}^{-1}$ for the case of $N = 3$.

Solution

$$\mathcal{W} = \begin{bmatrix} 1 & 1 & 1 \\ 1 & W & W^2 \\ 1 & W^2 & W^4 \end{bmatrix} \quad \text{and} \quad W = e^{-j2\pi/3} = -\frac{1}{2} - j\frac{\sqrt{3}}{2}$$

$$\mathcal{W}^{-1} = \frac{\begin{bmatrix} W^5 - W^4 & W^2 - W^4 & W^2 - W \\ W^2 - W^4 & W^4 - 1 & 1 - W^2 \\ W^2 - W & 1 - W^2 & W - 1 \end{bmatrix}}{W^5 + 2W^2 - W - 2W^4}$$

Dividing everything through by W^3, we obtain:

$$\mathcal{W}^{-1} = \frac{\begin{bmatrix} W^2 - W & W^{-1} - W & W^{-1} - W^{-2} \\ W^{-1} - W & W - W^{-3} & W^{-3} - W^{-1} \\ W^{-1} - W^{-2} & W^{-3} - W^{-1} & W^{-2} - W^{-3} \end{bmatrix}}{W^2 + 2W^{-1} - W^{-2} - 2W}$$

Now

$$W = -\frac{1}{2} - j\frac{\sqrt{3}}{2}, \qquad W^2 = -\frac{1}{2} + j\frac{\sqrt{3}}{2}, \qquad W^3 = 1$$

$$W^{-1} = -\frac{1}{2} + j\frac{\sqrt{3}}{2}, \qquad W^{-2} = -\frac{1}{2} - j\frac{\sqrt{3}}{2}, \qquad W^{-3} = 1$$

Substituting in these values, we have:

$$\mathcal{W}^{-1} = \frac{\begin{bmatrix} j\sqrt{3} & j\sqrt{3} & j\sqrt{3} \\ j\sqrt{3} & -\frac{3}{2} - j\frac{\sqrt{3}}{2} & \frac{3}{2} - j\frac{\sqrt{3}}{2} \\ j\sqrt{3} & \frac{3}{2} - j\frac{\sqrt{3}}{2} & -\frac{3}{2} - j\frac{\sqrt{3}}{2} \end{bmatrix}}{3j\sqrt{3}}$$

$$= \frac{1}{3} \begin{bmatrix} 1 & 1 & 1 \\ 1 & -\frac{1}{2} + j\frac{\sqrt{3}}{2} & -\frac{1}{2} - j\frac{\sqrt{3}}{2} \\ 1 & -\frac{1}{2} - j\frac{\sqrt{3}}{2} & -\frac{1}{2} + j\frac{\sqrt{3}}{2} \end{bmatrix}$$

and this expression can be written as:

$$\mathcal{W}^{-1} = \frac{1}{3} \begin{bmatrix} W^0 & W^0 & W^0 \\ W^0 & W^{-1} & W^{-2} \\ W^0 & W^{-2} & W^{-4} \end{bmatrix}$$

Now, using induction, we can show that the result of Example 9-3 can be generalized for any N as:

$$\mathcal{W}^{-1} = \frac{1}{N} \begin{bmatrix} W^0 & W^0 & W^0 & \cdots & W^{0(N-1)} \\ W^0 & W^{-1} & W^{-2} & \cdots & W^{-(N-1)} \\ W^0 & W^{-2} & W^{-4} & \cdots & W^{-2(N-1)} \\ \vdots & \vdots & \vdots & & \vdots \\ W^0 & W^{-(N-1)} & W^{-2(N-1)} & \cdots & W^{-(N-1)(N-1)} \end{bmatrix} \quad (9\text{-}21)$$

Comparing this inverse to the matrix in Equation 9-18, we note that the only differences are the change in sign in the exponent of W and the multiplicative factor of $1/N$. The change in sign actually corresponds to taking a complex conjugate. Thus in view of Equation 9-16 and the matrices $\mathcal{W}$ and $\mathcal{W}^{-1}$, we can write:

$$f(n) = \frac{1}{N} \sum_{k=0}^{N-1} F(k)\, e^{\,j(2\pi/N)kn} \quad (9\text{-}22)$$

This is the inverse DFT (IDFT). The DFT is expressed by Equation 9-16. The DFT pair can be indicated by the notation:

$$f(n) \leftrightarrow F(k) \quad (9\text{-}23)$$

Expressing Equation 9-23 in terms of W, we get:

$$F(k) = \sum_{n=0}^{N-1} f(n)\, W^{kn} \quad (9\text{-}24)$$

and

$$f(n) = \frac{1}{N} \sum_{k=0}^{N-1} F(k)\, W^{-kn} \quad (9\text{-}25)$$

The similarities are striking. The minus sign of W^{-kn}, again, indicates complex conjugation. Except for this and the factor of $1/N$, the forms of the DFT and the IDFT are identical.

9-2 ALIASING AND LEAKAGE PROBLEMS

So far we have illustrated the theory by using the fiction of a time-limited, band-limited signal. Actual physical signals, however, are time-limited. A signal from a radar trace, for example, must start at some finite time t_1 and end at some finite time t_2. Time-limited signals, in reality, are never band-limited. Thus $F(j\omega)$ might appear as in Figure 9-5(a) and, consequently, $F(k)$ would appear as in Figure 9-5(b). The overlap in $F(k)$ produces a phenomenon called *aliasing*. An alias is something that stands for something else, like an assumed name that takes the place of an actual name. For our purposes, **aliasing** refers to certain frequency components that substitute for other frequency components. Figure 9-5(b) reveals that at a frequency slightly below W, for instance, there is a frequency component standing in for the component that would have been there if no overlap occurred. The actual value at that frequency consists of a contribution from the original unaliased spectrum plus a term from a frequency component slightly above W. With aliasing, information is lost, signals are corrupted, and uncertainty is introduced. The effects of this unfortunate phenomenon can be minimized, as we have seen in Chapter 8, by sampling the original $f(t)$ at a sufficiently high sampling rate. The sampling theorem implied that we should sample at a frequency at least twice the highest frequency in $F(j\omega)$. If $F(j\omega)$ is band-limited, there is some finite sampling frequency that will avoid overlap. But, again, $F(j\omega)$ is, in reality, never band-limited. Thus we would need to sample at an infinite frequency to avoid overlap. What do we do? We sample as fast as possible within the hardware constraints and try to keep the inevitable aliasing to a minimum. We could also use a low-pass filter on the $f(t)$

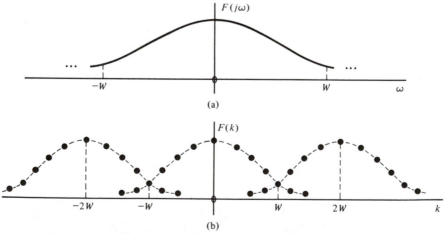

Figure 9-5 (a) Non-band-limited spectrum; (b) overlap in the DFT spectrum.

signal prior to sampling. This would reduce overlap by producing a spectrum with sharper roll-off.

Another DFT problem closely tied to the aliasing problem is that of *leakage*. Both these problems are rooted in the time-limited nature of real physical signals. The abrupt discontinuities in time-domain signals due to starting and stopping a data record produce frequency spectra that typically have main-lobes containing most of the spectral information, as well as side-lobes in which information is lost. Side-lobes are spurious frequency peaks that detract from information contained in the main-lobe. Minimizing the side-lobes will minimize the information that is lost in the side-lobes. The information lost in the side-lobes is called **leakage.** The standard method of side-lobe minimization employs the technique of *windowing*. **Windowing** smooths the abruptness of data record discontinuities.

How does the idea of windows fit here? A window not only limits a view but frames and shapes it as well. The data are always cut off or framed by a window. The phenomenon x under observation is observed from some time t_1 to some time t_2. In a typical DFT processing event, a signal x is available from t_1 to t_2: $x(t_1)$ to $x(t_2)$. In a discrete setting we can arrange these values as $x(n)$ from $n = 0$ to $N - 1$. Before delving too far into the discrete theory here, let us consider an example from continuous time. Continuous time representations seem to have a larger appeal to the uninitiated imagination. We will return shortly to the discrete world. This particular example is very simple, but nicely illustrates the general theory.

EXAMPLE 9-4

Given:

$$g(t) = \text{Cos } \omega_0 t \leftrightarrow G(j\omega) = \pi[\delta(\omega - \omega_0) + \delta(\omega + \omega_0)]$$

Let $w(t)$ be the unity gain square pulse:

$$w(t) = u(t + T) - u(t - T)$$

where
$$W(j\omega) = 2\,\frac{\text{Sin } \omega T}{\omega}$$

Solution. Then $f(t) = g(t)w(t)$ can be considered to be a "windowed" version of $g(t)$. $f(t)$ is a truncated cosine. This could be a data record over a finite time interval of a pure cosine function. Taking the Fourier transform of $f(t)$, we would convolve $G(j\omega)$ and $W(j\omega)$ to get:

$$F(j\omega) = \left[\frac{\text{Sin } (\omega - \omega_0)\,T}{\omega - \omega_0} + \frac{\text{Sin } (\omega + \omega_0)\,T}{\omega + \omega_0}\right]$$

Note the plots of $G(j\omega)$, $W(j\omega)$, and $F(j\omega)$ in Figure 9-6. The effect of the window is to replace the impulses of the original function with transforms shaped like the transform of the window (Sin x/x), except that they are located at the points where the impulses occur. Looking only at $F(j\omega)$, we can deduce the presence of a sinusoid at $\omega = \omega_0$ in this simple case. The

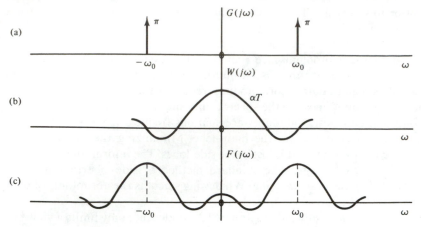

Figure 9-6   (a) Unwindowed Fourier transform; (b) transform of the window; (c) transform of the windowed function.

$F(j\omega)$ function peaks at exactly those frequencies, $\pm\omega_0$, where $G(j\omega)$ has its impulses. However, if $g(t)$ were composed of a sum of many sinusoids of many different frequencies, then $F(j\omega)$ would consist of many Sin x/x shapes centered at these different frequencies. The overlap problem becomes obvious. The peak value of one Sin x/x term, for example, might be so small that it gets lost in the side-lobes of a neighbor. Estimating the frequencies and amplitudes of the sinusoids in $g(t)$ from a data record $f(t)$ is called the **spectral estimation problem,** a large area of research within the field of signal processing. One straightforward tactic to increase the chance of estimating the sinusoids in $g(t)$ is to modify the window function $w(t)$ such that its transform is sharper or has lower side-lobes. For example, if $w(t)$ were chosen to be a triangle pulse instead of the square pulse, then its transform takes the form $(\text{Sin } x/x)^2$ which has lower side-lobes than Sin x/x. Many different windows have been proposed. Some of these will be considered shortly.

Now in the discrete world, let us assume we have samples of some time record $f(n)$, n = 0, 1, ..., N − 1. We could employ Equation 9-24 and compute the DFT without further ado. This would correspond to using a uniform or rectangular window $w(n) = 1$, $n = 0, 1, ..., N − 1$. Then $f(n) = g(n)w(n)$, where $g(n)$ is the *actual* time function that in principle could extend in time from minus to plus infinity. Data from an actual finite time record generally are gathered as $g(n)$ from $n = 0$ to $n = N − 1$. Without windowing, or with a uniform window, we would just let $f(n)$ be these $g(n)$ values. As Example 9-4 indicates, however, it is advantageous to employ nonuniform windows, especially if we want to know the spectral content of a signal. If a triangular window is used, for example, then the first points in the data record are *weighted* slightly, the middle points are weighted mostly, and the last points are also weighted slightly. To illustrate this weighting procedure, consider the following example.

EXAMPLE 9-5

Given:

$$g(n) = [\ldots, 5, 2, 1, \underset{\underset{n\,=\,0}{\uparrow}}{4}, 3, 4, 2, 2, 1, 1, 2, \underset{\underset{n\,=\,8}{\uparrow}}{1}, 1, 0, 2, \ldots]$$

is a data record to be processed from $n = 0$ to $n = 8$. Employ a triangular weighting

where $$w(n) = [0.00, 0.25, 0.50, 0.75, 1.00, 0.75, 0.50, 0.25, 0.00]$$
$$\underset{n\,=\,0}{\uparrow} \qquad\qquad\qquad\qquad\qquad\qquad\qquad\qquad \underset{n\,=\,8}{\uparrow}$$

Determine the values of $f(n)$ to be processed.

Solution.   The values of $f(n)$ to be processed are $f(n) = g(n)w(n)$.

Thus $$f(n) = [0.00, 0.75, 2.00, 1.50, 2.00, 0.75, 0.50, 0.50, 0.00]$$
$$\underset{n\,=\,0}{\uparrow} \qquad\qquad\qquad\qquad\qquad\qquad\qquad\qquad\qquad \underset{n\,=\,8}{\uparrow}$$

Now to consider some of the more common windows, assume that $w(n)$ is an even function with the origin as the point of symmetry. Since there is a point at $n = 0$, evenness will require an odd number of points for discrete window functions. However, we assume that $f(n)$ has an even number of points. This turns out to be convenient for the fast Fourier transform development that occurs later in the chapter. We also assume, in fact, that $N = 2^P$ where P is an integer. It seems problematic to have $w(n)$ with an odd number of points and $f(n)$ with an even number of points. This even–odd problem is resolved by recalling that $w(n)$ is a periodically extended function so that its first and last points are the same. Therefore we consider windows centered about the origin and having $N + 1$ points, where N is even. Then in an actual problem, since $w(n)$ is periodic, it can be shifted such that its left point coincides with the origin and the right point can be deleted. Remember that such a shift only contributes a phase shift term and leaves the magnitude of the frequency response unaffected.

In addition to the square and triangular windows, some of the other frequently used windows are: the Hann window, the Hamming window, the Gaussian window, the Blackman window, and the Dolph-Chebyschev window. There are also many varieties within these basic types. Since we must keep our discussion brief, we examine only a few of these.

9-2-1 The Hann Window

Analytically this window is described by the equation:

$$w(n) = 0.5 + 0.5 \cos\frac{2n\pi}{N}, \qquad n = \frac{-N}{2}, \ldots, -1, 0, 1, \ldots, \frac{N}{2} - 1 \qquad (9\text{-}26)$$

These kinds of windows are sometimes called "cosine on a pedestal" windows. The cosine is superimposed on a uniform or rectangular window. The result of this configuration is very low side-lobes in the frequency domain. Why this is the case can be seen if the Hann window is compared to the rectangular window. The Sin x/x pattern of the rectangular window has rather large side-lobes. The truncated cosine term in the Hann window produces a series of Sin x/x terms that are displaced from the origin in such a way that their peaks tend to cancel the side-lobe terms in the "pedestal" part of the Hann window. This is not magic but merely superposition.

EXAMPLE 9-6

Determine and plot the DFT for the Hann window if $N = 8$. Compare these results to the DFT for a corresponding rectangular window.

Solution. For the Hann window, let us first plot $w(n)$ shifted so that its left point coincides with the origin.
Let:

$$\overline{w}(n) = 0.5 + 0.5 \, \mathrm{Cos} \, \frac{n\pi}{4}, \qquad n = -4, -3, -2, -1, 0, 1, 2, 3,$$

which we plot as in Figure 9-7(a). Now shift $\overline{w}(n)$ to the right four units to form $w(n)$ as in Figure 9-7(b). Using $w(n)$, we compute the DFT $W(k)$:

$$W(k) = \sum_{n=0}^{7} w(n) W^{nk}, \qquad \text{where } W = e^{-j(2\pi/N)}$$

$$= e^{-j(\pi/4)} = \frac{1-j}{\sqrt{2}}$$

$$W(0) = 4.00 \qquad W(2) = W(3) = W(4) = W(5) = W(6) = 0$$

$$W(1) = -2.0 \quad \text{and} \quad W(7) = -2.0$$

And since these functions are periodic, $W(8) = 4$, and so on. We can plot $W(k)$ as in Figure 9-8. Now for the rectangular window, let $w_R(n) = 1$, $n = 0, 1, \ldots, 7$.

Then

$$W_R(k) = 8, \qquad k = 0$$

$$= 0, \qquad k = 1, 2, 3, 4, 5, 6, 7$$

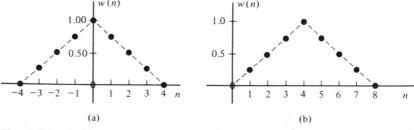

(a)   (b)

Figure 9-7   (a) Plot of Hann window for $N = 8$; (b) Hann window shifted.

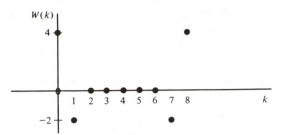

Figure 9-8   Plot of DFT for the Hann window.

We plot $W_R(k)$ as in Figure 9-9. Now observing the DFT plots in Figures 9-8 and 9-9, we seem to have results that contradict our discussion of the Hann window. From Figure 9-8, it appears that the Hann window exhibits some side-lobe behavior due to the values of -2.0 at $k = 1$ and $k = 7$. The rectangular window DFT in Figure 9-9 appears to exhibit no side-lobes at all. Why is there a discrepancy? The DFT of the rectangular window seems to have no side-lobes because the sample rate is such that we sample exactly at the zeros of the actual rectangular DFT. We know the rectangular window has a Sin x/x type transform. These patterns do have rather large side-lobe levels.

If instead of computing $W(k)$ for these two windows we were to first determine $W(\theta)$ and plot $W(\theta)$ versus θ, then we would see the side-lobes displayed in a continuous fashion. $W(\theta)$ for the Hann window and $W_R(\theta)$ for the rectangular window would appear as in Figure 9-10. Notice that the main-lobe for the Hann window is wider than the main-lobe for the rectangular window. The side-lobes for the Hann window, however, are very low.

Before we leave this Hann window to consider some other windows, a few general comments are in order. For small values of N (like $N = 8$) we should not expect $W(\theta)$ and $W(k)$ to be very similar. But since θ is sampled at $\theta = 2\pi k/N$, for very large values of N, the number of sample values of $W(\theta)$ in the range $-\pi < \theta < \pi$ becomes very large as well. Thus $W(k)$ should look more like $W(\theta)$ for large N. Now sometimes the side-lobes in these windows are reduced so much that they hardly appear at all on a linear plot of $W(k)$ or $W(\theta)$. For this reason a dB scale is often used to plot the DFT magnitudes. The dB level of the first or largest side-lobe is often used as a measure of window quality. But since lowering

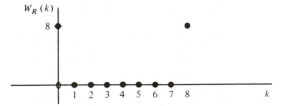

Figure 9-9   Plot of DFT for the rectangular window.

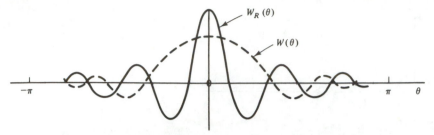

Figure 9-10   Plots of $W(\theta)$ and $W_R(\theta)$, the Hann and rectangle window transforms.

the side-lobes increases the width of the main-lobe, trade-offs are practically always necessary. The most useful windows in practice appear to be those exhibiting 60 dB or more side-lobe suppression while at the same time maintaining a main-lobe width that does not exceed about four times the width of the rectangular window's main-lobe. For comparison purposes, the rectangular window for large N exhibits a first side-lobe level of approximately 13 dB down from the main-lobe and the Hann window has a first side-lobe of 32 dB down.

9-2-2 The Gaussian Window

This window appears like a bell-shaped Gaussian distribution curve in the discrete time domain. The exact shape of the Gaussian window depends on a standard deviation parameter. One particular Gaussian window is described by the equation:

$$w(n) = \exp\left(-4.5(2n/N)^2\right), \qquad n = \frac{-N}{2}, \ldots, -1, 0, 1, \ldots, \frac{N}{2} - 1 \quad (9\text{-}27)$$

Compared to the Hann window, this Gaussian window has lower side-lobes— approximately 55 dB down—but has a wider main-lobe.

9-2-3 The Dolph-Chebyschev Window

This window is unique in that it provides uniform side-lobes and will have the minimum main-lobe width for a specified side-lobe level and specified number of data points. The use of this window in the signal processing area actually grew out of the radar design area. Radar designers have used ideas similar to windowing in the design of linear phased arrays. The current distribution across the face of a phased array forms a Fourier transform pair with the far-field pattern produced by the phased array radar. Dolph-Chebyschev concepts have been used in this context for many years.

Now the analytical description of the Dolph-Chebyschev window can be rather complicated because, in general, it employs Chebyschev polynomials. These polynomials can be expressed by the equation:

$$T_m(w) = \text{Cos}\,(m\,\text{Cos}^{-1}\,w) \qquad (9\text{-}28)$$

Then, for example:

$$T_4(w) = 1 - 8w^2 + 8w^4$$

For a particular case in which we want side-lobe levels of 30 dB down from the main-lobe and for which we have $N = 100$ data points, we can write the DFT of the Dolph-Chebyschev window as follows:

$$W(k) = \text{Cos}\left(100 \, \text{Cos}h^{-1}\left(\text{Cos}\,\frac{k\pi}{50}\right)\right) \qquad (9\text{-}29)$$

In general, the use of any of these windows will increase the chances of estimating the frequencies and amplitudes of sinusoids contained in a given data record. Which window should one use? There is no straightforward answer to this question.

Sometimes different windows are used with the same data record to see which window yields the minimal leakage. Occasionally, knowing beforehand what the data record is like will dictate the proper window. Because of these ambiguities, window selection is often considered more an art than a science. To conclude this window discussion, we note that the use of windowing occurs not only in DFT spectral estimation but in other areas as well, in particular, in the design and analysis of digital filters.

Aliasing and leakage are the most serious DFT problems. The ideal, of course, would be to have the DFT be equivalent to the continuous Fourier transform. We must content ourselves, however, with approximations. Aliasing and leakage corrupt our approximations. In this section we indicated the best way to deal with these corruptions. To eliminate the corruption due to aliasing, we need to sample the original signal at a rate greater than twice the highest frequency in the signal. Often this frequency is not known, in which case we sample at the highest practical rate. To deal with leakage, use windows. The more sophisticated windows require more computation time. As in most engineering problems, trade-offs are in order.

Assume now that we have $f(n)$ in hand. These data points have been gathered at the highest possible sampling rate and windowing has been done. We are ready then to return to Equations 9-24 and 9-25. To assist in the performance of the DFT and inverse DFT operations, we can employ a number of DFT properties. We will consider these next.

9-3 DFT PROPERTIES

Like the continuous Fourier transform of the previous chapter, there are a number of properties of the DFT that can facilitate some analytical tasks. The most important of the DFT properties are summarized in Table 9-1. Since a number of operations relating different time functions or different frequency functions are involved, we consider only two functions, $x(n)$ and $y(n)$, which have DFTs $X(k)$ and $Y(k)$, respectively.

TABLE 9-1 DFT PROPERTIES

Property	Data sequence representation	Transform sequence representation
Discrete Fourier transform	$x(n)$	$X(k)$
Linearity	$ax(n) + by(n)$	$aX(k) + bY(k)$
Periodicity of data and transform sequences	$x(n + lN)$, $l, m =$ $\ldots, -1, 0, 1, \ldots$	$X(k + mN)$
Horizontal axis sign change	$x(-n)$	$X(-k)$
Complex conjugation	$x^*(n)$	$X^*(-k)$
Data sequence sample shift	$x(n \pm n_0)$	$e^{\pm j2\pi k n_0/N} X(k)$
Single sideband modulation	$e^{\pm j2\pi k_0 n/N} x(n)$	$X(k \mp k_0)$
Double sideband modulation	$[\mathrm{Cos}\,(2\pi k_0 n)]\,x(n)$	$\frac{1}{2}[X(k + k_0) + X(k - k_0)]$
Data sequence circular convolution	$x(n) * y(n)$	$X(k)\,Y(k)$
Transform sequence circular convolution	$x(n)\,y(n)$	$X(k) * Y(k)$
Arithmetic correlation	$x(n) * y^*(-n)$	$X(k)\,Y^*(k)$
Arithmetic autocorrelation	$x(n) * x^*(-n)$	$\|X(k)\|^2$
Data sequence convolution	$\tilde{x}(n) * \tilde{y}(n)$ (augmented sequences)	$\tilde{X}(k)\,\tilde{Y}(k)$
Transform sequence convolution	$\tilde{x}(n)\,\tilde{y}(n)$	$\tilde{X}(k) * \tilde{Y}(k)$ (augmented sequences)
Symmetry	$(1/N)X(n)$	$x(-k)$
Parseval's theorem	$\displaystyle\sum_{n=0}^{N-1} \|x(n)\|^2$ equals	$\displaystyle\frac{1}{N}\sum_{k=0}^{N-1} \|X(k)\|^2$

Now since many of the DFT properties are very similar to the properties of the continuous Fourier transform presented in Chapter 8, we consider here only a few of the more important ones. These will be illustrated by examples.

EXAMPLE 9-7

Demonstrate the periodicity of the DFT.

Solution. Let:

$$X(k) = \sum_{n=0}^{N-1} x(n)\,W^{nk}$$

be the DFT of $x(n)$,

then

$$X(k + mN) = \sum_{n=0}^{N-1} x(n)\,W^{(k+mN)n} = \sum_{n=0}^{N-1} x(n)\,W^{nk}\,W^{mnN}$$

but

$$W^{mnN} = e^{-j(2\pi/N)mnN} = e^{-j2\pi mn}$$

which equals 1 as long as nm is an integer. Since n is an integer, mn is an integer if m is an integer.

Therefore $X(k + mN) = X(k)$,   for all $m = \ldots -1, 0, 1, \ldots$

that is, the DFT is periodic with period N.

EXAMPLE 9-8

Determine the DFT for $x(n) = [4, 3, 2, 1]$ and $N = 4$. Then demonstrate the horizontal axis sign change property and use it to determine the DFT for

$$y(n) = 4\delta(n) + 3\delta(n + 1) + 2\delta(n + 2) + \delta(n + 3) \quad \text{where} \quad N = 4.$$

Solution

$$x(n) \leftrightarrow X(k) = \sum_{n-0}^{N-1} x(n) W^{nk} = \text{DFT} \, [x(n)]$$

$$\text{DFT} \, [x(-n)] = \sum_{n-0}^{N-1} x(-n) W^{nk}$$

Letting $n = -m$, we get:

$$\text{DFT} \, [x(-n)] = \sum_{m-0}^{1-N} x(m) W^{-mk}$$

If we let $N = 8$ and expand this summation, we obtain:

$$\text{DFT} \, [x(-n)] = x(0) W^0 + x(-1) W^k + \cdots + x(-7) W^{7k}$$

But recall that $x(n)$ and W^{nk} are periodic:

$$W = e^{-j(2\pi/N)} = e^{-j(\pi/4)}$$

in this case

and $\qquad W^{7k} = W^{-k}, W^{6k} = W^{-2k}, \ldots, W^k = W^{-7k}$

Also, we can write $x(-1) = x(7), x(-2) = x(6), \ldots, x(-7) = x(1)$.

Therefore $\qquad \text{DFT} \, [x(-n)] = x(0) W^0 + x(1) W^{-k} + x(2) W^{-2k}$

$$+ \cdots + x(N - 1) W^{-k(N-1)}$$

$$= \sum_{n-0}^{N-1} x(n) W^{-nk}$$

and comparing this with $X(k)$, we write:

$$\text{DFT} \, [x(-n)] = X(-k)$$

therefore $\qquad x(-n) \leftrightarrow X(-k)$

Now for $\qquad x(n) = [4, 3, 2, 1] \quad \text{and} \quad N = 4,$

$$W = e^{-j(\pi/2)} = -j$$

$$X(k) = \sum_{n-0}^{3} x(n) W^{nk} = x(0) W^0$$

$$+ x(1) W^k + x(2) W^{2k} + x(3) W^{3k}$$

$$X(0) = 10, X(1) = 4 + 3(-j)$$
$$+ 2(-1) + j = 2 - 2j$$
$$X(2) = 4 + 3(-j)^2 + 2(-j)^4 + (-j)^6 = 2$$
$$X(3) = 4 + 3(-j)^3 + 2(-j)^6 + (-j)^9 = 2 + 2j$$

and since $X(k)$ is periodic, $X(4) = X(0) = 10$, and so on.

Also, $X(-1) = X(3) = 2 + 2j$, $X(-2) = X(2) = 2$,

$$X(-3) = X(1) = 2 - 2j$$

The function $y(n) = 4\delta(n) + 3\delta(n + 1) + 2\delta(n + 2) + \delta(n + 3)$ is $x(-n)$. According to the horizontal axis sign change property, we know $y(n) = x(-n) \leftrightarrow X(-k) = Y(k)$.

Therefore $Y(k) = 10$,   for $k = 0$

$$= 2 + 2j,   \text{for } k = 1$$

$$= 2,   \text{for } k = 2$$

$$= 2 - 2j,   \text{for } k = 3$$

We can check this by computing $Y(k)$ as:

$$Y(k) = \sum_{n=0}^{3} y(n) W^{nk} = y(0) W^0 + y(1) W^k + y(2) W^{2k} + y(3) W^{3k}$$

$$y(n) = [\underset{\underset{n=-3}{\uparrow}}{1}, 2, 3, 4]$$

so to get $y(0), \ldots, y(3)$, we invoke the periodic nature of $y(n)$, that is, $y(0) = 4, y(1) = 1, y(2) = 2, y(3) = 3$.

$$Y(0) = 10,   Y(1) = 4 + W + 2W^2 + 3W^3$$

$$= 4 - j - 2 + 3j = 2 + 2j$$

$$Y(2) = 4 + (-j)^2 + 2(-j)^4 + 3(-j)^6 = 2$$

and $Y(3) = 4 + (-j)^3 + 2(-j)^6 + 3(-j)^9 = 2 - 2j$

which checks with $Y(k)$ which is computed with the sign change property.

EXAMPLE 9-9

Demonstrate the convolution properties and use them to determine the convolution of $x(n) = [0, 1, 0, 1]$ and $y(n) = [0, 0, 0, 1]$ for $N = 4$.

Solution. There are two types of convolution employed with the DFT. **Circular convolution** (or periodic convolution) is used for periodic sequences, and **noncircular convolution** (or aperiodic convolution) is used for aperiodic sequences. Now the dilemma here is that from the DFT theory we have seen that our sequences are always periodic with period N, for example, $x(n + N) = x(n)$. This arises from the definition of $x(n)$ as the IDFT. The mechanisms of the DFT and IDFT give rise to periodic $x(n)$ and $X(k)$ functions. However, in reality, is the $x(n)$ we are dealing with periodic? Usually it is not. But if $x(n)$ is periodic and $y(n)$ is periodic, then $z(n) = x(n)*y(n)$ has a transform $Z(k) = X(k)Y(k)$, where $z(n)$, $Z(k)$, $X(k)$, $Y(k)$ are also all periodic with period N

and
$$z(n) = x(n)*y(n) = \frac{1}{N}\sum_{k-0}^{N-1} Z(k)W^{-nk}$$

$$= \frac{1}{N}\sum_{k-0}^{N-1} X(k)Y(k)W^{-nk}$$

$$= \frac{1}{N}\sum_{k-0}^{N-1}\left\{\sum_{n-0}^{N-1} x(n)W^{nk}\right\}\left\{\sum_{n-0}^{N-1} y(n)W^{nk}\right\}W^{-nk}$$

Change the indices on the second and third summations to p and q:

$$z(n) = \frac{1}{N}\sum_{k-0}^{N-1}\sum_{p-0}^{N-1}\sum_{q-0}^{N-1} x(p)y(q)W^{k(p+q)}W^{-nk}$$

$$= \frac{1}{N}\sum_{p-0}^{N-1}\sum_{q-0}^{N-1} x(p)y(q)\sum_{k-0}^{N-1} W^{k(p+q-n)}$$

but we know that:

$$\sum_{k-0}^{N-1} x^k = \frac{1-x^N}{1-x}$$

Therefore
$$\sum_{k-0}^{N-1} W^{k(p+q-n)} = \frac{1-W^{N(p+q-n)}}{1-W^{p+q-n}} = \frac{1-e^{-j2\pi(p+q-n)}}{1-e^{-j(2\pi/N)(p+q-n)}}$$

If $p + q - n \neq 0$,

then
$$1 - e^{-j2\pi(p+q-n)} = 0 \quad \text{and} \quad \sum_{k-0}^{N-1} = 0$$

But if $p + q - n = 0$,

then
$$\sum_{k-0}^{N-1} 1 = N$$

Thus
$$z(n) = \frac{N}{N}\sum_{p-0}^{N-1}\sum_{q-0}^{N-1} x(p)y(q),$$

but $p + q - n = 0$ or $q = n - p$ and we can eliminate the summation over q:

Therefore
$$z(n) = \sum_{p-0}^{N-1} x(p)y(n-p) = x(n)*y(n) \leftrightarrow X(k)Y(k)$$

This is periodic convolution. To illustrate, we let:

$$x(n) = [0, 1, 0, 1] \quad \text{and} \quad y(n) = [0, 0, 0, 1].$$

Then $z(n) = \displaystyle\sum_{p=0}^{3} x(p)y(n - p)$

$\quad = x(0)y(n) + x(1)y(n - 1)$

$\qquad + x(2)y(n - 2) + x(3)y(n - 3)$

$\quad = x(1)y(n - 1) + x(3)y(n - 3) = y(n - 1) + y(n - 3)$

$z(0) = y(-1) + y(-3) = 1, \qquad z(1) = y(0) + y(-2) = 0$

$z(2) = y(1) + y(-1) = 1, \qquad z(3) = y(2) + y(0) = 0$

$z(n) = [1, 0, 1, 0]$

Now for aperiodic convolution we must extend each function $x(n)$ and $y(n)$ by adding zero values in order to get no overlap in the intervals outside the given values. Imagine that the sequences are of length $N/2$, instead of N. Then form the sequences:

$$\hat{x}(n) = \left[x(0), x(1), \ldots, x\left(\frac{N}{2} - 1\right), 0, \ldots, 0\right]$$

$$\hat{y}(n) = \left[y(0), y(1), \ldots, y\left(\frac{N}{2} - 1\right), 0, \ldots, 0\right]$$

which are of length N. To both $x(n)$ and $y(n)$ we simply append $N/2$ zeros. Then $\hat{z}(n) = \hat{x}(n)*\hat{y}(n)$ is a periodic sequence of length N. Also, DFT $[\hat{z}(n)] = $ DFT $[\hat{x}(n)*\hat{y}(n)] = \hat{X}(k)\hat{Y}(k)$, where $\hat{X}(k)$ and $\hat{Y}(k)$ are the DFTs of $\hat{x}(n)$, $\hat{y}(n)$. All these functions, again, are periodic with period N. The periodic nature of these functions follows from the DFT and IDFT mechanisms. But by appending these zeros, we get a "true" convolution; that is, when the time sequences are shifted in the convolution operation, we get none of the overlap that periodic replication imposes. Let:

$\quad x(n) = [0, 1, 0, 1], y(n) = [0, 0, 0, 1] \quad$ and $\quad N = 8.$

Then $\hat{x}(n) = [0, 1, 0, 1, 0, 0, 0, 0], \hat{y}(n) = [0, 0, 0, 1, 0, 0, 0, 0]$

and $\hat{z}(n) = \displaystyle\sum_{p=0}^{7} \hat{x}(p)\hat{y}(n - p)$

$\quad = \hat{x}(0)\hat{y}(n) + \hat{x}(1)\hat{y}(n - 1) + \hat{x}(2)\hat{y}(n - 2)$

$\qquad + \hat{x}(3)\hat{y}(n - 3) + \hat{x}(4)\hat{y}(n - 4) + \hat{x}(5)\hat{y}(n - 5)$

$\qquad + \hat{x}(6)\hat{y}(n - 6) + \hat{x}(7)\hat{y}(n - 7)$

$\quad = \hat{y}(n - 1) + \hat{y}(n - 3)$

$\hat{z}(0) = \hat{y}(-1) + \hat{y}(-3) = 0, \qquad \hat{z}(1) = \hat{y}(0) + \hat{y}(-2) = 0$

$\hat{z}(2) = \hat{y}(1) + \hat{y}(-1) = 0, \qquad \hat{z}(3) = \hat{y}(2) + \hat{y}(0) = 0$

$\hat{z}(4) = \hat{y}(3) + \hat{y}(1) = 1, \qquad \hat{z}(5) = \hat{y}(4) + \hat{y}(2) = 0$

$$\hat{z}(6) = \hat{y}(5) + \hat{y}(3) = 1, \qquad \hat{z}(7) = \hat{y}(6) + \hat{y}(4) = 0$$

$$\hat{z}(n) = [0, 0, 0, 0, 1, 0, 1, 0]$$

Note how this is considerably different from $z(n)$.

We avoid convolution, of course, by doing multiplication of the DFTs. To obtain $z(n)$, get $Z(k) = X(k) Y(k)$ and invert.

$$X(k) = \sum_{n=0}^{3} x(n) W^{nk} = x(0) + x(1) W^{k}$$

$$+ x(2) W^{2k} + x(3) W^{3k}, \qquad W = -j$$

$$= (-j)^{k} + (-j)^{3k}$$

$$X(0) = 2, \quad X(1) = 0, \quad X(2) = -2, \quad X(3) = 0,$$

$$X(k) = [2, 0, -2, 0]$$

Also $\qquad Y(k) = y(3) W^{3k} = (-j)^{3k}$

$$Y(0) = 1, \quad Y(1) = j, \quad Y(2) = -1, \quad Y(3) = -j,$$

$$Y(k) = [1, j, -1, -j]$$

$$Z(k) = [2, 0, 2, 0] \quad \text{and} \quad z(n) = \tfrac{1}{4} \sum_{k=0}^{3} Z(k) W^{-nk}$$

$$z(n) = \tfrac{1}{4} [2 + 2 W^{-2n}] \rightarrow z(0) = 1,$$

$$z(1) = 0, \quad z(2) = 1, \quad z(3) = 0$$

$$z(n) = [1, 0, 1, 0]$$

which checks with the previous result. Finally:

$$\hat{Z}(k) = \hat{X}(k) \hat{Y}(k)$$

$$\hat{X}(k) = \sum_{n=0}^{7} \hat{x}(n) W^{nk} = W^{k} + W^{3k} \quad \text{and} \quad W = e^{-j\pi/4}$$

since $\qquad N = 8$

$$\hat{X}(0) = 2, \quad \hat{X}(1) = -j\sqrt{2}, \quad \hat{X}(2) = 0, \quad \hat{X}(3) = -j\sqrt{2},$$

$$\hat{X}(4) = -2, \quad \hat{X}(5) = j\sqrt{2}, \quad \hat{X}(6) = 0, \quad \hat{X}(7) = j\sqrt{2}$$

Therefore $\quad \hat{X}(k) = [2, -j\sqrt{2}, 0, -j\sqrt{2}, -2, j\sqrt{2}, 0, j\sqrt{2}]$

And $\qquad \hat{Y}(k) = W^{3k}.$

$$\hat{Y}(0) = 1, \quad \hat{Y}(1) = \frac{-1-j}{\sqrt{2}}, \quad \hat{Y}(2) = j,$$

$$\hat{Y}(3) = \frac{1-j}{\sqrt{2}}, \quad \hat{Y}(4) = -1,$$

$$\hat{Y}(5) = \frac{1+j}{\sqrt{2}}, \quad \hat{Y}(6) = -j, \quad \hat{Y}(7) = \frac{j-1}{\sqrt{2}}$$

Thus $\qquad \hat{Y}(k) = \left[1, \dfrac{-1-j}{\sqrt{2}}, j, \dfrac{1-j}{\sqrt{2}}, -1, \dfrac{1+j}{\sqrt{2}}, -j, \dfrac{j-1}{\sqrt{2}}\right]$

Therefore $\quad \hat{Y}(k)\hat{X}(k) = \hat{Z}(k) = [2, j - 1, 0, -1 - j,$
$$2, j - 1, 0, -1 - j]$$

$$\hat{z}(n) = \frac{1}{8}\sum_{k=0}^{7}\hat{Z}(k)W^{-nk} = \frac{1}{8}[2$$

$$+ (j - 1)W^{-n} + (-1 - j)W^{-3n}$$

$$+ 2W^{-4n} + (j - 1)W^{-5n} + (-1 - j)W^{-7n}]$$

and $\qquad \hat{z}(0) = \dfrac{1}{8}[2 + (j - 1) + (-1 - j)$

$$+ 2 + (j - 1) + (-1 - j)] = 0$$

$$\hat{z}(1) = \frac{1}{8}\left[2 + (j - 1)\frac{(1 + j)}{\sqrt{2}} + (-1 - j)\frac{(-1 + j)}{\sqrt{2}} - 2\right.$$

$$\left. + (j - 1)\frac{(-1 - j)}{\sqrt{2}} + \frac{(-1 - j)}{\sqrt{2}}(1 - j)\right] = 0$$

$$\hat{z}(2) = \frac{1}{8}[2 + (j - 1)j + (-1 - j)(-j) + 2$$

$$+ (j - 1)j + (-1 - j)(-j)] = 0$$

$$\hat{z}(3) = \frac{1}{8}\left[2 + (j - 1)\frac{(-1 + j)}{\sqrt{2}}\right.$$

$$+ (-1 - j)\frac{(1 + j)}{\sqrt{2}} + 2(-1)$$

$$\left. + (j - 1)\frac{(1 - j)}{\sqrt{2}} + (-1 - j)\frac{(-1 - j)}{\sqrt{2}}\right] = 0$$

$$\hat{z}(4) = \frac{1}{8}[2 + (j - 1)(-1) + (-1 - j)(-1)$$

$$+ 2(1) + (j - 1)(-1) + (-1 - j)(-1)] = 1$$

$$\hat{z}(5) = \frac{1}{8}\left[2 + (j - 1)\frac{(-1 - j)}{\sqrt{2}}\right.$$

$$+ (-1 - j)\frac{(1 - j)}{\sqrt{2}} + 2(-1)$$

$$\left. + (j - 1)\frac{(1 + j)}{\sqrt{2}} + (-1 - j)\frac{(j - 1)}{\sqrt{2}}\right] = 0$$

$$\hat{z}(6) = \frac{1}{8}[2 + (j-1)(-j) + (-1-j)(j)$$

$$+ 2(1) + (j-1)(j) + (-1-j)j] = 1$$

$$\hat{z}(7) = \frac{1}{8}\left[2 + (j-1)\frac{(1-j)}{\sqrt{2}}\right.$$

$$+ (-1-j)\frac{(-1-j)}{\sqrt{2}} + 2(-1)$$

$$\left. + (j-1)\frac{(j-1)}{\sqrt{2}} + (-1-j)\frac{(1+j)}{\sqrt{2}}\right] = 0$$

Therefore $\hat{z}(n) = [0, 0, 0, 0, 1, 0, 1, 0]$

which checks with the previous result.

Since most of the other properties are analogous to the continuous Fourier transform properties, we end our discussion here. Some other properties will be dealt with as problems at the end of the chapter. We turn now to consider the fast Fourier transform.

9-4 THE FAST FOURIER TRANSFORM

The fast Fourier transform (FFT) is an algorithm or a procedure with which the discrete Fourier transform can be computed using far fewer calculations. For many signal processing operations, computational requirements using the FFT can be reduced considerably. Although the FFT has had a long and interesting history (see Brigham, 1974, pp. 8–9), it was a paper by Cooley and Tukey (1965) that really put the FFT on the map. Since their famous work, which heralded a major technological breakthrough, there have been hundreds of papers written on the FFT. Fields as diverse as seismology, radar, astronomy, linear systems, optics, quantum physics, neurology, and communications have benefited from the FFT, mainly because the DFT requires N^2 complex multiply and add operations, whereas the FFT needs only aproximately $N/2 \log_2 N$ operations. For

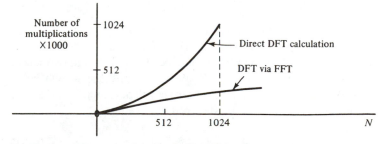

Figure 9-11 Computation for direct DFT and DFT via FFT.

low values of N these numbers are not very different, but for large N the difference becomes dramatic. (See Figure 9-11.)

Numerous FFT algorithms and variations on these algorithms are presented in Brigham's (1974) book and, more recently, at a more advanced level, in Elliott and Rao's (1982) work, both containing very extensive bibliographies. Our approach to the FFT will be limited to an introduction. We will consider what are called *power-of-2 FFT algorithms* and focus on the two types: *Decimation-in-time* algorithms and *decimation-in-frequency* algorithms.

9-4-1 The Decimation-in-Time Algorithm

Assume that the number of points in the data sequence is a power of 2: $N = 2^r$ where r is a positive integer. The number of points N will be repeatedly divided by 2 in order to facilitate the algorithm. Algorithms for arbitrary N values have been developed and are even more efficient than power-of-2 algorithms; however, they are more complicated and will not be explored here.

Let us write $F(k)$ as follows:

$$F(k) = \sum_{n=0}^{N-1} f(n) W^{nk}, \qquad k = 0, 1, \ldots, N - 1$$

$$= \underbrace{\sum_{n=0}^{N-1} f(n) W^{nk}}_{\text{even } n} + \underbrace{\sum_{n=0}^{N-1} f(n) W^{nk}}_{\text{odd } n} \qquad (9\text{-}30)$$

Then this expression can be written as:

$$F(k) = \sum_{n=0}^{N/2-1} f_1(n) W^{2kn} + \sum_{n=0}^{N/2-1} f_2(n) W^{k(2n+1)} \qquad (9\text{-}31)$$

where $f_1(n) = f(2n)$ and $f_2(n) = f(2n + 1)$ are, respectively, the even and odd components in $f(n)$, the original sequence to be transformed. Now let:

$$F_1(k) = \sum_{n=0}^{N/2-1} f_1(n) W^{2kn} \quad \text{and} \quad F_2(k) = \sum_{n=0}^{N/2-1} f_2(n) W^{2kn}$$

and we can write:

$$F(k) = F_1(k) + W^k F_2(k) \qquad (9\text{-}30)$$

which expresses the original N point DFT as a summation of two $N/2$ point DFTs with the second multiplied by W^k. The DFT $F(k)$ has N points $F(0)$, $F(1), \ldots, F(N - 1)$. Imagine breaking this sequence of N points into two sets or sequences, the first being $F(0), F(1), \ldots, F(N/2 - 1)$ and the second being $F(N/2), F(N/2 + 1), \ldots, F(N - 1)$. At this point in the development, the FFT advantage comes clearly to light. It can be shown that:

$$F\left(k + \frac{N}{2}\right) = F_1(k) - W^k F_2(k) \qquad (9\text{-}31)$$

Comparing Equations 9-30 and 9-31, we observe that, except for a minus sign, their right-hand sides are equal. The same information used in Equation 9-30 to calculate $F(k)$ for $k = 0, \ldots, N/2 - 1$ can be used in Equation 9-31 to calculate $F(k)$ for $k = N/2, N/2 + 1, \ldots, N - 1$, where the latter calculation is $F(k + N/2)$ for $k = 0, 1, \ldots, N/2 - 1$. Varying k only over the first half of its range $k = 0, 1, \ldots, N/2 - 1$, and using Equations 9-30 and 9-31, we can calculate all N points of $F(k)$.

EXAMPLE 9-10

Show how Equation 9-31 follows from Equation 9-30.

Solution.　From Equation 9-30 we can write $F(k + N/2) = F_1(k + N/2) + W^{k+N/2} F_2(k + N/2)$.

But　　$$F_1(k + N/2) = \sum_{n=0}^{N/2-1} f_1(n) W^{2n(k+N/2)} = \sum_{n=0}^{N/2-1} f_1(n) W^{2kn} W^{nN}$$

$$= \sum_{n=0}^{N/2-1} f_1(n) W^{2kn} = F_1(k)$$

since　　$$W^{nN} = e^{-j(2\pi/N)(nN)} = e^{-j2n\pi} = 1$$

Likewise:

$$F_2\!\left(k + \frac{N}{2}\right) = F_2(k)$$

Also,

$$W^{k+N/2} = W^k W^{N/2} = W^k e^{-j\pi} = -W^k$$

Substituting, we get Equation 9-31.

Now the operations involved in Equations 9-30 and 9-31 can be expressed in a convenient signal flow representation called a "butterfly," which is illustrated in Figure 9-12. This butterfly arrangement can be expanded gradually backward to form a "cocoon" type lattice. Similarly, each of $F_1(k)$ and $F_2(k)$ can be treated as was $F(k)$. Since $F_1(k)$ and $F_2(k)$ are $N/2$ point sequences, we will break each of these $N/2$ point sequences up into two $N/4$ point sequences:

$$F_1(k) = F_3(k) + W^{2k} F_4(k), \qquad k = 0, 1, \ldots, \frac{N}{4} - 1 \qquad (9\text{-}32)$$

$$F_1\!\left(k + \frac{N}{4}\right) = F_3(k) - W^{2k} F_4(k), \qquad k = 0, 1, \ldots, \frac{N}{4} - 1 \qquad (9\text{-}33)$$

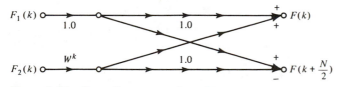

Figure 9-12　Butterfly computation scheme.

Then the $N/4$-point sequences $F_3(k)$ and $F_4(k)$ can be split in two, and so on, until we have only 1-point sequences remaining. An 8-point FFT can be used to illustrate this procedure.

EXAMPLE 9-11

Construct an 8-point decimation-in-time (DIT) FFT butterfly signal flow representation.

Solution

$$F(k) = F_1(k) + W^k F_2(k), \qquad k = 0, 1, 2, 3$$

$$F\left(k + \frac{N}{2}\right) = F_1(k) - W^k F_2(k), \qquad k = 0, 1, 2, 3$$

$F(k)$ is an 8-point DFT. $F_1(k)$ and $F_2(k)$ are both 4-point DFTs, each of which can be separately decomposed:

$$F_1(k) = F_3(k) + W^{2k} F_4(k), \qquad k = 0, 1$$

$$F_1\left(k + \frac{N}{4}\right) = F_3(k) - W^{2k} F_4(k), \qquad k = 0, 1$$

and

$$F_2(k) = F_5(k) + W^{2k} F_6(k), \qquad k = 0, 1$$

$$F_2\left(k + \frac{N}{4}\right) = F_5(k) - W^{2k} F_6(k), \qquad k = 0, 1$$

Now $F_3(k)$, $F_4(k)$, $F_5(k)$, $F_6(k)$ are each 2-point DFTs that can be expressed as sums and differences of the original data.

$$F_3(k) = \sum_{n=0}^{N/4-1} f_3(n) W^{4kn}, \qquad \text{where } f_3(n) = f_1(2n) = f(4n)$$

$$F_4(k) = \sum_{n=0}^{N/4-1} f_4(n) W^{4kn}, \qquad \text{where } f_4(n) = f_1(2n + 1) = f(4n + 2)$$

Because F_3 and F_4 both relate to F_1, the time functions f_3 and f_4 both relate to f_1. Recall that $f_1(n) = f(2n)$. In a similar manner:

$$F_5(k) = \sum_{n=0}^{N/4-1} f_5(n) W^{4kn}, \qquad F_6(k) = \sum_{n=0}^{N/4-1} f_6(n) W^{4kn}$$

where $f_5(n) = f_2(2n) = f(4n + 1)$

and

$$f_6(n) = f_2(2n + 1) = f(4n + 3)$$

Now if we carry out the expressions for F_3, F_4, F_5, and F_6, we obtain:

$$F_3(k) = f_3(0) W^0 + f_3(1) W^{4k}, \qquad \text{for } k = 0, 1$$
$$\text{and } W^4 = e^{-j\pi} = -1$$

Also

$$F_4(k) = f_4(0) W^0 + f_4(1) W^{4k}$$

$$F_5(k) = f_5(0) W^0 + f_5(1) W^{4k}$$

$$F_6(k) = f_6(0) W^0 + f_6(1) W^{4k}$$

Thus $F_3(0) = f(0) + f(4)$, $F_4(0) = f(2) + f(6)$

$F_3(1) = f(0) - f(4)$, $F_4(1) = f(2) - f(6)$

$F_5(0) = f(1) + f(5)$, $F_6(0) = f(3) + f(7)$

$F_5(1) = f(1) - f(5)$, $F_6(1) = f(3) - f(7)$

which can be viewed as four butterfly arrangements with gains of 1.0 and -1.0. The total 8-point FFT can be arranged as in Figure 9-13. Note that the F_1 and F_2 equations can take the form:

$$F_1(0) = F_3(0) + W^0 F_4(0)$$

$$F_1(1) = F_3(1) + W^2 F_4(1), \qquad \text{where } W^0 = 1$$
$$\text{and } W^2 = e^{-j\pi/2} = -j$$

$$F_1(2) = F_3(0) - W^0 F_4(0)$$

$$F_1(3) = F_3(1) - W^2 F_4(1)$$

and $F_2(0) = F_5(0) + W^0 F_6(0)$

$$F_2(1) = F_5(1) + W^2 F_6(1)$$

$$F_2(2) = F_5(0) - W^0 F_6(0)$$

$$F_2(3) = F_5(1) - W^2 F_6(1)$$

And, finally, the F equations take the form:

$$F(0) = F_1(0) + W^0 F_2(0) \qquad W^0 = 1$$

$$F(1) = F_1(1) + W^1 F_2(1) \qquad W^1 = 1/\sqrt{2} - j1/\sqrt{2}$$

$$F(2) = F_1(2) + W^2 F_2(2) \qquad W^2 = -j$$

$$F(3) = F_1(3) + W^3 F_2(3) \qquad W^3 = -1/\sqrt{2} - j1/\sqrt{2}$$

and $F(4) = F_1(0) - W^0 F_2(0)$

$$F(5) = F_1(1) - W^1 F_2(1)$$

$$F(6) = F_1(2) - W^2 F_2(2)$$

$$F(7) = F_1(3) - W^3 F_2(3)$$

The FFT computations start from the given time samples and proceed through r stages, where $2^r = N$. In Example 9-11 with $N = 8$, we had three butterfly stages. A 16-point FFT would require four butterfly stages; a 256-point FFT would need eight butterfly stages, and so on. After completion of each stage of computation, the results can be stored in the registers that held the previous stage's information. The inputs to each butterfly stage are used only once, which saves a lot of space in the computer's memory.

There is an interesting symmetry that occurs in the arrangement of the input data in these decimation-in-time FFT algorithms. Consider the input order for the 8-point FFT of the previous example: $f(0), f(4), f(2), f(6), f(1), f(5)$,

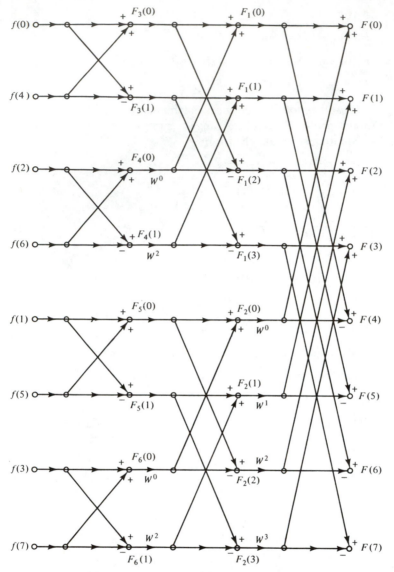

Figure 9-13   Butterfly signal flow graph for 8-point DIT FFT (all paths have unity gain unless otherwise indicated).

$f(3), f(7)$. This is the order in which the first butterfly stage calls for the data. It appears rather jumbled, the numbers in parentheses are not data values but locations of data in the data array. These locations are given addresses in computer memory. If we indicate these addresses in binary form, we can make the following correspondences:

$$f(0) \rightarrow 000$$
$$f(4) \rightarrow 100$$

$$f(2) \rightarrow 010$$
$$f(6) \rightarrow 110$$
$$f(1) \rightarrow 001$$
$$f(5) \rightarrow 101$$
$$f(3) \rightarrow 011$$
$$f(7) \rightarrow 111$$

In a typical signal processing situation, the data are gathered in a natural order, for example, $f(0), f(1), f(2), f(3), f(4), f(5), f(6), f(7)$. For this natural order we can make the following correspondences:

$$f(0) \rightarrow 000$$
$$f(1) \rightarrow 001$$
$$f(2) \rightarrow 010$$
$$f(3) \rightarrow 011$$
$$f(4) \rightarrow 100$$
$$f(5) \rightarrow 101$$
$$f(6) \rightarrow 110$$
$$f(7) \rightarrow 111$$

Now send the numbers in this data array reversed to the addresses indicated by the binary numbers. This technique of rearrangement is called **binary reversal.** After this reversal the data are in the proper order for the FFT processing.

EXAMPLE 9-12

Using the 8-point FFT for $f(n) = 1, n = 0, \ldots, 7$, determine F_3, F_4, F_5, F_6, then F_1, F_2 and, finally, $F(k)$ for $k = 0, \ldots, 7$.

Solution

$F_3(0) = 2$	$F_4(0) = 2$	$F_5(0) = 2$	$F_6(0) = 2$
$F_3(1) = 0$	$F_4(1) = 0$	$F_5(1) = 0$	$F_6(1) = 0$

then

$$F_1(0) = 4 \quad \text{and} \quad F_2(0) = 4$$
$$F_1(1) = 0 \qquad\qquad F_2(1) = 0$$
$$F_1(2) = 0 \qquad\qquad F_2(2) = 0$$
$$F_1(3) = 0 \qquad\qquad F_2(3) = 0$$

and $\qquad F(0) = 8$

$$F(1) = F(2) = F(3) = F(4) = F(5) = F(6) = F(7) = 0$$

The DFT pair $f(n) \leftrightarrow F(k)$ represents a mapping between N points in the discrete time domain and N points in the discrete frequency domain. But recall that in the DFT development, the actual functions $f(n)$ and $F(k)$ were two periodic discrete sequences. Tracing that development backward, we arrive at the aperiodic discrete time sequence $f(n)$ and the periodic continuous frequency function $F_E(\theta)$. Stepping back further still, through the duality property, we arrive at the periodic continuous time function $f_E(t)$ and the aperiodic discrete frequency function C_n. In light of this trace we can explain the results of Example 9-12 by viewing $f(n)$ as a periodically replicated and continuous $f_E(t)$ which becomes a constant: $f_E(t) = 1$ for all t. Such a function has a frequency domain c_n as a single function c_0 with all $c_n = 0$ for $n \neq 0$.

EXAMPLE 9-13

Using the 8-point FFT for $f(n) = [1, 1, 1, 1, 0, 0, 0, 0]$, determine $F(k)$ for $k = 0, \ldots, 7$.

Solution

$$F_3(0) = 1 \qquad F_4(0) = 1 \qquad F_5(0) = 1 \qquad F_6(0) = 1$$

$$F_1(0) = 2 \qquad F_2(0) = 2$$

$$F_3(1) = 1 \qquad F_4(1) = 1 \qquad F_5(1) = 1 \qquad F_6(1) = 1$$

$$F_1(2) = 0 \qquad F_2(2) = 0 \qquad\qquad F_1(1) = 1 - j \qquad F_2(1) = 1 - j$$

$$W^0 = 1, \qquad W^1 = 1/\sqrt{2} - j1/\sqrt{2}, \qquad W^2 = -j,$$

$$W^3 = -1/\sqrt{2} - j1/\sqrt{2}, \qquad F_1(3) = 1 + j, \qquad F_2(3) = 1 + j$$

$$F(0) = 4, \qquad F(1) = 1 - j\left(\frac{2}{\sqrt{2}} + 1\right), \qquad F(2) = 0,$$

$$F(3) = 1 + j(1 - 2/\sqrt{2})$$

$$F(4) = 0, \qquad F(5) = 1 + j\left(\frac{2}{\sqrt{2}} - 1\right), \qquad F(6) = 0,$$

$$F(7) = 1 + j(1 + 2/\sqrt{2})$$

Plotting the magnitude of these terms, we get a plot similar to that in the following sketch.

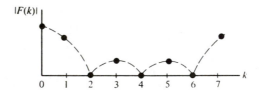

This is the familiar Sin x/x pattern that we had with the Fourier series coefficients when the time function was a pulse. Interestingly, if we determined $F(k)$ for $f(n) = [0, 0, 1, 1, 1, 1, 0, 0]$, we would get the same magnitude plot as shown here, although the phase of the $F(k)$ terms would change. This is explained by the time shift property of the DFT.

9-4-2 The Inverse FFT

If we return to the previous example and imagine that the complex conjugate of the outputs are inputs to an FFT algorithm, we will obtain some interesting results. We can write F^* as an array:

$$
\begin{bmatrix}
4 \\
1 + j(2/\sqrt{2} + 1) \\
0 \\
1 - j(1 - 2/\sqrt{2}) \\
0 \\
1 - j(2/\sqrt{2} - 1) \\
0 \\
1 - j(1 + 2/\sqrt{2})
\end{bmatrix}
$$

Let these function as the $f(n)$ inputs to our 8-point FFT; that is, $f(0) = 4, f(1) = 1 + j(2/\sqrt{2} + 1)$, and so on. Tracing through the butterflies in the flow graph of Figure 9-13, we get as outputs the following array:

$$
\begin{bmatrix}
8 \\
8 \\
8 \\
8 \\
0 \\
0 \\
0 \\
0
\end{bmatrix}
$$

Notice that if we divide this array by $N = 8$, we have as outputs of the FFT the array that was previously the inputs; that is, $f(n) = [1, 1, 1, 1, 0, 0, 0, 0]$. In this way the FFT algorithm can function as an inverse FFT (IFFT) algorithm. To be totally general, we would have to take the complex conjugate of the output. But this is seldom necessary since $f(n)$ is typically a real sequence of data points.

To formalize this result, take the complex conjugate of both sides of Equation 9-25

$$
f^*(n) = \left(\frac{1}{N} \sum_{k=0}^{N-1} F(k)W^{-kn} \right)^*
$$

$$
= \frac{1}{N} \sum_{k=0}^{N-1} F^*(k)W^{kn} \tag{9-34}
$$

Now change the k to n and the n to k:

$$
f^*(k) = \frac{1}{N} \sum_{n=0}^{N-1} F^*(n)W^{kn} \tag{9-35}
$$

Compare this summation to the DFT summation for $F(k)$ in Equation 9-24. Except for the N factor, they are the same if $F^*(n)$ replaces $f(n)$.

Then
$$f(k) = \left(\frac{1}{N} \sum_{n=0}^{N-1} F^*(n) W^{kn} \right)^* \qquad (9\text{-}36)$$

Thus to use the FFT to calculate the IFFT, that is, to determine the time sequence using FFT algorithms:

1. Apply the complex conjugate of the transform array to the FFT input.
2. Multiply the FFT outputs by $1/N$.
3. Take the complex conjugate (if necessary) of the output array.

9-4-3   The Decimation-in-Frequency Algorithm

Let us write $F(k)$ as follows:

$$F(k) = \sum_{n=0}^{N/2-1} f(n) W^{nk} + \sum_{n=N/2}^{N-1} f(n) W^{nk}, \qquad k = 0, 1, \ldots, N - 1 \qquad (9\text{-}37)$$

If we change the index of summation in the second term, we can write:

$$F(k) = \sum_{n=0}^{N/2-1} \left[f(n) W^{nk} + f\left(n + \frac{N}{2}\right) W^{(n+N/2)k} \right]$$

but
$$W^{(N/2)k} = e^{-j\pi k} = (-1)^k \qquad (9\text{-}38)$$

so we have:

$$F(k) = \sum_{n=0}^{N/2-1} \left(f(n) + f\left(n + \frac{N}{2}\right) \right) W^{nk}, \qquad \text{for } k = 0, 2, 4, 6, \ldots \qquad (9\text{-}39)$$

$$F(k) = \sum_{n=0}^{N/2-1} \left(f(n) - f\left(n + \frac{N}{2}\right) \right) W^{nk}, \qquad \text{for } k = 1, 3, 5, 7, \ldots \qquad (9\text{-}40)$$

It is from these equations that the "decimation-in-frequency" idea arises. We note that the frequency function $F(k)$ is broken up into even and odd components. In the "decimation-in-time" FFT considered earlier, recall that the time function $f(n)$ was broken up into even and odd

If we now make the definitions:

$$f_1(n) = f(n) + f\left(n + \frac{N}{2}\right) \qquad (9\text{-}41)$$

$$f_2(n) = \left[f(n) - f\left(n + \frac{N}{2}\right) \right] W^n \qquad (9\text{-}42)$$

then we can write:

$$F(2k) = \sum_{n=0}^{N/2-1} f_1(n) W^{2nk} \qquad (9\text{-}43)$$

$$F(2k + 1) = \sum_{n=0}^{N/2-1} f_2(n) W^{2nk} \qquad (9\text{-}44)$$

both of which are good for $k = 0, 1, \ldots, N/2 - 1$. Then we can write each of these equations as two summations by dividing the interval $n = 0$ to $N/2 - 1$ into two intervals $n = 0$ to $N/4 - 1$ and $n = N/4$ to $N/2 - 1$. We demonstrate this procedure, as before, with an 8-point FFT.

EXAMPLE 9-14

Solution

$$F(2k) = \sum_{n=0}^{3} f_1(n) W^{2nk}, \qquad k = 0, 1, 2, 3$$

$$F(2k + 1) = \sum_{n=0}^{3} f_2(n) W^{2nk}, \qquad k = 0, 1, 2, 3$$

Now write $F(2k) = \sum_{n=0}^{1} f_1(n) W^{2nk} + \sum_{n=2}^{3} f_1(n) W^{2nk}$

and $\quad F(2k + 1) = \sum_{n=0}^{1} f_2(n) W^{2nk} + \sum_{n=2}^{3} f_2(n) W^{2nk}$

where $\quad f_1(n) = f(n) + f(n + 4), \qquad f_2(n) = [f(n) - f(n + 4)] W^n$

Now we can write the second summation in $F(2k)$, $F(2k + 1)$ as summations from $n = 0$ to $n = 1$ so that:

$$F(2k) = \sum_{n=0}^{1} f_1(n) W^{2nk} + f_1(n + 2) W^{2(n+2)k}, \qquad k = 0, 1, 2, 3$$

$$F(2k + 1) = \sum_{n=0}^{1} f_2(n) W^{2nk} + f_2(n + 2) W^{2(n+2)k}, \qquad k = 0, 1, 2, 3$$

Now define:

$$f_3(n) = f_1(n) + f_1(n + 2)$$
$$f_4(n) = [f_1(n) - f_1(n + 2)] W^{2n}$$

and

$$f_5(n) = f_2(n) + f_2(n + 2)$$
$$f_6(n) = [f_2(n) - f_2(n + 2)] W^{2n}$$

and we can write:

$$F(4k) = \sum_{n=0}^{1} f_3(n) W^{4kn} = f_3(0) + f_3(1) W^{4k}, \qquad k = 0, 1$$

$$F(4k + 2) = \sum_{n=0}^{1} f_4(n) W^{4kn} = f_4(0) + f_4(1) W^{4k}, \qquad k = 0, 1$$

$$F(4k + 1) = \sum_{n=0}^{1} f_5(n) W^{4kn} = f_5(0) + f_5(1) W^{4k}, \qquad k = 0, 1$$

$$F(4k + 3) = \sum_{n=0}^{1} f_6(n) W^{4kn} = f_6(0) + f_6(1) W^{4k}, \qquad k = 0, 1$$

Thus
$$F(0) = f_3(0) + f_3(1)$$
$$F(4) = f_3(0) + f_3(1)W^4, \qquad W^4 = -1$$
$$= f_3(0) - f_3(1)$$
$$F(2) = f_4(0) + f_4(1)$$
$$F(6) = f_4(0) - f_4(1)$$
$$F(1) = f_5(0) + f_5(1)$$
$$F(5) = f_5(0) - f_5(1)$$
$$F(3) = f_6(0) + f_6(1)$$
$$F(7) = f_6(0) - f_6(1)$$

and
$$f_3(0) = f_1(0) + f_1(2), \qquad f_3(1) = f_1(1) + f_1(3)$$
$$f_4(0) = [f_1(0) - f_1(2)]W^0,$$
$$f_4(1) = [f_1(1) - f_1(3)]W^2, \qquad W^2 = -j$$
$$f_5(0) = f_2(0) + f_2(2), \qquad f_5(1) = f_2(1) + f_2(3)$$
$$f_6(0) = [f_2(0) - f_2(2)]W^0, \quad f_6(1) = [f_2(1) - f_2(3)]W^2$$

and
$$f_1(0) = f(0) + f(4)$$
$$f_1(1) = f(1) + f(5)$$
$$f_1(2) = f(2) + f(6)$$
$$f_1(3) = f(3) + f(7)$$

and
$$f_2(0) = [f(0) - f(4)]W^0$$
$$f_2(1) = [f(1) - f(5)]W$$
$$f_2(2) = [f(2) - f(6)]W^2$$
$$f_2(3) = [f(3) - f(7)]W^3$$

The total 8-point DIF FFT can be arranged as in Figure 9-14.

It is interesting to compare the DIT and DIF algorithms. The DIF algorithm has an advantage in that the data are input in their natural order, whereas the DIT algorithm requires that the data be put in the proper order for processing by performing a binary reversal. However, the DIF algorithm has its outputs in a jumbled order. These values can also be put in their natural order by binary reversal, if so desired.

EXAMPLE 9-15
Determine the DFT, then the DIF FFT, for the $f(n)$ used in Example 9-12. Compare the results.

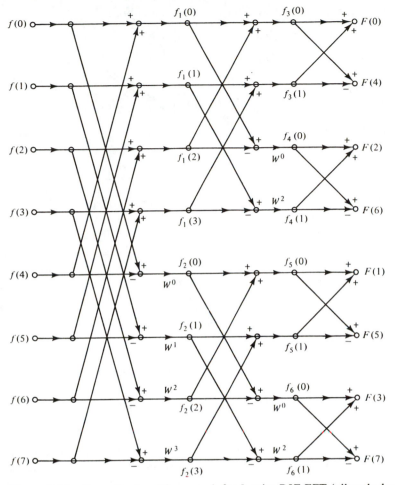

Figure 9-14 Butterfly signal flow graph for 8-point DIF FFT (all paths have unity gain unless otherwise indicated).

Solution

$$F(k) = \sum_{n=0}^{N-1} f(n) W^{kn} = \sum_{n=0}^{N-1} W^{kn} = \frac{1 - W^{kN}}{1 - W^k}$$

but

$$W^{kN} = e^{-j2\pi k} = 1.$$

Therefore the numerator $= 0$ and $F(k) = 0$. Except when? If $k = 0$, we get:

$$\Sigma = \sum_{n=0}^{N-1} W^0 = \sum_{n=0}^{N-1} (1)^n = N = 8$$

Therefore $\quad F(k) = [8, 0, 0, 0, 0, 0, 0, 0]$

Now tracing through the DIF FFT algorithm, we obtain:

$f_1(0) = 2$	$f_2(0) = 0$	$f_3(0) = 4$	$f_5(0) = 0$
$f_1(1) = 2$	$f_2(1) = 0$	$f_3(1) = 4$	$f_5(1) = 0$
$f_1(2) = 2$	$f_2(2) = 0$	$f_4(0) = 0$	$f_6(0) = 0$
$f_1(3) = 2$	$f_2(3) = 0$	$f_4(1) = 0$	$f_6(1) = 0$

and finally, we have:

$$F(0) = 8$$

$$F(4) = F(2) = F(6) = F(1) = F(5) = F(3) = F(7) = 0$$

Both these approaches yield the same result we obtained with the DIT FFT. The number of computations involved in these approaches is roughly similar. This is verified by the plot in Figure 9-11, which shows that the tremendous advantage of either DIF or DIT FFT algorithms over the direct DFT calculation becomes apparent at larger values of N. The value of $N = 8$ is too small to make much of a difference.

9-5 APPLICATION OF THE FFT

The FFT is typically applied in the modern world of engineering in almost every situation where Fourier transforms are employed. Thus for FFT applications consider the discussion of applications from the previous chapter and use the FFT in place of the continuous time Fourier transform. However, recall that we did not always want to compute Fourier transforms. Often we used Fourier transforms in a theoretical sense to describe concepts involved in the communication theory or in the filter theory. In those discussions the FFT would not be needed. Yet, when it comes to real-world engineering practices, we frequently are required to compute transforms rather than use the transform theory to develop or demonstrate abstract concepts. A knowledge of the basics of the FFT, then, seems more and more essential for today's engineering work.

One place where the FFT is useful is in digital filter design. We know that the output of a digital filter is expressed by the convolution summation:

$$y(n) = \sum_{i=-\infty}^{\infty} x(i)h(n-i) \tag{9-45}$$

where $x(n)$ is the input and $h(n)$ is the unit pulse response. If $x(n)$ and $h(n)$ are finite data sequences, nonzero from $n = 0$ to $N - 1$, then we can write:

$$y(n) = \sum_{i=0}^{N-1} x(i)h(n-i) \tag{9-46}$$

To implement this digital filter as a software package, we could simply calculate $y(n)$ from Equation 9-46. The FFT, however, would permit us to save a lot of computer time by computing $X(k) = \text{FFT}[x(n)]$ and $H(k) = \text{FFT}[h(n)]$. Then

$Y(k) = X(k)H(k)$ for $k = 0, \ldots, N - 1$ and $y(n) = \text{IDFT}[Y(k)]$, where the inverse discrete Fourier transform can also be computed with the FFT algorithm.

There is another useful procedure here that will save more computer time. Often the input sequence will be rather long, requiring a delay in real-time system processing, because all the $x(n)$ sequence is needed before evaluation of $y(n)$ from Equation 9-46 can be initiated. To shorten this delay, we can decompose the $x(n)$ sequence into a number of shorter segments, each of which can be processed individually with $y(n)$ then becoming a summation of partial convolutions. This idea is based on superposition. Instead of having a single input $x(n)$, we have a sum of inputs. Let $x(n) = x_1(n) + x_2(n) + \cdots + x_m(n)$. We can start the processing involved in Equation 9-46 by using $x_1(n)$ which is the data record gathered first. As this processing proceeds, the second data record $x_2(n)$ can be obtained. Then use $x_2(n)$ in Equation 9-46 as $x_3(n)$ is being collected, and so on.

In addition to its use in fast convolution, the FFT is also useful for fast correlation. Since the correlation operation is structurally similar to convolution, the sequences involved are often segmented into shorter sequences with which partial correlations are performed and the partial results are combined into total correlation functions. Correlation and autocorrelation techniques can be used for system identification. For a system with input $x(n)$, output $y(n)$, and impulse response $h(n)$, we can of course compute $X(k)$, $Y(k)$, and $H(k)$ using the FFT. If we have the input and output available, but not the system description, that is, $h(n)$ or $H(k)$, then we can write:

$$Y(k) = H(k)X(k) \tag{9-47}$$

$$Y(k)X^*(k) = H(k)X(k)X^*(k) \tag{9-48}$$

$$= H(k)\,|X(k)|^2$$

or
$$H(k) = Y(k)X^*(k)/|X(k)|^2 \tag{9-49}$$

The numerator is the DFT of the correlation of the input and the output. The denominator is the DFT of the autocorrelation of the input. Both these DFTs can be computed with FFT algorithms. Then the inverse FFT can be used to determine $h(n)$, that is, to identify the system.

Finally, spectral analysis is a popular area of FFT employment. The basic idea is to transform a time function to get its Fourier spectrum. The time function is discretized and the DFT or FFT is used to obtain the spectrum. Most of this work presupposes an extensive background in probability and random variable theory and is beyond our present scope. Some general comments: To obtain spectral estimates from a finite set of measurements typically employs the power spectral density function which is just the Fourier transform of the autocorrelation function of the data time sequence. The data sequences, though, are typically random variables. One interesting problem in this area comes from the fact that the determination of the power spectral density requires complete knowledge of the autocorrelation function, a function of generally infinite extent. Since we only have a finite data sequence available, we must make some estimates of the unavailable data. For instance, the autocorrelation function $r(n)$ is defined for all n, $-\infty < n < \infty$. However, in view of the finite nature of our data, we can only calculate $r(n)$ for $-k \le n \le< k$. The question is: How can we make reasonable estimates of the $r(n)$ values outside the range $-k \le n \le k$? A variety of spectral estimation techniques have been developed to deal with this problem. The interested reader is encouraged to consult the literature.

SUMMARY

In this chapter we have developed the discrete Fourier transform (DFT) and the inverse discrete Fourier transform (IDFT). We have shown the relationship of the discrete time Fourier transform to the Z transform. A number of examples illustrated this theory.

We considered the properties of the DFT, many of which were similar to the continuous Fourier transform properties. In particular, the convolution property deserved some extra attention due to the distinction that needed to be made between the periodic and aperiodic convolution.

Then we derived the fast Fourier transform (FFT) algorithms: The decimation-in-time (DIT) and the decimation-in-frequency (DIF) fast Fourier transforms. The similarities and the differences between these two approaches were indicated and the inverse FFT (IFFT) was developed. A few examples were worked to demonstrate the theory.

Finally, some FFT applications were discussed. These discussions were rather general in nature because many FFT applications are merely Fourier transform applications in which Fourier transforms are numerically calculated.

PROBLEMS

9-1. Let $f(n) = \text{Cos}\,(\pi/8\ n) + \text{Cos}\,(\pi/4\ n)$. Using the 8-point DIT FFT, determine the DFT of $f(n):F(k)$. Use samples of the $f(n)$ from $n = 0, \ldots, 7$.

9-2. Prove Parseval's theorem for the DFT, that is:

$$\sum_{n=0}^{N-1} |x(n)|^2 = \frac{1}{N} \sum_{k=0}^{N-1} |X(k)|^2$$

9-3. Prove the symmetry property for the DFT, that is:

$$\frac{1}{N}X(n) \leftrightarrow x(-k)$$

9-4. Find the discrete time Fourier transform for:
(a) $f(n) = u(n) - u(n - 3) + \delta(n - 4)$
(b) $f(n) = (\frac{1}{3})^n u(n)$
(c) $f(n) = (\frac{1}{3})^n \text{Cos}\ n\pi u(n)$
(d) $f(n) = (\frac{1}{3})^n u(n) - (\frac{1}{5})^n u(n - 1)$

9-5. Determine the Z transform for the functions in Problem 9.4. Then, using the Z transforms, find the discrete time Fourier transforms.

9-6. Determine the 8-point DFT for the sequences:
(a) $f(n) = [1, 0, 0, 1, 0, 0, 1, 0]$
(b) $f(n) = [1, 2, 3, 0, 0, 0, 0, 0]$
(c) $f(n) = [0, 0, 1, 1, 2, 2, 3, 3]$
(d) $f(n) = [10, 20, 10, 20, 10, 20, 10, 20]$

9-7. Determine the 8-point IDFT for:
(a) $F(k) = 1 - \text{Cos } \pi k/2$
(b) $F(k) = \text{Cos } \pi k + \text{Cos } \pi k/3$
(c) $F(k) = (k - 1)(\frac{1}{2})^k$

9-8. Construct the butterfly signal flow graph for the 16-point DIF FFT and the 16-point DIT FFT.

9-9. Compute the 8-point FFT (DIF or DIT) for:
(a) $f(n) = (\frac{1}{2})^n u(n)$
(b) $f(n) = u(n + 1) - u(n - 1)$
(c) $f(n) = u(n) \text{ Cos } n\pi/8$
(d) $f(n) = u(n) - u(n - 2)(\frac{1}{2})^n$

9-10. Consider the function $x(n) = (\frac{1}{3})^n u(n)$. Let $N = 6$.
(a) Determine the even and odd parts of $x(n)$: $x_e(n)$ and $x_0(n)$.
(b) Show that $X_e(k) = \text{Re}[X(k)]$ and $X_0(k) = j\text{Im}[X(k)]$.

9-11. A certain time sequence $x(n)$ has a DFT:

$$X(k) = [1, 1 - j, 0, 0, 0, 1 + j, j, -1], \qquad N = 8$$

Determine $x(n)$ and also the IDFT of $X(k - 4)$.

9-12. Let:

$$x(n) = [1, 0, 0, 1, j, -j, 1, 0]$$

Show that $x*(n) \leftrightarrow X*(-k)$.

9-13. Demonstrate Parseval's theorem for $x(n) \leftrightarrow X(k)$

where
$$x(n) = [8, 10, 4, 5], \qquad N = 4.$$

9-14. Using the symmetry property and knowing $x(n) \leftrightarrow X(k)$

where $x(n) = [\alpha_1, \alpha_2, \alpha_3, \alpha_4]$ and $X(k) = [\beta_1, \beta_2, \beta_3, \beta_4]$ and $N = 4$

Determine the DFT of $y(n) = [\beta_1, \beta_2, \beta_3, \beta_4]$.

9-15. Determine the "coordinate normalized Fourier transforms" $F_E(\theta)$ for the following time sequence:
(a) $f(n) = [1, 2, 3, 4, 3, 2, 1]$
(b) $f(n) = (\frac{1}{2})^n u(n) + 2^n u(-n)$
(c) $f(n) = \delta(n - 5) + \delta(n + 5) + 2\delta(n)$

9-16. Determine $f(n)$ using Equation 9-22 when $N = 4$ and
(a) $F(k) = [1, -1, 0, 1]$
(b) $F(k) = [0, 0, 1, -1]$
(c) $F(k) = [-1, 0, 0, 2]$

9-17. Determine the DFT of the Hamming window that is described by the equation $w(n) = 0.54 - 0.46 \cos [\pi(n + 0.5)/4]$. Let $N = 8$. Compare these results to the DFT for a corresponding rectangular window.

9-18. From Equation 9-25 we have:

$$f(n) = \frac{1}{N} \sum_{k=0}^{N-1} F(k) W^{-kn}$$

and from Equation 9-36 we can write:

$$f(n) = \left(\frac{1}{N} \sum_{k=0}^{N-1} F^*(k) W^{kn} \right)^*$$

(a) Show that these two results are the same.
(b) Explain why the second formulation is more appropriate than the first for IFFT computations.

9-19. Let:

$$f(n) = [1, 1, 1, 1, 0, 0, 0, 0], \qquad N = 8$$

We know that $|F(\theta)| = \alpha_1 \cos \alpha_2\theta \cos \alpha_3\theta$. Determine α_1, α_2, and α_3.

9-20. Demonstrate the linearity property of the DFT by determining $F(k)$ when

$$f(n) = 8(\tfrac{1}{2})^n + 12(\tfrac{1}{4})^n, \qquad N = 4, n = 0, \ldots, 3$$

9-21. Let:

$$x(n) = [1, 1, 0, 1] \quad \text{and} \quad y(n) = [0, 0, 1, 1], \qquad N = 4$$

Determine $X(k)$, $Y(k)$, and show that $z(n) = x(n)\, y(n) = [0, 0, 0, 1] \leftrightarrow X(k) * Y(k)$.

Chapter 10

State Variable Theory

INTRODUCTION

This final chapter on state variables provides an integration of much of the material in the text. State variable equations are solved with time-domain and transform methods. Our knowledge of differential and difference equation solutions comes into play, and we use the Laplace and Z transforms as well. We will consider the application of state variable ideas to some control system problems. State variable theory is one of the most important areas of systems theory and in control theory we have one of the most important applications of state variable theory.

State variable representations of linear systems are time-domain descriptions. They have certain advantages over the familiar difference or differential equation representations studied in Chapter 2. The state variable representation offers a concise and precise notation that lends itself well to digital computer processing. Multiple-input–multiple-output (MIMO) systems are easily managed within the state variable framework. In addition, extensions to time-varying and nonlinear systems—although not considered in this text—can be readily effected through the state variable representations.

The everyday idea of state is well-known: "the state of the Union," "the state of the art," "the state of one's health," and so on. In this sense, the concept of state refers to the conditions or attributes or circumstances of a person or thing. Within the linear systems world the idea of *state* is much more precise. The **state** of a dynamic system is the smallest set of numbers—called **state variables**—which if specified at some initial time t_0 or k_0 can be used to predict the system's behavior for all $t \geq t_0$ or $k \geq k_0$, provided that the input to the system is known for all $t \geq t_0$ or $k \geq k_0$. The condition of the system is exhibited in terms of the behavior of these state variables.

The discussion of linear systems in the previous chapters has focused on the input/output model. The input to the system, $x(t)$ or $x(k)$, yields an output, $y(t)$ or $y(k)$. The concern of this chapter will be not just to relate y and x but also to describe the internal system behavior by using the state variables. Internal to the system, let us assume that there are n state variables $q_1, q_2, \ldots, q_n$ needed to describe the system's behavior completely. These n state variables can be considered to be n components of a vector q. This vector q is called the **state vector.**

After obtaining the state variable descriptions of systems, we consider solving the state equations. Then we demonstrate the use of state variable ideas in a few areas of modern control theory. Throughout this development we will deal with both discrete and continuous systems. A facility with Z transforms and Laplace transforms will be assumed. Note also that we are using "k" rather than "n" as our discrete independent variable. This is because n has traditionally been reserved for the dimension of the state vector or the order of the system.

10-1 STATE VARIABLE REPRESENTATIONS

We begin the state variable approach by giving some simple examples.

EXAMPLE 10-1

Put the RC circuit of Figure 10-1 in the state variable form.

Solution. The input and output are related by the equation:

$$\dot{V}_{out}(t) + V_{out}(t) = V_{in}(t)$$

Let $q(t) = V_{out}(t)$ be the state variable and let $x(t) = V_{in}(t)$ be the input. Then we can write:

$$\dot{q}(t) = -q(t) + x(t)$$

Also, if we let $y(t) = V_{out}(t)$ be the output, we have:

$$y(t) = q(t)$$

The $\dot{q}$ equation and the y equation are called, respectively, the *state equation* and the *output equation*. Note that this is a first-order system

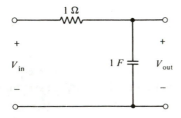

Figure 10-1 *RC* circuit for Example 10-1.

and, of course, is represented by a state variable equation of the first order.

EXAMPLE 10-2

Consider the second-order differential equation:

$$\frac{d^2y(t)}{dt^2} + 10\frac{dy(t)}{dt} + 20y(t) = x(t)$$

Assume $y(0) = 10$ and $\dot{y}(0) = 20$. Cast this equation into state variable form by letting $y(t)$ and $\dot{y}(t)$ be state variables.

Solution. Let:

$$q_1(t) = y(t) \quad \text{and} \quad q_2(t) = \frac{dy(t)}{dt}$$

Then

$$\ddot{y}(t) = x - 20q_1 - 10q_2 = \dot{q}_2 \quad \text{and} \quad \dot{q}_1 = q_2$$

Thus we can write:

$$\dot{q}(t) = \begin{bmatrix} 0 & 1 \\ -20 & -10 \end{bmatrix} q(t) + \begin{bmatrix} 0 \\ 1 \end{bmatrix} x(t)$$

$$y(t) = \begin{bmatrix} 1 & 0 \end{bmatrix} q(t) + \begin{bmatrix} 0 \end{bmatrix} x(t)$$

where

$$q(t) = \begin{bmatrix} q_1(t) \\ q_2(t) \end{bmatrix} \quad \text{and} \quad q(0) = \begin{bmatrix} 10 \\ 20 \end{bmatrix}$$

The $\dot{q}$ equation, involving a 2×2 matrix and a 2×1 matrix, is called the state equation. The $y(t)$ equation, involving a 1×2 matrix and a 1×1 zero matrix, is called the output equation. Note that a second-order differential equation has been represented by two first-order differential equations. Generally, the state variable formulation will cast an nth-order differential or difference equation into n first-order differential or difference equations.

These examples are single-input–single-output (**SISO**) systems. Hence the input $x(t)$ and the output $y(t)$ are scalars and can be considered as 1×1 matrices or one-dimensional vectors. More generally, however, a system can have multiple inputs (say, m) and multiple outputs (say, p). These are called **MIMO systems.** To generalize the state equation and the output equation from the preceding example, we can write the following equations in a vector-matrix format:

$$\dot{q}(t) = Aq(t) + Bx(t), \qquad q(t_0) = q_0 \tag{10-1}$$

$$y(t) = Cq(t) + Dx(t) \tag{10-2}$$

The A matrix is of dimension $n \times n$ where n is the order of the system. The B matrix is of dimension $n \times m$ where m is the number of inputs. The C matrix is of dimension $p \times n$ where p is the number of outputs. The D matrix is of dimension

$p \times m$. $q(t)$ is an n vector, $x(t)$ is an m vector, and $y(t)$ is a p vector. The matrices A, B, C, D are all assumed to be constant. This means that the systems we consider will be time-invariant. Equations 10-1 and 10-2 constitute the most general representation in the state variable format for any linear, lumped, time-invariant, deterministic, continuous system.

Now the A, B, C, D determination problem, which is called the *realization* problem, is not unique, and later in the chapter we examine two distinct approaches to this problem. One approach will be based on physical relationships in the system, defining the state variables as actual physical variables within the system. The other approach will be based on an existing input/output model, such as the differential equation of the second-order system considered earlier or perhaps a transfer function representation.

For discrete time systems the state variable formulation proceeds in a similar fashion. An example will illustrate the formulation.

EXAMPLE 10-3

Consider the two-input–two-output system represented by the following difference equations:

$$y_1(k + 2) + y_2(k + 1) + y_1(k + 1) + y_2(k) = x_1(k) + x_2(k)$$

$$y_2(k + 2) + 2y_2(k + 1) + 2y_2(k) - y_1(k) = x_1(k)$$

Assume all initial conditions are zero, that is, $y_1(0) = y_2(0) = y_1(1) = y_2(1) = 0$. Cast these equations into the state variable format. Let:

$$y_1(k) = q_1(k)$$

$$y_1(k + 1) = q_2(k) = q_1(k + 1)$$

$$y_2(k) = q_3(k)$$

$$y_2(k + 1) = q_4(k) = q_3(k + 1)$$

Solution

$$y_1(k + 2) = q_2(k + 1) = x_1(k) + x_2(k) - q_3(k) - q_2(k) - q_4(k)$$

and

$$y_2(k + 2) = q_4(k + 1) = x_1(k) + q_1(k) - 2q_3(k) - 2q_4(k)$$

Then collecting all $q(k + 1)$ equations, we can write:

$$q_1(k + 1) = q_2(k)$$

$$q_2(k + 1) = -q_2(k) - q_3(k) - q_4(k) + x_1(k) + x_2(k)$$

$$q_3(k + 1) = q_4(k)$$

$$q_4(k + 1) = q_1(k) - 2q_3(k) - 2q_4(k) + x_1(k)$$

Thus we have:

$$q(k+1) = \begin{bmatrix} 0 & 1 & 0 & 0 \\ 0 & -1 & -1 & -1 \\ 0 & 0 & 0 & 1 \\ 1 & 0 & -2 & -2 \end{bmatrix} q(k) + \begin{bmatrix} 0 & 0 \\ 1 & 1 \\ 0 & 0 \\ 1 & 0 \end{bmatrix} x(k)$$

$$y(k) = \begin{bmatrix} 1 & 0 & 0 & 0 \\ 0 & 0 & 1 & 0 \end{bmatrix} q(k) + \begin{bmatrix} 0 & 0 \\ 0 & 0 \end{bmatrix} x(k)$$

where $\quad q(k) = \begin{bmatrix} q_1(k) \\ q_2(k) \\ q_3(k) \\ q_4(k) \end{bmatrix}, \quad q(0) = \begin{bmatrix} 0 \\ 0 \\ 0 \\ 0 \end{bmatrix},$

$$x(k) = \begin{bmatrix} x_1(k) \\ x_2(k) \end{bmatrix}, \quad y(k) = \begin{bmatrix} y_1(k) \\ y_2(k) \end{bmatrix}$$

As in the continuous case, the $q(k+1)$ equation is called the **state equation** and the $y(k)$ equation is called the **output equation.**

To generalize from this discrete example, we can write:

$$q(k+1) = Aq(k) + Bx(k), \qquad q(k_0) = q_0 \tag{10-3}$$

$$y(k) = Cq(k) + Dx(k) \tag{10-4}$$

The dimensions of these vectors and matrices are the same as those of the continuous time systems. The A, B, C, D matrices here could be distinguished from those in Equations 10-1 and 10-2 by using subscripts. This, however, clutters the notation and will be avoided whenever possible. In any particular problem that uses the A, B, C, D matrices, the reference to a continuous time system or to a discrete time system should be apparent from the context.

10-2 BLOCK DIAGRAMMATICS

Before considering various realizations of A, B, C, D, we investigate some block diagram ideas that will help to illustrate those realizations. Some repetition from earlier chapters will be in the service of this illustration. A **block diagram** is a shorthand, graphical way to demonstrate the relationships of variables within a physical system. A single block with one input and one output is the simplest form of the block diagram (see Figure 10-2). Note that information flows, as indicated by the arrows, through the block in only one direction. Contained within the interior of the block is some description, usually mathematical, of the system at issue. An example will illustrate the basic idea.

Figure 10-2   Linear system block diagram with input and output.

EXAMPLE 10-4

Construct the simplest block diagram representation of the equation from Example 10-2 assuming zero-initial conditions.

Solution.   The equation is:

$$\frac{d^2y(t)}{dt^2} + 10\frac{dy(t)}{dt} + 20y(t) = x(t)$$

Laplace transforming this equation, assuming zero-initial conditions, we get the transfer function:

$$H(s) = \frac{Y(s)}{X(s)} = \frac{1}{s^2 + 10s + 20}$$

Then to exhibit this transfer function, we can construct the block diagram of Figure 10-3.

The input and output variables that are referenced to blocks in a block diagram may be represented as time-domain variables or frequency-domain variables. Generally, if the block expresses a frequency-domain operation, then the variables are frequency-domain variables; and if the block expresses a time-domain operation, then the variables are time-domain variables. Other examples of what often appears interior to a block are: "$G(s)$," indicating a transfer function of some kind or other; "unit delay," indicating a discrete system whose input is a variable, say, $x(k + 1)$, and whose output, $x(k)$, is a delayed version of the input; and the symbol "$\int$" or "$1/s$," indicating an integration operation.

In addition to blocks, we use three other elements to complete our diagramatics. These are the summing device, the takeoff point, and the constant multiplier, symbolized in Figure 10-4. The output of the summing device is the algebraic sum of the input. There may be any number of inputs and each may carry either a plus or minus sign. The output of the constant multiplier is just the input multiplied by the constant term α. A takeoff point indicates a variable that branches off unchanged into two or more paths. We will use the circle notation for multipliers, although some texts use the block notation with the α interior to the block. The choice is arbitrary. Only consistency matters. There is another block diagramatic element, the multiplying device, that is not the same thing as

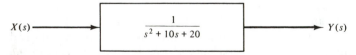

Figure 10-3   Block diagram for a second-order system.

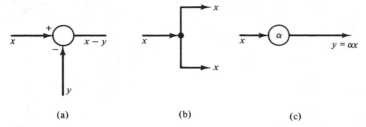

(a)　　　　　　　　　　　(b)　　　　　　　　　　　(c)

Figure 10-4　Symbols for (a) summing device; (b) takeoff point; (c) constant multiplier.

the multiplier. The multiplying device takes as inputs two or more variables and yields their product at the output. We have seen these in our discussion of modulation in Chapter 8. However, for linear system block diagramatics, multipliers are seldom used and will not be considered here.

Block diagrams, then, are constructed from various combinations of blocks, summing devices, constant multipliers, and takeoff points. Block diagrams mimic or simulate or represent equations. These equations in turn represent real physical systems. Given a block diagram representation of a system, we can derive a unique set of corresponding equations. However, given a set of equations, there are many block diagrams that will adequately represent these equations. Which block diagram will be most revealing for a particular problem depends on the circumstances. This is where the realization problem—considered in the next section—comes into play.

Let us consider some examples.

EXAMPLE 10-5

Consider the block diagram of Figure 10-5. Write the equation that the block diagram simulates.

Solution. The system output $y(k)$ is also the output of the second unit delay. The input to this delay is $y(k + 1)$. And the input to the first delay is $y(k + 2)$. But $y(k + 2)$ is also the output of the summing device. Usually the easiest way to write the equations represented by a block diagram is to write the equations indicated by all the summing devices in the system. In this case:

$$y(k + 2) = x(k) - 2y(k + 1) - 2y(k)$$

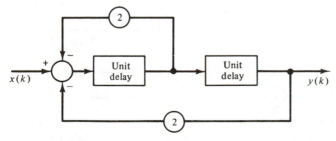

Figure 10-5　Block diagram for a second-order discrete system.

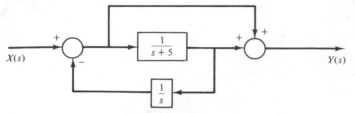

Figure 10-6   Control system example.

From this we form the standard difference equation:

$$y(k + 2) + 2y(k + 1) + 2y(k) = x(k)$$

EXAMPLE 10-6

For the control system represented in Figure 10-6, determine the overall system transfer function.

Solution.   Let $V_1(s)$ be the output of the left summing device. Let $V_2(s)$ be the output of the $1/(s + 5)$ block.

Then $V_1(s) = X(s) - \dfrac{1}{s} V_2(s)$ and $Y(s) = V_1(s) + V_2(s)$

Also, $V_1(s) \dfrac{1}{s + 5} = V_2(s)$

Eliminating V_1 and V_2, we obtain the system transfer function:

$$\frac{Y(s)}{X(s)} = \frac{s(s + 6)}{s^2 + 5s + 1} = H(s)$$

EXAMPLE 10-7

Write the state variable equations for the triple integrator system of Figure 10-7.

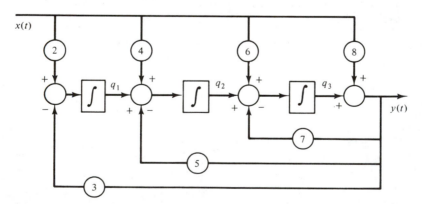

Figure 10-7   A triple integrator system.

Solution. The outputs of the three integrators are the state variables q_1, q_2, and q_3. Choosing integrator outputs as state variables generally leads to an appropriate state variable formulation. The inputs to the integrators $\dot{q}_1$, $\dot{q}_2$, and $\dot{q}_3$ are outputs of the summing devices and can be written as:

$$\dot{q}_1 = 2x - 3y$$

$$\dot{q}_2 = q_1 + 4x - 5y$$

$$\dot{q}_3 = q_2 + 6x - 7y$$

To get rid of $y(t)$ in these equations we need only observe that $y = q_3 + 8x$. Substituting and collecting terms, we can write:

$$\dot{q}(t) = \begin{bmatrix} 0 & 0 & -3 \\ 1 & 0 & -5 \\ 0 & 1 & -7 \end{bmatrix} q(t) + \begin{bmatrix} -22 \\ -36 \\ -50 \end{bmatrix} x(t)$$

and

$$y(t) = \begin{bmatrix} 0 & 0 & 1 \end{bmatrix} q(t) + [8]x(t)$$

These equations are in the standard state variable form of Equations 10-1 and 10-2. The A, B, C, D matrices are readily identified.

EXAMPLE 10-8

Consider a difference equation from which two separate realizations can be derived. The original equation is:

$$y(k + 2) + 3y(k + 1) + 2y(k) = x(k + 1) + 2x(k)$$

As we will see later, this equation can be written in the following state variable format:

$$q(k + 1) = \begin{bmatrix} 0 & 1 \\ -2 & -3 \end{bmatrix} q(k) + \begin{bmatrix} 0 \\ 1 \end{bmatrix} x(k)$$

$$y(k) = \begin{bmatrix} 2 & 1 \end{bmatrix} q(k)$$

and the original difference equation can be cast into the alternative format:

$$\hat{q}(k + 1) = \begin{bmatrix} 0 & -2 \\ 1 & -3 \end{bmatrix} \hat{q}(k) + \begin{bmatrix} 2 \\ 1 \end{bmatrix} x(k)$$

$$y(k) = \begin{bmatrix} 0 & 1 \end{bmatrix} \hat{q}(k)$$

In these formulations $q(k)$ and $\hat{q}(k)$ represent two different two-dimensional state vectors. The relationship between $x(k)$ and $y(k)$, however, remains the same in both formulations. Show how each formulation yields the same second-order difference equation. Construct the block diagram for each formulation.

Solution

$$q_1(k + 1) = q_2(k), \qquad q_2(k + 1) = -2q_1(k) - 3q_2(k) + x(k)$$
$$y(k) = 2q_1(k) + q_2(k)$$

Z-transform these equations and assume zero-initial conditions.

$$zQ_1(z) = Q_2(z)$$
$$zQ_2(z) = -2Q_1(z) - 3Q_2(z) + X(z)$$
$$Y(z) = 2Q_1(z) + Q_2(z)$$

Thus
$$z^2Q_1(z) = -2Q_1(z) - 3zQ_1(z) + X(z)$$
$$(z^2 + 3z + 2)Q_1(z) = X(z)$$

$$Y(z) = \frac{2X(z)}{z^2 + 3z + 2} + z\frac{X(z)}{z^2 + 3z + 2}$$

$$H(z) = \frac{Y(z)}{X(z)} = \frac{z + 2}{z^2 + 3z + 2}$$

Converting this to a difference equation, we have:

$$y(k + 2) + 3y(k + 1) + 2y(k) = x(k + 1) + 2x(k)$$

For the second formulation we obtain:

$$zQ_1(z) = -2Q_2(z) + 2X(z)$$
$$zQ_2(z) = Q_1(z) - 3Q_2(z) + X(z)$$
$$Y(z) = Q_2(z)$$

$$Q_1(z) = \frac{-2}{z}Q_2(z) + \frac{2}{z}X(z)$$

and
$$zQ_2(z) = \frac{-2}{z}Q_2(z) + \frac{2}{z}X(z) - 3Q_2(z) + X(z)$$

$$\left(z + \frac{2}{z} + 3\right)Q_2(z) = \left(\frac{2}{z} + 1\right)X(z) = \left(\frac{z^2 + 3z + 2}{z}\right)Y(z)$$

$$= \left(\frac{2 + z}{z}\right)X(z)$$

Then
$$H(z) = \frac{Y(z)}{X(z)} \equiv \frac{z + 2}{z^2 + 3z + 2}$$

as before.

This transfer function yields the same difference equation. Block diagram realizations of these two formulations are presented in Figures 10-8 and 10-9. Note that these block diagrams are constructed from

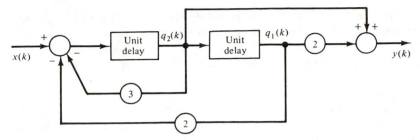

Figure 10-8   Block diagram of the $q(k)$ state system.

exactly the same elements; however, the elements are structured very differently.

10-3 REALIZATIONS OF *A, B, C, D*

The **realization problem** is defined as the problem of determining A, B, C, D given the system's transfer functions. But we could just as well proceed from a knowledge of the system's difference or differential equations, since in previous chapters we saw how the transfer function follows from the differential or difference equation and vice versa. Once A, B, C, D are found, the system can be built or "realized." Before any actual hardware is constructed, however, a simulation is usually performed. This provides a way of testing the system's performance and stability in a purely abstract fashion at very little cost.

10-3-1 Physical System Realizations

Our first task is to realize A, B, C, D from physical system considerations. The state variables are chosen to be physical system variables, for example, the position of an actuator in a mechanical system, the pitch angle of an aircraft, the

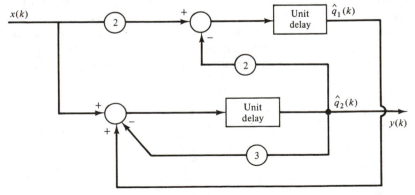

Figure 10-9   Block diagram of the $\hat{q}(k)$ state system.

capacitor voltage in an *RLC* circuit. To realize *A, B, C, D* from physical variables requires that we have knowledge of the details of the physical system and not just a knowledge of the system transfer function. Some of the problems encountered with physical system realizations are illustrated in the following example.

EXAMPLE 10-9

Consider the *RLC* circuit of Figure 10-10. Assume that the circuit is initially quiescent. Let $x(t)$ be the input current in amperes. Let $y(t)$ be the output voltage in volts. Represent this circuit in a state variable format.

Solution.   From node equations, we get:

$$x = v_1 + \dot{v}_1 - \dot{v}_2,$$

$$\ddot{v}_1 - \ddot{v}_2 = \dot{v}_2 + v_2 - v_3$$

$$\dot{v}_3 = v_2 - v_3$$

The output $y(t)$ is $v_3(t)$. Eliminating v_1 and v_2 by using the Laplace transform, we obtain:

$$H(s) = \frac{y(s)}{x(s)} = \frac{\frac{1}{2}s}{s^2 + 2s + 1}$$

From this transfer function we can write the equation:

$$\ddot{y}(t) + 2\dot{y}(t) + y(t) = \tfrac{1}{2}\dot{x}(t)$$

Since the system was assumed to be quiescent, $y(0) = \dot{y}(0) = 0$ and this differential equation can be solved quite easily if $x(t)$ is specified. However, the equation is not in a state variable format. Also—and this is the heart of the issue—if the system is not initially quiescent, then we would need to determine values for $y(0)$ and $\dot{y}(0)$.

If these values are not known, then they must be determined from whatever knowledge we have about the circuit. Generally, we know something about the initial energy stored in the circuit. In *RLC* circuits this energy is expressed in terms of the initial capacitor voltages and initial inductor currents. To express $y(0)$ and $\dot{y}(0)$ in terms of capacitor voltages and inductor currents is often no trivial task. (In this particular case, however, it can be done and will be done shortly.) Because of the difficulty involved in determining $y(0)$ and $\dot{y}(0)$, the question about alternatives

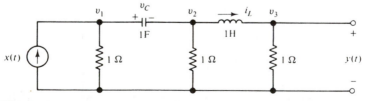

Figure 10-10   *RLC* circuit example.

comes to light. Can we avoid $y(0)$ and $\dot{y}(0)$ altogether and use the initial capacitor voltages and inductor currents directly? The answer is yes, if we define $v_c(t)$ and $i_L(t)$ to be state variables.

Consider again the circuit in Figure 10-10. Using mesh and node equations and eliminating v_1 and v_2, we can write:

$$\dot{v}_c = -\tfrac{1}{2}v_c + \tfrac{1}{2}i_L + \tfrac{1}{2}x$$

$$\dot{i}_L = -\tfrac{1}{2}v_c - \tfrac{3}{2}i_L + \tfrac{1}{2}x$$

and $y = i_L$. These equations can be cast into the familiar state variable format by letting $q_1 = v_c$ and $q_2 = i_L$.

$$\dot{q}(t) = \begin{bmatrix} -\tfrac{1}{2} & \tfrac{1}{2} \\ -\tfrac{1}{2} & -\tfrac{3}{2} \end{bmatrix} q(t) + \begin{bmatrix} \tfrac{1}{2} \\ \tfrac{1}{2} \end{bmatrix} x(t)$$

$$y(t) = [0 \quad 1]q(t) + [0]x(t)$$

We have, then, cast the given *RLC* circuit into a state variable format wherein the state variables are real physical variables within the system.

Now to express $y(0)$ and $\dot{y}(0)$ in terms of $v_c(0)$ and $i_L(0)$, we know that

$$y(t) = (1 \ \Omega)i_L(t)$$

Thus $\qquad\qquad y(0) = i_L(0)$

Also, $\qquad\qquad \dot{v}_3 = \dot{y} = v_2 - v_3 = v_2 - y$

and $\qquad\qquad x = v_1 + \dot{v}_1 - \dot{v}_2 = v_1 + v_2 + i_L$

where $\qquad\qquad v_2 = v_1 - v_c, \qquad i_L = y$

thus $\qquad\qquad x = v_1 + v_1 - v_c + y = 2v_1 - v_c + y$

from which $\qquad\quad v_1 = (x + v_c - y)/2.$

Thus $\qquad\qquad\qquad \dot{y} = v_2 - y$

$$= v_1 - v_c - y$$

$$= \tfrac{1}{2}x - \tfrac{1}{2}v_c - \tfrac{3}{2}y$$

and when $t = 0$, $\qquad \dot{y} = \tfrac{1}{2}x(0) - \tfrac{1}{2}v_c(0) - \tfrac{3}{2}y(0)$

Therefore $\qquad\qquad y(0) = i_L(0)$

$$\dot{y}(0) = \tfrac{1}{2}x(0) - \tfrac{1}{2}v_c(0) - \tfrac{3}{2}i_L(0)$$

Considerable work is required just to get these initial conditions.

In general, for *RLC* circuits, by choosing capacitor voltages and inductor currents to be state variables, we can write first-order differential equations of the form of Equations 10-1 and 10-2. There are some exceptions to this general approach. In particular, the capacitor voltages and inductor currents must all be independent variables. Sometimes, for more complicated circuits, the determination of dependence or independence among the variables is difficult. In practice, however, the selection of appropriate state variables from capacitor voltages and inductor currents is seldom questionable. For that reason such problematic situations will not be discussed further here.

For simple mechanical systems it is typical to consider independent velocities and positions to be state variables. This choice of state variables will lead to differential equations of the form of Equations 10-1 and 10-2.

Other types of systems do not generally yield to any accessible systematic method that will allow determination of system state variables corresponding to real physical variables. Often, however, by simple consideration of the system and its representations we can pinpoint certain physical variables to serve as state variables. An example will illustrate this.

EXAMPLE 10-10

In the study of aircraft autopilots the equations that describe the lateral motions are:

$$\dot{\beta} + r = k_1\beta + k_2\phi$$

$$\dot{r} + k_3\dot{p} = k_4\beta + k_5r + k_6p + x_1$$

$$\dot{p} + k_7\dot{r} = k_8p + k_9\beta + k_{10}r + x_2$$

$$\dot{\phi} = p$$

$$\dot{\psi} = r$$

where x_1 and x_2 are inputs to the system, r is the angular yaw velocity, ψ is the yaw angle, p is the angular roll velocity, ϕ is the angle of roll, and β is the angle of sideslip. Cast this system into the state variable format.

Solution.   There are five state variables here and two inputs. We can arrange this system such that each state variable derivative is equal to some linear combination of the state variables and the inputs. To do this, solve the second and third equations for $\dot{r}$ and $\dot{p}$ to get:

$$\dot{p} = k_{11}\beta + k_{12}r + k_{13}p + k_{14}x_1 + k_{15}x_2$$

$$\dot{r} = k_{16}\beta + k_{17}r + k_{18}p + k_{19}x_1 + k_{20}x_2$$

k_{11} through k_{20} are various combinations of k values from k_3 through k_{10}. For example:

$$k_{11} = \frac{k_7k_4 - k_9}{k_3k_7 - 1}$$

Once these equations are put into this form, we can collect all five state equations as follows:

$$
\begin{bmatrix} \dot{\beta} \\ \dot{p} \\ \dot{r} \\ \dot{\phi} \\ \dot{\psi} \end{bmatrix} = \begin{bmatrix} k_1 & 0 & -1 & k_2 & 0 \\ k_{11} & k_{13} & k_{12} & 0 & 0 \\ k_{16} & k_{18} & k_{17} & 0 & 0 \\ 0 & 1 & 0 & 0 & 0 \\ 0 & 0 & 1 & 0 & 0 \end{bmatrix} \begin{bmatrix} \beta \\ p \\ r \\ \phi \\ \psi \end{bmatrix} + \begin{bmatrix} 0 & 0 \\ k_{14} & k_{15} \\ k_{19} & k_{20} \\ 0 & 0 \\ 0 & 0 \end{bmatrix} \begin{bmatrix} x_1 \\ x_2 \end{bmatrix}
$$

This is the familiar $\dot{q} = Aq + Bx$ form.

Certainly, these physical variable state equation models are desirable because the variables in the model of the system correspond directly to variables in the system. The main problem is that these models are often difficult to determine. Also, in many cases, we will be given only an input/output model of the system, for example, a difference equation or a transfer function. In these cases we can develop a number of different state variable forms called canonical forms, which we discuss next.

10-3-2 Canonical Form Realizations

The word *canonical* is an adjective meaning authorized, recognized, or accepted. **Canonical form realizations** are those A, B, C, D determinations that are special in the sense of being authorized, recognized, or accepted. Consider three canonical form realizations:

1. the controllability canonical form
2. the observability canonical form
3. the Jordan canonical form

The first two forms are related to the concepts of controllability and observability which will be taken up later in the chapter. For now, just think of these as names we assign to two distinct and revealing types or forms of the state variable representation. The Jordan form is a very useful form into which any other A, B, C, D realization can be transformed. In what follows, for the sake of simplicity, we restrict ourselves to single-input–single-output systems. Extensions to MIMO systems are straightforward but would be confusing to this elementary presentation.

Consider the transfer function:

$$
H(s) = \frac{b_n s^n + b_{n-1} s^{n-1} + \cdots + b_1 s + b_0}{s^n + a_{n-1} s^{n-1} + \cdots + a_1 s + a_0} \tag{10-5}
$$

which can also be written as:

$$H(s) = \frac{c_{n-1}s^{n-1} + \cdots + c_1 s + c_0}{s^n + a_{n-1}s^{n-1} + \cdots + a_1 s + a_0} + b_n \tag{10-6}$$

From these equations we can deduce that:

$$b_{n-1} = a_{n-1}b_n + c_{n-1}$$

$$b_{n-2} = a_{n-2}b_n + c_{n-2}$$

$$\vdots \tag{10-7}$$

$$b_0 = a_0 b_n + c_0$$

The same transfer function form is assumed for discrete systems with z replacing s: $H(z)$. Also, for physically realizable systems it is necessary that the order of the numerator polynomials be at most n.

This transfer function can be cast into the state variable controllability canonical form realization as follows:

$$\dot{q}(t) = \begin{bmatrix} 0 & 1 & 0 & \cdots & 0 \\ 0 & 0 & 1 & \cdots & 0 \\ \vdots & & & & \vdots \\ 0 & 0 & 0 & \cdots & 1 \\ -a_0 & -a_1 & -a_2 & \cdots & -a_{n-1} \end{bmatrix} q(t) + \begin{bmatrix} 0 \\ 0 \\ \vdots \\ 1 \end{bmatrix} x(t) \tag{10-8}$$

$$y(t) = [c_0 c_1 \quad \cdots \quad c_{n-1}] q(t) + b_n x(t) \tag{10-9}$$

The A, B, C, D matrices are readily distinguishable here. We call these A_c, B_c, C_c, D_c to indicate the controllability form. For discrete systems the equations become:

$$q(k + 1) = A_c q(k) + B_c x(k) \tag{10-10}$$

$$y(k) = C_c q(k) + D_c x(k) \tag{10-11}$$

EXAMPLE 10-11

Using the Laplace transform method show that Equations 10-8 and 10-9 imply Equations 10-5 and 10-6 for the case of $n = 3$. Assume zero-initial conditions.

Solution. The equations to be transformed are:

$$\dot{q}_1 = q_2$$

$$\dot{q}_2 = q_3$$

$$\dot{q}_3 = -a_0 q_1 - a_1 q_2 - a_2 q_3 + x$$

and
$$y = c_0 q_1 + c_1 q_2 + c_2 q_3 + b_3 x$$

$$sQ_1(s) = Q_2(s)$$

$$sQ_2(s) = Q_3(s)$$

$$sQ_3(s) = -a_0 Q_1(s) - a_1 Q_2(s) - a_2 Q_3(s) + X(s)$$

Thus
$$Q_1 = \frac{Q_2}{s}, \qquad Q_2 = \frac{Q_3}{s}, \qquad Q_1 = Q_3/s^2$$

and
$$sQ_3 = -a_0 \frac{Q_3}{s^2} - a_1 \frac{Q_3}{s} - a_2 Q_3 + X$$

$$(s^3 + a_2 s^2 + a_1 s + a_0) Q_3 = s^2 X, \qquad Q_3 = s^2 X / (s^3 + a_2 s^2 + a_1 s + a_0)$$

$$= s^2 X / \Delta$$

Thus
$$Y(s) = \frac{c_0 \, s^2 X(s)}{s^2 \quad \Delta} + \frac{c_1 \, s^2 X(s)}{s \quad \Delta} + \frac{c_2 s^2 X(s)}{\Delta} + b_3 X(s)$$

$$H(s) = \frac{Y(s)}{X(s)} = \frac{c_0 + c_1 s + c_2 s^2}{s^3 + a_2 s^2 + a_1 s + a_0} + b_3$$

$$= \frac{b_3 s^3 + (a_2 b_3 + c_2)s^2 + (a_1 b_3 + c_1)s + (a_0 b_3 + c_0)}{s^3 + a_2 s^2 + a_1 s + a_0}$$

$$= \frac{b_3 s^3 + b_2 s^2 + b_1 s + b_0}{s^3 + a_2 s^2 + a_1 s + a_0}$$

EXAMPLE 10-12

Cast the discrete system represented by the following difference equation into the state variable controllability canonical form.

$$y(k + 3) + 2y(k + 1) + y(k) = x(k + 1) + x(k)$$

Solution. First form the transfer function that this equation yields.

$$H(z) = \frac{z + 1}{z^3 + 2z + 1}$$

From this we can read off the coefficients:

$$b_3 = 0$$
$$a_2 = 0$$
$$b_2 = 0$$
$$a_1 = 2$$
$$b_1 = 1$$
$$a_0 = 1$$
$$b_0 = 1$$

Then we can calculate:

$$c_0 = 1$$
$$c_1 = 1$$
$$c_2 = 0$$

$$\text{Then} \qquad q(k+1) = \begin{bmatrix} 0 & 1 & 0 \\ 0 & 0 & 1 \\ -1 & -2 & 0 \end{bmatrix} q(k) + \begin{bmatrix} 0 \\ 0 \\ 1 \end{bmatrix} x(k)$$

$$y(k) = \begin{bmatrix} 1 & 1 & 0 \end{bmatrix} q(k) + \begin{bmatrix} 0 \end{bmatrix} x(k)$$

A second canonical form realization called the observability canonical form realization also takes off from the transfer functions of Equations 10-5 and 10-6. These transfer functions can be cast into the observability canonical form realization as follows:

$$\dot{q}(t) = \begin{bmatrix} 0 & 0 & 0 & \cdots & 0 & -a_0 \\ 1 & 0 & 0 & \cdots & 0 & -a_1 \\ 0 & 1 & 0 & \cdots & 0 & -a_2 \\ \vdots & \vdots & \vdots & & & \vdots \\ 0 & 0 & 0 & \cdots & 1 & -a_{n-1} \end{bmatrix} q(t) + \begin{bmatrix} c_0 \\ c_1 \\ \vdots \\ c_{n-1} \end{bmatrix} x(t) \qquad (10\text{-}12)$$

$$y(t) = \begin{bmatrix} 0 & 0 & \cdots & 1 \end{bmatrix} q(t) + b_n x(t) \qquad (10\text{-}13)$$

The A, B, C, D matrices here will be called A_0, B_0, C_0, D_0. For discrete systems the equations become:

$$q(k+1) = A_0 q(k) + B_0 x(k) \qquad (10\text{-}14)$$

$$y(k) = C_0 q(k) + D_0 x(k) \qquad (10\text{-}15)$$

Again, the significance of these two canonical realizations will be discussed later in the chapter when we consider what controllability and observability mean when applied to linear systems in general. For now just note that there is a very interesting relationship between these two canonical forms. From inspection of the matrices involved in these representations we observe:

$$A_c = A_0^T$$
$$B_c = C_0^T$$
$$C_c = B_0^T \qquad (10\text{-}16)$$

and
$$D_c = D_0$$

EXAMPLE 10-13

Put the system of Example 10-12 into the observability canonical form.

Solution. We can proceed in two different ways here. First, we can take

$H(z) = (z + 1)/(z^3 + 2z + 1)$ and reading off the coefficients construct the following from Equations 10-14 and 10-15:

$$q(k + 1) = \begin{bmatrix} 0 & 0 & -1 \\ 1 & 0 & -2 \\ 0 & 1 & 0 \end{bmatrix} q(k) + \begin{bmatrix} 1 \\ 1 \\ 0 \end{bmatrix} x(k)$$

and
$$y(k) = \begin{bmatrix} 0 & 0 & 1 \end{bmatrix} q(k) + [0] x(k)$$

Alternatively, we can employ Equations 10-16 to deduce:

$$A_0 = A_c^T = \begin{bmatrix} 0 & 0 & -1 \\ 1 & 0 & -2 \\ 0 & 1 & 0 \end{bmatrix}$$

$$B_0 = C_c^T = \begin{bmatrix} 1 \\ 1 \\ 0 \end{bmatrix}, \quad C_0 = B_c^T = \begin{bmatrix} 0 & 0 & 1 \end{bmatrix}, \quad \text{and} \quad D_0 = D_c = 0$$

The observability and controllability realizations, then, permit us to take any physically realizable transfer function and to transform it into a state variable description. This of course includes any difference or differential equation description because these are readily converted to transfer functions. Recall that for a differential equation like $y'' + y' + y = x$, we obtained a state variable form by letting $y = q_1$ and $\dot{y} = q_2$. If, however, we have a differential equation like $y'' + y' + y = x' + x$, then getting the state variable form is more difficult because we cannot simply let $y = q_1$ and $\dot{y} = q_2$. Yet, this difficulty vanishes if we convert the differential equation to its transfer function and then employ the observability or the controllability forms.

The third canonical form representation is actually first in the sense of utility. The Jordan canonical form realization is most useful when we actually have to solve the state equation for $q(t)$ or $q(k)$, a task we will deal with in the next section. Before defining the Jordan form in an explicit fashion, we note that the typical Jordan form has an A matrix that is diagonal. However, even in those cases of a nondiagonal A matrix, the Jordan form considerably simplifies the solution of the state variable equation. The next example illustrates this.

The Jordan canonical form employs the notion of *eigenvalues*. Eigenvalues are simply the roots of the characteristic equation of the system. The characteristic equation is obtained by setting the denominator of the transfer function equal to zero. The utility of the Jordan canonical form lies in the fact that the A matrix directly exhibits the eigenvalues on its main diagonal. While avoiding the details of solving the state equation in general, a simple example will illustrate some of these ideas.

EXAMPLE 10-14

Assume we have a continuous system with

$$A = \begin{bmatrix} -1 & 1 & 0 \\ 0 & -1 & 0 \\ 0 & 0 & -2 \end{bmatrix}$$

This happens to be a Jordan canonical form for the A matrix. Assume the input $x(t)$ is zero and assume $q_1(0) = q_2(0) = q_3(0) = \alpha$. Determine $q(t)$.

Solution.   We have:

$$\dot{q}_1 = -q_1 + q_2$$

$$\dot{q}_2 = -q_2$$

$$\dot{q}_3 = -2q_3$$

Thus

$$q_3(t) = k_3 e^{-2t} = \alpha e^{-2t}$$

$$q_2(t) = k_2 e^{-t} = \alpha e^{-t}$$

and

$$\dot{q}_1(t) + q_1(t) = \alpha e^{-t}$$

$$q_{1H}(t) = k_1 e^{-t}$$

$$q_{1p}(t) = k_5 t e^{-t}$$

$$\dot{q}_{1p} = k_5 e^{-t} - k_5 t e^{-t}$$

and

$$-k_5 t e^{-t} + k_5 e^{-t} + k_5 t e^{-t} = \alpha e^{-t}$$

Therefore

$$k_5 = \alpha$$

Thus

$$q_1(t) = q_{1H} + q_{1p} = k_1 e^{-t} + \alpha t e^{-t}$$

$$q_1(0) = k_1 = \alpha$$

Therefore

$$q(t) = \begin{bmatrix} \alpha(1 + t)e^{-t} \\ \alpha e^{-t} \\ \alpha e^{-2t} \end{bmatrix}$$

Comments: Systems in the Jordan canonical form can usually be solved from the bottom up, that is, obtain $q_3(t)$ first, then $q_2(t)$, and so on. The eigenvalues of this system are -1, -1, -2. Therefore $(s + 1)(s + 1)(s + 2) = 0$ is the characteristic equation. Also, $s^3 + 4s^2 + 5s + 2$ is the denominator of the system transfer function. This example is a case of repeated roots. When the system has no repeated roots, the A matrix is purely diagonal and the solution of the state equation is even simpler.

The obvious question is: How do we put a given system into this Jordan canonical form? At the center of the matter lies the theory of coordinate transformations. In the canonical forms considered previously, we had two different state vectors, q, which could be used to describe exactly the same system. We can distinguish these as q_c and q_0, for the controllability and observability forms, respectively. There are actually an infinite number of ways to define the state vector for a given system. It is convenient to view different ways of defining a state vector as transformations of the coordinates of the system. This transformation is effected through a matrix multiplication:

$$\hat{q} = Pq \qquad (10\text{-}17)$$

If q and $\hat{q}$ are n-dimensional vectors, then P will be an $n \times n$ nonsingular constant matrix. For instance, in Equation 10-17 we might have $q = q_0$ and $\hat{q} = q_c$. But our problem here is to find P when $\hat{q}$ is any given state vector and q is the state vector of the Jordan canonical form.

A system is said to be in the **Jordan canonical form** if its $n \times n$ A matrix appears as follows:

$$A = \begin{bmatrix} A_1 & & & & 0 \\ & A_2 & & & \\ & & \ddots & & \\ 0 & & & & A_s \end{bmatrix} \qquad (10\text{-}18)$$

The A_j terms are matrices of dimension $m_j \times m_j$

where

$$A_j = \begin{bmatrix} \lambda & 1 & 0 & \cdots & 0 & 0 \\ 0 & \lambda & 1 & \cdots & 0 & 0 \\ \vdots & \vdots & \vdots & & \vdots & \vdots \\ 0 & 0 & 0 & \cdots & \lambda & 1 \\ 0 & 0 & 0 & \cdots & 0 & \lambda \end{bmatrix} \qquad (10\text{-}19)$$

These matrices are called *Jordan blocks*. Since it is standard practice, we use the λ notation for the roots of the characteristic equation when these roots are functioning as eigenvalues. In Example 10-14, for instance, we would have $\lambda_1 = -1, \lambda_2 = -1$, and $\lambda_3 = -2$. The λ's in each A_j are usually, but not always, different. Also, the following equation holds:

$$\sum_{j=1}^{s} m_j = n \qquad (10\text{-}20)$$

In Example 10-14 we had

$$A_1 = \begin{bmatrix} -1 & 1 \\ 0 & -1 \end{bmatrix} \quad \text{and} \quad A_2 = [-2] \quad \text{and} \quad m_1 = 2 \quad \text{and} \quad m_2 = 1.$$

The Jordan blocks are typically enclosed by dashed lines, so A might appear as:

$$A = \begin{bmatrix} -1 & 1 & \vdots & 0 \\ 0 & -1 & \vdots & 0 \\ \hdashline 0 & 0 & \vdots & -2 \end{bmatrix} = \begin{bmatrix} A_1 & \vdots & 0 \\ \hdashline 0 & \vdots & A_2 \end{bmatrix}$$

Assume we are given the linear system:

$$\dot{\hat{q}} = \hat{A}\hat{q} + \hat{B}x \qquad (10\text{-}21)$$

$$y = \hat{C}\hat{q} + \hat{D}x \qquad (10\text{-}22)$$

We want to transform it into the Jordan canonical form

$$\dot{q} = Aq + Bx \qquad (10\text{-}23)$$

$$y = Cq + Dx \qquad (10\text{-}24)$$

Using the transformation of Equation 10-17, $\hat{q} = Pq$, we have:

$$\frac{d\hat{q}}{dt} = P\frac{dq}{dt} = \hat{A}Pq + \hat{B}x \qquad (10\text{-}25)$$

$$y = \hat{C}Pq + \hat{D}x \qquad (10\text{-}26)$$

from which we deduce:

$$\begin{aligned} A &= P^{-1}\hat{A}P \\ B &= P^{-1}\hat{B} \\ C &= \hat{C}P \\ D &= \hat{D} \end{aligned} \qquad (10\text{-}27)$$

We have the "hat" system in hand. Now we are seeking the A, B, C, D realization. Obviously, we need P. Eigenvalues and eigenvectors are involved in this determination. We already defined eigenvalues, λ_i, to be roots of the system's characteristic equation. However, often only the $\hat{A}$, $\hat{B}$, $\hat{C}$, $\hat{D}$ matrices and not the system's transfer function are given. To avoid calculation of the system's transfer function, we use the following equation which is sufficient to solve for the eigenvalues of $\hat{A}$.

$$\det(\lambda I - \hat{A}) = 0 \qquad (10\text{-}28)$$

But these are also eigenvalues of A.

EXAMPLE 10-15
Prove that the eigenvalues of A are the eigenvalues of $\hat{A}$.

Solution

$$\det(\lambda I - A) = \det(\lambda I - P^{-1}\hat{A}P) = \det[P^{-1}(\lambda PP^{-1} - \hat{A})P]$$

$$= \det P^{-1} \det(\lambda I - \hat{A}) \det P = \det(P^{-1}P) \det(\lambda I - \hat{A})$$

$$= \det I \det(\lambda I - \hat{A})$$

$$= \det(\lambda I - \hat{A}) = 0$$

Eigenvectors are *n*-dimensional vectors p_i which satisfy:

$$(\lambda_i I - \hat{A})p_i = 0 \qquad (10\text{-}29)$$

where λ_i is the *i*th eigenvalue of $\hat{A}$. To each eigenvalue there is a corresponding eigenvector.

Our problem, again, is to find *P*. We proceed by first considering the simplest kind of Jordan canonical form: the case of nonrepeated roots where *A* is purely diagonal. From this particular problem we jump to the general Jordan form determinations.

The case of nonrepeated roots is sometimes called the **diagonalization problem.** Given any $\hat{A}$ matrix, find *P* such that $A = P^{-1}\hat{A}P$ is purely diagonal with the terms on the diagonal being the eigenvalues.

$$A = \begin{bmatrix} \lambda_1 & 0 & \cdots & 0 \\ 0 & \lambda_2 & & \\ \vdots & & \ddots & \\ 0 & & & \lambda_n \end{bmatrix} \qquad (10\text{-}30)$$

The standard approach to this problem is to let *P* be the matrix whose columns are eigenvectors of *A*

$$P = [p_1 \vdots p_2 \vdots --- \vdots p_n] \qquad (10\text{-}31)$$

The determination of p_i follows from Equation 10-29. An example illustrates this.

EXAMPLE 10-16

Let

$$\hat{A} = \begin{bmatrix} -1 & 1 \\ 0 & -2 \end{bmatrix}$$

Find *A*.

Solution. First we need the eigenvalues. Using Equation 10-28, we obtain:

$$\det(\lambda I - \hat{A}) = \det\left[\begin{pmatrix} \lambda & 0 \\ 0 & \lambda \end{pmatrix} - \begin{pmatrix} -1 & 1 \\ 0 & -2 \end{pmatrix}\right]$$

$$= \det\begin{bmatrix} \lambda + 1 & -1 \\ 0 & \lambda + 2 \end{bmatrix} = (\lambda + 1)(\lambda + 2) = 0$$

Therefore $\lambda_1 = -1$ $\lambda_2 = -2$

Now using Equation 10-29, we have:

$$(\lambda_1 I - \hat{A})p_1 = 0$$

Let:

$$p_1 = \begin{bmatrix} p_{11} \\ p_{21} \end{bmatrix}$$

$$\left[\begin{pmatrix} -1 & 0 \\ 0 & -1 \end{pmatrix} - \begin{pmatrix} -1 & 1 \\ 0 & -2 \end{pmatrix}\right]\begin{bmatrix} p_{11} \\ p_{21} \end{bmatrix} = 0$$

$$\begin{bmatrix} 0 & -1 \\ 0 & 1 \end{bmatrix}\begin{bmatrix} p_{11} \\ p_{21} \end{bmatrix} = \begin{bmatrix} 0 \\ 0 \end{bmatrix}, \qquad \begin{matrix} -p_{21} = 0 \\ p_{21} = 0 \end{matrix}$$

But this tells us nothing about p_{11}. What should we do next? In solving eigenvector problems of this type we have a certain freedom of choice. This freedom often causes uneasiness in our search for clear and distinct ideas. What it really means, though, in this context, is that we can choose p_{11} to be anything other than zero. (A zero choice would make the eigenvector p_1 trivial). We can choose $p_{11} = 5.876$. We can choose $p_{11} = 10.87 + 983.49j$. To make life simple, let $p_{11} = 1$.

Then $$p_1 = \begin{bmatrix} 1 \\ 0 \end{bmatrix}$$

For the second eigenvalue:

$$(\lambda_2 I - \hat{A})p_2 = 0 = \left[\begin{pmatrix} -2 & 0 \\ 0 & -2 \end{pmatrix} - \begin{pmatrix} -1 & 1 \\ 0 & -2 \end{pmatrix}\right]\begin{bmatrix} p_{12} \\ p_{22} \end{bmatrix} = \begin{bmatrix} 0 \\ 0 \end{bmatrix}$$

$$\begin{pmatrix} -1 & -1 \\ 0 & 0 \end{pmatrix}\begin{bmatrix} p_{12} \\ p_{22} \end{bmatrix} = \begin{bmatrix} 0 \\ 0 \end{bmatrix}, \qquad -p_{12} - p_{22} = 0, \quad \text{or} \quad p_{22} = -p_{12}$$

Choose $p_{22} = -1$, then $p_{12} = 1$,

and $$p_2 = \begin{bmatrix} 1 \\ -1 \end{bmatrix}$$

(As an additional exercise choose $p_{22} = 1$ and see if you get the same results. Remember that the choice of p_{22} is arbitrary.) We get:

$$P = [p_1 \mid p_2] = \begin{bmatrix} 1 & 1 \\ 0 & -1 \end{bmatrix}$$

Also

$$P^{-1} = \begin{bmatrix} 1 & 1 \\ 0 & -1 \end{bmatrix}$$

Thus

$$A = P^{-1}\hat{A}P = \begin{bmatrix} 1 & 1 \\ 0 & -1 \end{bmatrix}\begin{bmatrix} -1 & 1 \\ 0 & -2 \end{bmatrix}\begin{bmatrix} 1 & 1 \\ 0 & -1 \end{bmatrix}$$

$$= \begin{bmatrix} -1 & 0 \\ 0 & -2 \end{bmatrix} = \begin{bmatrix} \lambda_1 & 0 \\ 0 & \lambda_2 \end{bmatrix}$$

Before doing another example, we should note that we apparently only need to compute λ_i and not P for these problems because it is just necessary to fill in the diagonal terms to get A. This is true as far as A is concerned. However, we also need B, C, D for the complete Jordan canonical form. The computation of B and C requires knowledge of the P matrix.

EXAMPLE 10-17

Let:

$$\hat{A} = \begin{bmatrix} 2 & -4 & 4 \\ -4 & 6 & 0 \\ 4 & 0 & 6 \end{bmatrix}$$

Find A.

Solution

$$\det(\lambda I - \hat{A}) = \begin{bmatrix} \lambda & 0 & 0 \\ 0 & \lambda & 0 \\ 0 & 0 & \lambda \end{bmatrix} - \begin{bmatrix} 2 & -4 & 4 \\ -4 & 6 & 0 \\ 4 & 0 & 6 \end{bmatrix} = \begin{bmatrix} \lambda - 2 & 4 & -4 \\ 4 & \lambda - 6 & 0 \\ -4 & 0 & \lambda - 6 \end{bmatrix}$$

$$= (\lambda - 2)(\lambda^2 - 12\lambda + 36) - 4(4)(\lambda - 6) - 4(4)(\lambda - 6)$$

$$= \lambda^3 - 14\lambda^2 + 28\lambda + 120 = 0$$

Guess $\lambda_1 = -2$. Check $-8 - 56 - 56 + 120 = 0$. Then $\lambda^2 - 16\lambda + 60 = 0$.

Therefore $\lambda_2 = 6, \qquad \lambda_3 = 10$

The eigenvectors are obtained as follows

$$(\lambda_1 I - \hat{A})p_1 = 0 \quad \text{or} \quad \begin{bmatrix} -4 & 4 & -4 \\ 4 & -8 & 0 \\ -4 & 0 & -8 \end{bmatrix} \begin{bmatrix} p_{11} \\ p_{21} \\ p_{31} \end{bmatrix} = \begin{bmatrix} 0 \\ 0 \\ 0 \end{bmatrix}$$

Letting $p_{31} = 1$, we get $p_{21} = -1$ and $p_{11} = -2$. Likewise for λ_2 and p_2, letting $p_{22} = 1$, we obtain $p_{12} = 0$, $p_{22} = 1$, $p_{32} = 1$. For λ_3 and p_3 we have $p_{33} = 1$, $p_{23} = -1$, $p_{13} = 1$.

Thus
$$P = \begin{bmatrix} -2 & 0 & 1 \\ -1 & 1 & -1 \\ 1 & 1 & 1 \end{bmatrix}$$

Inverting, we obtain:

$$P^{-1} = \frac{\begin{bmatrix} 2 & 0 & -2 \\ 1 & -3 & 2 \\ -1 & -3 & -2 \end{bmatrix}^T}{-6}$$

$$P^{-1} = \begin{bmatrix} -\dfrac{1}{3} & -\dfrac{1}{6} & \dfrac{1}{6} \\ 0 & \dfrac{1}{2} & \dfrac{1}{2} \\ \dfrac{1}{3} & -\dfrac{1}{3} & \dfrac{1}{3} \end{bmatrix}$$

To check, compute:

$$A = P^{-1}\hat{A}P = \begin{bmatrix} \lambda_1 & 0 & 0 \\ 0 & \lambda_2 & 0 \\ 0 & 0 & \lambda_3 \end{bmatrix}$$

Usually when the characteristic equation has repeated roots, the $\hat{A}$ matrix cannot be diagonalized. However, we can still write $A = P^{-1}\hat{A}P$, except that now the A matrix will be in what is called the Jordan canonical form. The diagonal form for A is just a special case of this. In the Jordan blocks mentioned previously the diagonal elements for a given block are all the same. The elements immediately above the diagonal terms of a given block will be 1's.

Our problem, as before, is to find P. Assume $\hat{A}$ has r distinct eigenvalues among the n total eigenvalues. Eigenvectors of these first-order eigenvalues are found as before:

$$(\lambda_i I - \hat{A})p_i = 0, \qquad i = 1, 2, \ldots, r \tag{10-32}$$

Now if a repeated root is repeated s times, then there will be $s - 1$ additional eigenvectors corresponding to this root. Call the s eigenvectors for this repeated root $\tilde{p}_1, \tilde{p}_2, \ldots, \tilde{p}_s$. These can be calculated from the following equations, assuming that the eigenvalue at issue is λ_j:

$$(\lambda_j I - \hat{A})\tilde{p}_1 = 0 \qquad \text{(This is just like the nonrepeated root case.)}$$
$$(\lambda_j I - \hat{A})\tilde{p}_2 = -\tilde{p}_1$$
$$\vdots$$
$$(\lambda_j I - \hat{A})\tilde{p}_s = -\tilde{p}_{s-1} \tag{10-33}$$

Examples will illustrate this rather complicated procedure.

EXAMPLE 10-18

Cast the following matrix into its Jordan canonical form:

$$\hat{A} = \begin{bmatrix} 2.5 & 2 & 0 \\ 0 & 0.5 & 0 \\ -2 & 2 & 0.5 \end{bmatrix}$$

Solution

$$\det(\lambda I - \hat{A}) = (\lambda - 2.5)(\lambda - 0.5)^2 = 0$$

thus

$$\lambda_1 = 2.5, \qquad \lambda_2 = 0.5 \qquad \lambda_3 = 0.5$$

In this case there are $r = 2$ distinct eigenvalues.

$$(\lambda_1 I - \hat{A})p_1 = 0 = \begin{bmatrix} 0 & -2 & 0 \\ 0 & 2 & 0 \\ 2 & -2 & 2 \end{bmatrix} \begin{bmatrix} p_{11} \\ p_{21} \\ p_{31} \end{bmatrix} = \begin{bmatrix} 0 \\ 0 \\ 0 \end{bmatrix}$$

Therefore $\qquad -2p_{21} = 0, \qquad 2p_{21} = 0, \qquad 2p_{11} - 2p_{21} + 2p_{31} = 0$

Then $\qquad p_{21} = 0, \quad \text{choose} \quad p_{11} = 1$

Thus $\qquad p_{31} = -1 \quad \text{and} \quad p_1 = \begin{bmatrix} 1 \\ 0 \\ -1 \end{bmatrix}$

$$(\lambda_2 I - \hat{A})p_2 = 0 = \begin{bmatrix} -2 & -2 & 0 \\ 0 & 0 & 0 \\ 2 & -2 & 0 \end{bmatrix} \begin{bmatrix} p_{12} \\ p_{22} \\ p_{32} \end{bmatrix} = \begin{bmatrix} 0 \\ 0 \\ 0 \end{bmatrix}$$

Therefore $-2p_{12} - 2p_{22} = 0$ and $2p_{12} - 2p_{22} = 0$

Thus $p_{12} = p_{22} = 0$

and we must choose $p_{32} \neq 0$. Let $p_{32} = 1$.

Then
$$p_2 = \begin{bmatrix} 0 \\ 0 \\ 1 \end{bmatrix}$$

Now the repeated root, 0.5, is repeated $s = 2$ times. Thus there will be $s - 1 = 1$ additional eigenvectors to compute. Referring to Equation 10-33, we can write:

$$(\lambda_j I - \hat{A})\tilde{p}_1 = 0$$
$$(\lambda_j I - \hat{A})\tilde{p}_2 = -\tilde{p}_1$$

The first of these has already been employed to calculate

$$p_2 = \begin{bmatrix} 0 \\ 0 \\ 1 \end{bmatrix}.$$

That is, $p_2 = \tilde{p}_1$. Thus we must solve:

$$(\lambda_2 I - \hat{A})\tilde{p}_2 = -\begin{bmatrix} 0 \\ 0 \\ 1 \end{bmatrix}$$

for the third eigenvector $\tilde{p}_2 = p_3$:

$$\begin{bmatrix} -2 & -2 & 0 \\ 0 & 0 & 0 \\ 2 & -2 & 0 \end{bmatrix} \begin{bmatrix} p_{13} \\ p_{23} \\ p_{33} \end{bmatrix} = \begin{bmatrix} 0 \\ 0 \\ -1 \end{bmatrix}$$

therefore $-2p_{13} - 2p_{23} = 0$, $2p_{13} - 2p_{23} = -1$, and $p_{33} =$ anything. Thus $-4p_{23} = -1$ or $p_{23} = 0.25$. Then $p_{13} = -0.25$. For simplicity, choose $p_{33} = 0$:

Thus
$$P = \begin{bmatrix} 1 & 0 & -0.25 \\ 0 & 0 & 0.25 \\ -1 & 1 & 0 \end{bmatrix}$$

from which it follows that:

$$A = \left[\begin{array}{c|cc} 2.5 & 0 & 0 \\ \hline 0 & 0.5 & 1 \\ 0 & 0 & 0.5 \end{array} \right] = P^{-1}\hat{A}P$$

EXAMPLE 10-19

Cast the matrix:

$$\hat{A} = \begin{bmatrix} 0 & 0.5 & 0 \\ 0 & 0 & 0.5 \\ 4 & -6 & 3 \end{bmatrix}$$

into the Jordan form.

Solution

$$\det(\lambda I - \hat{A}) = \lambda^3 - 3\lambda^2 + 3\lambda - 1 = (\lambda - 1)^3 = 0; \qquad \lambda_1 = \lambda_2 = \lambda_3 = 1$$

Therefore we have $r = 1$ distinct eigenvalue. Also, $s = 3$; thus $(\lambda I - \hat{A})p_1 = 0$.

$$\begin{bmatrix} 1 & -0.5 & 0 \\ 0 & 1 & -0.5 \\ -4 & 6 & -2 \end{bmatrix} \begin{bmatrix} p_{11} \\ p_{21} \\ p_{31} \end{bmatrix} = \begin{bmatrix} 0 \\ 0 \\ 0 \end{bmatrix} \qquad \begin{array}{l} p_{11} - 0.5p_{21} = 0 \\ p_{21} - 0.5p_{31} = 0 \\ -4p_{11} + 6p_{21} - 2p_{31} = 0 \end{array}$$

Let $p_{21} = 2$:

Then $\qquad p_{11} = 1 \quad$ and $\quad p_{31} = 4 \quad$ and $\quad p_1 = \begin{bmatrix} 1 \\ 2 \\ 4 \end{bmatrix}$

Then we need:

$$(\lambda I - \hat{A})\tilde{p}_2 = -\tilde{p}_1$$

$$(\lambda I - \hat{A})\tilde{p}_3 = -\tilde{p}_2$$

$$\begin{bmatrix} 1 & -0.5 & 0 \\ 0 & 1 & -0.5 \\ -4 & 6 & -2 \end{bmatrix} \begin{bmatrix} p_{12} \\ p_{22} \\ p_{32} \end{bmatrix} = \begin{bmatrix} -1 \\ -2 \\ -4 \end{bmatrix}$$

$$p_{12} - 0.5p_{22} = -1 \qquad \text{Let } p_{22} = 2; \text{ then } p_{12} = 0 \quad \text{and} \quad p_{32} = 8$$

$$p_{22} - 0.5p_{32} = -2$$

$$-4p_{12} + 6p_{22} - 2p_{32} = -4$$

Thus
$$\tilde{p}_2 = p_2 = \begin{bmatrix} 0 \\ 2 \\ 8 \end{bmatrix}$$

Finally:

$$\begin{bmatrix} 1 & -0.5 & 0 \\ 0 & 1 & -0.5 \\ -4 & 6 & -2 \end{bmatrix} \begin{bmatrix} p_{13} \\ p_{23} \\ p_{33} \end{bmatrix} = \begin{bmatrix} 0 \\ -2 \\ -8 \end{bmatrix}$$

$$p_{13} - 0.5p_{23} = 0 \qquad \text{Let } p_{13} = 1; \text{ then } p_{23} = 2 \quad \text{and} \quad p_{33} = 8$$

$$p_{23} - 0.5p_{33} = -2$$

$$-4p_{13} + 6p_{23} - 2p_{33} = -8$$

Thus
$$\tilde{p}_3 = p_3 = \begin{bmatrix} 1 \\ 2 \\ 8 \end{bmatrix}$$

$$P = \begin{bmatrix} 1 & 0 & 1 \\ 2 & 2 & 2 \\ 4 & 8 & 8 \end{bmatrix}$$

The Jordan canonical form is:

$$A = \begin{bmatrix} 1 & 1 & 0 \\ 0 & 1 & 1 \\ 0 & 0 & 1 \end{bmatrix}$$

which is equal to $P^{-1}\hat{A}P$.

There are many special cases and special theorems related to these Jordan canonical forms that we are unable to consider. But although our treatment here is brief, the previous developments should be sufficient to transform most low-order systems into the Jordan canonical form. Computer software is available to treat higher-order systems and is usually employed in real engineering problems. However, learning how to handle low-order systems analytically will provide insight into these complex software packages.

Thus far we have considered the controllability canonical form, the observability canonical form, and the Jordan canonical form. Although many other canonical forms exist, these three are very widely employed. We now move on to other considerations and discuss next the block diagramatic realizations of these three canonical forms.

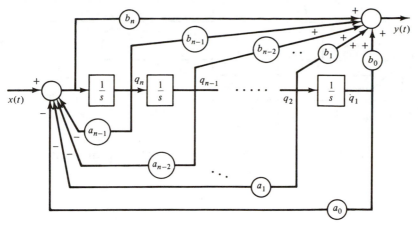

Figure 10-11 Controllability canonical form realization.

10-3-3 Canonical Block Diagramatics

The general block diagrams for the controllability canonical form and for the observability canonical form are presented in Figures 10-11 and 10-12. These diagrams are for the continuous case. For the discrete case we need only replace the integrators by unit delays.

EXAMPLE 10-20

A certain system has a transfer function $H(s) = s^2 + s + 1/s^3 + s + 1$. Construct the controllability canonical form realization of this system in block diagram format.

Solution. From the general transfer function we can read off the coefficients: $a_{n-1} = a_2 = 0$, $a_1 = 1$, $a_0 = 1$ and $b_n = b_3 = 0$, $b_2 = 1$, $b_1 = 1$, $b_0 = 1$. Then from the block diagram of Figure 10-11 we can construct the block

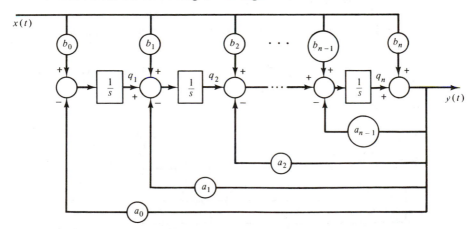

Figure 10-12 Observability canonical form realization.

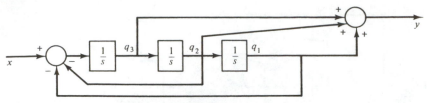

Figure 10-13   Block diagram for $H(s) = s^2 + s + 1/s^3 + s + 1$.

diagram of Figure 10-13. From this block diagram we can construct the
state variable equation as follows:

$$\dot{q}_3 = x - q_2 - q_1$$

$$\dot{q}_2 = q_3$$

$$\dot{q}_1 = q_2 \qquad \text{and} \qquad y = q_3 + q_2 + q_1$$

From which

$$A = \begin{bmatrix} 0 & 1 & 0 \\ 0 & 0 & 1 \\ -1 & -1 & 0 \end{bmatrix}, \quad B = \begin{bmatrix} 0 \\ 0 \\ 1 \end{bmatrix}, \quad C = [1 \quad 1 \quad 1] \text{ and } D = [0]$$

These check with the general form of the A, B, C, D canonical realization
equations, Equations 10-8 and 10-9.

EXAMPLE 10-21

Given the block diagram of Figure 10-14, construct the observability
canonical form state equations.

Solution.   The figure shows that we have a third-order system (the
number of integrators—or unit delays—in the system is the order of the
system). Label the outputs of the integrators from left to right q_1, q_2, and
q_3. Then we can write:

$$\dot{q}_1 = 2x - 2q_3, \qquad \dot{q}_2 = q_1 - 2q_3, \qquad \dot{q}_3 = q_2 + 4x \quad \text{and} \quad y = q_3$$

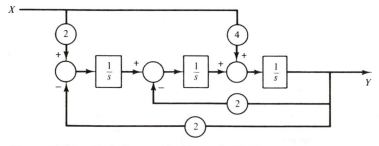

Figure 10-14   Block diagram for Example 10-21.

These equations can be cast into the general format:

$$\dot{q} = \begin{bmatrix} 0 & 0 & -2 \\ 1 & 0 & -2 \\ 0 & 1 & 0 \end{bmatrix} q + \begin{bmatrix} 2 \\ 0 \\ 4 \end{bmatrix} x, \qquad y = [0 \quad 0 \quad 1]q + [0]x$$

By comparing these equations with Equations 10-12 and 10-13, we can read off the coefficients of the transfer function:

$$a_0 = 2, \quad a_1 = 2, \quad a_2 = 0, \quad \text{and,} \quad b_0 - a_0 b_n = 2$$

$$b_1 - a_1 b_n = 0$$

$$b_2 - a_2 b_n = 4$$

But since $b_n = b_3 = 0$, $b_0 = 2$, $b_1 = 0$, and $b_2 = 4$. From these we can construct the transfer function:

$$H(s) = \frac{4s^2 + 2}{s^3 + 2s + 2}$$

EXAMPLE 10-22

Given the block diagram of Figure 10-15 which is in the controllability format, transform the representation to the observability format.

Solution. Compare Figure 10-15 with Figure 10-11 where the $1/s$ integrators are replaced by unit delays. From the comparison we see that $n = 4$, $b_4 = 3$, $b_3 = b_2 = b_1 = 0$, $b_0 = 1$, and $a_3 = 2$, $a_2 = 0$, $a_1 = 2$, $a_0 = 3$. Now, following the general block diagram format for the observability canonical form realization, Figure 10-12, we can construct the block diagram of Figure 10-16.

Block diagrams for the Jordan forms of continuous systems are based on transfer functions of the form $(s - \lambda_i)^{-1}$, where λ_i represents eigenvalues of the system. This transfer function form can be decomposed into an integrator with a

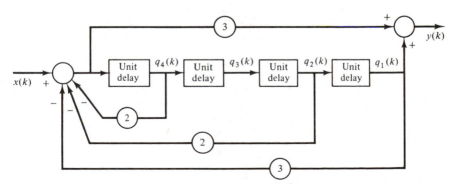

Figure 10-15   Discrete system in controllability form.

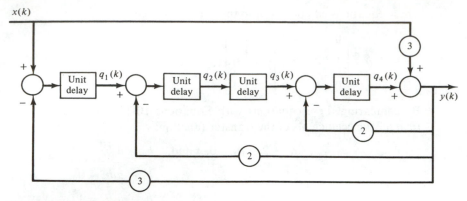

Figure 10-16 Block diagram for Example 10-22.

feedback loop as in Figure 10-17. For discrete systems we could use this feedback form with the integrator replaced by a unit delay. Although this unit delay form is most typical, discrete systems can also be handled with transfer functions of the form $(z - \lambda_i)^{-1}$ which is analogous to the $(s - \lambda_i)^{-1}$ form for continuous systems. Examples will illustrate the block diagrammatics involved with the Jordan canonical form realizations.

EXAMPLE 10-23

Construct the block diagram representation of the following diagonalized system:

$$\dot{q}(t) = \begin{bmatrix} -1 & 0 \\ 0 & -2 \end{bmatrix} q(t) + \begin{bmatrix} 3 \\ 2 \end{bmatrix} x(t)$$

$$y(t) = [4 \quad 3] q(t) + [0] x(t)$$

Solution.   We can write $\dot{q}_1 = -q_1 + 3x$ and $\dot{q}_2 = -2q_2 + 2x$. Laplace transform these equations assuming zero-initial conditions. $sQ_1 = -Q_1 + 3X$ and $sQ_2 = -2Q_2 + 2X$.

Then $$Q_1(s) = \frac{3X(s)}{s + 1} \quad \text{and} \quad Q_2(s) = \frac{2X(s)}{s + 2}$$

Also, $$Y(s) = 4Q_1(s) + 3Q_2(s)$$

A block diagram for these equations can be constructed using $(s - \lambda_1)^{-1}$ blocks. Here we have $\lambda_1 = -1$ and $\lambda_2 = -2$ (see Figure 10-18).

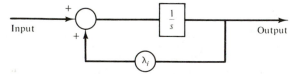

Figure 10-17 Decomposition of the $(s - \lambda_i)^{-1}$ term.

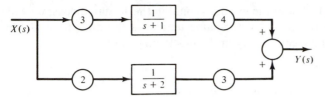

Figure 10-18   Block diagram for Example 10-23.

EXAMPLE 10-24

Construct the block diagram representation of the Jordan form equations:

$$\dot{q}(t) = \begin{bmatrix} -1 & 1 & 0 \\ 0 & -1 & 0 \\ 0 & 0 & -2 \end{bmatrix} q(t) + \begin{bmatrix} 3 \\ 2 \\ 1 \end{bmatrix} x(t)$$

and $y(t) = [4 \quad 5 \quad 8] q(t) + [0] x(t)$

Solution

$$sQ_3 = -2Q_3 + X$$

$$sQ_2 = -Q_2 + 2X$$

$$sQ_1 = -Q_1 + Q_2 + 3X$$

and $Y = 4Q_1 + 5Q_2 + 8Q_3$

Therefore $Q_3(s) = \dfrac{X(s)}{s + 2}, \qquad Q_2(s) = \dfrac{2X(s)}{s + 1},$

and $Q_1(s) = \dfrac{Q_2(s)}{s + 1} + \dfrac{3X(s)}{s + 1}$

From these equations we can construct the block diagram of Figure 10-19.

Comments: In the diagonal case one $(s - \lambda_i)^{-1}$ block corresponds to each eigenvalue λ_i. The block diagram is arranged in a parallel fashion. In the

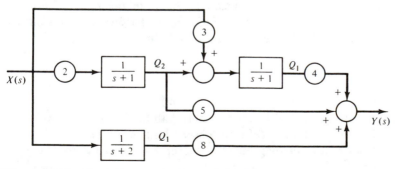

Figure 10-19   Block diagram for Example 10-24.

repeated root case with 1's above the main diagonal in the Jordan blocks, we will have a string of $(s - \lambda_i)^{-1}$ blocks of the same length as the dimension of the block. These strings will appear serially in the block diagram.

EXAMPLE 10-25

Construct the block diagram for the system:

$$q(k + 1) = \begin{bmatrix} 2 & 1 & 0 & 0 \\ 0 & 2 & 0 & 0 \\ 0 & 0 & 2 & 0 \\ 0 & 0 & 0 & 3 \end{bmatrix} q(k) + \begin{bmatrix} 2 \\ 3 \\ 3 \\ 2 \end{bmatrix} x(k)$$

and $y(k) = [0 \quad 1 \quad 0 \quad 1]q(k) + [2]x(k)$

Solution. In these discrete problems we could use the Z transform to write, for example, $zQ_4(z) = 3Q_4(z) + 2X(z)$. Then $Q_4(z) = 2X(z)/(z - 3)$ and we could use blocks of $(z - \lambda_i)^{-1}$ instead of $(s - \lambda_i)^{-1}$. However, the unit delay devices can also be employed by replacing the s^{-1} term in Figure 10-17 with a unit delay. Let us write out the equations involved here:

$$q_1(k + 1) = 2q_1(k) + q_2(k) + 2x(k)$$
$$q_2(k + 1) = 2q_2(k) + 3x(k)$$
$$q_3(k + 1) = 2q_3(k) + 3x(k)$$
$$q_4(k + 1) = 3q_4(k) + 2x(k)$$

and $y(k) = q_2(k) + q_4(k) + 2x(k)$

These equations can be simulated as in Figure 10-20 using four unit delay devices.

There are many aspects of the realization problem that will not be considered here. Typical questions that are beyond our present scope are: Which realization is best and what do we mean by best? in the sense of minimal sensitivity? in the sense of the least amount of required hardware? and so on. The next section in this chapter considers the solution of the state equations.

Drill Set: State Variables

1. Let $H(s) = s^2 + 10s + 30/s^4 + 20s^3 + 40$.
   (a) Determine A, B, C, D for the controllability canonical form.
   (b) Determine A, B, C, D for the observability canonical form.
   (c) Construct block diagrams for the representations in parts (a) and (b).

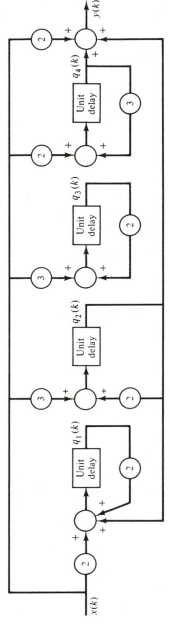

Figure 10-20 Block diagram for Example 10-25.

2. Let $H(z) = z + 1/z^2 + z + 1$. Represent this system in the Jordan canonical form and construct the block diagram.

3. Represent the following circuit in the state variable format. Let $V_{c1} = q_1$, $V_{c2} = q_2$, and $i_L = q_3$. Let $x_1(t) = V_s(t)$ and $x_2(t) = i_s(t)$. Assume $y(t) = V_0(t)$.

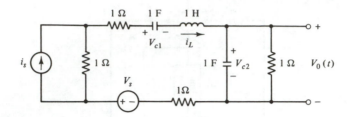

4. Consider the following discrete system:

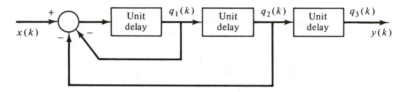

   (a) Determine A, B, C, D of the state variable representation.
   (b) Diagonalize A.
   (c) Construct the block diagram of the diagonal system using unit delay devices.

5. Diagonalize the following A matrices:

   (a) $A = \begin{bmatrix} 1 & 2 & 1 \\ 0 & 1 & 1 \\ 0 & 0 & 1 \end{bmatrix}$

   (b) $A = \begin{bmatrix} -1 & -1 \\ -1 & -2 \end{bmatrix}$

   (c) $A = \begin{bmatrix} -1 & 0 & 1 \\ 0 & -2 & 1 \\ 0 & 0 & -3 \end{bmatrix}$

   (d) $A = \begin{bmatrix} 100 & 50 \\ 75 & 150 \end{bmatrix}$

10-4 SOLUTION OF THE STATE EQUATIONS

The analysis problem seeks the output of a linear system when the system description is known and the input is specified. Referring to the general state and output equations, we note that to obtain $y(t)$ or $y(k)$, we need C, D, and q and x. If the system description is known, then C and D are known. If the input is specified, then x is known. That leaves only q to be obtained. To find q, we need to solve the state equation. This is the task at hand.

The state equation is a set of first-order difference or differential equations. The solution of these kinds of equations is well known. What is different about the approach now is that these equations are expressed in vector-matrix notation. Some concepts of vector-matrix algebra will need to be employed. As with scalar difference or differential equations, we solve the state equation by using both the classical time-domain approach and the transform approach.

10-4-1 The Discrete State Equation Solution

Start with the difference equation:

$$q(k + 1) = Aq(k) + Bx(k) \tag{10-3}$$

Assume we know $q(0)$ and $x(k)$ for $k = 0, 1, \ldots, N$. Then, letting $k = 0$, we get:

$$q(1) = Aq(0) + Bx(0)$$

Then
$$q(2) = Aq(1) + Bx(1) = A^2q(0) + ABx(0) + Bx(1)$$

$$q(3) = Aq(2) + Bx(2) = A^3q(0) + A^2Bx(0) + ABx(1) + Bx(2)$$

From these we can induce the following:

$$q(k) = A^kq(0) + A^{k-1}Bx(0) + A^{k-2}Bx(1) + \cdots$$
$$+ A^0Bx(k - 1)$$
$$= A^kq(0) + \sum_{m=1}^{k} A^{k-m}Bx(m - 1)$$

or
$$q(k) = A^kq(0) + \sum_{m=0}^{k-1} A^{k-m-1}Bx(m), \qquad k > 0 \tag{10-34}$$

In this equation, A, B, $q(0)$, and $x(m)$ are assumed known. The only problem here is to find A^k and A^{k-m-1}. This is a minor problem as long as the powers of A are small. If the powers get very large, however, like A^{100}, then it would be nice to be able to avoid multiplying the matrix A by itself one hundred times. Fortunately, a circumvention is available. It is based on the Cayley-Hamilton theorem which says simply that the A matrix satisfies its characteristic equation.

EXAMPLE 10-26

Prove the Cayley-Hamilton theorem.

Solution. The characteristic equation of a linear system A, B, C, D can be written as:

$$\det(\lambda I - A) = 0$$

and this equation can be expressed as:

$$g(\lambda) = \lambda^n + a_1\lambda^{n-1} + \cdots + a_{n-1}\lambda + a_n = 0$$

The Cayley-Hamilton theorem says that:

$$g(A) = A^n + a_1 A^{n-1} + \cdots + a_{n-1}A + a_n I = 0$$

This is a matrix equation as distinguished from the characteristic equation which is scalar. Let us assume in this proof that A is diagonalizable. For the case of repeated roots the Cayley-Hamilton theorem is still true, although the proof is somewhat more involved [cf. Bellman (1968)].

$$g(\lambda_i) = 0 \qquad \text{for all } \lambda_i, i = 1, 2, \cdots, n$$

Let $\Lambda = P^{-1}AP = \text{diag}[\lambda_1, \lambda_2, \ldots, \lambda_n]$. Thus $A = P\Lambda P^{-1}$. We know that $\det(\lambda I - A) = \det(\lambda I - \Lambda) = g(\lambda) = 0$, the characteristic equation of the system.

$$\text{Now}\quad g(A) = A^n + a_1 A^{n-1} + \cdots + a_{n-1}A + a_n I$$

$$= (P\Lambda P^{-1})^n + a_1(P\Lambda P^{-1})^{n-1} + \cdots + a_{n-1}(P\Lambda P^{-1}) + a_n I$$

$$= P\Lambda^n P^{-1} + a_1 P\Lambda^{n-1}P^{-1} + \cdots + a_{n-1}P\Lambda P^{-1} + a_n PP^{-1}$$

$$= P(\Lambda^n + a_1\Lambda^{n-1} + \cdots + a_{n-1}\Lambda + a_n I)P^{-1}$$

The term in parentheses is an $n \times n$ diagonal matrix:

$$\begin{bmatrix} \lambda_1^n + a_1\lambda_1^{n-1} + \cdots + a_{n-1}\lambda_1 + a_n & 0 & & \cdots & 0 \\ 0 & \lambda_2^n + a_1\lambda_2^{n-1} + \cdots + a_{n-1}\lambda_2 + a_n & & \cdots & 0 \\ \vdots & \vdots & & & \vdots \\ 0 & 0 & & & \lambda_n^n + a_1\lambda_n^{n-1} + \cdots + a_{n-1}\lambda_n + a_n \end{bmatrix}$$

Each of these diagonal terms is zero because $g(\lambda) = 0$ for all $\lambda_i, i = 1, \ldots, n$.

Thus $$g(A) = P0P^{-1} = 0$$

EXAMPLE 10-27

Consider the matrix

$$A = \begin{bmatrix} 10 & 40 \\ 40 & 10 \end{bmatrix}$$

Determine A^4 using the Cayley-Hamilton theorem.

Solution

$$g(\lambda) = \det(\lambda I - A) = \begin{vmatrix} \lambda - 10 & -40 \\ -40 & \lambda - 10 \end{vmatrix}$$

$$= \lambda^2 - 20\lambda - 1500 = 0$$

From the Cayley-Hamilton theorem, we can write:

$$A^2 - 20A - 1500I = 0$$

Therefore $A^2 = 20A + 1500I$ and

$$A^3 = 20A^2 + 1500A$$

$$= 20(20A + 1500I) + 1500A$$

$$= 400A + 30{,}000I + 1500A$$

$$= 1900A + 30{,}000I$$

And $A^4 = A(A^3) = 1900A^2 + 30{,}000A$

$$= 1900(20A + 1500I) + 30{,}000A$$

$$= 68{,}000A + 2{,}850{,}000I$$

Thus $$A^4 = \begin{bmatrix} 3.53 \times 10^6 & 2.72 \times 10^6 \\ 2.72 \times 10^6 & 3.53 \times 10^6 \end{bmatrix}$$

Note that we were able to get A^4 without doing any matrix multiplication. The fourth power of A is written as a linear combination of A and I.

As a generalization of the previous example, it can be shown that for any integer $m > 0$, A^{n+m} is a linear combination of $I, A, A^2, \ldots, A^{n-1}$. This is a very powerful result. If we are seeking A^{100}, for example, and A is a 4 × 4 matrix, then we can write $A^{100} = \beta_0 I + \beta_1 A + \beta_2 A^2 + \beta_3 A^3$. The problem here is to determine the β terms. To do this we use the obvious fact that the eigenvalues also satisfy the characteristic equation. From this fact we can deduce that any matrix polynomial expressed in terms of $I, A, A^2, \ldots, A^{n-1}$ is also satisfied by the scalar equation that results from replacing A by λ and I by 1. We illustrate this deduction with the next example.

EXAMPLE 10-28
Determine A^5 when

$$A = \begin{bmatrix} 1 & 2 & 1 \\ 0 & -1 & -2 \\ 0 & 0 & 2 \end{bmatrix}$$

Solution

$$\det(\lambda I - A) = (\lambda - 1)(\lambda + 1)(\lambda - 2)$$

$$= \lambda^3 - 2\lambda^2 - \lambda + 2 = 0$$

Therefore $\lambda_1 = 1, \qquad \lambda_2 = -1, \quad \text{and} \quad \lambda_3 = 2$

From the Cayley-Hamilton theorem it follows that:

$$A^5 = \beta_0 I + \beta_1 A + \beta_2 A^2$$

From the fact that the eigenvalues satisfy the characteristic equation we have:

$$\lambda^3 = 2\lambda^2 + \lambda - 2$$

$$\lambda^4 = 2(2\lambda^2 + \lambda - 2) + \lambda^2 - 2\lambda$$

$$= 5\lambda^2 - 4$$

and $$\lambda^5 = 5\lambda^3 - 4\lambda$$

$$= 5(2\lambda^2 + \lambda - 2) - 4\lambda$$

$$= 10\lambda^2 + \lambda - 10$$

Since these equations are also true with λ replaced by A and 1 replaced by I, by comparing coefficients we can deduce that $\beta_1 = 1$, $\beta_2 = 10$, and $\beta_0 = -10$. Now instead of using this iteration procedure, we can postulate that:

$$\lambda^5 = \beta_0 + \beta_1 \lambda + \beta_2 \lambda^2$$

where the λ's are eigenvalues of A. This follows from the Cayley-Hamilton result that $A^5 = \beta_0 I + \beta_1 A + \beta_2 A^2$.

Let: $\lambda_1 = 1 \quad \text{and} \quad 1^5 = \beta_0 + \beta_1 + \beta_2 = 1$

$\lambda_2 = -1 \quad \text{and} \quad (-1)^5 = \beta_0 - \beta_1 + \beta_2 = -1$

$\lambda_3 = 2 \quad \text{and} \quad 2^5 = \beta_0 + 2\beta_1 + 4\beta_2 = 32$

From these equations $\beta_1 = 1$, $\beta_2 = 10$ and $\beta_0 = -10$

Therefore $$A^5 = -10I + A + 10A^2$$

$$= \begin{bmatrix} 1 & 2 & -9 \\ 0 & -1 & -22 \\ 0 & 0 & 32 \end{bmatrix}$$

The case of repeated eigenvalues can be handled by differentiating with respect to λ the scalar equation that is analogous to the matrix equation that arises from the Cayley-Hamilton theorem. Differentiate once for each time the root is repeated.

EXAMPLE 10–29

Determine A^5 when

$$A = \begin{bmatrix} 1 & 2 & 1 \\ 0 & 1 & 2 \\ 0 & 0 & 1 \end{bmatrix}$$

Solution

$$\det(\lambda I - A) = (\lambda - 1)^3 = 0$$

where

$$\lambda_1 = \lambda_2 = \lambda_3 = 1$$

$$A^5 = \beta_0 I + \beta_1 A + \beta_2 A^2$$

Also,

$$\lambda^5 = \beta_0 + \beta_1 \lambda + \beta_2 \lambda^2$$

Plugging in $\lambda_1 = 1$, we get $1 = \beta_0 + \beta_1 + \beta_2$. We need two more equations to solve for all three unknowns. Differentiate with respect to λ twice and we get:

$$5\lambda^4 = \beta_1 + 2\beta_2 \lambda \quad \text{and} \quad 20\lambda^3 = 2\beta_2$$

Plugging $\lambda = 1$ into these equations, we obtain:

$$5(1)^4 = \beta_1 + 2\beta_2 = 5$$

$$20(1)^3 = 2\beta_2 = 20$$

where

$$\beta_2 = 10, \qquad \beta_1 = -15, \qquad \beta_0 = 6$$

Thus

$$A^5 = 6I - 15A + 10A^2$$

$$= \begin{bmatrix} 1 & 10 & 45 \\ 0 & 1 & 10 \\ 0 & 0 & 1 \end{bmatrix}$$

From our Cayley-Hamilton discussion let us return to Equation 10-34. Any A matrix to any power can now be adequately handled. Instead of A being set to a specific power, Equation 10-34 calls for A^k or A^{k-m-1}. These determinations are obtained in exactly the same way as before.

EXAMPLE 10-30

Assume the input to a discrete system is zero

but $\quad q(0) = \begin{bmatrix} 1 \\ 1 \end{bmatrix}$

Let:

$$A = \begin{bmatrix} -1 & -1 \\ -1 & 1 \end{bmatrix}, \qquad B = \begin{bmatrix} 0 \\ 1 \end{bmatrix}, \qquad C = [1 \quad 1], \qquad D = 0$$

Determine $q(k)$ for all $k > 0$.

Solution.   From Equation 10-34, since $x(k) = 0$, we need:

$$q(k) = A^k q(0) = A^k \begin{bmatrix} 1 \\ 1 \end{bmatrix}$$

Our problem is to find A^k.

$$\det(\lambda I - A) = \lambda^2 - 2 = 0$$

where
$$\lambda = \pm \sqrt{2}, \qquad \lambda_1 = \sqrt{2}, \qquad \lambda_2 = -\sqrt{2}$$

We can write $A^k = \beta_0 I + \beta_1 A$. Also,

$$\lambda^k = \beta_0 + \beta_1 \lambda$$

Plugging in the λ's, we get:

$$\sqrt{2}^k = \beta_0 + \sqrt{2}\beta_1$$
$$(-\sqrt{2})^k = \beta_0 - \sqrt{2}\beta_1$$

Thus
$$\beta_0 = \frac{(\sqrt{2})^k + (-\sqrt{2})^k}{2} \quad \text{and} \quad \beta_1 = \frac{(\sqrt{2})^k - (-\sqrt{2})^k}{2\sqrt{2}}$$

Then
$$A^k = \begin{bmatrix} \beta_0 & 0 \\ 0 & \beta_0 \end{bmatrix} + \begin{bmatrix} -\beta_1 & -\beta_1 \\ -\beta_1 & \beta_1 \end{bmatrix} = \begin{bmatrix} \beta_0 - \beta_1 & -\beta_1 \\ -\beta_1 & \beta_0 + \beta_1 \end{bmatrix}$$

and
$$q(k) = A^k \begin{bmatrix} 1 \\ 1 \end{bmatrix} = \begin{bmatrix} \beta_0 - 2\beta_1 \\ \beta_0 \end{bmatrix}$$

or
$$q_1(k) = \beta_0 - 2\beta_1 = 1.207(-\sqrt{2})^k - 0.207(\sqrt{2})^k, \qquad k > 0$$
$$q_2(k) = \beta_0 = 0.5(\sqrt{2})^k + 0.5(-\sqrt{2})^k, \qquad k > 0$$

Thus far, we have focused on the solution of the discrete state equation using an iterative time-domain procedure. Induction was employed to arrive at the solution for $q(k)$ in Equation 10-34. The problem of determining A^k was solved by using the Cayley-Hamilton theorem. In addition to the classical time-domain approach for the solution of difference equations, we also have available the transform approach, that is, the Z transform in the case of discrete equations.

The Z transform approach to solving Equation 10-3 will now be discussed.

Taking the Z transform of Equation 10-3, we obtain:

$$zQ(z) - zq(0) = AQ(z) + BX(z)$$

or

$$Q(z) = (I - z^{-1}A)^{-1}q(0) + (I - z^{-1}A)^{-1}z^{-1}BX(z) \qquad (10\text{-}35)$$

Then $q(k)$ is the inverse Z transform of $Q(z)$.

$$q(k) = Z^{-1}\{(I - z^{-1}A)^{-1}\}q(0) + Z^{-1}\{(I - z^{-1}A)^{-1}z^{-1}BX(z)\} \qquad (10\text{-}36)$$

Comparing this equation with Equation 10-34, we can note an alternative to the Cayley-Hamilton approach to A^k:

$$A^k = Z^{-1}\{(I - z^{-1}A)^{-1}\} \qquad (10\text{-}37)$$

Thus, to get A^k, form $I - z^{-1}A$ which is an $n \times n$ matrix, next take the inverse of this matrix, and then take the inverse Z transform of this inverted matrix.

EXAMPLE 10-31

Determine A^k using the Z transform method when:

$$A = \begin{bmatrix} 0 & 1 \\ -2 & -3 \end{bmatrix}$$

Solution

$$I - z^{-1}A = \begin{pmatrix} 1 & 0 \\ 0 & 1 \end{pmatrix} - \begin{pmatrix} 0 & z^{-1} \\ -2z^{-1} & -3z^{-1} \end{pmatrix} = \begin{bmatrix} 1 & -z^{-1} \\ 2z^{-1} & 1 + 3z^{-1} \end{bmatrix}$$

The inverse of this matrix is:

$$\frac{\begin{pmatrix} 1 + 3z^{-1} & z^{-1} \\ -2z^{-1} & 1 \end{pmatrix}}{2z^{-2} + 3z^{-1} + 1} = \begin{bmatrix} \dfrac{z(z + 3)}{(z + 2)(z + 1)} & \dfrac{z}{(z + 2)(z + 1)} \\ \dfrac{-2z}{(z + 2)(z + 1)} & \dfrac{z^2}{(z + 2)(z + 1)} \end{bmatrix}$$

Finally, taking the inverse Z transform of this matrix in a term-by-term fashion we get:

$$A^k = \begin{bmatrix} 2(-1)^k - (-2)^k & (-1)^k - (-2)^k \\ 2(-2)^k - 2(-1)^k & 2(-2)^k - (-1)^k \end{bmatrix}$$

10-4-2 The Continuous State Equation Solution

As the next logical step in our presentation, consider the solution of the continuous version of the state equation:

$$\dot{q}(t) = Aq(t) + Bx(t), \qquad q(t_0) = q_0 \qquad (10\text{-}1)$$

Since any solution of a differential equation that satisfies the initial condition is *the* solution, we will propose a solution and then test it by substitution. Consider the scalar version of Equation 10-1.

$$\dot{q}(t) = aq(t) + bx(t), \qquad q(t_0) \tag{10-38}$$

Assume $q(t) = q_H(t) + q_p(t)$, the homogeneous solution and the particular solution. By inspection:

$$q_H(t) = e^{at}\beta \tag{10-39}$$

where β is the constant to be determined.

To determine $q_p(t)$, use the method of variation of parameters and let:

$$q_p(t) = e^{at}\beta(t) \tag{10-40}$$

Here β is assumed to vary with time. It follows that:

$$\dot{q}_p = ae^{at}\beta + e^{at}\dot{\beta}$$

$$= aq + bx$$

$$= ae^{at}\beta + bx$$

Then $\qquad\qquad e^{at}\dot{\beta} = bx$

and $\qquad\qquad \dot{\beta} = e^{-at}bx$

Thus $\qquad\qquad \beta(t) = \int_{-\infty}^{t} e^{-a\lambda}bx(\lambda)\, d\lambda$

and $\qquad\qquad q_p(t) = \int_{-\infty}^{t} e^{a(t-\lambda)}bx(\lambda)\, d\lambda$

Thus $\qquad\qquad q(t) = e^{at}\beta + \int_{-\infty}^{t} e^{a(t-\lambda)}bx(\lambda)\, d\lambda \tag{10-41}$

and solving for β (the constant), we use $q(t_0)$

$$q(t_0) = e^{at_0}\beta + \int_{-\infty}^{t_0} e^{at_0}e^{-a\lambda}bx(\lambda)\, d\lambda$$

Thus $\qquad\qquad \beta = e^{-at_0}q(t_0) - \int_{-\infty}^{t_0} e^{-a\lambda}bx(\lambda)\, d\lambda$

and $\qquad\qquad q(t) = e^{a(t-t_0)}q(t_0) - \int_{-\infty}^{t_0} e^{a(t-\lambda)}bx(\lambda)\, d\lambda + \int_{-\infty}^{t} e^{a(t-\lambda)}bx(\lambda)\, d\lambda$

which can be written as:

$$q(t) = e^{a(t-t_0)}q(t_0) + \int_{t_0}^{t} e^{a(t-\lambda)}bx(\lambda)\, d\lambda \tag{10-42}$$

Keeping the form of this scalar equation, we postulate for the solution of Equation 10-1:

$$q(t) = e^{A(t-t_0)}q(t_0) + \int_{t_0}^{t} e^{A(t-\lambda)}Bx(\lambda)\, d\lambda \tag{10-43}$$

Now an exponential to the power of an $n \times n$ matrix is itself an $n \times n$ matrix:

$$e^M = I + M + \frac{M^2}{2!} + \frac{M^3}{3!} + \cdots \qquad (10\text{-}44)$$

If $M = At$, where A and M are $n \times n$ matrices and t is a scalar, then we can differentiate $e^M = e^{At}$ as follows:

$$\frac{d}{dt} e^{At} = \frac{d}{dt}\left(I + At + \frac{A^2 t^2}{2!} + \frac{A^3 t^3}{3!} + \cdots \right)$$

$$= A + A^2 t + \frac{A^3 t^2}{2!} + \cdots$$

$$= A\left(I + At + \frac{A^2 t^2}{2!} + \cdots \right)$$

Therefore $\qquad \dfrac{d}{dt} e^{At} = A e^{At} \qquad\qquad\qquad (10\text{-}45)$

Let us now verify Equation 10-43 by substituting it into Equation 10-1. First, determine $\dot{q}(t)$:

$$\dot{q}(t) = A e^{A(t-t_0)} q(t_0) + e^{A(t-t)} Bx(t) + \int_{t_0}^{t} A e^{A(t-\lambda)} Bx(\lambda)\, d\lambda$$

$$= A\left[e^{A(t-t_0)} q(t_0) + \int_{t_0}^{t} e^{A(t-\lambda)} Bx(\lambda)\, d\lambda \right] + Bx(t)$$

But the term in brackets is just $q(t)$. Thus Equation 10-43 does satisfy Equation 10-1 and is *the* solution of the state equation. Observing Equation 10-43, we note that $q(t_0)$ is given, $x(t)$ the input is known, and A and B are known. That leaves only $e^{A(t-t_0)}$ and $e^{A(t-\lambda)}$ to be determined. It will suffice to determine e^{At}. Then let t be $t - t_0$ or let t be $t - \lambda$ as needed. This problem is analogous to the A^k determination problem considered for discrete systems. In fact, using the Cayley-Hamilton theorem, we can write the infinite series of Equation 10-44 as a finite series:

$$e^{At} = \beta_0 I + \beta_1 A + \cdots + \beta_{n-1} A^{n-1} \qquad (10\text{-}46)$$

where n is the order of the A matrix. Then we can use the fact that the eigenvalues satisfy the characteristic equation to deduce that:

$$e^{\lambda t} = \beta_0 + \beta_1 \lambda + \cdots + \beta_{n-1} \lambda^{n-1} \qquad (10\text{-}47)$$

for all the eigenvalues λ_i, $i = 1, \ldots, n$. For the case of repeated roots we can differentiate Equation 10-47 with respect to λ.

EXAMPLE 10-32

Determine e^{At} if

$$A = \begin{bmatrix} -1 & 1 \\ 0 & -2 \end{bmatrix}$$

Solution

$$\det(\lambda I - A) = \begin{vmatrix} \lambda + 1 & -1 \\ 0 & \lambda + 2 \end{vmatrix} = (\lambda + 1)(\lambda + 2) = 0$$

Thus $\lambda_1 = -1$ and $\lambda_2 = -2$

Since $n = 2$ we can write:

$$e^{At} = \beta_0 I + \beta_1 A$$

Also, $e^{\lambda t} = \beta_0 + \beta_1 \lambda$

or $e^{-t} = \beta_0 - \beta_1$ and $e^{-2t} = \beta_0 - 2\beta_1$

Therefore $\beta_1 = e^{-t} - e^{-2t}$ and $\beta_0 = 2e^{-t} - e^{-2t}$

Thus $e^{At} = \beta_0 \begin{pmatrix} 1 & 0 \\ 0 & 1 \end{pmatrix} + \beta_1 \begin{pmatrix} -1 & 1 \\ 0 & -2 \end{pmatrix}$

$$= \begin{bmatrix} \beta_0 - \beta_1 & \beta_1 \\ 0 & \beta_0 - 2\beta_1 \end{bmatrix} = \begin{bmatrix} e^{-t} & e^{-t} - e^{-2t} \\ 0 & e^{-2t} \end{bmatrix}$$

EXAMPLE 10-33
   Determine e^{At} if:

$$A = \begin{bmatrix} 1 & 1 & 1 \\ 0 & 1 & 2 \\ 0 & 0 & 2 \end{bmatrix}$$

Solution

$$\det(\lambda I - A) = (\lambda - 1)^2(\lambda - 2) = 0$$
$$\lambda_1 = 1, \qquad \lambda_2 = 1, \qquad \text{and} \qquad \lambda_3 = 2$$
$$e^{At} = \beta_0 I + \beta_1 A + \beta_2 A^2$$

$e^{\lambda t} = \beta_0 + \beta_1 \lambda + \beta_2 \lambda^2$, which we can differentiate with respect to λ to yield $te^{\lambda t} = \beta_1 + 2\beta_2 \lambda$ and substituting the eigenvalues into these equation, we get:

$$e^t = \beta_0 + \beta_1 + \beta_2 \quad \text{and} \quad te^t = \beta_1 + 2\beta_2 \quad \text{and} \quad e^{2t} = \beta_0 + 2\beta_1 + 4\beta_2$$

From these equations we obtain:

$$\beta_0 = e^{2t} - 2te^t, \qquad \beta_1 = 3te^t + 2e^t - 2e^{2t}, \quad \text{and} \quad \beta_2 = e^{2t} - e^t - te^t$$

$$\text{Thus} \quad e^{At} = \begin{bmatrix} \beta_0 & 0 & 0 \\ 0 & \beta_0 & 0 \\ 0 & 0 & \beta_0 \end{bmatrix} + \begin{bmatrix} \beta_1 & \beta_1 & \beta_1 \\ 0 & \beta_1 & 2\beta_1 \\ 0 & 0 & 2\beta_1 \end{bmatrix} + \begin{bmatrix} \beta_2 & 2\beta_2 & 5\beta_2 \\ 0 & \beta_2 & 6\beta_2 \\ 0 & 0 & 4\beta_2 \end{bmatrix}$$

$$= \begin{bmatrix} e^t & te^t & 3e^{2t} - 3e^t - 2te^t \\ 0 & 0 & 2e^{2t} - 2e^t \\ 0 & 0 & e^{2t} \end{bmatrix}$$

Thus far in the solution to the continuous state equation we have used classical time-domain procedures. We arrived at the solution for $q(t)$ in Equation 10-43. However, the transform approach is also available. As we used the Z transform to solve the discrete state equation, similarly, we can use the Laplace transform to solve the continuous state equation.

The Laplace transform solution to Equation 10-1 will now be investigated. Taking the Laplace transform of Equation 10-1, we get:

$$sQ(s) - q(0) = AQ(s) + BX(s)$$

or
$$Q(s) = (sI - A)^{-1}q(0) + (sI - A)^{-1}BX(s) \qquad (10\text{-}48)$$

Then $q(t)$ is the inverse Laplace transform of $Q(s)$:

$$q(t) = \mathcal{L}^{-1}\{(sI - A)^{-1}\}q(0) + \mathcal{L}^{-1}\{(sI - A)^{-1}BX(s)\} \qquad (10\text{-}49)$$

Comparing this equation with Equation 10-43, with $t_0 = 0$, we note that:

$$e^{At} = \mathcal{L}^{-1}\{(sI - A)^{-1}\} \qquad (10\text{-}50)$$

This is an alternative to the Cayley-Hamilton approach to e^{At}. Now if $t_0 \neq 0$, we can form the right-hand side of Equation 10-49 as a function of t and $q(0)$. Then evaluate $q(t)$ at $t = t_0$ and solve for $q(0)$ in terms of $q(t_0)$ which is assumed known.

In the realization discussion earlier in the chapter we stressed the Jordan canonical forms because of their utility in solving the state equations. The basic problems, again, are to find e^{At} and A^k. If A is a diagonal matrix

$$A^k = \begin{bmatrix} \lambda_1^k & 0 & \cdots & 0 \\ 0 & \lambda_2^k & \cdots & 0 \\ \vdots & \vdots & & \vdots \\ 0 & 0 & \cdots & \lambda_n^k \end{bmatrix} \qquad (10\text{-}51)$$

and
$$e^{At} = \begin{bmatrix} e^{\lambda_1 t} & 0 & \cdots & 0 \\ 0 & e^{\lambda_2 t} & \cdots & 0 \\ \vdots & \vdots & & \vdots \\ 0 & 0 & \cdots & e^{\lambda_n t} \end{bmatrix} \qquad (10\text{-}52)$$

If A is not diagonal but consists of Jordan blocks, then the A^k and e^{At} determinations without resorting to the Cayley-Hamilton theorem or the Laplace or Z transforms are more difficult and will not be considered further in this elementary presentation.

In general, the transform approach to solving for $q(t)$ or $q(k)$ is very straightforward but matrix inversion is involved. If n becomes large, matrix inversion can become very tedious. Time-domain approaches are typically employed in computer solutions of higher-order systems. In order to develop good insight into state variable solutions of linear systems, one should master the fundamentals of both time-domain and frequency-domain methods. In a typical industrial setting there will be many digital computer software packages available to perform the numerous tasks outlined in this chapter. Blindly calling the subroutines, however, though it may provide the right answer, often masks the intuition which is provided by mastering fundamentals. Such intuition is essential for deeper understanding and interpretation of the systems at issue.

10-4-3 Transfer Functions

Transfer function determination, the final topic in this section, is closely tied to the transform solution of the state equations. For the continuous system we know that:

$$Y(s) = CQ(s) + DX(s) \qquad (10\text{-}53)$$

which is the Laplace transform of the output equation. We have determined $Q(s)$ in Equation 10-48. Letting $q(0) = 0$ and substituting $Q(s)$ into Equation 10-53, we obtain:

$$Y(s) = \{C(sI - A)^{-1}B + D\}X(s) \qquad (10\text{-}54)$$

The term in brackets is the system transfer function that in general is a matrix of dimension $p \times m$. In the single-input–single-output case we have the scalar:

$$H(s) = C(sI - A)^{-1}B + D \qquad (10\text{-}55)$$

For discrete systems we have:

$$Y(z) = CQ(z) + DX(z) \tag{10-56}$$

Setting $q(0) = 0$ in Equation 10-35 and plugging $Q(z)$ into Equation 10-56, we get:

$$Y(z) = [C(I - z^{-1}A)^{-1}z^{-1}B + D]X(z) \tag{10-57}$$

The term in brackets is the discrete system transfer function, generally a matrix of dimension $p \times m$. In the single-input–single-output case we have the scalar:

$$H(z) = C(I - z^{-1}A)^{-1}z^{-1}B + D \tag{10-58}$$

10-5 CONTROLLABILITY

Assuming we have adequately described a dynamic system in the state variable format, the question of what to do with such a description becomes important. Generally, we like to have that particular system do something for us. Often we prefer to take the state at some time t_0 and transfer it to a different but specified state at a later time t_1. Can the system do that? Can it take $q(t_0)$ to $q(t_1)$ where $t_1 - t_0$ is finite? If so, the system is said to be *controllable*. Whether or not a system is controllable can be determined explicitly by computing a certain matrix and then considering what is called the rank of that matrix. Before stating the precise definition of controllability, let us consider the concept of *rank*.

What is rank? Specifically, what is the rank of a matrix, any matrix, say, A? The **rank** of A, $r(A)$, is the number of linearly independent rows or columns of A. This is equivalent to saying that $r(A)$ is the order of the largest nonsingular submatrix that A contains. This may be A itself. If A is an $n \times m$ matrix, then $r(A)$ will be less than or equal to the smaller of n and m. To test for nonsingularity, we can check rows or columns to see if they are linearly independent. This can be done by elementary operations on the rows or columns or by looking for the largest square matrix whose determinant is nonzero.

EXAMPLE 10-34

Determine the rank of the following matrices:

(a) $A = \begin{bmatrix} 1 & 0 & 1 & 1 \\ 1 & 0 & 0 & 0 \\ 0 & 1 & 0 & 0 \end{bmatrix}$
(b) $A = \begin{bmatrix} 1 & 1 & 2 \\ 1 & 2 & 2 \\ 1 & 1 & 2 \end{bmatrix}$

(c) $A = \begin{bmatrix} 1 & 0 \\ 0 & 2 \\ 2 & 2 \end{bmatrix}$
(d) $A = \begin{bmatrix} 2 & 2 & 2 & 0 & 1 \\ 2 & 0 & 1 & 0 & 2 \\ 2 & 1 & 2 & 2 & 0 \end{bmatrix}$

Solution

(a) $r(A) \leq 3$ for this matrix:

$$\det \begin{bmatrix} 1 & 0 & 1 \\ 1 & 0 & 0 \\ 0 & 1 & 0 \end{bmatrix} = 1$$

therefore $r(A) = 3$.

(b) $\det A = 1(4 - 2) - 1(2 - 2) + 2(1 - 2) = 2 - 2 = 0$

Thus at most $r(A) = 2$. Note that columns 1 and 3 are linearly dependent, therefore, again, at most $r(A) = 2$. Now is there any 2×2 submatrix with a nonzero determinant? Yes:

$$\det \begin{bmatrix} 1 & 1 \\ 1 & 2 \end{bmatrix} = 1, \quad \text{thus} \quad r(A) = 2$$

(c) Since this matrix is 3×2, $r(A)$ at most is 2.

And $\det \begin{bmatrix} 1 & 0 \\ 0 & 2 \end{bmatrix} = 2,$

therefore $r(A) = 2$

(d) Since this matrix is 3×5, $r(A)$ at most is 3. Check:

$$\det \begin{bmatrix} 2 & 2 & 2 \\ 2 & 0 & 1 \\ 2 & 1 & 2 \end{bmatrix} = 2(-1) - 2(4 - 2) + 2(2) = -2$$

thus $r(A) = 3$

There are actually two types of controllability: state controllability and output controllability. We consider both of these. To facilitate our discussion we will cover here—and in the last few sections—only continuous time systems. The relationship of these results to discrete time systems is quite straightforward. However, there are some subtle points of difference, a consideration of which would be distracting to this introductory presentation.

Completely State Controllable   The system A, B, C, D is **completely state controllable** if for some initial time t_0 we can construct a control $x(t)$ that will transfer a given initial state $q(t_0)$ to a specified final state $q(t_f)$ in a finite time interval, $t_0 \leq t \leq t_f$.

Completely Output Controllable   The system A, B, C, D is **completely output controllable** if for some initial time t_0 we can construct a control $x(t)$ that will transfer a given initial output $y(t_0)$ to a specified final state $y(t_f)$ in a finite time interval, $t_0 \le t \le t_f$.

Now to test a given system to see if it is or is not controllable, a number of criteria have been developed. We will consider only the most simple ones. Let us define the matrices P_s and P_0 as follows:

$$P_s \equiv [B \vdots AB \vdots \cdots \vdots A^{n-1}B]_{n \times nm} \qquad (10\text{-}59)$$

and

$$P_0 \equiv [CB \vdots CAB \vdots \cdots \vdots CA^{n-1}B \vdots D]_{px(n+1)m} \qquad (10\text{-}60)$$

Recall the dimensionality of the vectors: The state $q(t)$ is an n vector, the control $x(t)$ is an m vector, and the output $y(t)$ is a p vector.

To test for controllability of the system, we test the rank of P_s and P_0. The system A, B, C, D is completely state controllable if and only if $r(P_s) = n$ and completely output controllable if and only if $r(P_0) = p$. The proof of these contentions is beyond our present scope but can be found in more advanced texts in the control theory area. We will focus on the application of these criteria to test the controllability of given systems.

EXAMPLE 10-35

Determine if the following systems are state and output controllable:

(a) $A = \begin{pmatrix} 1 & 1 & 1 \\ 0 & 1 & 0 \\ 0 & 1 & 1 \end{pmatrix}$, $\quad B = \begin{pmatrix} 0 & 1 \\ 0 & 2 \\ 2 & 1 \end{pmatrix}$,

$C = \begin{pmatrix} 1 & 0 & 1 \\ 2 & 2 & 1 \end{pmatrix}$, $\quad D = \begin{pmatrix} 0 & 0 \\ 0 & 0 \end{pmatrix}$

(b) $A = \begin{pmatrix} 1 & 1 \\ 2 & 1 \end{pmatrix}$, $\quad B = \begin{pmatrix} 2 & 2 \\ 1 & 0 \end{pmatrix}$, $\quad C = \begin{pmatrix} 0 & 1 \\ 1 & 1 \end{pmatrix}$, $\quad D = \begin{pmatrix} 0 & 0 \\ 0 & 0 \end{pmatrix}$

(c) $A = \begin{pmatrix} 2 & 0 & 1 \\ 0 & 0 & 0 \\ 0 & 1 & 3 \end{pmatrix}$, $\quad B = \begin{pmatrix} 1 \\ 0 \\ 0 \end{pmatrix}$, $\quad C = (2 \quad 3 \quad 0)$, $\quad D = (0)$

(d) $A = \begin{pmatrix} 0 & 1 & 0 & 0 \\ 0 & 0 & 1 & 0 \\ 0 & 0 & 0 & 1 \\ 2 & 1 & 2 & 1 \end{pmatrix}$, $\quad B = \begin{bmatrix} 0 \\ 0 \\ 0 \\ 1 \end{bmatrix}$,

$C = (1 \quad 0 \quad 1 \quad 2)$, $\quad D = (0)$

Solutions

(a) $P_s = \begin{bmatrix} 0 & 1 & \vdots & 2 & 4 & \vdots & 4 & 9 \\ 0 & 2 & \vdots & 0 & 2 & \vdots & 0 & 2 \\ 2 & 1 & \vdots & 2 & 3 & \vdots & 2 & 5 \end{bmatrix}$, $P_0 = \begin{bmatrix} 2 & 2 & \vdots & 4 & 7 & \vdots & 6 & 14 & \vdots & 0 & 0 \\ 2 & 7 & \vdots & 6 & 15 & \vdots & 10 & 27 & \vdots & 0 & 0 \end{bmatrix}$

where $n = 3$, $m = 2$, and $p = 2$. Thus, is $r(P_s) = 3$ and $r(P_0) = 2$?
Check:

$$\det \begin{pmatrix} 0 & 1 & 2 \\ 0 & 2 & 0 \\ 2 & 1 & 2 \end{pmatrix} = -8$$

therefore $\qquad\qquad r(P_s) = 3$

Check:

$$\det \begin{pmatrix} 2 & 2 \\ 2 & 7 \end{pmatrix} = 10$$

Therefore $\qquad\qquad r(P_0) = 2$

Thus this system is both state and output controllable.

(b) $\qquad P_s = \begin{bmatrix} 2 & 2 & \vdots & 3 & 2 \\ 1 & 0 & \vdots & 5 & 4 \end{bmatrix}$ and $r(P_s) = 2$

$\qquad P_0 = \begin{bmatrix} 1 & 0 & \vdots & 5 & 4 & \vdots & 0 & 0 \\ 3 & 2 & \vdots & 8 & 6 & \vdots & 0 & 0 \end{bmatrix}$ and $r(P_0) = 2$

Thus, since $n = m = p = 2$, this system is both state and output controllable.

(c) $P_s = \begin{bmatrix} 1 & \vdots & 2 & \vdots & 4 \\ 0 & \vdots & 0 & \vdots & 0 \\ 0 & \vdots & 0 & \vdots & 0 \end{bmatrix}$, $P_0 = [2 \quad 4 \quad 8 \quad 0]$ and $r(P_s) = r(P_0) = 1$

but $n = 3$, $m = p = 1$. Therefore this system is completely output controllable but not completely state controllable.

(d) $\qquad P_s = \begin{bmatrix} 0 & \vdots & 0 & \vdots & 0 & \vdots & 1 \\ 0 & \vdots & 0 & \vdots & 1 & \vdots & 1 \\ 0 & \vdots & 1 & \vdots & 1 & \vdots & 3 \\ 1 & \vdots & 1 & \vdots & 3 & \vdots & 6 \end{bmatrix}$ and $P_0 = [2 \vdots 3 \vdots 7 \vdots 16 \vdots 0]$

Since $\det P_s = 1$, $r(P_s) = 4$, and $r(P_0) = 1$, and $n = 4$, $m = p = 1$, this system is both completely state and output controllable.

Now notice that the last part of the preceding example dealt with a system in its controllability canonical form. There is an interesting fact that *any* SISO system represented in its controllability canonical form is, in fact, completely state controllable.

EXAMPLE 10-36

Demonstrate the complete state controllability of a general SISO system in its controllability canonical form.

Solution

$$P_s = [B_c \vdots A_c B_c \vdots \cdots \vdots A_c^{n-1} B_c]_{m=p=1}$$

$$B_c = \begin{bmatrix} 0 \\ 0 \\ \vdots \\ 0 \\ 0 \\ 1 \end{bmatrix} \quad \text{and}$$

$$A_c = \begin{bmatrix} 0 & 1 & 0 & 0 & 0 \\ 0 & 0 & 1 & 0 & 0 \\ \vdots & \vdots & \vdots & \vdots & \vdots \\ 0 & 0 & 0 & 1 & 0 \\ 0 & 0 & 0 & 0 & 1 \\ -a_0 & -a_1 & -a_2 \cdots & -a_{n-2} & -a_{n-1} \end{bmatrix}$$

$$\text{Then} \qquad P_s = \begin{bmatrix} 0 & 0 & 0 & \cdots & 0 & 1 \\ 0 & 0 & 0 & \cdots & 1 & -a_{n-1} \\ \vdots & \vdots & \vdots & & \vdots & \vdots \\ 0 & 0 & 1 & \cdots & \alpha_1 & \alpha_2 \\ 0 & 1 & -a_{n-1} & \cdots & \alpha_3 & \alpha_4 \\ 1 & -a_{n-1} & -a_{n-2} + a_{n-1}^2 & \cdots & \alpha_5 & \alpha_6 \end{bmatrix}$$

The terms $\alpha_1, \alpha_2, \ldots, \alpha_6$ are rather complicated combinations of $a_0, a_1,$ $\ldots, a_{n-1}$. But exactly what they are does not matter, because all the reverse diagonal terms are 1's. This means that det $P_s = 1$ or -1. Thus $r(P_s) = n$ and the general system is completely state controllable.

This example brings out a very important point: The state controllability of a SISO system is not dependent on the system as such, but rather on the representation or the realization of the system. Thus if state controllability is essential for any given SISO system, what we can do is to test P_s for its rank, and if $r(P_s) < n$, then we can transform the representation into its controllability canonical form. In terms of what we have done in this chapter, the best approach to transform the noncontrollable form into the controllable form is to obtain $H(s)$ from Equation 10-55 and then read off the coefficients to obtain A_c, B_c, C_c, D_c.

EXAMPLE 10-37

Consider the system

$$A = \begin{pmatrix} -1 & -1 \\ 0 & -2 \end{pmatrix}, \qquad B = \begin{pmatrix} 2 \\ 2 \end{pmatrix}, \qquad C = (1 \quad 2), \qquad D = [0]$$

Show that $r(P_s) < 2$ and transform the system into a realization for which $r(P_s) = 2$.

Solution

$$P_s = [B \quad AB] = \begin{bmatrix} 2 & \vdots & -4 \\ 2 & \vdots & -4 \end{bmatrix} \qquad \text{for which } r(P_s) = 1$$

Then $H(s) = C[(sI - A)^{-1}]B$

$$sI - A = \begin{pmatrix} s & 0 \\ 0 & s \end{pmatrix} - \begin{pmatrix} -1 & -1 \\ 0 & -2 \end{pmatrix} = \begin{pmatrix} s+1 & 1 \\ 0 & s+2 \end{pmatrix}$$

which has the inverse:

$$\frac{\begin{pmatrix} s+2 & 0 \\ -1 & s+1 \end{pmatrix}^{\mathrm{T}}}{(s+1)(s+2)} = \begin{bmatrix} \dfrac{1}{s+1} & \dfrac{-1}{(s+1)(s+2)} \\ 0 & \dfrac{1}{s+2} \end{bmatrix}$$

Thus

$$H(s) = (1 \quad 2) \begin{pmatrix} \dfrac{1}{s+1} & \dfrac{-1}{(s+1)(s+2)} \\ 0 & \dfrac{1}{s+2} \end{pmatrix} \begin{pmatrix} 2 \\ 2 \end{pmatrix}$$

$$\begin{bmatrix} \dfrac{1}{s+1} & \dfrac{2s+1}{(s+1)(s+2)} \end{bmatrix} \begin{bmatrix} 2 \\ 2 \end{bmatrix} = \frac{2}{s+1} + \frac{4s+2}{(s+1)(s+2)}$$

$$H(s) = \frac{6s+6}{(s+1)(s+2)} = \frac{6s+6}{s^2+3s+2}$$

$c_1 = 6, c_0 = 6, a_1 = 3, a_0 = 2, b_2 = 0$

Thus
$$A_c = \begin{pmatrix} 0 & 1 \\ -2 & -3 \end{pmatrix}, \quad B_c = \begin{pmatrix} 0 \\ 1 \end{pmatrix},$$

$$C_c = [6 \quad 6], \quad \text{and} \quad D_c = [0]$$

Check:

$$P_s = [B_c \quad A_c B_c] = \begin{bmatrix} 0 & \vdots & 1 \\ 1 & \vdots & -3 \end{bmatrix}, \quad \text{for which } r(P_s) = 2$$

Also, $P_0 = [6 \vdots -12 \vdots 0]$ for the first system and $r(P_0) = p = 1$ and $P_0 = [6 \vdots -12 \vdots 0]$ for the system in its controllability canonical form and, again, $r(P_0) = 1$. Thus, for either representation, the system is completely output controllable.

For the general MIMO system, there is no necessary relationship between state and output controllability. For SISO systems, as we have seen, if $H(s)$ exists as a physically realizable system function, we can transform any realization into its completely state controllable controllability canonical form. As far as output controllability is concerned for SISO systems, note that $r(P_0)$ need only be one. This is almost always attainable in realistic systems.

Once we have a controllable system, what do we do with it? Remember, controllability only says that a control exists, it does not say what it is. How do we obtain a proper control for a controllable system? This question opens up the whole field of control systems design. We may not only want to transfer $q(t)$ or $y(t)$ from some initial to some final value, but we may also want to do so in minimum time or by using minimum energy. There are many other constraints that arise in the general control problem. In the next section we investigate one kind of control problem that takes off from a given system assumed to be controllable: state variable feedback. Many other modern control topics are discussed in the numerous texts available in the field as well as in the extensive periodical literature.

10-6 STATE VARIABLE FEEDBACK

Since this topic itself has an extensive literature, we focus only on the basics. The technique of state variable feedback uses the states as information fed back to the input in order to form part of or all the input $x(t)$. Each state q_i is fed back through an amplifier whose gain is k_i. The k_i's are called the *feedback coefficients*. Recall

$$\dot{q}(t) = Aq(t) + Bx(t) \tag{10-1}$$

$$y(t) = Cq(t) + Dx(t) \tag{10-2}$$

In the general case, we can let:

$$x(t) = \bar{x}(t) - kq(t) \tag{10-61}$$

where k is of dimension $m \times n$ and is called the *matrix of feedback coefficients*. For SISO systems, $m = 1$ and we get:

$$k = [k_1 k_2 \cdots k_n] \tag{10-62}$$

Now $\bar{x}(t)$ is the new input that can be zero or some reference input toward which we might like to drive the state or the output of the system. Substituting, we obtain the new system:

$$\dot{q}(t) = (A - Bk)q(t) + B\bar{x}(t) \tag{10-63}$$

$$y(t) = (C - Dk)q(t) + D\bar{x}(t) \tag{10-64}$$

If the original system is completely state controllable, this new system is also completely state controllable. If the original system is not completely state controllable, then this new system cannot be made completely state controllable no matter how we choose k. This fact highlights the need to begin with a controllable A, B, C, D.

There are many interesting things we can do with our new system. We focus on one of them, namely, how we can modify the roots of the characteristic equation of the system. The roots of the characteristic equation of the original system are the roots of the polynomial:

$$\det(sI - A) = 0 \tag{10-65}$$

The roots of the characteristic equation of the new system are the roots of the polynomial:

$$\det(sI - A + Bk) = 0 \tag{10-66}$$

Let s_i, $i = 1, \ldots, n$ be the roots of Equation 10-65 and $\bar{s}_i$, $i = 1, \ldots, n$ be the roots of Equation 10-66. Some examples will illustrate these ideas.

EXAMPLE 10-38

Let:

$$A = \begin{bmatrix} 1 & 2 \\ 0 & 3 \end{bmatrix}, \quad B = \begin{bmatrix} 0 \\ 1 \end{bmatrix}, \quad C = (1 \quad 0),$$

$$D = [0], \quad \text{and} \quad k = [3 \quad 2]$$

Determine the poles of the system with and without state variable feedback. Check controllability.

Solution

$$P_s = [B \vdots AB] = \begin{bmatrix} 0 & 2 \\ 1 & 3 \end{bmatrix} \quad \text{and} \quad r(P_s) = 2$$

Thus this system is originally completely stable controllable.

$$P_0 = [CB \quad CAB \quad D] = [0 \quad 2 \quad 0] \quad \text{and} \quad r(P_0) = 1$$

Therefore this system is originally completely output controllable.

$$\det(sI - A) = \left| \begin{pmatrix} s & 0 \\ 0 & s \end{pmatrix} - \begin{pmatrix} 1 & 2 \\ 0 & 3 \end{pmatrix} \right|$$

$$= \left| \begin{pmatrix} s-1 & -2 \\ 0 & s-3 \end{pmatrix} \right| = (s-1)(s-3) = 0$$

Thus $\qquad s_1 = 1, s_2 = 3$

$$\det(sI - A + Bk) = \left| \begin{pmatrix} s & 0 \\ 0 & s \end{pmatrix} - \begin{pmatrix} 1 & 2 \\ 0 & 3 \end{pmatrix} + \begin{pmatrix} 0 \\ 1 \end{pmatrix} (3 \quad 2) \right|$$

$$= \left| \begin{pmatrix} s-1 & -2 \\ 0 & s-3 \end{pmatrix} + \begin{pmatrix} 0 & 0 \\ 3 & 2 \end{pmatrix} \right|$$

$$= \left| \begin{pmatrix} s-1 & -2 \\ 3 & s-1 \end{pmatrix} \right| = s^2 - 2s + 1 + 6$$

$$= s^2 - 2s + 7 = 0 \rightarrow \bar{s}_1 = 1 + j2.45$$

and $\qquad \bar{s}_2 = 1 - j2.45$

To check controllability of the new system, let $A - Bk$ be the new A matrix.

$$A - Bk = \begin{pmatrix} 1 & 2 \\ 0 & 3 \end{pmatrix} - \begin{pmatrix} 0 & 0 \\ 3 & 2 \end{pmatrix} = \begin{pmatrix} 1 & 2 \\ -3 & 1 \end{pmatrix} = A_{new}$$

Then $\qquad P_s = \begin{pmatrix} 0 & 2 \\ 1 & 1 \end{pmatrix}$ and $r(P_s) = 2.$

Note that $C - Dk = C$, therefore $P_0 = [CB \vdots CA_{new}B \vdots D] = [0 \vdots 2 \vdots 0]$ and $r(P_0) = 1$ as expected.

Example 10-38 was an example of analysis, that is, given some k matrix, check the consequences of using it. The design problem is to determine k for some desired result. In the previous example, by using $k = [3 \quad 2]$, we are able to move the poles from 1, 3 to $1 \pm j2.45$. All these poles are in the right half of the S plane, indicating an unstable system. The more typical problem in the state feedback area is to find a k matrix for a given system that will yield a desired pole configuration.

EXAMPLE 10-39

For the system:

$$A = \begin{pmatrix} -2 & 1 \\ 0 & -3 \end{pmatrix}, \qquad B = \begin{pmatrix} 1 \\ 1 \end{pmatrix}, \qquad C = (0 \quad 1), \qquad D = 0$$

Determine $k = [k_1 \quad k_2]$ such that the new system has poles at $\bar{s} = -1 \pm 2j$.

Solution

$$\det(sI - A + Bk) = \left| \begin{pmatrix} s & 0 \\ 0 & s \end{pmatrix} - \begin{pmatrix} -2 & 1 \\ 0 & -3 \end{pmatrix} + \begin{pmatrix} 1 \\ 1 \end{pmatrix} (k_1 \quad k_2) \right|$$

$$= \left| \begin{pmatrix} s + 2 & -1 \\ 0 & s + 3 \end{pmatrix} + \begin{pmatrix} k_1 & k_2 \\ k_1 & k_2 \end{pmatrix} \right|$$

$$= \begin{vmatrix} s + 2 + k_1 & k_2 - 1 \\ k_1 & s + 3 + k_2 \end{vmatrix}$$

$$= s^2 + 2s + k_1 s + 3s + 6 + 3k_1$$

$$\quad + sk_2 + 2k_2 + k_1 k_2 + k_1 = 0$$

$$= s^2 + s(5 + k_1 + k_2) + 6 + 4k_1 + 2k_2 = 0$$

And the desired poles $\rightarrow (s + 1 + 2j)(s + 1 - 2j) = 0$

$$\rightarrow s^2 + 2s + 5 = 0$$

Thus $5 + k_1 + k_2 = 2$

$$6 + 4k_1 + 2k_2 = 5 \rightarrow k_1 = 2.5 \quad \text{and} \quad k_2 = -5.5$$

Note that the poles of the original system are at s_1, s_2, where $\det(sI - A) = 0 \rightarrow s_1 = -2, s_2 = -3$. So the effect of the state variable feedback is to undampen the system slightly and make it faster responding.

This problem of choosing k to generate a specific pole pattern, the pole-placement problem, is a very general problem. The pole-placement problem is subsumed by a number of famous classical control issues, in particular, the proportional integral derivative (PID) control problem and the rate feedback control problem. Exploration of these issues, although not within our present scope, would be further evidence of the importance of state variable feedback. In practical applications, however, state variable feedback can be problematic for at least two reasons: (1) If n is large, implementing a state variable feedback design can be very expensive, requiring a large number of sensing devices to make $q_1, \ldots, q_n$ available and a large number of amplifiers with gains $k_1, \ldots, k_n$. (2) All the state variables are not always physically available. A capacitor, for instance, may be encased in a piece of hardware and we may not be able to measure the voltage across the capacitor. If that voltage is a state variable, we need to estimate it. How can estimation of system state variables be accomplished? *Observers* are devices used to construct estimates of unavailable states. The use of observers presupposes that the system we are dealing with is *observable*. Observability and observers are the issues taken up in the last two sections of this chapter.

10-7 OBSERVABILITY

This is the companion concept to controllability. Often a system must be both observable and controllable before any advanced control or optimal control techniques can be used in the design of the system. A system is observable if the states are observable. This means that we can determine the states by knowledge of the inputs and measurements of the outputs. The explicit definition of observability, like those of controllability, employs finite time intervals, $t_0 \leq t \leq t_f$.

Definition. The system A, B, C, D is *completely observable* if the initial state $q(t_0)$ can be determined from knowledge of the input $x(t)$ and from measurements of the output $y(t)$ over a finite time interval, $t_0 \leq t \leq t_f$. The complete record of $q(t)$ can be obtained by incrementation, letting $t = t_0$, $t_0 + \Delta$, $t_0 + 2\Delta, \ldots, t_f$.

As in the case of controllability, there are a number of ways to test the observability of a given system. We use an approach similar to controllability testing. It involves checking the rank of a matrix. We define:

$$P_B = [C^T \vdots A^T C^T \vdots \cdots \vdots (A^T)^{n-1} C^T]_{n \times np} \tag{10-67}$$

The system A, B, C, D is completely observable if and only if $r(P_B) = n$. Some applications of this criteria follow:

EXAMPLE 10-40
Determine if the following systems are completely observable:

(a) $A = \begin{pmatrix} -1 & 0 \\ 0 & -2 \end{pmatrix}$, $\qquad B = \begin{pmatrix} 1 \\ 1 \end{pmatrix}$, $\qquad C = (2 \quad 0)$, $\qquad D = (0)$

(b) $A = \begin{pmatrix} 2 & 1 & 0 \\ 1 & 1 & 0 \\ 0 & 0 & 1 \end{pmatrix}$, $\quad B = \begin{pmatrix} 1 & 0 \\ 0 & 1 \\ 1 & 1 \end{pmatrix}$, $\quad C = \begin{pmatrix} 1 & 0 & 1 \\ 0 & 0 & 2 \end{pmatrix}$, $\quad D = \begin{pmatrix} 0 & 0 \\ 0 & 0 \end{pmatrix}$

(c) $A = \begin{pmatrix} 1 & 1 & 0 \\ 0 & 1 & 0 \\ 0 & 0 & 2 \end{pmatrix}$, $\quad B = \begin{pmatrix} 1 \\ 0 \\ 1 \end{pmatrix}$, $\qquad C = (1 \quad 0 \quad 1)$, $\qquad D = (0)$

(d) $A = \begin{pmatrix} 0 & 0 & -1 \\ 1 & 0 & -2 \\ 0 & 1 & -3 \end{pmatrix}$, $\quad B = \begin{pmatrix} 1 \\ 1 \\ 0 \end{pmatrix}$, $\qquad C = (0 \quad 0 \quad 1)$, $\qquad D = (0)$

Solution

(a) $P_B = \begin{bmatrix} 2 & \vdots & -2 \\ 0 & \vdots & 0 \end{bmatrix}$ and $r(P_B) = 1$

therefore it is not observable.

(b) $P_B = \begin{bmatrix} 1 & 0 & \vdots & 2 & 0 & \vdots & 5 & 0 \\ 0 & 0 & \vdots & 1 & 0 & \vdots & 3 & 0 \\ 1 & 2 & \vdots & 1 & 2 & \vdots & 1 & 2 \end{bmatrix}$ and $r(P_B) = 3$

thus it is observable.

(c) $P_B = \begin{bmatrix} 1 & \vdots & 1 & \vdots & 1 \\ 0 & \vdots & 1 & \vdots & 2 \\ 1 & \vdots & 2 & \vdots & 4 \end{bmatrix}$ and $r(P_B) = 3$

therefore it is observable.

(d) $P_B = \begin{bmatrix} 0 & \vdots & 0 & \vdots & 1 \\ 0 & \vdots & 1 & \vdots & -3 \\ 1 & \vdots & -3 & \vdots & 7 \end{bmatrix}$ and $r(P_B) = 3$

therefore it is observable.

Analogous to the controllability discussion, notice that the last part of the last example dealt with a system in its observability canonical form. *Any* SISO system represented in its observability canonical form is, in fact, completely observable.

EXAMPLE 10-41

Demonstrate the complete observability of a general SISO system in its observability canonical form.

Solution

$$P_B = [C_0^T \vdots A_0^T C_0^T \vdots \cdots \vdots (A_0^T)^{n-1} C_0^T]_{n \times n}, \qquad m = p = 1$$

$$C_0 = [0 \quad 0 \quad \cdots \quad 1] \quad \text{and} \quad A_0 = \begin{bmatrix} 0 & 0 & 0 & \cdots & 0 & 0 & -a_0 \\ 1 & 0 & 0 & \cdots & 0 & 0 & -a_1 \\ 0 & 1 & 0 & \cdots & 0 & 0 & -a_2 \\ \vdots & & & & & & \\ 0 & 0 & 0 & & 1 & 0 & -a_{n-2} \\ 0 & 0 & 0 & & 0 & 1 & -a_{n-1} \end{bmatrix}$$

Then
$$C_0^T = \begin{bmatrix} 0 \\ 0 \\ \cdot \\ \cdot \\ \cdot \\ 1 \end{bmatrix}$$

and
$$A_0^T = \begin{bmatrix} 0 & 1 & 0 & \cdots & 0 & 0 \\ 0 & 0 & 1 & & 0 & 0 \\ \cdot & \cdot & \cdot & & \cdot & \cdot \\ \cdot & \cdot & \cdot & & \cdot & \cdot \\ \cdot & \cdot & \cdot & & \cdot & \cdot \\ 0 & 0 & 0 & & 1 & 0 \\ 0 & 0 & 0 & & 0 & 1 \\ -a_0 & -a_1 & -a_2 & & -a_{n-2} & -a_{n-1} \end{bmatrix}$$

But note that $C_0^T = B_C$ and $A_0^T = A_c$

Thus
$$P_B = [B_c \vdots A_c B_c \vdots \cdots \vdots A_c^{n-1} B_c]_{nxn} = P_s$$

for which, from Example 10-36, $r(P_B) = n$. Therefore the general system is completely observable.

Thus the observability of a SISO system depends on the realization, that is, on how the state variables are defined. To cast an unobservable system into an observable form, obtain $H(s)$ from Equation 10-55 and then read off the coefficients to obtain A_0, B_0, C_0, D_0.

EXAMPLE 10-42
Put the system of Example 10-40(a) into an observable form.

Solution

$$H(s) = C(sI - A)^{-1}B$$

$$= (2 \quad 0) \left[\begin{pmatrix} s & 0 \\ 0 & s \end{pmatrix} - \begin{pmatrix} -1 & 0 \\ 0 & -2 \end{pmatrix} \right]^{-1} \begin{pmatrix} 1 \\ 1 \end{pmatrix}$$

$$\begin{pmatrix} s+1 & 0 \\ 0 & s+2 \end{pmatrix}^{-1} = \frac{\begin{pmatrix} s+2 & 0 \\ 0 & s+1 \end{pmatrix}}{(s+1)(s+2)}$$

$$\rightarrow H(s) = (2 \quad 0) \begin{pmatrix} \dfrac{s+2}{(s+1)(s+2)} \\ \dfrac{s+1}{(s+1)(s+2)} \end{pmatrix}$$

Therefore $\qquad H(s) = \dfrac{2s + 4}{s^2 + 3s + 2} \longrightarrow c_0 = 4, \qquad c_1 = 2,$

$$a_1 = 3, \qquad a_0 = 2,$$

$$b_2 = 0$$

Thus $\qquad A_0 = \begin{pmatrix} 0 & -2 \\ 1 & -3 \end{pmatrix}, \qquad B_0 = \begin{pmatrix} 4 \\ 2 \end{pmatrix},$

$$C_0 = (0 \quad 1), \qquad D_0 = 0$$

and $\qquad P_B = \begin{bmatrix} 0 & \vdots & 1 \\ 1 & \vdots & -3 \end{bmatrix}$

for which $r(P_B) = 2$ and the new representation is completely observable.

Closer investigation of the concepts of controllability and observability reveals a number of interesting relationships between these concepts. We restrict our remarks here to SISO systems. Again, these remarks will be in the service of clarity rather than completeness. We know from Equation 10-55 that every SISO system has a transfer function representation. If $H(s)$ is factored and written as a product of zero terms divided by a product of pole terms, then it may happen that some zero term is identical to a pole term. We can, of course, cancel these terms, which results in a system whose order is one less than the order of the original system. Note that the $H(s)$'s from Examples 10-42 and 10-37 can be written $H(s) = 2/(s + 1)$ and $H(s) = 6/(s + 2)$, respectively, after pole zero cancellation. What is the significance of this pole zero cancellation? There are three main ideas here:

1. If no cancellation occurs, then the system can always be expressed in a representation that is *both* completely observable *and* completely state controllable.
2. If cancellation is possible and is effected, the order of the system is reduced by one for each cancellation and the reduced order system can always be expressed in a representation that is *both* completely observable *and* completely state controllable.
3. If cancellation is possible but is not effected, then the system can always be expressed in a representation that is completely observable *or* completely state controllable, but not both.

Now let us consider some implications of these ideas. The third point actually expresses a necessary and sufficient condition. In other words, if a system turns up *either* not observable *or* not state controllable, then pole zero cancellation is possible. But could we not discern the possibility of pole zero cancellation from a glance at $H(s)$? No. Not necessarily. $H(s)$ may not be in factored form. The numerator and denominator may be high-order polynomials in s which are not conveniently factored. If a possible pole zero cancellation occurs, when should it be effected? There is no easy answer to this question. The reduced order system would surely be easier to handle mathematically. But, in practice, we may lose access to state variables, a knowledge of which is needed for some reason or other. On the other hand, a pole zero cancellation might indicate a way to simplify the hardware, for instance, by replacing a third-order filter with a second-order filter. Usually, though, $H(s)$ is not just a single system but a combination of, say, a plant transfer function and a feed-back transfer function. In any event, care should be taken every time a possible pole zero cancellation arises.

EXAMPLE 10-43

Investigate the controllability and observability of the system:

$$\dot{q}_1 = q_1 + x, \qquad \dot{q}_2 = -2q_2 + 2x, \qquad \dot{q}_3 = -q_3 + q_2, \quad \text{and} \quad y = q_3.$$

Solution. First construct the block diagram as in Figure 10-21. Note that no matter how many measurements we make of $y(t)$, even with a knowledge of $x(t)$, we will not be able to determine $q_1(t)$. Thus, intuitively, the system is not observable.

$$A = \begin{bmatrix} 1 & 0 & 0 \\ 0 & -2 & 0 \\ 0 & 1 & -1 \end{bmatrix}, \qquad B = \begin{bmatrix} 1 \\ 2 \\ 0 \end{bmatrix}, \qquad C = (0 \quad 0 \quad 1), \qquad D = (0)$$

$$P_B = [C^T \vdots A^T C^T \vdots (A^T)^2 C^T] = \begin{bmatrix} 0 & 0 & 0 \\ 0 & 1 & -3 \\ 1 & -1 & 1 \end{bmatrix} \quad \text{and} \quad r(P_B) = 2$$

Thus the system is not observable and it will have a pole zero cancellation.

$$P_s = [B \vdots AB \vdots A^2 B] = \begin{bmatrix} 1 & 1 & 1 \\ 2 & -4 & 8 \\ 0 & 2 & -6 \end{bmatrix} \quad \text{and} \quad r(P_s) = 3$$

and $\qquad P_0 = [0 \vdots 2 \vdots -6 \vdots 0] \quad \text{and} \quad r(P_0) = 1$

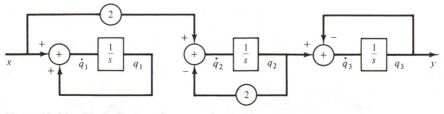

Figure 10-21 Block diagram for an unobservable system.

Therefore the system is both completely state and output controllable. We can transform the given realization into the observability canonical form that could make the system completely observable but not completely state controllable. This might be important if observability is required but controllability is not. In any event a pole zero cancellation is possible in this system. Now:

$$H(s) = C[(sI - A)^{-1}]B$$

$$= (0 \quad 0 \quad 1) \begin{bmatrix} s-1 & 0 & 0 \\ 0 & s+2 & 0 \\ 0 & -1 & s+1 \end{bmatrix}^{-1} \begin{bmatrix} 1 \\ 2 \\ 0 \end{bmatrix}$$

where the inverse

$$= \frac{\begin{bmatrix} (s+2)(s+1) & 0 & 0 \\ 0 & (s-1)(s+1) & 0 \\ 0 & s-1 & (s-1)(s+2) \end{bmatrix}}{(s-1)(s+1)(s+2)}$$

and

$$H(s) = \frac{2(s-1)}{(s-1)(s+1)(s+2)} = \frac{2}{(s+1)(s+2)}$$

which represents a completely observable and completely state controllable second-order system.

As a final point regarding pole zero cancellation, note in the last example that an unstable pole is canceled by the zero term $s - 1$. This is a very delicate situation because an instability in the system can be masked by a mathematical simplification. Unless the pole and zero are *exactly* at $s = 1$, serious problems can arise. If the pole, say, occurs at $s = 0.99$ and the zero at, say $s = 1.01$, then the actual system will be unstable. What should one do in this type of situation? There are many sophisticated approaches for controlling an unstable system. However, the most obvious thing to do first is to return to the original system and try to stabilize it by, for example, using classical control compensation techniques.

If a system turns out to be both observable and controllable, what can we do with it? We saw before that controllability was an existence theorem because the fact that a system is controllable does not tell us how it should be controlled. Then we briefly investigated state variable feedback as a particular way to control a system. The use of state variable feedback presupposes that we have the states $q(t)$ available. If they are not directly available, can we determine them or observe them? This is the observability question. Observability is also an existence theorem because the fact that a system is observable does not tell us how to determine or observe the states $q(t)$. This is the topic of the next section on observers.

10-8 OBSERVERS

We are given a system A, B, C, D which is known to be completely observable. Therefore with a knowledge of the input $x(t)$ and measurements of the output $y(t)$, we should be able to determine the states $q(t)$. We would like to construct a system like the one in Figure 10-22. Such a system is called *an observer*. Let us form the following system, the observer system:

$$\dot{r}(t) = Nr(t) + My(t) + Lx(t) \tag{10-68}$$

where $r(t)$ is the state of the observer and $y(t)$, $x(t)$ are inputs. Let the output of the observer be the state $r(t)$. From these considerations and in view of Figure 10-22, we would like $r(t)$ to be such that:

$$r(t) \approx q(t) \tag{10-69}$$

Then the observer output would be the desired state vector of the original system. The question here is how to chose N, M, and L such that this is the case.

Recalling the original state variable description:

$$\dot{q}(t) = Aq(t) + Bx(t) \tag{10-1}$$

$$y(t) = Cq(t) + Dx(t) \tag{10-2}$$

we can subtract Equation 10-1 from Equation 10-68, substituting $y(t)$ from Equation 10-2 to get:

$$\dot{r} - \dot{q} = Nr - Aq + M(Cq + Dx) + (L - B)x \tag{10-70}$$

$$= Nr + (MC - A)q + (MD + L - B)x \tag{10-71}$$

Now let us define a state error signal:

$$e = r - q \tag{10-72}$$

and then let:

$$MC - A = -N \tag{10-73}$$

and

$$MD + L - B = 0 \tag{10-74}$$

from which we get:

$$\dot{e} = Ne \tag{10-75}$$

So, what has this achieved? Remember we are trying to find N, M, and L. We know A, B, C, and D and would ideally like to have $e(t) = 0$. However, we settle for:

$$\lim_{t \to \infty} e(t) = 0 \tag{10-76}$$

Figure 10-22   Block diagram of a general observer.

Looking at the differential equation in Equation 10-75, we can get the solution $e(t)$ to decay exponentially to zero very rapidly if the eigenvalues of N are chosen to be very negative. Since we need to determine the matrices N, M, and L, this choice of eigenvalues will give us some constraints with which to work. Also, we have the constraints of Equations 10-73 and 10-74. There are some results from matrix theory that say we must also have $MC \neq 0$ and the matrices A and N must have no eigenvalues in common.

A good general approach to these observer problems is first to select values of the M matrix that yield appropriate eigenvalues for N. Then substitute M into Equation 10-74 and solve for L.

EXAMPLE 10-44

Construct an observer for the following completely observable system.

$$A = \begin{bmatrix} 0 & -5 \\ 1 & -2 \end{bmatrix}, \qquad B = \begin{bmatrix} 2 \\ 0 \end{bmatrix}, \qquad C = [0 \quad 1], \qquad D = (2)$$

Solution

From Equation 10-73, we obtain:

$$M_{2\times1}[0 \quad 1]_{1\times2} - \begin{pmatrix} 0 & -5 \\ 1 & -2 \end{pmatrix} = -(N)_{2\times2}$$

Thus
$$N = \begin{pmatrix} 0 & -5 \\ 1 & -2 \end{pmatrix} - \begin{pmatrix} m_{11} \\ m_{12} \end{pmatrix}(0 \quad 1) = \begin{pmatrix} 0 & -5 \\ 1 & -2 \end{pmatrix} - \begin{pmatrix} 0 & m_{11} \\ 0 & m_{12} \end{pmatrix}$$

$$= \begin{pmatrix} 0 & -5 - m_{11} \\ 1 & -2 - m_{12} \end{pmatrix}$$

To check eigenvalues, write $\det(\lambda I - N) = 0$.

$$\begin{vmatrix} \lambda & 5 + m_{11} \\ -1 & \lambda + 2 + m_{12} \end{vmatrix} = \lambda^2 + 2\lambda + m_{12}\lambda + 5 + m_{11} = 0$$

$$\text{or } \lambda^2 + \lambda(2 + m_{12}) + 5 + m_{11} = 0$$

$$\text{and } \lambda = \frac{-(2 + m_{12}) \pm \sqrt{(2 + m_{12})^2 - 4(5 + m_{11})}}{2}$$

And $\det(\lambda I - A) = 0$ gives us the eigenvalues of A:

$$\begin{vmatrix} \lambda & 5 \\ -1 & \lambda + 2 \end{vmatrix} = \lambda^2 + 2\lambda + 5 = 0$$

$$\rightarrow \lambda = \frac{-2 \pm \sqrt{4 - 20}}{2}$$

$$= -1 \pm j\sqrt{2}$$

If we select $m_{12} = 20$ and $m_{11} = 116$, we will get eigenvalues for N at $\lambda = $

-11, -11, which are both very negative and both different from the eigenvalues of A.

Therefore

$$M = \begin{bmatrix} 116 \\ 20 \end{bmatrix}$$

Also,

$$MC = \begin{pmatrix} 116 \\ 20 \end{pmatrix} (0 \quad 1) = \begin{pmatrix} 0 & 116 \\ 0 & 20 \end{pmatrix} \neq 0$$

Then

$$N = \begin{bmatrix} 0 & -121 \\ 1 & -22 \end{bmatrix}$$

and

$$L = B - MD$$

$$= \begin{pmatrix} 2 \\ 0 \end{pmatrix} - \begin{pmatrix} 116 \\ 20 \end{pmatrix} 2 = \begin{bmatrix} -230 \\ -40 \end{bmatrix}$$

Having the states available at the output of the observer, we are then in a position to proceed into a variety of modern control procedures. The state variable feedback, mentioned earlier, is one of these.

The observer developed in this brief presentation is called the *full-order-observer*. If none of the states in the state vector $q(t)$ is directly available for measurement, then this observer is typically used. If all states are available, then we need no observer. However, what if some states are available and some are not? If so, we can use what is called the *reduced-order-observer*. The order of a full-order-observer is the same as the order of the system being observed. This is true for the observer considered earlier, although there are more sophisticated full-order-observers whose order can be slightly reduced. We will not consider these. Nor will we consider the reduced-order-observer, except to mention that for higher-order systems this observer can provide a tremendous reduction in computation if only a few states need to be observed.

Topics like observer theory and state variable feedback are playing a large role in the modern design of sophisticated control systems. Enmeshed with these issues are the concepts of controllability and observability. This cluster of ideas grows out of the state variable theory. A working knowledge of state variable time-domain ideas is becoming as essential in the modern engineering world as the frequency-domain Laplace, Z, and Fourier transforms. It is hoped that this chapter will provide impetus to students to pursue developments in some of the broad areas of systems and controls for which this fundamental state variable theory provides a point of departure.

SUMMARY

This chapter has considered the representation of linear continuous and discrete systems in the state variable format. Some block diagram ideas were investigated as an aid to a circumspective overview of systems. The realization problem considered:

1. physical system realizations
2. canonical form realizations
   (a) controllability form
   (b) observability form
   (c) Jordan form

Transformations to the Jordan form were stressed because of the ease with which the resulting state equations are solved. Block diagrams for the three canonical forms were considered. Then the solutions of the state equations were developed and transfer functions in terms of A, B, C, D were derived.

Next, we turned toward a few state variable topics that are important in the theory of automatic control. Controllability was taken up and a criterion to test controllability was studied. State variable feedback was considered briefly. Then, observability was dealt with and a criterion to test observability was examined. Finally, the topic of observers was considered.

PROBLEMS

10-1. Put the following equations into the controllability canonical form:
  (a) $y(k + 3) + 2y(k + 2) + 3y(k + 1) + 4y(k) = x(k)$
  (b) $y(k + 3) + y(k) = x(k + 1) + 2x(k)$
  (c) $\dddot{y}(t) + 2\dot{y}(t) + 5y(t) = \dot{x}(t)$
  (d) $y'''(t) + y''(t) = x''(t) + x'(t) + 2x(t)$

10-2. Put the following equations into the observability canonical form:
  (a) $y''(t) = x(t)$
  (b) $y'''(t) = x''(t) - y'(t) + 5x'(t) + y(t) + 10x(t)$
  (c) $y(k + 6) + y(k) = x(k)$
  (d) $y(k + 3) - y(k) = x(k + 2) - 10x(k)$

10-3. Write state variable equations for the following systems:
  (a)

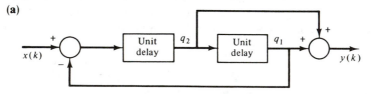

  (b)

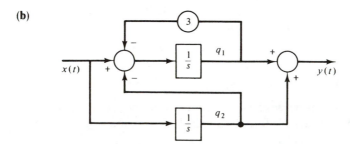

(c)

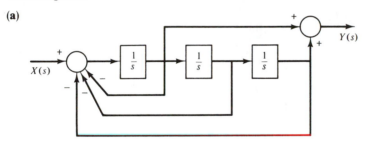

10-4. Put the following equations into the Jordan form:

(a) $\dot{q} = \begin{bmatrix} 1 & 0 & 0 \\ 1 & 1 & 0 \\ 1 & 1 & 1 \end{bmatrix} q + \begin{bmatrix} 1 \\ 0 \\ 1 \end{bmatrix} x, \qquad y = [0 \quad 0 \quad 1]q + [0]x$

(b) $q(k + 1) = \begin{bmatrix} 10 & 20 \\ 30 & 40 \end{bmatrix} q(k) + \begin{bmatrix} 50 \\ 60 \end{bmatrix} x(k), \qquad y(k) = [10 \quad 10]q(k)$

10-5. Find the transfer functions for the systems of Problem 10-4.

10-6. Determine the transfer functions for the systems represented by the following block diagrams:

(a)

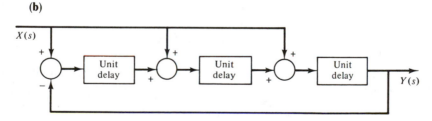

(b)

10-7. Construct block diagrams for the following systems:

(a) $\dot{q} = \begin{bmatrix} 1 & 1 & 0 \\ 0 & 1 & 0 \\ 0 & 0 & -1 \end{bmatrix} q + \begin{bmatrix} 0 \\ 1 \\ 1 \end{bmatrix} x, \qquad y = [1 \quad 1 \quad 1]q + [0]x$

(b) $q(k + 1) = \begin{bmatrix} -2 & 1 & 0 \\ 0 & -2 & 1 \\ 0 & 0 & -2 \end{bmatrix} q(k) + \begin{bmatrix} 1 \\ 1 \\ 1 \end{bmatrix} x(k),$

$\qquad y(k) = [0 \quad 0 \quad 1]q(k) + [0]x(k)$

(c) $\dot{q} = \begin{bmatrix} -1 & 0 & 0 \\ 0 & -2 & 0 \\ 0 & 0 & -3 \end{bmatrix} q + \begin{bmatrix} 2 \\ 1 \\ 2 \end{bmatrix} x, \qquad y = \begin{bmatrix} 1 & 2 & 1 \end{bmatrix} q + [0] x$

10-8. Put the following systems into the Jordan form:

(a) $\dot{q} = \begin{bmatrix} 1 & 2 \\ 2 & 2 \end{bmatrix} q + \begin{bmatrix} 1 \\ 1 \end{bmatrix} x, \qquad y = \begin{bmatrix} 0 & 1 \end{bmatrix} q + [5] x$

(b) $q(k + 1) = \begin{bmatrix} 1 & 1 & 1 \\ 0 & 1 & 1 \\ 0 & 0 & 1 \end{bmatrix} q(k) + \begin{bmatrix} 1 \\ 0 \\ 1 \end{bmatrix} x(k), \qquad y(k) = \begin{bmatrix} 0 & 1 & 0 \end{bmatrix} q(k)$

(c) $q(k + 1) = \begin{bmatrix} 20 & 30 \\ 10 & 20 \end{bmatrix} q(k) + \begin{bmatrix} 1 \\ 1 \end{bmatrix} x(k), \qquad y(k) = \begin{bmatrix} 1 & 2 \end{bmatrix} q(k) + [5] x(k)$

10-9. Using the Cayley-Hamilton method, determine A^k for the following A matrices:

(a) $A = \begin{bmatrix} 8 & 0 \\ 0 & 1 \end{bmatrix}$ \qquad (b) $A = \begin{bmatrix} 8 & 5 \\ 3 & 1 \end{bmatrix}$

(c) $A = \begin{bmatrix} 1 & 0 & 0 \\ 1 & 2 & 0 \\ 2 & 1 & 1 \end{bmatrix}$ \qquad (d) $A = \begin{bmatrix} -2 & 1 & 0 \\ 0 & -2 & 1 \\ 0 & 0 & -2 \end{bmatrix}$

10-10. Using the Z transform method, determine A^k for the matrices of Problem 10-9.

10-11. Using the Laplace transform method, determine e^{At} for the matrices of Problem 10-9.

10-12. Using the Cayley-Hamelton method, determine e^{At} for the matrices of Problem 10-9.

10-13. For the following system, $x(t) = e^{-t} u(t)$ and $q(0) = \begin{bmatrix} -1 & -2 \end{bmatrix}^T$, determine $q(t)$ for all $t \geq 0$:

$$\dot{q} = \begin{bmatrix} 0 & 1 \\ -2 & -3 \end{bmatrix} q + \begin{bmatrix} 1 \\ 2 \end{bmatrix} x$$

10-14. For the following system:

$$x(t) = \begin{bmatrix} x_1 \\ x_2 \end{bmatrix} = \begin{bmatrix} u(t) \\ e^{-t} u(t) \end{bmatrix}$$

Determine $q(t)$ for all $t \geq 0$.

$$\dot{q} = \begin{bmatrix} 0 & 2 \\ -1 & -2 \end{bmatrix} q + \begin{bmatrix} 1 & -1 \\ 2 & 2 \end{bmatrix} x, \qquad \text{where } q(0) = \begin{bmatrix} 0 \\ 0 \end{bmatrix}$$

10-15. Determine $y(k)$, $k > 0$, if $x(k) = 0$

and
$$q(k + 1) = \begin{bmatrix} 5 & -1 \\ 1 & 0 \end{bmatrix} q(k) + \begin{bmatrix} 1 \\ 2 \end{bmatrix} x(k), \qquad q(0) = \begin{bmatrix} -2 \\ -3 \end{bmatrix}$$

and
$$y(k) = [2 \quad -3] q(k)$$

10-16. Determine $H(z)$ for the system:

$$A = \begin{bmatrix} 0 & 1 & 1 \\ 0 & 1 & 2 \\ 1 & 1 & 0 \end{bmatrix}, \qquad B = \begin{bmatrix} 1 \\ 0 \\ 0 \end{bmatrix}, \qquad C = [0 \quad 0 \quad 1], \qquad D = [0]$$

10-17. Determine $H(s)$ for the system:

$$A = \begin{bmatrix} 1 & 0 \\ -1 & 2 \end{bmatrix}, \qquad B = \begin{bmatrix} 2 & 0 \\ 1 & 1 \end{bmatrix}, \qquad C = \begin{bmatrix} 4 & 2 \\ 1 & 0 \end{bmatrix}, \qquad D = \begin{bmatrix} 0 & 0 \\ 0 & 0 \end{bmatrix}$$

10-18. Determine $H(z)$ for the system:

$$A = \begin{bmatrix} 0 & 1 & 0 & 0 \\ 0 & 0 & 1 & 0 \\ 0 & 0 & 0 & 1 \\ -1 & -1 & -1 & -1 \end{bmatrix}, \qquad B = \begin{bmatrix} 0 \\ 0 \\ 0 \\ 1 \end{bmatrix}, \qquad C = [1 \quad 1 \quad 0 \quad 1], \qquad D = [2]$$

10-19. Determine $H(s)$ for the system:

$$A = \begin{bmatrix} 0 & 0 & 0 & -2 \\ 1 & 0 & 0 & -3 \\ 0 & 1 & 0 & -1 \\ 0 & 0 & 1 & -4 \end{bmatrix}, \qquad B = \begin{bmatrix} 1 \\ 2 \\ 2 \\ 3 \end{bmatrix}, \qquad C = [0 \quad 0 \quad 0 \quad 1], \qquad D = [2]$$

10-20. Represent the following differential equation in a Jordan canonical form:

$$\frac{d^3 y(t)}{dt^3} + \frac{4 \, d^2 y(t)}{dt^2} + \frac{5 \, dy(t)}{dt} + 2y(t) = 5x(t)$$

10-21. Given the following state variable representation, construct the corresponding scalar difference equation relating input and output

$$q(k + 1) = \begin{bmatrix} 0 & 1 & 0 \\ 0 & 0 & 1 \\ 0 & -2 & 2 \end{bmatrix} q(k) + \begin{bmatrix} 2 \\ 1 \\ -1 \end{bmatrix} x(k), \qquad y(k) = [0 \quad 1 \quad 0] q(k)$$

10-22. Find a matrix P such that $\hat{q} = Pq$,

where
$$\frac{d\hat{q}}{dt} = \begin{bmatrix} 1 & 2 \\ 2 & 3 \end{bmatrix} \hat{q} + \begin{bmatrix} 1 \\ 0 \end{bmatrix} x$$

and
$$\frac{dq}{dt} = \begin{bmatrix} 2 & 3 \\ 1 & 0 \end{bmatrix} q + \begin{bmatrix} 1 \\ -1 \end{bmatrix} x$$

10-23. For a discrete system $x(k) = 0$ and

$$q(3) = \begin{bmatrix} 2 \\ 3 \\ 2 \end{bmatrix}$$

Determine $q(0)$ if:

$$A = \begin{bmatrix} -3 & 0 & 0 \\ 1 & -3 & 0 \\ 0 & 1 & -2 \end{bmatrix}$$

10-24. Given the system:

$$\dot{q} = \begin{bmatrix} -2 & 0 \\ 2 & -5 \end{bmatrix} q + \begin{bmatrix} 2 & 0 \\ 2 & 2 \end{bmatrix} x$$

with
$$q(0) = \begin{bmatrix} 0 \\ 0 \end{bmatrix},$$

find $x(t)$ such that:

$$q_1(t) = 1 - e^{-5t} \quad \text{and} \quad q_2(t) = 1 - e^{-2t} + e^{-5t} - e^{-7t}, \qquad \text{for } t \geq 0.$$

10-25. Consider the system:

$$q(k + 1) = \begin{bmatrix} 0 & 1 \\ -3 & -2 \end{bmatrix} q(k) + \begin{bmatrix} 0 & 1 \\ 1 & 2 \end{bmatrix} x(k)$$

Let $x(k) = Gq(k)$

where
$$G = \begin{bmatrix} g_{11} & g_{12} \\ g_{21} & g_{22} \end{bmatrix}$$

Then we get the "closed-loop" system $q(k + 1) = \tilde{A}q(k)$. Determine G such that the closed-loop system has poles at $z = -1$ and $z = -0.5$.

10-26. For the system represented in the block diagram:
(a) Determine $H(s)$.
(b) Determine e^{At}.
(c) Find $y(t)$ if $x(t) = u(t)$ and $q(0) = 0$.

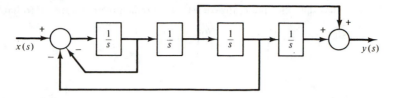

10-27. Let $q_1 = e^{-t} - e^{-2t}, q_2 = 2 - e^{-3t}, q_3 = 1 - e^{-5t} - e^{-6t}$ for $t \geq 0$. Assume $x(t) = 0$, and assume:

$$A = \begin{bmatrix} a_{11} & 3 & 0 \\ 0 & -2 & a_{32} \\ 0 & 0 & a_{33} \end{bmatrix}$$

Determine a_{11}, a_{32}, and a_{33}.

10-28. Determine if the following systems are completely state controllable and/or completely output controllable.

(a) $\quad A = \begin{pmatrix} 1 & 1 & 1 \\ 0 & 0 & 1 \\ 1 & 1 & 0 \end{pmatrix}, \quad B = \begin{pmatrix} 1 \\ 1 \\ 1 \end{pmatrix}, \quad C = (0 \quad 1 \quad 0), \quad D = [0]$

(b) $\quad A = \begin{pmatrix} 1 & 2 \\ 3 & 5 \end{pmatrix}, \quad B = \begin{pmatrix} 8 & 0 \\ 2 & 2 \end{pmatrix}, \quad C = \begin{pmatrix} 2 & 1 \\ 1 & 3 \end{pmatrix}, \quad D = \begin{pmatrix} 5 & 4 \\ 2 & 0 \end{pmatrix}$

(c) $\quad A = \begin{bmatrix} 1 & 0 & 0 & 0 \\ 0 & -1 & 0 & 0 \\ 1 & 0 & 2 & 0 \\ 0 & 0 & 0 & -2 \end{bmatrix}, \quad B = \begin{bmatrix} 1 \\ 0 \\ 0 \\ 0 \end{bmatrix}, \quad C = [0 \quad 2 \quad 0 \quad 0], \quad D = [0]$

(d) $A = \begin{bmatrix} -1 & -1 & 2 \\ 2 & 0 & 2 \\ 0 & 0 & -2 \end{bmatrix}, \quad B = \begin{bmatrix} 0 & 1 \\ 1 & 0 \\ 0 & 0 \end{bmatrix}, \quad C = \begin{bmatrix} 0 & 0 & 1 \\ 1 & 0 & 0 \end{bmatrix}, \quad D = \begin{pmatrix} 1 & 2 \\ 0 & 2 \end{pmatrix}$

10-29. Let:

$$A = \begin{pmatrix} 1 & -2 \\ 0 & 3 \end{pmatrix}, \quad B = \begin{bmatrix} b_1 \\ 2 \end{bmatrix}, \quad C = [2 \quad 1], \quad D = [0]$$

Determine b_1 such that this system is completely output controllable but not completely state controllable.

10-30. A system has the following realization that is not completely state controllable:

$$A = \begin{pmatrix} 1 & 1 & 1 \\ 0 & 0 & 1 \\ 0 & 1 & 0 \end{pmatrix}, \qquad B = \begin{pmatrix} 0 \\ 0 \\ 1 \end{pmatrix}, \qquad C = [2 \ \ 1 \ \ 0], \qquad D = [0]$$

Transform this realization into one that is completely state controllable.

10-31. For the following system, use state variable feedback to transform the system into one that has poles at $s = -5, -10, -15$. Determine $K = [k_1 \ \ k_2 \ \ k_3]$.

$$A = \begin{bmatrix} 1 & -1 & -1 \\ -1 & -1 & -1 \\ 1 & 0 & -1 \end{bmatrix}, \qquad B = \begin{bmatrix} 2 \\ 0 \\ -1 \end{bmatrix}, \qquad C = [1 \ \ 0 \ \ 1], \qquad D = [0]$$

10-32. Determine if the following systems are completely observable:

(a) $\qquad A = \begin{pmatrix} 0 & 1 \\ 1 & 1 \end{pmatrix}, \qquad B = \begin{pmatrix} 1 & 0 \\ 1 & 1 \end{pmatrix}, \qquad C = \begin{pmatrix} 1 & 1 \\ 1 & 0 \end{pmatrix}, \qquad D = \begin{pmatrix} 1 & 1 \\ 0 & 1 \end{pmatrix}$

(b) $\quad A = \begin{pmatrix} 1 & 1 & 0 \\ 1 & 0 & 1 \\ 1 & 1 & 1 \end{pmatrix}, \qquad B = \begin{pmatrix} 0 & 1 \\ 0 & 0 \\ 1 & 0 \end{pmatrix}, \qquad C = \begin{pmatrix} 0 & 1 & 0 \\ 0 & 0 & 1 \end{pmatrix}, \qquad D = \begin{pmatrix} 0 & 0 \\ 0 & 0 \end{pmatrix}$

(c) $\qquad A = \begin{pmatrix} 2 & 0 & 0 \\ 0 & 1 & 2 \\ 2 & 1 & 0 \end{pmatrix}, \qquad B = \begin{pmatrix} 0 \\ 1 \\ 0 \end{pmatrix}, \qquad C = (-1 \ \ 0 \ \ 0), \qquad D = [0]$

(d) $\qquad\qquad A = \begin{pmatrix} 1 & -1 & -2 \\ 0 & -1 & 0 \\ 2 & 0 & 0 \end{pmatrix}, \qquad B = \begin{pmatrix} 0 & 0 \\ 0 & 0 \\ 0 & 1 \end{pmatrix},$

$$C = \begin{pmatrix} 1 & 0 & 0 \\ 0 & 0 & 0 \end{pmatrix}, \qquad D = \begin{pmatrix} 1 & 2 \\ 0 & 1 \end{pmatrix}$$

10-33. Consider the following system:

$$A = \begin{pmatrix} 1 & 0 & 0 \\ 0 & 0 & 1 \\ 0 & 1 & 0 \end{pmatrix}, \qquad B = \begin{pmatrix} 1 \\ 1 \\ 0 \end{pmatrix}, \qquad C = [2 \ \ 1 \ \ c_3], \qquad D = [0]$$

Determine c_3 such that this system is not completely observable. Then using this value of c_3, transform the representation into one that is completely observable.

10-34. Construct observers for the following systems:

(a)
$$A = \begin{pmatrix} 0 & -1 \\ 1 & -2 \end{pmatrix}, \qquad B = \begin{pmatrix} 1 \\ 1 \end{pmatrix}, \qquad C = (1 \quad 2), \qquad D = [3]$$

(b)
$$A = \begin{pmatrix} 0 & 0 & -1 \\ 0 & 1 & 0 \\ 2 & 0 & 1 \end{pmatrix}, \qquad B = \begin{pmatrix} 0 \\ 1 \\ 0 \end{pmatrix}, \qquad C = (1 \quad 1 \quad 1), \qquad D = [2]$$

Appendix: The Formulas of Complex Variables

We will now enumerate in logical sequence of development the formulas of complex variable that we use when evaluating inverse Laplace and Z transforms.

Definition of Integration

Given a complex function $f(z) = u(x, y) + jv(x, y)$, its integral over a path C from point $P_A = z_A$ where $z_A = x_A + jy_A$ to a point $P_B = z_B$ where $z_B = x_B + jy_B$ is defined as:

$$\int_{z_A}^{z_B} f(z) \, dz \triangleq \lim_{\Delta z_i \to 0} \sum_{i=1}^{N} f(z_i') \, \Delta z_i \qquad (A\text{-}1)$$

where z_i' and Δz_i are defined for the curve C in Figure A-1. If C is defined by $y = g(x)$, then A-1 becomes,

$$\int_{C}^{z_B}{}_{z_A} f(z) \, dz = \left[\int_{C} u(x, y) \, dx - v(x, y) \, dy \right] + j \left[\int_{C} v(x, y) \, dx + u(x, y) \, dy \right]$$

and putting $y = g(x)$ and $dy = g'(x) \, dx$ this becomes,

$$\int_{C}^{z_B}{}_{z_A} f(z) \, dz = \int_{x_A}^{x_B} u(x, g(x)) \, dx - \int_{x_A}^{x_B} v(x, g(x))g'(x) \, dx$$

$$+ j \left[\int_{x_A}^{x_B} v(x, g(x)) \, dx + \int_{x_A}^{x_B} u(x, g(x)) \, g'(x) \, dx \right]$$

EXAMPLE A-1

Given:

$$f(z) = 3x + j4y$$

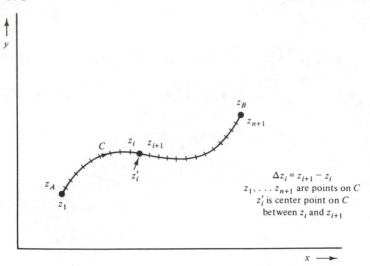

Figure A-1   A curve C used in defining complex integration.

Evaluate:

(a) $\displaystyle\int_C^{2+j2} f(z)\, dz,$ where C is the straight line path joining the end points.

(b) $\displaystyle\int_{C_1^0+C_2}^{2+j2} f(z)\, dz,$ where $C_1: y = 0, 0 < x < 2$ and $C_1: x = 2, 0 < y < 2$

Solution.   Strictly in terms of z we could write $f(z) = 3z + jIm(z)$, where Im stands for the "imaginary part of."

(a)

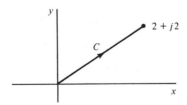

The path C is defined by $y = x, 0 < x < 2$ and $dy = dx$

Thus $\displaystyle\int_C (3x + j4y)(dx + j\, dy) = \left[\int_0^2 (3x\, dx - 4x\, dx)\right]$

$$+ j\left[\int_0^2 (4x\, dx + 3x\, dx)\right]$$

$$= -2 + j14 \qquad \text{with some work}$$

(b)

$C_1: y = 0, dy = 0, 0 < x < 2$

$C_2: x = 2, dx = 0, 0 < y < 2$

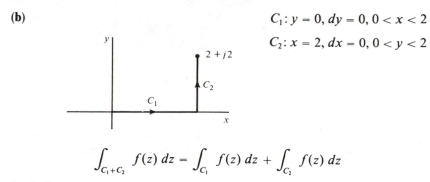

$$\int_{C_1+C_2} f(z)\, dz = \int_{C_1} f(z)\, dz + \int_{C_2} f(z)\, dz$$

Path C_1

From our thumbnail sketch we have:

$$\int_{C_1} (3x + j4y)(dx + j\, dy) = \int_0^2 3x\, dx + j0$$

$$= 6$$

Path C_2

$$\int_{C_2} (3x + j4y)(dx + j\, dy) = -\int_0^2 4y\, dy + j \int_0^2 6\, dy$$

$$= -8 + j12$$

and finally,

$$\int_{C_1+C_2} (3x + j4y)\, dz = -2 + j12$$

We note that the answers from parts (a) and (b) are different and so the integral is path dependent.

Complex Integrals That Are Independent of Path

The derivative of a complex function $f(z)$ at $z = z$ is defined as:

$$f(z) = \lim_{\Delta z \to 0} \frac{f(z + \Delta z) - f(z)}{\Delta z} \tag{A-2}$$

and $f'(z)$ exists at $z = z$ if the Cauchy-Riemann conditions for $f(z) = u(x, y) + jv(x, y)$

$$\frac{\delta u}{\delta x} = \frac{\delta v}{\delta y}$$

and

$$\frac{\delta v}{\delta x} = \frac{\delta u}{\delta y} \tag{A-3}$$

are satisfied and all second partials are continuous at $z = z$. If $f(z)$ possesses a derivative or is analytic at $z = z$, two formulas for $f'(z)$ are:

$$f'(z) = \frac{\delta u}{\delta x} + j\frac{\delta v}{\delta x}$$

or

$$f'(z) = \frac{\delta v}{\delta y} - j\frac{\delta u}{\delta y} \qquad \text{(A-4)}$$

For polynomial functions of z plus products and quotients of polynomials all the derivative formulas from calculus carry over to complex functions. In addition, the exponential, trigonometric, and hyperbolic functions defined with their derivatives are:

$$e^z = e^x \underline{|y}$$
$$= e^x \text{Cos } y + je^x \text{Sin } y$$

and

$$\frac{d}{dz}(e^z) = e^z$$

$$\text{Cos } z = \frac{1}{2}(e^{jz} + e^{-jz})$$

$$\text{Sin } z = \frac{-j}{2}(e^{jz} - e^{-jz})$$

and

$$\frac{d}{dz}\text{Sin } z = \text{Cos } z, \qquad \frac{d}{dz}\text{Cos } z = -\text{Sin } z$$

$$\text{Cosh } z = \frac{1}{2}(e^z + e^{-z})$$

$$\text{Sinh } z = \frac{1}{2}(e^z - e^{-z})$$

and

$$\frac{d}{dz}\text{Cosh } z = \text{Sinh } z, \qquad \frac{d}{dz}\text{Sinh } z = \text{Cosh } z$$

Cauchy's Theorem. If a function $f(z)$ is analytic on and inside a contour

then

$$\oint_C f(z)\, dz = 0 \qquad \text{(A-5a)}$$

and

$$\int_{C_1} f(z)\, dz = \int_{C_2} f(z)\, dz \qquad \text{(A-5b)}$$

where C_1 and C_2 make up a closed contour C.

This is illustrated in Figure A-2(a).

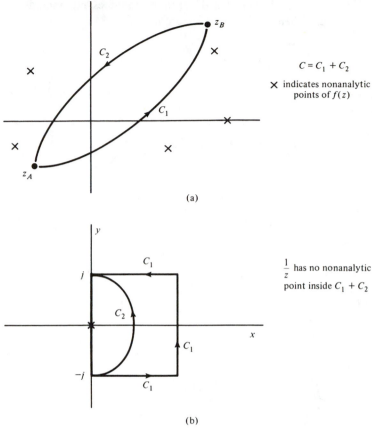

$C = C_1 + C_2$

$\times$ indicates nonanalytic
  points of $f(z)$

$\dfrac{1}{z}$ has no nonanalytic
point inside $C_1 + C_2$

Figure A-2 (a) The contours C, C_1, and C_2 for Cauchy's theorem; (b) the contours C_1 and C_2 for Example A-2.

EXAMPLE A-2
   Evaluate:

(a) $\displaystyle\oint_{C_1+C_2} \frac{1}{z}\, dz$

(b) and $\displaystyle\int_C \frac{1}{z}\, dz$

where C_1 and C_2 make up a closed contour C. This theorem is illustrated in Figure A-2(a).

   Solution

   (a) The denominator is zero at $z = 0$ and is not analytic at $z = 0$. Since this point is not included in the closed contour $C_1 + C_2$, the answer is zero by Cauchy's theorem.

(b) Since C_1 and C_2 make up a closed contour that does not include a nonanalytic point, then by Cauchy's theorem:

$$\int_{C_1} \frac{1}{z} \, dz = \int_{C_2} \frac{1}{z} \, dz$$

C_2 is a much simpler contour than C_1 and is defined by $z(\theta) = 2e^{j\theta}$ and $dz/d\theta = j2 \, e^{j\theta}$, $-\pi/2 < \theta < \pi/2$.

$$\int_{C_2} \frac{1}{z} dz = \int_{-\pi/2}^{\pi/2} j \frac{2e^{j\theta}}{2e^{j\theta}} \, d\theta$$

$$= j\pi$$

The Fundamental Theorem of Integration

If a complex function $f(z)$ is analytic everywhere

then $\qquad \displaystyle\int_{C^{z_A}}^{z_B} f(z) \, dz = F(z_B) - F(z_A), \qquad$ where C is a simple path

and $\qquad \displaystyle\frac{d}{dz} F(z) = f(z)$

For example:

$$\int_0^{j4} z \, dz = (0.5z^2 + A) \, \Big|_0^{j4} = -8$$

A simple path is one that does not intersect itself (such as C_1 or C_2 in Figure A-2 (a) and (b)).

The Residue Theorem

If a function $g(z)$ is analytic inside a closed contour except at a finite number of points called *poles,* where, for example, an *p*th-order pole at $z = z_0$ is defined by:

- $(z - z_0)^m \, g(z)$ is not analytic $m = 1, 2, \cdot \cdot \cdot (p - 1)$ at z_0
- but $(z - z_0)^p \, g(z)$ is analytic at z_0

then $\qquad \displaystyle\oint_C g(z) \, dz = 2\pi j \Sigma$ [residues of the poles of $g(z)$ inside C] $\qquad$ (A-5)

where the residue of a *p*th-order pole of $g(z)$ at $z = z_0$ is:

$$[\text{residue of } p\text{th-order pole}] = \frac{1}{(p-1)!} \left[\frac{d^{p-1}}{dz^{p-1}} (z - z_0)^p g(z) \right]_{z=z_0} \qquad (\text{A-6})$$

where "!" is the factorial notation, [e.g., $4! = 4 \times 3 \times 2 \times 1 = 24$].

Example A-3

Evaluate for a fixed t using the residue theorem:

(a) $\dfrac{1}{2\pi j} \displaystyle\oint_C \dfrac{ze^{zt}}{(z-1)(z-2)} \, dz,$ $\qquad$ where C is defined by $1.5e^{j\theta}$

(b) $\dfrac{1}{2\pi j} \displaystyle\oint_C \dfrac{e^{zt}}{z(z-1)^2} \, dz,$ $\qquad$ where C is defined by $3e^{j\theta}$

Solution. Sketch the curves for parts (a) and (b)

(a) Since only the pole at $z = 1$ is included inside the contour, the answer is:

$$\frac{e^t}{(1-2)^2} = e^t \qquad \text{for any } t$$

(b) Since both the first-order pole at $z = 0$ and the second-order pole at $z = 1$ are included inside C, the answer is:

$$\frac{e^{0t}}{(-1)^2} + \frac{d}{dz}\left(\frac{e^{zt}}{z}\right)\bigg|_{\text{at } z=1} = 1 + te^t - e^t$$

for any fixed t.

The Inside–Outside Theorem

If the integrand for a closed contour C is of the form $g(z) = N(z)/D(z)$ where $D(z)$ is of order more than one higher than $N(z)$

then $\qquad \displaystyle\oint_C g(z)\, dz = 2\pi j \, \Sigma \, [\text{residues of the poles of } g(z) \text{ inside } C]$

or $\qquad = -2\pi j \, \Sigma \, [\text{residues of the poles of } g(z) \text{ outside } C]$

This theorem may be proved by constructing the contour C_1, defined by $z(\theta) = Re^{j\theta}$ where R approaches infinity, and showing that:

$$\oint_{C_1} g(z)\, dz \to 0$$

Then by contour subdivision:

$$\oint_C g(z)\, dz = \oint_C g(z)\, dz + \oint_{C_1} g(z)\, dz$$
$$= -2\pi j \, \Sigma \, [\text{residues of the poles outside } C]$$

These contours are shown in Figure A-3.

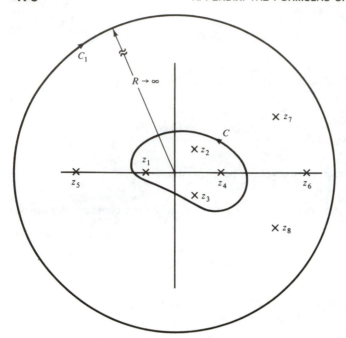

Figure A-3   A function $g(z)$ with four poles inside and outside a contour C.

EXAMPLE A-4

Use the inside–outside theorem, if possible, to evaluate:

$$\frac{1}{2\pi j} \oint_C \frac{2z + 1}{(z - 1)^2(z - 2)} \, dz, \qquad \text{where } C \text{ is defined by:}$$

(a) $|z| = 0.5$
(b) $|z| = 1.5$
(c) $|z| = 3$

Solution

(a) Since the integrand contains no poles inside C the answer is zero.
(b) The order of the denominator is three and the numerator is of order one. Therefore we may use the inside–outside theorem to advantage to obtain:

$$\frac{1}{2\pi j} \oint_C g(z) \, dz = - \, [\text{residue of the pole at } z = 2]$$
$$= -5$$

This is much faster than finding the answer by the residue theorem as the residue of the second-order pole at $z = 1$.

(c) The inside–outside theorem says the answer is zero since there are no poles outside the curve $|z| = 3$.

The Laurent Series

If $F(z) = N(z)/D(z)$, then it may be expressed in a Laurent series:

$$F(z) = \sum_{n=-\infty}^{\infty} A_n z_n \qquad (A\text{-}7)$$

where the Laurent coefficients are:

$$A_n = \frac{1}{2\pi j} \oint_C \frac{F(z)}{z^{n+1}} \, dz \qquad (A\text{-}8)$$

and the series will converge in an annulus $\rho_1 < |z| < \rho_2$, where there are consecutive poles of $F(z)$ located at $|z| = \rho_1$ and $|z| = \rho_2$ as indicated in Figure A-4. We will now solve an example using Equations A-7 and A-8 and then discuss some general properties of the Laurent series.

EXAMPLE A-5
  Express $F(z) = 1/(z - 1)(z - 2)$ in a Laurent series for the regions:

(a) $|z| < 1$
(b) $1 < |z| < 2$
(c) $|z| > 3$

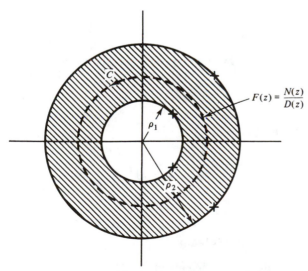

Figure A-4   Given that $F(z)$ is analytic, $\sigma_1 < |z| < \sigma_2$, it can be expressed in a Laurent series.

Solution

(a) Laurent Series for $|z| < 1$

$$A_n = \frac{1}{2\pi j} \oint_C \frac{1}{z^{n+1}(z-1)(z-2)} \, dz$$

may be interpreted for $n < 0$ and $n \geq 0$

Case $n \geq 0$
It is advantageous to use the inside–outside theorem and to find A_n as:

$$A_n = -[\text{residue at } z = 1 + \text{residue at } z = 2]$$

$$= -\left(\frac{1}{1(-1)} + \frac{1}{2^{n+1}}\right)$$

$$= 1 - \left(\frac{1}{2}\right)^{n+1}$$

Case $n < 0$

$$A_n = \frac{1}{2\pi j} \oint_C \frac{1}{z^{n+1}(z-1)(z-2)} \, dz$$

$$= \frac{1}{2\pi j} \oint_C \frac{z^{|n|-1}}{(z-1)(z-2)} \, dz$$

$$= 0$$

by the residue theorem since there are no poles inside C. We may now write the Laurent series for $0 < |z| < 1$ as:

$$\frac{1}{(z-1)(z-2)} = \frac{1}{2} + \frac{3}{4}z + \frac{7}{8}z^2 + \cdots + \left(1 - \left(\frac{1}{2}\right)^{n+1}\right)z^n + \cdots$$

This Laurent series and region of convergence are shown in Figure A-5(a).

(b) Laurent Series $1 < |z| < 2$
We now consider evaluating:

$$A_n = \frac{1}{2\pi j} \oint_C \frac{1}{z^{n+1}(z-1)(z-2)} \, dz$$

in this region for $n \geq 0$ or $n < 0$.

For $n \geq 0$
Using the inside–outside theorem, we obtain:

$$A_n = -[\text{residue of the pole at } z = 2]$$

$$= -\left(\frac{1}{2}\right)^{n+1}$$

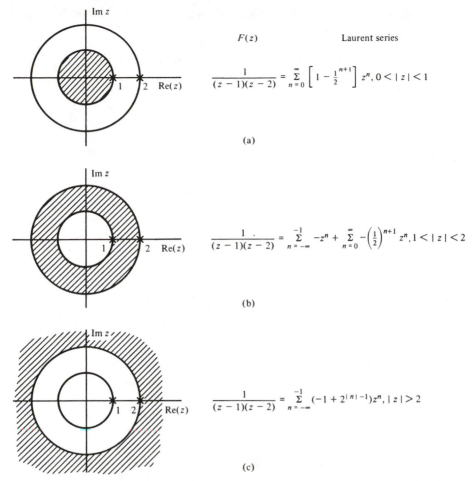

$$\frac{1}{(z-1)(z-2)} = \sum_{n=0}^{\infty} \left[1 - \frac{1}{2}^{n+1} \right] z^n, 0 < |z| < 1$$

(a)

$$\frac{1}{(z-1)(z-2)} = \sum_{n=-\infty}^{-1} -z^n + \sum_{n=0}^{\infty} -\left(\frac{1}{2}\right)^{n+1} z^n, 1 < |z| < 2$$

(b)

$$\frac{1}{(z-1)(z-2)} = \sum_{n=-\infty}^{-1} (-1 + 2^{|n|-1}) z^n, |z| > 2$$

(c)

$F(z)$ Laurent series

Figure A-5   The Laurent series for Example A-5; (a) for $|z| < 1$; (b) for $1 < |z| < 2$; (c) for $|z| > 2$.

For $n < 0$

$$A_n = \frac{1}{2\pi j} \oint_C \frac{z^{|n|-1}}{(z-1)(z-2)} \, dz$$

$$= [\text{residue of the pole at } z = 1]$$

$$= -1$$

We may now write the Laurent series for $1 < |z| < 2$ as:

$$\frac{1}{(z-1)(z-2)} = \sum_{n=0}^{\infty} \left(\frac{1}{-2}\right)^{n+1} z^n + \sum_{n=-\infty}^{-1} -z^n$$

$$= \left(-\frac{1}{2} + \frac{1}{4}z - \frac{1}{8}z^2 + \cdots \right)$$

$$- (z^{-1} + z^{-2} + z^{-3} + \cdots)$$

This Laurent series and region of convergence are shown in Figure A-5(b).

(c) Proceeding as in parts (a) and (b) we can easily find:

For n ≥ 0

$$A_n = 0$$

For n < 0

$$A_n = -1 + \frac{1}{2^{n+1}}$$

$$= -1 + 2^{|n|-1}$$

The Laurent series for $|z| > 2$ is:

$$\frac{1}{(z-1)(z-2)} = \sum_{-\infty}^{-1}(-1 + 2^{|n|-1})z^n$$

$$= z^{-2} + 3z^{-3} + 7z^{-4} + 15z^{-5} + \cdots$$

This series and the region of convergence are shown in Figure A-5(c).

The solution of this problem should be carefully studied since the evaluation of inverse Z transforms is almost identical to obtaining the Laurent series except for some initial confusion because of the assignment of negative powers of z to the values of a discrete function for positive n. It should be noted that: (1) the Laurent series for the region inside all the poles contains only positive powers of z since $f(z)$ is analytic (a MacLauren's series); (2) the Laurent series between two poles contains both positive and negative powers of z, and (3) the Laurent series in the region outside all the poles contains only negative powers of z.

Applications of Complex Variables

In the text we are interested in two main applications of complex integration:

1. the evaluation of inverse Z transforms
2. the evaluation of inverse Laplace transforms

Inverse Z transforms   The Z transform $F(z)$ of a discrete function $f(n)$ is defined as:

$$F(z) = \sum_{-\infty}^{\infty} f(n)z^{-n} \tag{A-9}$$

$$= (f(0) + f(1)z^{-1} + \cdots + f(n)z^{-n}) + (f(-1)z$$

$$+ \cdots + f(-n)z^n)$$

and if $F(z)$ exists it will do so for some $\rho_1 < |z| < \rho_2$. If $F(z)$ is known in the form $F(z) = N(Z)/D(Z)$, $\rho_1 < |z| < \rho_2$, then by the previous theory we find its Laurent coefficients A_{-n} or the values of $f(n)$ as:

$n > 0$

$$f(n) = A_{-n} = \frac{1}{2\pi j} \oint_C F(z) z^{n-1}\, dz$$

$$= \Sigma \,[\text{residues of the poles of } F(z)z^{n-1}]$$

$n \leq 0$

$$f(n) = A_{-n} = \frac{1}{2\pi j} \oint_C \frac{F(z)}{z^{|n|+1}}\, dz$$

$$= - \Sigma \,[\text{residues of the poles of } F(z)z^{n-1} \text{ outside } C]$$

For example, given:

$$F(z) = \frac{1}{(z-1)(z-2)}, \qquad 1 < |z| < 2$$

then

$$f(n) = -\left(\frac{1}{2}\right)^n, \qquad n \leq 0$$

$$= -1, \qquad n > 0$$

and $F(z)$ and $f(n)$ are shown in Figure A-6.

Inverse Laplace Transforms   The bilateral Laplace transform of a function $f(t)$ is:

$$F(s) = \int_{-\infty}^{\infty} f(t)e^{-st}\, dt$$

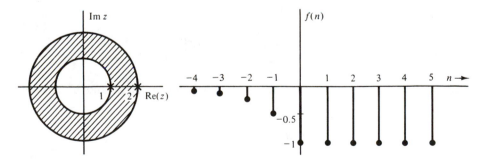

$$F(z) = \frac{1}{(z-1)(z-2)}, 1 < |z| < 2 \longleftrightarrow f(n) = \frac{1}{2\pi j} \oint z^{n-1} F(z)\, dz$$

$$= -\left(\tfrac{1}{2}\right)^n \; n \leqslant 0$$

$$= -1 \; n > 0$$

Figure A-6   A Z transform and its inverse, the discrete time function $f(n)$.

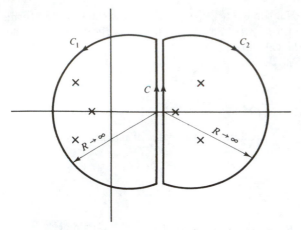

Figure A-7   Closing the contour C for the inverse Laplace transform.

and if $F(s)$ exists, it does so for $\sigma_1 < \text{Re}(s) < \sigma_2$. The inverse Laplace transform is:

$$f(t) = \frac{1}{2\pi j} \int_{C}^{\sigma+j\infty}_{\sigma-j\infty} F(s)e^{+st}\, ds$$

If $F(s) = N(s)/D(s)$ and the order of $D(s)$ is at least one higher than $N(s)$, then we close the contour C with another contour C_1 or C_2, chosen such that $\int_{C_1} F(s)e^{st}\, ds = 0$ or $\int_{C_2} F(s)e^{st}\, dt = 0$ Figure A-7 shows the two situations that arise for positive or negative time.

   Case $t < 0$

If $t > 0$, $F(s)e^{st} \to 0$ on C_1 where C is closed to the left and by Jordan's lemma, $\int_{C_1} F(s)\, e^{st}\, ds = 0$ and therefore:

$$f(t) = \frac{1}{2\pi j}\oint_{C+C_1} F(s)e^{st}\, ds$$

$$= \Sigma\,[\text{residues of the poles of } F(s)\, e^{st} \text{ to the left of } \sigma]$$

   Case $t < 0$

If $t < 0$, $F(s)e^{st} \to 0$ on C_2 where C is closed to the right and by Jordan's lemma:

$$\int_{C_2} F(s)e^{st}\, ds = 0$$

and therefore $f(t) = \frac{1}{2\pi j}\oint_{C+C_2} F(s)e^{st}\, ds$

$$= -\,\Sigma\,[\text{residues of the poles of } F(s) \text{ to the}$$

$$\text{right of } \sigma]$$

A good discussion of Jordan's lemma is given by many texts and the reader is referred to *Transform and State Variable Methods in Linear Systems* by Gupta, John Wiley & Sons, 1966, pp. 368–371.

EXAMPLE A-6

Find the inverse Laplace transform of:

$$F(s) = \frac{s+2}{(s-1)(s+3)}, \qquad -3 < \sigma < +1$$

Solution

$$f(t) = \frac{1}{2\pi j} \int_{\sigma-j\infty}^{\sigma+j\infty} \frac{s+2}{(s-1)(s+3)} e^{st}\, ds$$

For t > 0

$$f(t) = [\text{residue of the pole at } s = -3]$$

$$= \frac{-1}{-4} e^{-3t}$$

$$= 0.25 e^{-3t}$$

For t < 0

$$f(t) = [-\text{ residue of the pole at } s = +1]$$

$$= -0.75 e^{t}$$

and

$$f(t) = 0.25 e^{-3t} u(t) - 0.75 e^{t} u(-t).$$

$F(s)$ and $f(t)$ are shown for this example in Figure A-8.

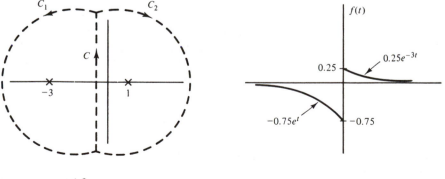

$$F(s) = \frac{s+2}{(s-1)(s+3)}; \quad -3 < \sigma < +1 \qquad\qquad f(t) = 0.25 e^{-3t} u(t) - 0.75 e^{t} u(-t)$$

Figure A-8 The Laplace transform and its inverse, the time function $f(t)$ for Example A-6.

Bibliography

Bellman, R. E. *Matrix Analysis,* 2d Ed. New York: McGraw-Hill, 1968.

Brigham, E. O. *The Fast Fourier Transform.* Englewood Cliffs, NJ: Prentice-Hall, 1974.

Cooley, J. W., and J. W. Tukey. "An Algorithm for Machine Calculation of Complex Fourier Series," *Math. Computation,* 19 (April 1965): 297–301.

Elliott, D. F., and K. R. Rao. *Fast Transforms: Algorithms, Analyses, Applications.* New York: Academic Press, 1982.

Gabel, R. A., and R. A. Roberts. *Signals and Linear Systems,* 2d Ed. New York: Wiley, 1980.

Kailath, T. *Linear Systems.* Englewood Cliffs, NJ: Prentice-Hall, 1980.

McGillem, C. D., and G. R. Cooper. *Continuous and Discrete Signal and System Analysis,* 2d Ed. New York: Holt, Rinehart and Winston, 1984.

O'Flynn, Michael. *Probabilities, Random Variables, and Random Processes.* New York: Harper & Row, 1982.

Oppenheim, A. V., A. S. Willsky, and I. T. Young. *Signals and Systems.* Prentice-Hall Signal Processing Series. Englewood Cliffs, NJ: Prentice-Hall, 1983.

Papoulis, A. *Circuits and Systems.* Holt, Rinehart and Winston Series in Electrical Engineering, Electronics, and Systems. New York: Holt, Rinehart and Winston, 1980.

Zadeh, L. A., and C. A. Desoer. *Linear System Theory.* New York: McGraw-Hill, 1963.

Answers to Drill Sets

DRILL SET: OPERATING ON DISCRETE FUNCTIONS, page 14

1. (d) $n \leqslant -3 \cup n \geqslant 3$ (e) $n \leqslant 1 \cup n \geqslant 7$

DRILL SET: SINGULARITY FUNCTIONS, page 29

1. $f'(t) = 16t \sqcap (t) + 2\delta(t + \frac{1}{2}) - 2\delta(t - \frac{1}{2})$
   $f''(t) = 16 \sqcap (t) + 2\delta'(t + \frac{1}{2}) - 2\delta'(t - \frac{1}{2}) - 8\delta(t + \frac{1}{2}) - 8\delta(t - \frac{1}{2})$
   $f'''(t) = 2\delta''(t + \frac{1}{2}) - 8\delta'(t + \frac{1}{2}) + 16\delta(t + \frac{1}{2}) - 2\delta''(t - \frac{1}{2})$
   $- 8\delta'(t - \frac{1}{2}) - 16\delta(t - \frac{1}{2})$

2. $F(t) = 2u(t + 2) - (t - 2) \sqcap (0.5t - 1.5) - 3\delta(t - 2) - 2u(t - 4)$
   $G(t) = 2(t + 2)\{u(t + 2) - u(t - 2)\} - \frac{1}{2}(t - 4)^2 \sqcap (0.5t - 1.5) + 7u(t - 2)$

DRILL SET: DISCRETE PULSE FUNCTION $\delta(n)$, page 31

1. (a) 0.25   (b) 0.125   (c) $F(-4) = 0.25$, $F(0) = 4.0$, $F(3) = 32.0$, $F(n) = 4.2^n$

DRILL SET: THE IMPULSE RESPONSE, page 48

1. $y''(t) + 3y'(t) + 2y(t) = 5 \cos t - 15 \sin t$; $y(0) = 7$, $\dot{y}(0) = -5$
2. $y'(t) + y(t) = 2x(t) - x'(t)$; $H(s) = -s + 2/s + 1$

DRILL SET: CONVOLUTION INTEGRALS, page 65

1. (a) $y(t) = 0, \quad t < 0$
$\qquad\qquad = 2 - 2e^{-t}, \quad 0 < t < 1$
$\qquad\qquad = 2e^{-t}(e - 1), \quad t > 1$

(b) $y(t) \approx 0.10e^{-(t-0.025)}u(t - 0.025)$

2. (a) $r(t) = 0, \quad t < 5.5$
$\qquad\qquad = 2t - 11, \quad 5.5 < t < 6.5$
$\qquad\qquad = 2, \quad 6.5 < t < 10.5$
$\qquad\qquad = -2t - 23, \quad 10.5 < t < 11.5$
$\qquad\qquad = 0, \quad t > 11.5$

(b) $(t - 2)u(t - 2)$

(c) $\frac{1}{3}e^t \quad$ for $t < 0$
$\qquad \frac{1}{3}e^{-2t} \quad$ for $t > 0$

DRILL SET: DIFFERENCE EQUATIONS, page 75

1. (a) $y(n) + 0.5y(n - 1) - 0.5y(n - 2) = 2x(n) + x(n - 1)$
(b) $y_{F0}(n) = 2.22(2^n)$ **(c)** $y(n) = 9 + \frac{8}{3}\{(-1)^n - (\frac{1}{2})^n\}$

DRILL SET: DISCRETE CONVOLUTION, page 91

1. (a) $y(n) = (-0.7)^n u(n) - 2(-0.7)^{n-1}u(n - 1) + 3(-0.7)^{n-4}u(n - 4)$
(b) $y(n) = [0.59 + 0.41(-0.7)^n]u(n)$
2. (a) $\delta(n + 1)$
(b) $(n - 2)u(n - 3)$
(c) $r(n) = [0.8(2)^{n-2} + 0.2(-0.5)^{n-2}]u(n-2)$

DRILL SET: LAPLACE TRANSFORMS AND THEOREMS, page 161

1. (a) $\dfrac{\alpha}{s^2 - \alpha^2}, \quad \dfrac{s}{s^2 - \alpha^2}$

(b) $\dfrac{6s^2 - 2}{(s^2 + 1)^3}, \quad e^{-4s}\left(\dfrac{2}{s^3} + \dfrac{8}{s^2} + \dfrac{19}{s}\right)$

DRILL SET: INVERSE LAPLACE TRANSFORMS, page 170

1. (a) $6 \cos 3t + \sin 3t$ **(b)** $6 \cosh 3t + \sinh 3t$
(c) $(6 - 3t)e^{-t}u(t)$ **(d)** $(5.14e^{-1.7t} + 0.86 e^{-0.3t})u(t)$
(e) $(6 \cos \sqrt{7} t - 1.13 \sin \sqrt{7}t)e^{-t}u(t)$

DRILL SET: INVERSE Z TRANSFORMS, page 241

1. (a) $(3 + 5.5n)(0.8)^n u(n)$ (b) $(3.44n^2 + 0.31n)(0.8)^n u(n)$
(c) $(0.44 + 1.66n - 0.44(-2)^n)u(n)$

DRILL SET: FOURIER SERIES, page 300

2. $d_1 = -0.188$, $d_2 = 0.845$, $t_1 = 3.0$, $\alpha_1 = 1.5$, $\alpha_2 = 0.845$
3. $f(t) = 0.5 \cos \pi t - 0.25 \cos 3\pi t - 0.25 \cos 5\pi t$
5. $C_n = 1/T$

DRILL SET: FOURIER TRANSFORMS, page 326

1. $\text{FT} \left(\dfrac{\sin t}{t} \right)^2 = \pi \left(1 - \dfrac{|\omega|}{2} \right)$, $|\omega| < 2$

$$= 0, \quad |\omega| > 2$$

2. $\text{FT} t^3 e^{-5t} u(t) = \dfrac{-6}{(j\omega + 5)^4}$

3. $F(j\omega) = 2\pi \displaystyle\sum_{n=-\infty}^{\infty} C_n \delta(\omega - n\omega_0)$,   where $\omega_0 = \dfrac{2\pi}{3}$

and $C_n = \dfrac{1}{2n\pi j} (e^{jn(2\pi/3)} - 2e^{-jn(4\pi/3)})$

DRILL SET: STATE VARIABLES, page 426

1. (a) $A_c = \begin{bmatrix} 0 & 1 & 0 & 0 \\ 0 & 0 & 1 & 0 \\ 0 & 0 & 0 & 1 \\ -40 & 0 & 0 & -20 \end{bmatrix}$, $B_c = \begin{bmatrix} 0 \\ 0 \\ 0 \\ 1 \end{bmatrix}$, $C_c = [30 \quad 10 \quad 1 \quad 0]$, $D_c = [0]$

(b) $A_0 = \begin{bmatrix} 0 & 0 & 0 & -40 \\ 1 & 0 & 0 & 0 \\ 0 & 1 & 0 & 0 \\ 0 & 0 & 1 & -20 \end{bmatrix}$, $B_0 = \begin{bmatrix} 30 \\ 10 \\ 1 \\ 0 \end{bmatrix}$, $C_0 = [0 \quad 0 \quad 0 \quad 1]$, $D_0 = [0]$

2. $A = \begin{bmatrix} \lambda_1 & 0 \\ 0 & \lambda_2 \end{bmatrix}$, $B = \dfrac{2j}{3\sqrt{3}} \begin{bmatrix} \lambda_1 \\ \lambda_2 \end{bmatrix}$, $C = 2j \dfrac{\sqrt{3}}{3} [1 \quad 1]$, $D = [0]$, where $\lambda_1 = -\frac{1}{2} + j\frac{\sqrt{3}}{2}$ and $\lambda_2 = -\frac{1}{2} - j\frac{\sqrt{3}}{2}$

4. $A = \begin{bmatrix} 0 & 0 & 1 \\ 0 & -1 & 1 \\ -1 & -1 & -3 \end{bmatrix}$, $B = \begin{bmatrix} 0 & 0 \\ 0 & 0 \\ 1 & 1 \end{bmatrix}$, $C = [0 \quad 1 \quad 0]$, $D = [0 \quad 0]$

Answers to Selected Problems

CHAPTER 1

2. (a) $x(t - 4) = e^{t-4}$, $t < 4$
$$= t - 4, \quad 4 < t < 6$$

   (d) $x(-4t - 2) = -4t - 2$, $0 < t < \frac{1}{2}$
$$= e^{-4t-2}, \quad t > \frac{1}{2}$$

3. (b) $f(n + 5)u(n) = (n + 5)(n + 4)(0.5)^{n+3}\, u(n)$

5. (b) $y(n) = \delta(n)$

6. (e) $h(n) = 0$ for $k \le -3 \cup k \ge 3$
   (f) $1(n) = 0$ for $k \le 1 \cup k \ge 7$

8. (a) $\pi/2$ **(b)** 0   **(c)** 76.77   **(d)** 13.5

14. (a) $F(n) = 8u(n - 3)$ **(b)** $G(n) = F(n)$ **(c)** $H(n) = \infty$

CHAPTER 2

1. (a) $y(t) = 1.25e^{-2t} + 2te^{-t} - \frac{1}{2}t + \frac{3}{4}$
   (b) $y(t) = e^{-t} + te^{-t} + t \,\text{Sin}\, t - \text{Sin}\, t + \text{Cos}\, t$

2. $y(0) = 6$, $\dot{y}(0) = -11$

3. (a) $h(t) = \frac{7}{3}e - \frac{5}{3}tu(t)$ **(b)** $h(t) = \frac{2}{3}e - \frac{5}{3}tu(t) + \delta(t)$
   (d) $h(t) = e^{-t}(3 \,\text{Cos}\, t - \text{Sin}\, t)u(t)$

4. (b) $H(s) = \dfrac{s + 2}{2s^2 + 6s + 1}$

8. (c) $6e^{-2t}u(t) - 6e^{-t}u(t)$

 (e) $t < 0 \rightarrow 2e^t$ and $t > 0 \rightarrow 2$

10. (b) $y(n) = 0.259(\frac{1}{3})^n - 0.681(-\frac{2}{3})^n + 0.2$

11. (c) $h(n) = [-1.45(-0.54)^n + 0.55(0.21)^n]u(n)$

CHAPTER 3

1. (a) undefined   **(b)** $e^{2t}u(-t) + e^t u(t)$ **(d)** $(1 - e^{-t})u(t)$

 (e) $-t u(-t)$ **(f)** $t u(t)$

5. (a) $0.77(0.6)^{-n} u(-n-1) + 0.77 (-0.5)^n u(n)$ **(d)** $(2 - (0.5)^n) u(n)$

 (e) $(n + 1) u(n)$

14. (a) $\tilde{x}(t) = 0$, $\tilde{x}^2(t) = 4$, $\sigma_x^2 = 4$ **(b)** $R_{xx}(\tau) = 4 - 1.33 |\tau|$, $0 < |\tau| < 3$; $= 0$,
 otherwise   **(c)** $R_{yy}(\tau) = 8 - 2.67 |\tau|$, $0 < |\tau| < 3$; $= 0$, otherwise, $R_{yy}(\tau) = 2R_{xx}(\tau)$, $R_{xy}(\tau) = R_{yx}(\tau) = 0$ **(d)** $R_{zz}(\tau) = R_{yy}(\tau)$. $R_{xy}(\tau) = R_{yx}(\tau) = 0$.

17. $C_{hh}^{(n)} = 1.52 (-0.6)^{|n|}$, $R_{yy}(n) = 6.1 (-0.6)^{|n|}$, P_{av} at input is 4, P_{av} at output is 6.

CHAPTER 4

1. (e) $\dfrac{4.6s + 7.7}{s^2 + 4}$ **(i)** $\dfrac{10}{s^3} + \dfrac{3}{s^2}$

 (k) $\left(\dfrac{6}{s^3} + \dfrac{4}{s^2} + \dfrac{3}{s}\right)e^{-2s}$ **(m)** $\dfrac{120}{(s + 2)^6}$ **(q)** $\dfrac{2}{s(s + 1)}$

2. (c) $e^{-3t}(2 \cos 1.414t - 3.54 \sin 1.414t)u(t)$

 (g) $(t^2 - \frac{7}{6}t^3)e^{-4t}u(t)$ **(i)** $\dfrac{(t - 5)^5}{7680} e^{-1/2(t-5)}u(t - 5)$

4. (b) $h(t) = 0.33e^{-1.17t}u(t)$ **(d)** $y(t) = [2.24e^{-1.17t} + 0.28t - 0.24]u(t)$

7. $y(t) = (0.91e^{-3/8t} + 0.091e^t)u(t)$

CHAPTER 5

1. (a) $F_1(s) = \dfrac{5s^2 + s - 3}{s^2(s - 1)}$; $0 < \sigma < 1$ **(d)** $F_4(s) = \dfrac{-10!}{(s + 2)^{11}}$; $\sigma < -2$

 (e) $F_6(s) = \dfrac{-8}{s^2 - 4}$; $-2 < \sigma < 2$

3. (a) $-2e^{-2t}u(-t)$ **(b)** $-2te^{-3t}u(-t)$ **(c)** $(-0.5e^{-t} - 1.5e^{-3t})u(-t)$

6. (b) $\delta(t - b + \alpha)$ **(d)** $(4e^{-t} - 4e^{-2t}) u(t)$ **(e)** $(1.33e^{-t} - 1.33e^{2t})u(-t)$

12. (a) $T(s) = \dfrac{-4}{s^2 - 25}$, $S_{yy}(s) = \dfrac{-16}{s^2 - 25}$, $S_{xy}(s) = \dfrac{8}{s + 5}$

 (b) $y^2(t) = 1.6$

CHAPTER 6

1. (b) $\dfrac{3z}{(z + 0.6)^2}$ **(f)** $\dfrac{-2z^3 + 15z^2 - 9z}{(z - 3)^3}$ **(g)** $\dfrac{0.25z^{-2}}{z - 2}$

(j) $\dfrac{-0.512z^{-2}}{(z + 0.8)}$ **(k)** $\dfrac{0.34z^{-1}}{(z - 0.7)^2}$

2. (c) $\dfrac{n(n - 1)}{2} 2^n u(n - 2)$ **(e)** $3(n - 3)(-1)^n u(n - 3)$

3. (a) $[1.4(2)^n - 0.4(0.6)^n] u(n)$

4. (a) $H(z) = \dfrac{2z}{z + 0.6}$ **(c)** $[1.25 + 0.75(-0.6)^n] u(n)$

(e) $[0.909(0.5)^n + 1.09(-0.6)^n] u(n)$

5. (b) $h(n) = [8(2)^n - 4] u(n)$
(f) $y(n) = [1.33(0.5)^n - 11.98 + 22.66(2)^n] u(n)$

CHAPTER 7

1. (a) $F_1(z) = \dfrac{3z^2 - 1}{z}$ **(c)** undefined

(d) $F_4(z) = (z^3 - 4z^2 - 1.33z + 0.33)/[z - 0.33)(z - 1)^2]$
(h) denominator $D(z) = (z + 0.5)^3(z - 0.7)^2, 0.5 < |z| < 0.7$

4. (a) $f_1(n) = 2\delta(n + 1) + 3\delta(n) + \delta(n - 2),$ **(d)** $[-0.68(2)^n 1.4(-3)^n$
$+ 4.4n(-3)^{n-1}]$ **(e)** $0.68(2)^n u(n) + [-1.4(-3)^n + 4.4(-3)^{n-1}] u(-n - 1)$

5. (a) $(3(2)^n - 2(3)^n) u(-n)$ **(b)** $-2.2 (2)^{n-1} u(n - 1) - 1.2(\tfrac{1}{3})^n u(-n)$

10. (a) $g(n) = 5(-1)^n u(n) - 4(-0.8)^n u(n), S_{mm}(z) = \dfrac{-7.5z}{(z + 0.8)(z + 1.25)}$

$S_{nm}(z) = \dfrac{6z}{z + 0.8}, S_{mn}(z) = \dfrac{7.5z}{z + 1.25}$

(c) $\widetilde{m^2}(n) = 13.7$.

CHAPTER 8

1. $a_0 = \dfrac{4\pi^2}{3}, \quad a_n = \dfrac{4}{n^2}, \quad b_n = \dfrac{-4\pi}{n}$

2. $\hat{f}(t) = (2.4e^{-2t} - 0.6e^{-t}) u(t), \quad E = 0.667, \quad \hat{E} = 0.660$

4. $y(t) = 0.75 + 0.3 \sin(\pi t - 17.44°) + 0.076 \sin(3\pi t - 43.30°) + 0.034 \cdot \sin(5\pi t - 57.50°)$

9. (a) $A = 2$ **(b)** $A = 6$ **(c)** $c_2 = 0$

10. $f_1(t) = 5 + 4 \cos 2\pi f_0 t$ and $f_2(t) = 4 \cos 2\pi f_0 t$

13. $A = B = 0.5$, $t_1 = 0$, $t_2 = 2$

14. (b) $G(j\omega) = 2j \sin \dfrac{\omega}{2} (e^{-j(\omega/2)} + e^{-j(5\omega/2)})$

17. $G(j\omega) = \dfrac{1 + j\omega}{(1 + j\omega)^2(4 + j\omega)(2 + j\omega)}$

20. (b) $F_1(j\omega) = \dfrac{1}{2j\omega} - \dfrac{1}{2j\omega + 4} + \dfrac{\pi}{2}\delta(\omega)$

21. $k = 0.636$

22. 0.333

CHAPTER 9

6. (a) $F(0) = 3$, $F(1) = 0.29 + j0.29$, $F(2) = j$, $F(3) = 1.7 - j1.7$, $F(4) = 1$, $F(5) = 1.7 + j1.7$, $F(6) = -j$, $F(7) = 0.29 - j0.29$

(d) $F(0) = 120$, $F(1) = F(2) = F(3) = 0$, $F(4) = -40$, $F(5) = F(6) = F(7) = 0$

7. (a) $f(n) = \{1, 0, -\frac{1}{2}, 0, 0, 0, -\frac{1}{2}, 0\}$

8. (c) $F(0) = 1$, $F(1) = 1 - j3.24$, $F(2) = 1 - j1.08$, $F(3) = 1 - j0.4$, $F(4) = 1$, $F(5) = 1 + j0.4$, $F(6) = 1 + j1.08$, $F(7) = 1 + j3.24$

CHAPTER 10

1. (d) $H(s) = \dfrac{s^2 + s + 2}{s^3 + s^2}$

2. (c) $H(z) = \dfrac{1}{z^6 + 1}$ **(d)** $H(z) = \dfrac{z^2 - 10}{z^3 - 1}$

3. (a) $A = \begin{bmatrix} 1 & 1 & 0 \\ 0 & 1 & 1 \\ 0 & 0 & 1 \end{bmatrix}$, $B = \begin{bmatrix} 1 \\ 1 \\ 1 \end{bmatrix}$, $C = \{1 \quad 0 \quad 0\}$, $D = 0$

5. (a) $H(s) = \dfrac{s^2 - s + 1}{(s - 1)^3}$

8. (a) $A = \begin{bmatrix} 3.56 & 0 \\ 0 & -0.56 \end{bmatrix}$, $B = \begin{bmatrix} 1.10 \\ -0.11 \end{bmatrix}$, $C = \{1 \quad 1\}$, $D = \{0\}$

(c) $A = \begin{bmatrix} 37.32 & 0 \\ 0 & 2.63 \end{bmatrix}$, $B = \begin{bmatrix} 0.69 \\ -0.183 \end{bmatrix}$, $C = [4.31 - 0.31]$, $D = \{5\}$

9. (c) $A^k = \begin{bmatrix} 1 & 0 & 0 \\ 2^k - 1 & 2^k & 0 \\ 2^k + k - 1 & 2^k - 1 & 1 \end{bmatrix}$

10. (a) $A^k = \begin{bmatrix} 8^k & 0 \\ 0 & 1 \end{bmatrix}$

11. (a) $e^{At} = \begin{bmatrix} e^{8t}u(t) & 0 \\ 0 & e^{t}u(t) \end{bmatrix}$ (d) $e^{At} = \begin{bmatrix} e^{-2t} & 0 & 0 \\ te^{-2t} & e^{-2t} & 0 \\ t^2 e^{-2t} & te^{-2t} & e^{-2t} \end{bmatrix} u(t)$

15. $y(k) = -2(4.79)^k + 7(0.21)^k$

16. $H(z) = \dfrac{z - 1}{z^3 - z^2 - 3z - 1}$

18. $H(z) = \dfrac{3z^4 + 2z^3 + 2z^2 + z + 1}{z^4 + z^3 + z^2 + 1}$

20. $\dot{q}(t) = \begin{bmatrix} -1 & 1 & 0 \\ 0 & -1 & 0 \\ 0 & 0 & -2 \end{bmatrix} q(t) + \begin{pmatrix} 1 \\ -1 \\ 1 \end{pmatrix} x(t), \quad y(t) = \begin{bmatrix} -5 & 0 & 5 \end{bmatrix} q(t)$

24. $x_1(t) = 1 + 1.5e^{-5t}, \quad x_2(t) = 0.5 - 1.5e^{-2t} - 0.5e^{-5t} + e^{-7t}$

26. (c) $y(t) = te^{t}u(t)$

Index

NOTATION IN ORDER OF ITS APPEARANCE

(Continued from inside front cover.)

$x(n)$ — The inverse Z transform, defined as:

$$x(n) = \frac{1}{2\pi j} \oint X(z)z^{n-1}\, dz$$

$$= \Sigma(\text{residues of the poles inside } C), \qquad n > 0$$

$$= -\Sigma(\text{residues of the poles outside } C), \qquad n \leq 0$$

$H(s), H(z)$ — The system function:

$$H(s) = \frac{Y(s)}{X(s)} \qquad \text{with } y^p(0^-) = 0, \qquad p \geq 0$$

$$\text{and} \qquad H(z) = \frac{Y(z)}{X(z)} \qquad \text{with } y(n) = 0, \qquad n \leq 0$$

$Y(s), Y(z)$ — The transform of the output of a LTIC system, given as:

$$Y(s) = H(s)X(s) \quad \text{or} \quad Y(z) = H(z)X(z)$$

$S_{xx}(s), S_{xx}(z)$ — The power spectral density of an ergodic random process $x(t)$ or $x(n)$ given as the transform of $R_{xx}(\tau)$ or $R_{xx}(n)$.

$S_{xy}(s), S_{xy}(z)$ — The cross-spectral density defined as the transform of $R_{xy}(\tau)$ or $R_{xy}(n)$, given as $S_{xx}(s)H(s)$ or $S_{xx}(z)H(z)$.

$S_{yy}(s), S_{yy}(z)$ — The output power spectral density for the random input:

$$S_{yy}(s) = S_{xx}(s)H(s)H(-s), \quad S_{yy}(z) = S_{xx}(z)H(z)H(z^{-1})$$

$T(s), T(z)$ — The power transfer function $H(s)H(-s)$ or $H(z)H(z^{-1})$:

$$T(s) = \overline{C_{hh}(\tau)} \quad \text{and} \quad T(z) = \overline{C_{hh}(n)}$$

$H_d(z)$ — The system function for a discrete system obtained by replacing a continuous system $H(s)$ by one with $h_d(n) = h(t)$ with $t = n\tau$:

$$H_d(z) = \Sigma\left(\text{residues of poles of } Y(p) \text{ in } \frac{Y(p)}{1 - e^{\tau p}z^{-1}}\right)$$

Chapter 8

a_n, b_n, a_0 — The Fourier series coefficients for the trigonometric Fourier series.

T, ω_0 — The period of a periodic waveform and the fundamental angular frequency, related by $\omega_0 = 2\pi/T$.

$f_e(t), f_0(t)$ — The even and odd parts of a given $f(t)$, where: $f(t) = f_e(t) + f_0(t)$

$\hat{f}(t)$ — The approximation to a given $f(t)$ in terms of basis functions $\phi_n(t)$.

α_i — Generalized Fourier series coefficients.

$E, \hat{E}$ — The energy in a given signal and the energy in an approximation to a given signal.

$F(j\omega)$ — The Fourier transform of $f(t)$:

$$F(j\omega) = \int_{-\infty}^{\infty} f(t)e^{-j\omega t}\, dt$$

$|F(j\omega)|^2$ — The energy spectral density.

$w(t)$ — The unit step response:

$$w(t) = \int_{-\infty}^{t} h(x)\, dx$$

where $h(t)$ is the impulse response.